GEOMETRIC MODELING:
ALGORITHMS AND NEW TRENDS

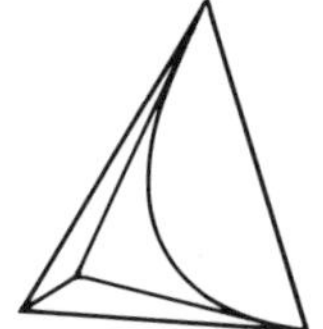 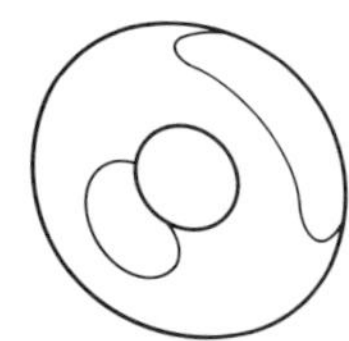 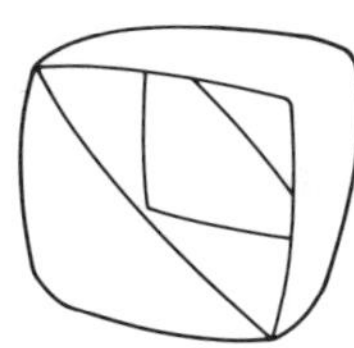 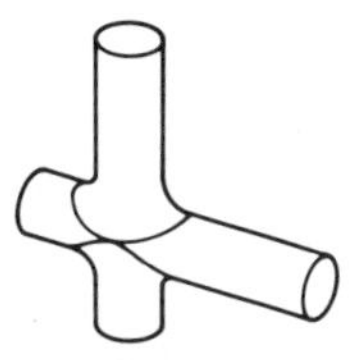

Geometric Modeling:
Algorithms and New Trends

EDITED BY GERALD E. FARIN

siam

Gerald E. Farin
Arizona State University
Department of Computer Science
Tempe, Arizona 85287

The conference from which some of the papers in this book were derived was supported by the U.S. Army Research Office, under grant DAAG29-85-G-0093.

Library of Congress Catalog Card Number 86-61772
ISBN 0-89871-206-8

Preface

This text is intended for CAD/CAM systems developers as well as for university and industry researchers in the fields of curve/surface design and solid modeling. It contains a mixture of survey-like articles and of research papers and should therefore contain material that is of interest both for beginners and experts in the field of geometric modeling.

The starting point for this book was the SIAM Conference on Geometric Modeling and Robotics, held at Albany, New York from July 15–19, 1985. It seemed that many of the papers that dealt with geometric modeling could serve as a basis for a reference text, possibly after adding in some other papers to complete each of the intended sections. That plan was further pursued, and so about 60% of the articles in this volume are largely based on conference presentations, while the remaining 40% are invited.

As for the included material, the first two sections—Curves and Surfaces and Solid Modeling and Multivariate Surfaces—consist mainly of expository articles, some being surveys and some giving new presentations of established methods. The last two sections—Shape and Intersection Algorithms—reflect the formation of two trends in geometric modeling, both in the research community and in industry. Each section starts with a separate introduction.

I wish to thank Harry McLaughlin from Rensselaer Polytechnic Institute and Dave Ferguson from Boeing Computer Services for their help in initiating this book. Thanks also go to the Utah Math CAGD group for valuable discussions.

G. Farin,
Salt Lake City, Utah

Contributors

P. Alfeld, *University of Utah, Salt Lake City, Utah*

F. Almgren, *Princeton University and The Institute for Advanced Study, Princeton, New Jersey*

R. Barnhill, *Arizona State University, Tempe, Arizona*

W. Boehm, *Technical University of Braunschweig, Braunschweig, Federal Republic of Germany*

K. Bosworth, *Utah State University, Logan, Utah*

V. Chandru, *Purdue University, West Lafayette, Indiana*

C. de Boor, *University of Wisconsin, Madison, Wisconsin*

R. Farouki, *IBM Thomas J. Watson Research Center, Yorktown Heights, New York*

D. Field, *General Motors Research Laboratories, Warren, Michigan*

F. Fritsch, *Lawrence Livermore National Laboratory, Livermore, California*

R. Goldman, *Control Data Corporation, Arden Hills, Minnesota*

G. Herron, *Colorado State University, Fort Collins, Colorado*

C. Hoffmann, *Purdue University, West Lafayette, Indiana*

K. Höllig, *University of Wisconsin, Madison, Wisconsin*

J. Hopcroft, *Cornell University, Ithaca, New York*

T. Jensen, *Evans and Sutherland Computer Corporation, Salt Lake City, Utah*

A. Jones, *Boeing Computer Services, Tukwila, Washington*

B. Kochar, *Purdue University, West Lafayette, Indiana*

E. Lee, *Boeing Commercial Airplane Company, Seattle, Washington*

L. Lichten, *California State University, Northridge, California*

G. Makatura, *Informatics General Corporation, Palo Alto, California*

F. Munchmeyer, *University of New Orleans, New Orleans, Louisianna*

G. Nielson, *Arizona State University, Tempe, Arizona*

J. Owen, *Shape Data Ltd., Cambridge, England*

C. Petersen, *Evans and Sutherland Computer Corporation, Salt Lake City, Utah*

B. Piper, *University of Utah, Salt Lake City, Utah*

K. Rescorla, *University of Utah, Salt Lake City, Utah*

A. Rockwood, *Evans and Sutherland Computer Corporation, Salt Lake City, Utah*

M. Samek, *Abel Image Research, Hollywood, California*
A. Schwartz, *University of Michigan, Ann Arbor, Michigan*
T. Sederberg, *Brigham Young University, Provo, Utah*
S. Stead, *NASA Ames Research Center, Moffett Field, California*
A. Worsey, *University of North Carolina, Wilmington, North Carolina*

I am grateful to the following people whose comments and suggestions greatly improved the quality of the articles that were accepted for inclusion in this volume:

P. Alfeld, *University of Utah, Salt Lake City, Utah*
R. Barnhill, *Arizona State University, Tempe, Arizona*
W. Boehm, *Technical University of Braunschweig, Braunschweig, Federal Republic of Germany*
C. de Boor, *University of Wisconsin, Madison, Wisconsin*
D. Ferguson, *Boeing Computer Services, Tukwila, Washington*
D. Field, *General Motors Research Laboratories, Warren, Michigan*
R. Franke, *Naval Postgraduate School, Monterey, California*
R. Goldman, *Control Data Corporation, Arden Hills, Minnesota*
G. Herron, *Colorado State University, Fort Collins, Colorado*
K. Höllig, *University of Wisconsin, Madison, Wisconsin*
T. Jensen, *Evans and Sutherland Computer Corporation, Salt Lake City, Utah*
A. Jones, *Boeing Computer Services, Tukwila, Washington*
M. Lucian, *Boeing Computer Services, Tukwila, Washington*
G. Nielson, *Arizona State University, Tempe, Arizona*
B. Piper, *University of Utah, Salt Lake City, Utah*
A. Requicha, *University of Rochester, Rochester, New York*
R. Sarraga, *General Motors Research Laboratories, Warren, Michigan*
R. Scott, *University of Chicago, Chicago, Illinois*
L. Schumaker, *Texas A & M University, College Station, Texas*
T. Sederberg, *Brigham Young University, Provo, Utah*
W. Tiller, *Structural Dynamics Research Corporation, Milford, Ohio*
A. Worsey, *University of North Carolina, Wilmington, North Carolina*

Contents

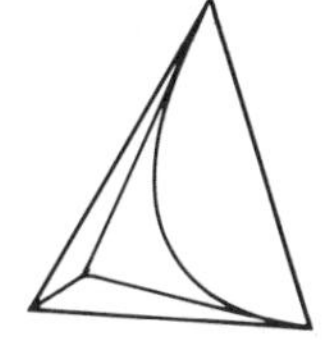

Part I:
Curves and Surfaces

The papers in this section are either surveys or describe known material from a new viewpoint.

The fact that conic sections may be written in rational Bézier form is well known; however, no detailed account of this connection has appeared in the literature so far. The article by E. Lee closes that gap by first giving the underlying theory and then obtaining results about conics in the Bézier formulation.

Another fundamental curve scheme is that of B-splines. While the classical derivation uses divided differences of truncated power functions, the paper by C. de Boor and K. Höllig uses a different approach, namely, that of starting with the recursion formula for B-splines. This leads to a "cleaner" way to derive the basic results about B-splines.

A field that is steadily gaining momentum in geometric modeling is concerned with algebraic geometry methods—the fact that most papers in the "Intersection Methods" section deal with implicit surfaces is a clear indication of this. The paper by T. Sederberg is an introduction to these methods and a survey of work done in the field thus far.

Although minimal surfaces are well understood in differential geometry, so far there have been almost no modeling applications, despite the fact that minimal surfaces have several very interesting "physical" properties that make them potential rivals of the existing surface methods. The article by F. Almgren proposes a method whose major application is the determination of minimal surfaces.

Bézier curves and surfaces, which are fundamental to geometric modeling, may be developed through the use of a difference calculus (see also the paper by C. de Boor in the next section). This difference approach has been used by several authors; the paper by A. Schwartz reviews these approaches and gives a new formulation that eliminates some inaccuracies present in previous formulations.

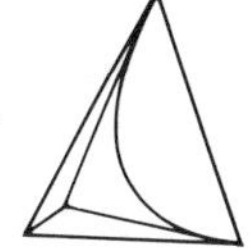

The Rational Bézier Representation for Conics

EUGENE T. Y. LEE

Abstract. A self-contained development of general properties of conics in the rational quadratic Bézier representation is given, together with examples of parameterizations. Formulas are obtained expressing various conic characteristics (center, vertices, etc.) directly in terms of the control points and weights.

Conics have been widely used in industry, due mainly to their familiar classical properties. Rational quadratic parametric representations for conics, particularly in the Bernstein–Bézier form, have recently become popular in CAGD (computer aided geometric design) circles. A number of works dealing with these are already available, among them Boehm et al. [1], Farin [2], Faux and Pratt [3], Forrest [4], [5], Pavlidis [8] and Piegl [9].

Here we present another exposition, in a somewhat different manner from each of the above references. We stay entirely in Euclidean 3-space; no concepts or theorems involving projective space are used, as, for example, in Forrest [4], [5]. (In particular, homogeneous coordinates will not even be mentioned. It is our experience that such concepts often confuse many practitioners in CAGD.) More importantly, we show that it is possible (and easy) to start with an arbitrary rational quadratic Bézier form and to deduce from it its general properties as a conic; this contrasts with the approach in Faux and Pratt [3], where a specific parameterization for the conic is imposed at the outset. We feel that this elementary exposition would be of interest for those CAGD workers not already familiar with the subject.

The general properties are developed in § 1, of which §§ 1.1 and 1.2 form the core. In § 1.1, we show that a rational quadratic parametric curve is a conic; in § 1.2, we obtain the type classification of conics. Section 2 contains some examples showing how a parameterization is fixed by choosing the "weights." Together these sections can be regarded as a comprehensive survey. In § 3, we obtain various geometric characteristics

3

of conics (center, vertices, etc.) directly in terms of the control points and weights. The formulas seem to be new.

Throughout the exposition, we assume the reader is familiar with the basic properties of polynomial Bézier curves. The subdivision algorithm for Bézier curves will be used several times; a review of this is contained in § 1.4.

1. Rational Quadratic Bézier Curves.

1.1. Let ϕ_i denote the quadratic Bernstein basis

$$\phi_0(t) = (1 - t)^2, \quad \phi_1(t) = 2t(1 - t), \quad \phi_2(t) = t^2,$$

and let

$$(1) \qquad P(t) = \sum w_i P_i \phi_i(t) \Big/ \sum w_i \phi_i(t)$$

be a rational quadratic curve in Bézier form. The following notation will be used throughout: capital Roman letters will denote points (vectors) in 3-space, lower case letters real numbers (scalars); $|X|$ will denote the Euclidean norm of X. Terminology: the P_i's are the control points, the w_i's the weights.

We will restrict our attention to the standard case where $w_i > 0$ for all i. Then the denominator

$$w(t) = \sum w_i \phi_i(t)$$

is positive on $[0, 1]$, and so $P[0, 1]$ is bounded. In fact,

$$P(t) = \sum P_i \psi_i(t), \quad \text{with}$$

$$(2)$$

$$\psi_i(t) = \frac{w_i \phi_i(t)}{w(t)} .$$

Since also $\psi_i \geq 0$ on $[0, 1]$ and $\sum \psi_i = 1$, we have

$$P[0, 1] \subset \Delta(P_0, P_1, P_2);$$

that is, the segment $P(t)$, $0 \leq t \leq 1$, lies in the convex hull of the control points. The $\psi_i(t)$'s are the barycentric coordinates of $P(t)$ with respect to P_0, P_1, P_2. We shall call the segment that lies in $\Delta(P_0, P_1, P_2)$ the standard segment, and the rest of the curve (including the endpoints P_0, P_2) the complementary "segment," even though the use of the term "segment" in the latter case is not strictly correct. We observe that the denominator $w(t)$ itself can be regarded as a planar Bézier curve $(t, w(t)) = \sum (i/2, w_i)\phi_i(t)$, since $2t = \sum i\phi_i(t)$.

It is easy to see that

$$P(0) = P_0, \qquad P(1) = P_2,$$

$$P'(0) = \frac{2w_1}{w_0}(P_1 - P_0), \qquad P'(1) = \frac{2w_1}{w_2}(P_2 - P_1);$$

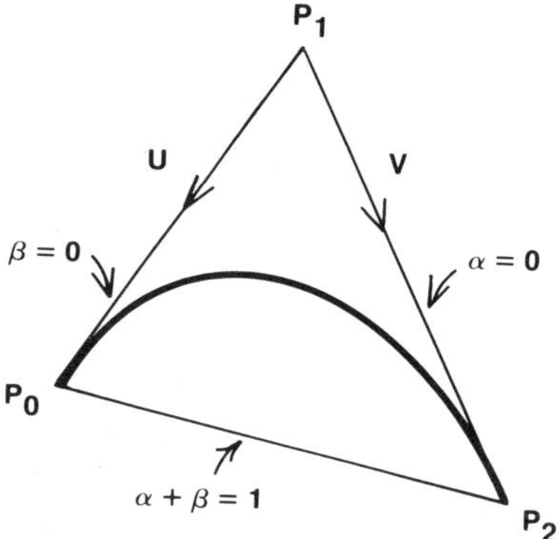

Figure 1.

thus the tangents at the endpoints $P(0)$, $P(1)$ meet at P_1. We exclude the degenerate case of P_0, P_1, P_2 being collinear. Then

$$U = P_0 - P_1, \qquad V = P_2 - P_1$$

are linearly independent. Any point X in the plane of P_0, P_1, P_2 can be written as

$$X = P_1 + \alpha U + \beta V$$
$$= \alpha P_0 + (1 - \alpha - \beta)P_1 + \beta P_2,$$

which shows the relationship of the coordinates α, β of X in the oblique coordinate system $(P_1; U, V)$ with the barycentric coordinates of X with respect to P_0, P_1, P_2. (See Fig. 1.) From (2), the oblique coordinates of $P(t)$ are $\alpha(t) = \psi_0(t)$, $\beta(t) = \psi_2(t)$. Note the happy coincidence that $\phi_0\phi_2 = \phi_1^2/4$; thus for any t,

$$\alpha(t)\beta(t) = \frac{w_0 w_2}{w^2(t)}\, \phi_0(t)\phi_2(t) = \frac{w_0 w_2}{4w^2(t)}\, \phi_1^2(t) = \frac{w_0 w_2}{4w_1^2}\, \psi_1^2(t).$$

Hence the curve satisfies the (oblique) implicit equation

$$(3) \qquad\qquad \alpha\beta = k(1 - \alpha - \beta)^2$$

where $k = w_0 w_2/4w_1^2$. Since (3) is quadratic, we can conclude that the entire curve $P(t)$ is a conic.

It will be seen in § 2 that various representations of the same curve segment in the form of (1) are possible, resulting in different values of $w_0 : w_1 : w_2$. However, since the coordinate system $(P_1; U, V)$ is fixed, the implicit equation (3) remains the same, hence $w_0 w_2/4w_1^2 = k$ is invariant under all such parameterizations.

As a second remark, we note that it is easy to convert (3) into one in rectangular coordinates. This is due to the important simple fact that barycentric coordinates are areas. For completeness, we give a proof of this fact here. Suppose $X = P_1 + \alpha U + \beta V$.

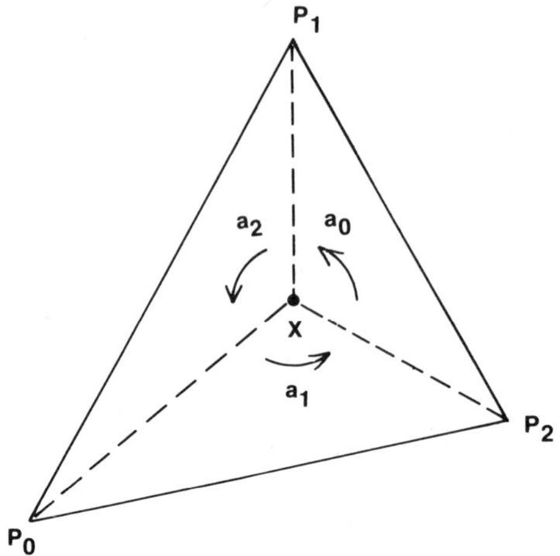

Figure 2.

Let N be a normal to the plane of P_0, P_1, P_2, and let

$$a = (P_0 - P_1) \times (P_2 - P_1) \cdot N,$$
$$a_0 = (P_2 - X) \times (P_1 - X) \cdot N/a,$$
$$a_1 = (P_0 - X) \times (P_2 - X) \cdot N/a,$$
$$a_2 = (P_1 - X) \times (P_0 - X) \cdot N/a.$$

Thus a_i is the (signed) area of the triangle with the vertex P_i replaced by X, weighted so that $\sum a_i = 1$. (See Fig. 2. Note that X need not be inside $\Delta(P_0, P_1, P_2)$.) Taking cross products of

$$0 = \alpha(P_0 - X) + (1 - \alpha - \beta)(P_1 - X) + \beta(P_2 - X)$$

with $P_2 - X$ and with $P_0 - X$, we obtain

$$\alpha a_1 = (1 - \alpha - \beta)a_0 \quad \text{and} \quad \beta a_1 = (1 - \alpha - \beta)a_2.$$

Hence we have the fact

$$(4) \qquad\qquad \alpha = a_0, \quad 1 - \alpha - \beta = a_1, \quad \beta = a_2.$$

Thus, if the plane of P_0, P_1, P_2 is the x-y plane and if $X = (x, y)$, $P_i = (x_i, y_i)$, then the oblique coordinates α, β of X are, in longhand,

$$\alpha = \frac{(y - y_1)(x_2 - x_1) - (x - x_1)(y_2 - y_1)}{(y_0 - y_1)(x_2 - x_1) - (x_0 - x_1)(y_2 - y_1)},$$
$$\beta = \frac{(y - y_1)(x_0 - x_1) - (x - x_1)(y_0 - y_1)}{(y_2 - y_1)(x_0 - x_1) - (x_2 - x_1)(y_0 - y_1)};$$

substitution into (3) gives the rectangular Cartesian equation.

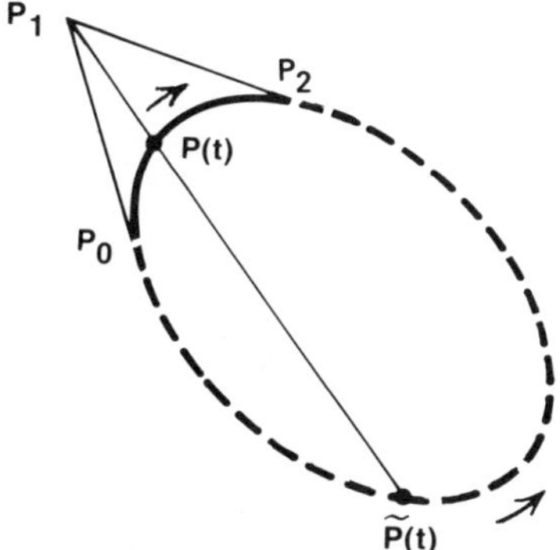

Figure 3.

1.2. Still with all $w_i > 0$, define the curve $\tilde{P}$ by changing the sign of the mid-weight:

(5)
$$\tilde{P}(t) = (w_0 P_0 \phi_0(t) - w_1 P_1 \phi_1(t) + w_2 P_2 \phi_2(t))/\bar{w}(t), \quad \text{with}$$
$$\bar{w}(t) = w_0 \phi_0(t) - w_1 \phi_1(t) + w_2 \phi_2(t).$$

The coordinates $\tilde{\alpha}$, $\tilde{\beta}$ of $\tilde{P}$ are $\tilde{\alpha} = w_0 \phi_0/\bar{w}$, $\tilde{\beta} = w_2 \phi_2/\bar{w}$, and $1 - \tilde{\alpha} - \tilde{\beta} = -w_1 \phi_1/\bar{w}$. Thus the same implicit equation (3) is satisfied by $(\tilde{\alpha}, \tilde{\beta})$, i.e., $\tilde{P}$ describes the same conic. Furthermore, (5) shows that

(6)
$$\tilde{P}(t) - P_1 = \frac{w(t)}{\bar{w}(t)} (P(t) - P_1)$$

whenever $\bar{w}(t) \neq 0$; in other words,

$$P_1, P(t), \tilde{P}(t) \text{ are collinear.}$$

Thus the complementary segment is given by $\tilde{P}(t)$, $0 \leq t \leq 1$. (See Fig. 3. Note the reversal of traverse.)

The roots of $\bar{w}$, if they exist, lie in $(0, 1)$, since $\bar{w}$ has weights alternating in sign. From $\bar{w}(t) = (w_0 + 2w_1 + w_2)t^2 - 2(w_0 + w_1)t + w_0$, the roots t_1, t_2 are given by

(7)
$$t_i = \frac{w_0 + w_1 \pm w_1 \sqrt{1 - 4k}}{w_0 + 2w_1 + w_2}.$$

We can now obtain the type classification of $P(t)$ from the behavior of $\tilde{P}$ on $[0, 1]$:

- $4k > 1$: $\bar{w}$ has no real roots, so it is bounded away from zero, and $\tilde{P}[0, 1]$ is bounded. This is the elliptic case.
- $4k = 1$: $t_1 = t_2$ but $\bar{w}$ does not change sign. Hence $\tilde{P}[0, 1]$ lies in the first quadrant of the coordinate system $(P_1; U, V)$, but is unbounded. This is the parabolic case. $P_1 P(t_1)$ gives the direction of the axis.
- $4k < 1$: $\bar{w}$ has distinct roots $t_1 < t_2$. As t goes through t_1, $\tilde{P}(t)$ flips from the first quadrant to the third (see (6)); it flips back to the first quadrant after t_2.

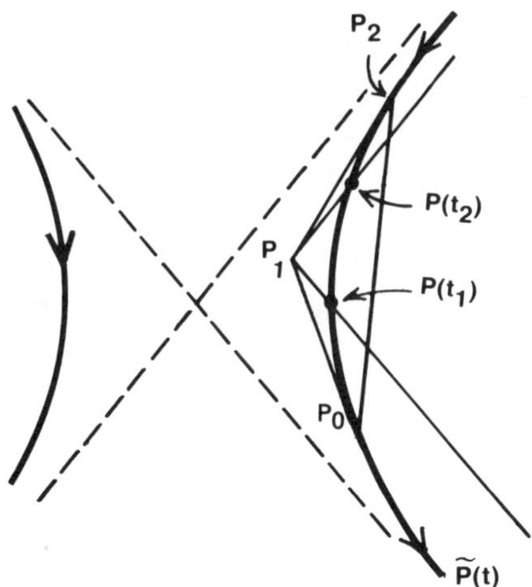

Figure 4.

This is the hyperbolic case. $P_1 P(t_i)$ meets the curve only once at $P(t_i)$. These two directions are the asymptotic directions. See Fig. 4.

Remark. We can easily recover the parameter t for any point $X = P(t)$ on the curve. Let the line joining P_1 and X intersect $P_0 P_2$ at $Y = (1 - \lambda)P_0 + \lambda P_2$. ($\lambda = |Y - P_0|/|P_2 - P_0|$.) Then $P(t) - P_1$ is a multiple of $Y - P_1$, and a comparison of the coordinates shows that

$$(8) \qquad \frac{1 - \lambda}{w_0 \phi_0(t)} = \frac{\lambda}{w_2 \phi_2(t)},$$

giving the roots

$$(9) \qquad \tau_1 = \frac{\sqrt{\lambda w_0}}{\sqrt{\lambda w_0} + \sqrt{(1 - \lambda)w_2}}, \qquad \tau_2 = \frac{\sqrt{\lambda w_0}}{\sqrt{\lambda w_0} - \sqrt{(1 - \lambda)w_2}}.$$

Clearly, if X is on the standard segment, $X = P(\tau_1)$, and if X is on the complementary segment, $X = P(\tau_2)$. Since $P(\tau_2) = \tilde{P}(\tau_1)$ by collinearity, and since $\tau_1 + \tau_2 = 2\tau_1\tau_2$ by (9), we have the following correspondence:

$$(10) \qquad \tilde{P}(t) = P\left(\frac{t}{2t - 1}\right).$$

1.3. One advantage of the rational Bézier form is its invariance under some commonly used transformations in CAGD. If T is an affine transformation of the 3-space (i.e., a linear transformation followed by a translation, $T(X) = L(X) + A$), then $T(P(t)) = \Sigma L(P_i)\psi_i(t) + A \Sigma \psi_i(t) = \Sigma T(P_i)\psi_i(t)$, that is, $T(P(t)) = \Sigma w_i T(P_i)\phi_i(t)/w(t)$. An example is a parallel projection. The new control points are projections of the old.

A transformation that is not affine is the perspective or central projection $X \to S(X) \in \mathcal{P}$ from a center C onto a plane $\mathcal{P}$ not containing C, $\mathcal{P}$ having a normal N and passing through a point Q:

$$S(X) = (1 - \rho)X + \rho C$$

with

$$\rho = \frac{(X - Q) \cdot N}{(X - C) \cdot N}.$$

Simple algebra shows that S also preserves the rational Bézier form (however, with changed weights):

$$S(P(t)) = \frac{\sum w_i' S(P_i)\phi_i(t)}{\sum w_i' \phi_i(t)}$$

with

$$w_i' = w_i(P_i - C) \cdot N.$$

As a simple application, the weights and control points of a planar intersection with a cone based on $P(t)$ can be readily computed.

These remarks pertain, of course, to any rational Bézier curves. For quadratics, the above relationship between the weights also shows that, given a conic, we can easily find a center C and a plane $\mathcal{P}$ so that the perspective image $S(P(t))$ from C is, say, a parabola. We simply pick any C not in the plane of P_0, P_1, P_2. Then $A_i = w_i(P_i - C)$ are linearly independent. Hence the system

$$A_i \cdot N = 1, \qquad i = 0, 1, 2$$

is nonsingular and can be solved for the required normal N.

1.4. We have required at the beginning that $w_i > 0$ for all i. This is really no restriction. In the general case,

$$P(t) = \sum V_i\phi_i(t) \Big/ \sum w_i\phi_i(t).$$

Since $P(0) = V_0/w_0$, $P(1) = V_2/w_2$, we should require at least that $w_0 \neq 0$, $w_2 \neq 0$. The properties of $P(t)$ are determined on any interval of t. For sufficiently small t_0, the curve $P(t)$, $0 \leq t \leq t_0$, when linearly reparameterized to $[0, 1]$, yields a rational Bézier curve with its new weights all of the same sign. The reparameterization is effected by subdivision at t_0.

The subdivision algorithm for Bézier curves is well known. (See Hosaka and Kimura [6], Lane and Riesenfeld [7], or Boehm et al. [1], where it is attributed to de Casteljau.) To subdivide a rational curve

$$P(t) = \sum V_i\phi_i(t) \Big/ \sum w_i\phi_i(t)$$

at $t_0 \in (0, 1)$, simply split the numerator and the denominator. We review the scheme. With $V_{0,i} = V_i$, $w_{0,i} = w_i$, form

$$
\begin{array}{ccccccc}
V_{0,0} & V_{0,1} & V_{0,2} & \quad & w_{0,0} & w_{0,1} & w_{0,2} \\
& V_{1,1} & V_{1,2} & & & w_{1,1} & w_{1,2} \\
& & V_{2,2} & & & & w_{2,2}
\end{array}
$$

where $V_{i,j} = (1 - t_0)V_{i-1,j-1} + t_0 V_{i-1,j}$, and similarly for $w_{i,j}$. The left part of the curve, i.e., $P(t_0 t)$ for $0 \le t \le 1$, has weights $w_{0,0}$, $w_{1,1}$, $w_{2,2}$ and hence control points $V_{0,0}/w_{0,0}$, $V_{1,1}/w_{1,1}$, $V_{2,2}/w_{2,2}$, if the new weights are nonzero. Note the new weights will be of the same sign if t_0 is small: simply look at the graph of the denominator as a planar Bézier curve. Similarly the right part of the curve, i.e., $P(t_0 + (1 - t_0)t)$ for $0 \le t \le 1$, will have weights $w_{2,2}$, $w_{1,2}$, $w_{0,2}$ and control points (if the new weights are nonzero), $V_{2,2}/w_{2,2}$, $V_{1,2}/w_{1,2}$, $V_{0,2}/w_{0,2}$. Note that the tangent at $P(t_0)$ is in the direction of

$$
\frac{V_{1,2}}{w_{1,2}} - \frac{V_{1,1}}{w_{1,1}}.
$$

2. Weights and Parameterization. To fix a particular member among the family of conics through P_0 and P_2 with tangent intercept P_1, it is necessary and sufficient to specify another point P_* on the conic, usually called a shoulder point. (See (3). This fixes k.) However, to fix a parameterization (i.e. to determine the weights $w_0 : w_1 : w_2$), one needs also to specify a parameter value t_* and determine the weights by requiring that

$$
P(t_*) = P_*.
$$

Let a_0, a_1, a_2 be the areas determined by P_* and P_0, P_1, P_2, as described in § 1.1. The barycentric coordinates of $P(t_*)$ are $\alpha = \psi_0(t_*)$, $1 - \alpha - \beta = \psi_1(t_*)$, $\beta = \psi_2(t_*)$, as in (2). Thus by (4), we have

$$
(11) \qquad \frac{w_0}{a_0/\phi_0(t_*)} = \frac{w_1}{a_1/\phi_1(t_*)} = \frac{w_2}{a_2/\phi_2(t_*)}.
$$

Clearly for P_* in the standard segment, a choice of $t_* \in (0, 1)$ will give all $w_i > 0$. The particular choice $t_* = \frac{1}{2}$, $(\phi_0 : \phi_1 : \phi_2 = 1 : 2 : 1)$, shows that we may simply take

$$
(12) \qquad w_0 : w_1 : w_2 = a_0 : \tfrac{1}{2}a_1 : a_2.
$$

Note, whatever t_*, $k = w_0 w_2/4w_1^2$ is always $a_0 a_2/a_1^2$.

Example 1. Consider a quarter of the unit circle, with $P_0 = (1, 0)$, $P_1 = (1, 1)$, $P_2 = (0, 1)$ (see Fig. 5). Taking $P_* = (3/5, 4/5)$, simple geometry shows that $a_0 : a_1 : a_2 = 1 : 2 : 2$, so that with $t_* = \frac{1}{2}$, we have $w_0 : w_1 : w_2 = 1 : 1 : 2$ and

$$
P(t) = \binom{x(t)}{y(t)} = \frac{\binom{1}{0}(1 - t)^2 + 2\binom{1}{1}t(1 - t) + 2\binom{0}{1}t^2}{(1 - t)^2 + 2t(1 - t) + 2t^2}, \qquad 0 \le t \le 1,
$$

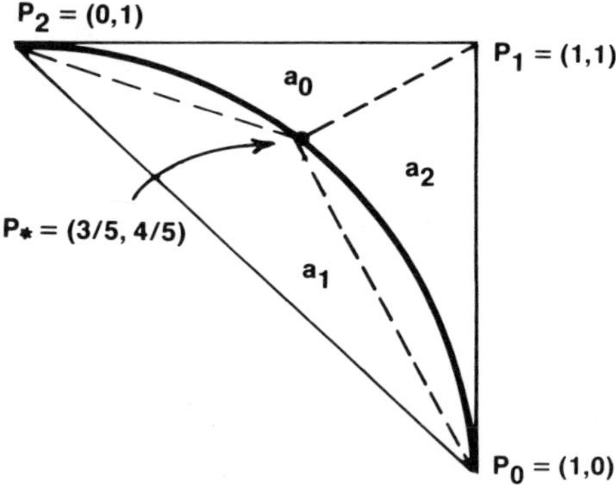

Figure 5.

which is the usual parameterization

$$x(t) = \frac{1-t^2}{1+t^2}, \qquad y(t) = \frac{2t}{1+t^2}.$$

(Note here and elsewhere that we do not make a distinction between "row vectors" and "column vectors.") Application of (9) shows that the point $(4/5, 3/5)$ is attained at $t = \frac{1}{3}$. $\square$

An often used parameterization, sometimes called the rho-conic parameterization, is given by the following special choice. Here we take $t_* = \frac{1}{2}$ and $P_* = \bar{P}$, an intersection of $P_1 P_m$ with the conic, $P_m = (P_0 + P_2)/2$ being the midpoint of the chord. See Fig. 6. Again we need not restrict $\bar{P}$ to be in the standard segment. The so-called rho-value (or the conic shape factor in Forrest [5]), ρ, is defined by

$$(13) \qquad\qquad \bar{P} - P_m = \rho(P_1 - P_m).$$

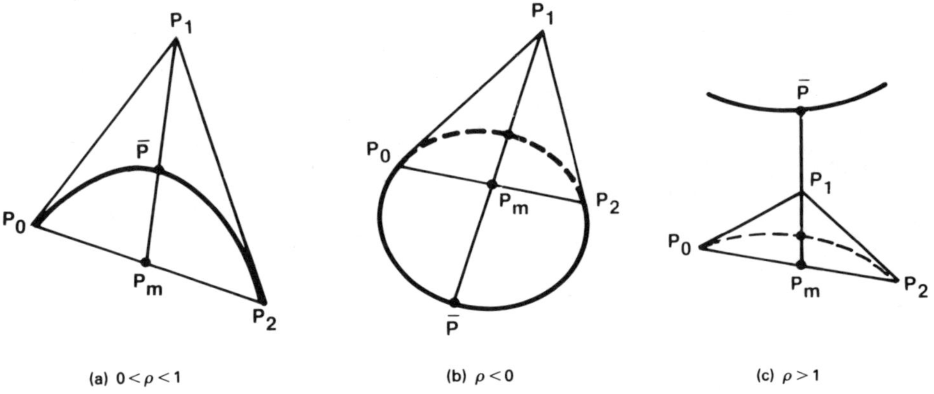

Figure 6.

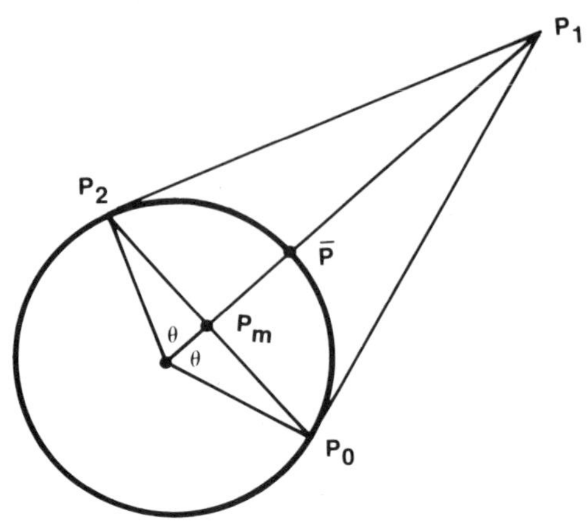

Figure 7.

In terms of ρ, we have

(14)
$$w_0 : w_1 : w_2 = 1 - \rho : \rho : 1 - \rho.$$

For $0 < \rho < 1$, i.e., $\bar{P}$ in $\Delta(P_0, P_1, P_2)$, (14) is easy to see by simple geometry. ($a_0 : a_1 : a_2 = 1 - \rho : 2\rho : 1 - \rho$.) In general, it can also be derived by simple vector algebra, which we omit here.

Note that $k = w_0 w_2 / 4w_1^2 = (1 - \rho)^2 / 4\rho^2$. Solving for ρ, we have

(15)
$$\rho_1 = \frac{1}{1 + 2\sqrt{k}}, \qquad \rho_2 = \frac{1}{1 - 2\sqrt{k}}.$$

ρ_1, always in $(0, 1)$, pertains to $\bar{P} = \bar{P}^1$ inside $\Delta(P_0, P_1, P_2)$. $\rho_1 = \frac{1}{2}$ for parabola, $\rho_1 < \frac{1}{2}$ for ellipse and $\rho_1 > \frac{1}{2}$ for hyperbola. ρ_2 pertains to $\bar{P} = \bar{P}^2$ outside the triangle; $\rho_2 < 0$ for ellipse, $\rho_2 > 1$ for hyperbola, but does not exist in the parabolic case. Note that (15) affords a simple way of changing any given parameterization to the rho-conic parameterization (with the same P_0, P_1, P_2); see example below.

Example 2.   Consider a circular arc, unit radius, subtending an angle of 2θ, $0 < \theta < \pi/2$ (see Fig. 7). Then $|\bar{P} - P_m| = 1 - \cos\theta$, $|P_1 - P_m| = |P_1 - P_0|\sin\theta = \sin^2\theta / \cos\theta$. Thus $\rho = |\bar{P} - P_m| / |P_1 - P_m| = \cos\theta / (1 + \cos\theta)$, $1 - \rho = 1/(1 + \cos\theta)$, and we may take $w_0 : w_1 : w_2 = 1 : \cos\theta : 1$ in the rho-conic representation:

$$P(t) = \frac{P_0(1 - t)^2 + 2\cos\theta P_1 t(1 - t) + P_2 t^2}{(1 - t)^2 + 2(\cos\theta)t(1 - t) + t^2}, \qquad 0 \le t \le 1.$$

For the arc in Example 1, $\theta = \pi/4$, so

$$P(t) = \frac{\binom{1}{0}(1 - t)^2 + \sqrt{2}\binom{1}{1}t(1 - t) + \binom{0}{1}t^2}{(1 - t)^2 + \sqrt{2}t(1 - t) + t^2}, \qquad 0 \le t \le 1.$$

This second form can be obtained directly from Example 1 through (15), since from Example 1, $w_0:w_1:w_2 = 1:1:2$ and $k = \frac{1}{2}$. Hence $\rho_1/(1 - \rho_1) = 1/(2\sqrt{k}) = 1/\sqrt{2}$.   □

We also include an example illustrating the use of subdivision.

Example 3.   Let $\mathscr{C}$ be the quarter circle described by P in Example 1, and $\tilde{\mathscr{C}}$ be its complement described by $\tilde{P}$. Let us split $\tilde{P}$ at parameter value $\frac{1}{2}$. The numerator and denominator of $\tilde{P}$ are subdivided according to the scheme described in § 1.4:

$$\begin{pmatrix} 1 \\ 0 \end{pmatrix} \quad \begin{pmatrix} -1 \\ -1 \end{pmatrix} \quad \begin{pmatrix} 0 \\ 2 \end{pmatrix} \qquad\qquad 1 \quad -1 \quad 2$$

$$\begin{pmatrix} 0 \\ -\frac{1}{2} \end{pmatrix} \quad \begin{pmatrix} -\frac{1}{2} \\ \frac{1}{2} \end{pmatrix} \qquad\qquad\qquad 0 \quad\quad \tfrac{1}{2}$$

$$\begin{pmatrix} -\frac{1}{4} \\ 0 \end{pmatrix} \qquad\qquad\qquad\qquad\qquad \tfrac{1}{4}$$

obtaining a left portion $\tilde{\mathscr{C}}_{\mathrm{L}}$ and a right portion $\tilde{\mathscr{C}}_{\mathrm{R}}$:

$$\tilde{\mathscr{C}}_{\mathrm{L}}: \quad \frac{\begin{pmatrix} 1 \\ 0 \end{pmatrix}\phi_0 - \dfrac{1}{2}\begin{pmatrix} 0 \\ 1 \end{pmatrix}\phi_1 + \dfrac{1}{4}\begin{pmatrix} -1 \\ 0 \end{pmatrix}\phi_2}{\phi_0 + \dfrac{1}{4}\phi_2},$$

$$\tilde{\mathscr{C}}_{\mathrm{R}}: \quad \frac{\dfrac{1}{4}\begin{pmatrix} -1 \\ 0 \end{pmatrix}\phi_0 + \dfrac{1}{2}\begin{pmatrix} -1 \\ 1 \end{pmatrix}\phi_1 + 2\begin{pmatrix} 0 \\ 1 \end{pmatrix}\phi_2}{\dfrac{1}{4}\phi_0 + \dfrac{1}{2}\phi_1 + 2\phi_2}$$

(see Fig. 8). Note $\tilde{\mathscr{C}}_{\mathrm{L}}$ has its mid-weight w_1 zero. This means that the tangents at the endpoints of $\tilde{\mathscr{C}}_{\mathrm{L}}$ are parallel. We do not have space to elaborate on the parallel tangent case here. However, a further subdivision of $\tilde{\mathscr{C}}_{\mathrm{L}}$ at parameter $\frac{1}{2}$ gives

$$\tilde{\mathscr{C}}_{\mathrm{LL}}: \quad \frac{\begin{pmatrix} 1 \\ 0 \end{pmatrix}\phi_0 + \dfrac{1}{2}\begin{pmatrix} 1 \\ -\frac{1}{2} \end{pmatrix}\phi_1 + \dfrac{5}{16}\begin{pmatrix} \frac{3}{5} \\ -\frac{4}{5} \end{pmatrix}\phi_2}{\phi_0 + \dfrac{1}{2}\phi_1 + \dfrac{5}{16}\phi_2},$$

$$\tilde{\mathscr{C}}_{\mathrm{LR}}: \quad \frac{\dfrac{5}{16}\begin{pmatrix} \frac{3}{5} \\ -\frac{4}{5} \end{pmatrix}\phi_0 + \dfrac{1}{8}\begin{pmatrix} -1 \\ -2 \end{pmatrix}\phi_1 + \dfrac{1}{4}\begin{pmatrix} -1 \\ 0 \end{pmatrix}\phi_2}{\dfrac{5}{16}\phi_0 + \dfrac{1}{8}\phi_1 + \dfrac{1}{4}\phi_2}.$$

Were we to subdivide $\tilde{\mathscr{C}}_{\mathrm{L}}$ at parameter value $\frac{2}{3}$, we would get two quarter circles instead. (Why $\frac{2}{3}$? By the parameter recovery procedure and (9), one finds that the point $(0, -1)$ is $P(-1)$, which is $\tilde{P}(\frac{1}{3})$ by (10). Note also, there is a rescaling $t \to 2t$ from the domain of $\tilde{P}$ to that of the representation of $\tilde{\mathscr{C}}_{\mathrm{L}}$ above.)

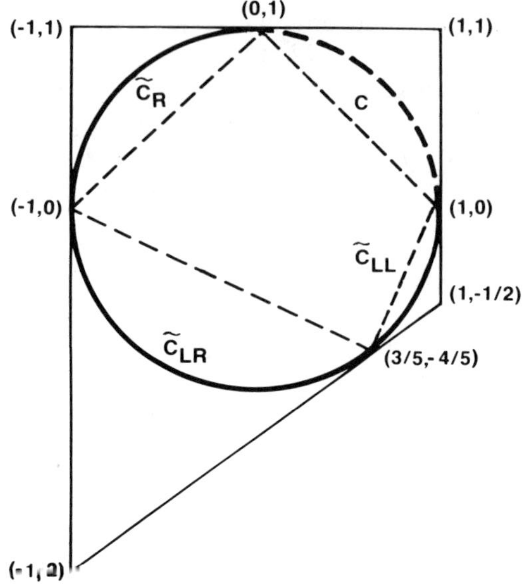

Figure 8.

Similarly if $\mathscr{C}$ is the circular arc subtending an angle of $2\theta = 2\pi/3$, parameterized by $P(t)$ in Example 2, then $P\bar{P}$ intersects the circle in a second point $\bar{P}(\frac{1}{2})$. Hence, splitting P at $\frac{1}{2}$ would give C as two arcs of $2\pi/3$ each.   □

We have given examples for the circle, so as to have nice numbers. The reader is invited to consider further examples for noncircular segments.

Remark. In Faux and Pratt [3, § 5.2.1], a conic segment is initially parameterized in the following way. A point P on the curve is described by the two intercepts A, B of the tangent at P with P_0P_1, P_2P_1, respectively, that is, by the numbers g_1, g_2:

$$A - P_1 = g_1(P_0 - A), \qquad B - P_1 = g_2(P_2 - B).$$

See Fig. 9. Suppose the conic is also given by $P(t)$ as in (1). It is easy to see the

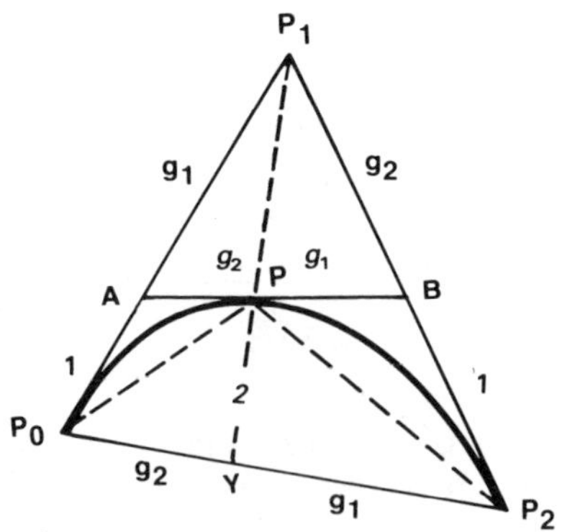

Figure 9.

relationship between the two parameterizations. Let $P = P(t)$ have the intercept ratios g_1, g_2. Subdivision as described in § 1.4 shows that $A = V_{1,1}/w_{1,1}$, $B = V_{1,2}/w_{1,2}$. That is,

$$(16) \qquad A = \frac{(1-t)w_0 P_0 + tw_1 P_1}{(1-t)w_0 + tw_1}, \qquad B = \frac{(1-t)w_1 P_1 + tw_2 P_2}{(1-t)w_1 + tw_2}.$$

Forming $A - P_1$, $P_0 - A$, etc., we immediately read off

$$g_1 = \frac{w_0(1-t)}{w_1 t} = \frac{2w_0 \phi_0(t)}{w_1 \phi_1(t)}, \qquad g_2 = \frac{w_2 t}{w_1(1-t)} = \frac{2w_2 \phi_2(t)}{w_1 \phi_1(t)},$$

which describes the reparameterization. By (11), the areas determined by (i.e., the barycentric coordinates of) the point $P = P(t)$ are

$$\frac{a_0}{g_1} = \frac{a_1}{2} = \frac{a_2}{g_2},$$

as can be also read off from (5.23) of Faux and Pratt [3]. Alternatively (from (8)), P is the intersection of AB with $P_1 Y$, where $Y = (1 - \lambda)P_0 + \lambda P_2$, with $(1 - \lambda):\lambda = g_1:g_2$.

Incidentally (16) also shows that, given any u_0, u_1, u_2, with $u_0 u_2/u_1^2 = w_0 w_2/w_1^2$, one can reparameterize (1) so that u_0, u_1, u_2 are the weights, that is,

$$\sum w_i P_i \phi_i(t) \Big/ \sum w_i \phi_i(t) = \sum u_i P_i \phi_i(s) \Big/ \sum u_i \phi_i(s).$$

The reparameterization is simply given by (see (16))

$$(17) \qquad \frac{w_0(1-t)}{w_1 t} = \frac{u_0(1-s)}{u_1 s} \quad \text{or} \quad \frac{w_2 t}{w_1(1-t)} = \frac{u_2 s}{u_1(1-s)}.$$

This can of course also be observed directly from (4).

3. Conic Characteristics.   The rational Bézier form for conics has the advantage of being coordinate independent. To further emphasize this point, we now wish to show that all the important geometric characteristics of a conic can be expressed in terms of the control points P_0, P_1, P_2 and $k = w_0 w_2/4w_1^2$. We start off with a formula for the center of a central conic:

$$(18) \qquad \text{Center:} \qquad C = P_1 + \frac{2k}{4k - 1}(U + V),$$

where $U = P_0 - P_1$, $V = P_2 - P_1$ as before (see Fig. 10).

In preparation for a proof, we first note a simple Lemma: Let $\bar{P}$ be as before, (i.e., an intersection of the conic with $P_1 P_m$). Then the tangent at $\bar{P}$ is parallel to $P_0 P_2$. This is easily seen through subdivision. We split the rho-conic representation at $\bar{P} = P(\frac{1}{2})$. In the notation of § 1.4, we get $2V_{1,1} = (1-\rho)P_0 + \rho P_1$, $2V_{1,2} = (1-\rho)P_2 +$

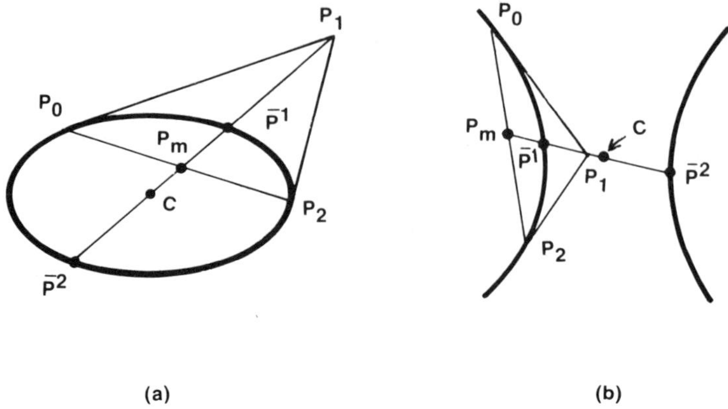

Figure 10.

ρP_1 and $w_{1,1} = w_{1,2} = \frac{1}{2}$. Hence the tangent at $\bar{P}$ is in the direction of

$$\frac{V_{1,2}}{w_{1,2}} - \frac{V_{1,1}}{w_{1,1}} = (1 - \rho)(P_2 - P_0).$$

By definition, the center C lies on the bisectors of parallel chords. The above lemma says that the tangents at $\bar{P}^1$ and $\bar{P}^2$ are limiting members of a family of parallel chords, hence C must be the midpoint of $\bar{P}^1$ and $\bar{P}^2$. (See also Note at the end of the section.) In the parabolic case, $\bar{P}^2$ does not exist, hence no center. Otherwise, by (13) and (15),

$$C = \frac{\bar{P}_1 + \bar{P}_2}{2} = P_m + \frac{\rho_1 + \rho_2}{2}(P_1 - P_m)$$

$$= P_1 + \frac{U + V}{2}\left(1 - \frac{\rho_1 + \rho_2}{2}\right) = P_1 + \frac{2k}{4k - 1}(U + V).$$

Below is a list of symbols, together with their definitions and geometric meanings, that will be reserved for the further formulas:

$$\varepsilon = 2k/(4k - 1),$$
$$\alpha = |U|^2, \quad \beta = U \cdot V, \quad \gamma = |V|^2,$$
$$\delta = \alpha\gamma - \beta^2 = |U \times V|^2,$$
$$\zeta = \alpha + \gamma + 2\beta = |U + V|^2,$$
$$\eta = \alpha + \gamma - 2\beta = |U - V|^2.$$

For the parabola, we have the following formulas:

(19)   Focus of parabola: $F = P_1 + \dfrac{\gamma U + \alpha V}{\zeta}.$

(20)   Vertex of parabola: $X = P_1 + \left(\dfrac{\gamma + \beta}{\zeta}\right)^2 U + \left(\dfrac{\alpha + \beta}{\zeta}\right)^2 V.$

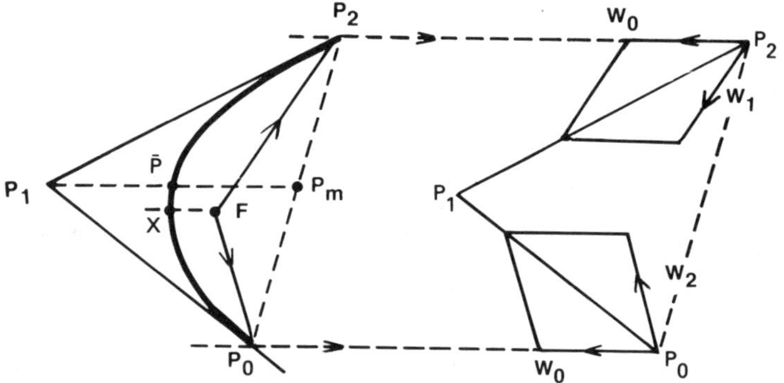

Figure 11. Illustrating the proof of (19).

We will use the reflection property of the parabola to find the focus F. The axis is parallel to $P_1\bar{P}$, since $P_1\bar{P}$ meets the parabola only once. Let $W_0 = 2(P_1 - P_m) = -(U + V)$, and determine $W_1 = xU + yV$ so that

$$(21) \qquad |W_1| = |W_0| \quad \text{and} \quad W_1 + W_0 = \lambda(P_1 - P_2).$$

This means that W_0 and W_1 make equal angles with the tangent at P_2, and consequently the line through P_2 in the direction of W_1 will contain F. See Fig. 11. The second equation in (21) determines $x = 1$, $y = 1 - \lambda$. Then, since $|W_0|^2 = \alpha + 2\beta + \gamma$, $|W_1|^2 = \alpha + 2\beta(1 - \lambda) + \gamma(1 - \lambda)^2$, their equality leads to the nontrivial solution

$$\lambda = \frac{2(\beta + \gamma)}{\gamma}.$$

(The trivial solution $\lambda = 0$ is inadmissible unless $W_0 \cdot (P_1 - P_2) = (U + V) \cdot V = \beta + \gamma = 0$, but this is already contained in the above formula.) Thus

$$W_1 = U + \left(1 - 2\frac{\beta + \gamma}{\gamma}\right)V.$$

Similarly one can determine W_2 such that $|W_2| = |W_0|$ and $W_2 + W_0 = \mu(P_1 - P_0)$. A similar procedure as above leads to

$$W_2 = \left(1 - 2\frac{\alpha + \beta}{\alpha}\right)U + V.$$

Now, according to the reflection property mentioned above, $F = P_2 + sW_1$ for some s, which is determined by $(F - P_0) \times W_2 = 0$. A little algebra shows that $s = \gamma/\zeta$, and this gives (19) immediately.

Next, the vertex X of the parabola is at

$$X = F + rW_0 = P_1 + \left(\frac{\gamma}{\zeta} - r\right)U + \left(\frac{\alpha}{\zeta} - r\right)V$$

for some r. For X to be on the conic, its coordinates must satisfy the implicit equation

(3), with $k = \frac{1}{4}$:

$$4\left(\frac{\gamma}{\zeta} - r\right)\left(\frac{\alpha}{\zeta} - r\right) = \left(\frac{\alpha + \gamma}{\zeta} - 2r - 1\right)^2,$$

which reduces to $\zeta^2 r = \delta$. With this r, we get (20).

We now turn to finding vertices of the ellipse and the hyperbola. These are points on the conic whose distances from the center C achieve extrema. Thus, let $W = xU + yV$ be a unit vector (i.e., $\alpha x^2 + 2\beta xy + \gamma y^2 = 1$) and consider the point

$$C + sW = P_1 + (\varepsilon + sx)U + (\varepsilon + sy)V$$

on the conic. The implicit equation

$$(\varepsilon + sx)(\varepsilon + sy) = k(2\varepsilon + sx + sy - 1)^2$$

reduces to

$$2(k(x + y)^2 - xy)s^2 = \varepsilon.$$

Thus $C \pm sW$ is on the conic with

(22) $$s = \sqrt{\varepsilon/(2k(x + y)^2 - 2xy)},$$

and we must therefore find the extreme values of

$$f(x, y) = 2k(x + y)^2 - 2xy$$

subject to the constraint $\alpha x^2 + 2\beta xy + \gamma y^2 = 1$.

The standard way to do this is through the use of Lagrange multipliers, leading to an eigenvalue problem. The eigenvalues are the extreme values of f; the corresponding eigenvectors give the principal directions. We will not carry out the routine details here, but merely state the results.

The resulting characteristic equation for the eigenvalue λ is

(23) $$\delta\lambda^2 - 2(k\eta + \beta)\lambda + (4k - 1) = 0.$$

Let the eigenvalues be $\lambda_1 \leq \lambda_2$. In the elliptic case, $\lambda_1\lambda_2 = (4k - 1)/\delta > 0$, and $\lambda_1 + \lambda_2 = 2(k\eta + \beta)/\delta > 0$ irrespective of the sign of β, since also $k\eta + \beta = k\zeta - (4k - 1)\beta$ (see the list of symbols). Thus $\lambda_1 > 0$. In the hyperbolic case, $\lambda_1\lambda_2 < 0$, so $\lambda_1 < 0 < \lambda_2$; note here $\varepsilon < 0$, also. Hence by (22), the semi-axis lengths are:

$$\text{ellipse:} \quad d_1 = \sqrt{\varepsilon/\lambda_1} \geq d_2 = \sqrt{\varepsilon/\lambda_2},$$
$$\text{hyperbola:} \quad d_1 = \sqrt{\varepsilon/\lambda_1}.$$

Let X_1, X_2 be the two vertices on the hyperbola, or the two vertices on the major axis of the ellipse. Let $\bar{x}$, $\bar{y}$ be given by any one of the following sets:

$$\bar{x} = \beta\lambda_1 - 2k + 1, \qquad \bar{y} = 2k - \alpha\lambda_1$$

or

$$\bar{x} = 2k - \gamma\lambda_1, \qquad \bar{y} = \beta\lambda_1 - 2k + 1;$$

and let

$$x_0 = \frac{\bar{x}}{\rho}, \quad y_0 = \frac{\bar{y}}{\rho}$$

with

$$\rho = \alpha\bar{x}^2 + 2\beta\bar{x}\bar{y} + \gamma\bar{y}^2.$$

Then the vertices X_1, X_2 are given by

$$X_{1,2} = P_1 + (\varepsilon \pm d_1 x_0)U + (\varepsilon \pm d_1 y_0)V.$$

It is of course unnecessary to give a similar formula for the vertices on the minor axis of the ellipse, since they are at distance d_2 from C and since the two axes are orthogonal. The foci of the ellipse can be obtained from the center and the vertices.

Example 4. Let us check the semi-axis lengths for the case of the circular arc given in Example 2. Here $k(1 - \rho)^2/4\rho^2 = 1/(4\cos^2\theta)$, and $\varepsilon = 2k/(4k - 1) = 1/(2\sin^2\theta)$. Finding α, β, γ, δ, η by simple geometry (see Fig. 7), one sees that (23) is

$$4(\sin^4\theta)\lambda^2 - 4(\sin^2\theta)\lambda + 1 = 0,$$

having equal roots $\lambda = 1/(2\sin^2\theta) = \varepsilon$. Thus $d_1 = d_2 = 1$ as expected. $\square$

Note. We are indebted to the referee for a simpler proof of the formula (18). The implicit equation (3) is symmetric in α and β. If $Q = P_1 + \alpha U + \beta V$ is a point on the conic, so is $R = P_1 + \beta U + \alpha V$. Clearly, such segments $\overline{QR}$ are all parallel (to $P_0 P_2$), and are bisected by the line $P_1 P_m$. Hence C must be the midpoint of $\bar{P}^1$ and $\bar{P}^2$. (We have kept the lemma in the text, however, as its proof illustrates another application of the subdivision technique.)

Acknowledgments. The author wishes to thank Miriam Lucian for her interest and several inspiring conversations, and Robert Blomgren for his recognition of the usefulness of this work and his constant urgings to put it into print. A previous version containing additional materials was written and circulated in 1983.

REFERENCES

[1]   W. BOEHM, G. FARIN AND J. KAHMANN, *A survey of curve and surface methods in* CAGD, Computer Aided Geometric Design, 1 (1984), pp. 1–60.

[2]   G. FARIN, *Algorithms for rational Bézier curves*, Computer Aided Design, 15 (1983), pp. 73–77.

[3]   I. E. FAUX AND M. J. PRATT, *Computational Geometry for Design and Manufacture*, Ellis Horwood, Chichester, 1979.

[4]   A. R. FORREST, *Curves and surfaces for computer-aided design*, Ph.D. thesis, Univ. Cambridge, Cambridge, 1968.

[5]   ———, *The twisted cubic curve: a computer aided geometric design approach*, Computer Aided Design, 12 (1980), pp. 165–172.

[6]   M. HOSAKA AND F. KIMURA, *A theory and methods for three dimensional free form shape construction*, J. Inform. Process., 3 (1980), pp. 140–151.

[7]   J. M. LANE AND R. F. RIESENFELD, *A theoretical development for the computer generation and display of piecewise polynomial surfaces*, IEEE Trans. Pattern Anal. Machine Intelligence, 2 (1980), pp. 35–46.

[8]   T. PAVLIDIS, *Curve fitting with conic splines*, ACM Trans. Graphics, 2 (1983), pp. 1–31.

[9]   L. PIEGL, *Defining curves containing conic segments*, Computer and Graphics, 8 (1984), pp. 177–182.

B-Splines Without Divided Differences

CARL DE BOOR AND KLAUS HÖLLIG

Abstract. This note develops the basic B-spline theory without using divided differences. Instead, the starting point is the definition of B-splines via recurrence relations. This approach yields very simple derivations of basic properties of spline functions and algorithms.

1. Basic Properties. The standard definition of the univariate B-spline uses divided differences, and the calculus of divided differences has been heavily used in developing the univariate spline theory. In this note we propose an alternative approach. We choose the well-known B-spline recurrence relations [2], [5] as the starting point and derive from them the main properties of B-splines and the basic algorithms in a simple way.

Let $\mathbf{t} := \cdots, t_{-1}, t_0, t_1, \cdots$ be a nondecreasing, bi-infinite sequence with $\lim_{i \to \pm\infty} t_i = \pm\infty$. The **B-splines** corresponding to the "knot sequence" $\mathbf{t}$ are defined by the recurrence relation

$$(1a) \qquad B_{i,k} := \omega_{i,k} B_{i,k-1} + (1 - \omega_{i+1,k}) B_{i+1,k-1},$$

with

$$(1b) \qquad B_{i,1}(t) := \begin{cases} 1 & \text{if } t_i \le t < t_{i+1}, \\ 0 & \text{otherwise,} \end{cases}$$

$$\omega_{i,k}(t) := \begin{cases} \dfrac{t - t_i}{t_{i+k-1} - t_i} & \text{if } t_i < t_{i+k-1}, \\ 0 & \text{otherwise.} \end{cases}$$

This gives $B_{i,k}$ in the form

$$(2) \qquad B_{i,k} = \sum_{j=i}^{i+k-1} b_{j,i} B_{j,1},$$

with each $b_{j,i}$ a polynomial degree $<k$ since it is the sum of products of $k-1$ linear polynomials. From this we read off that $B_{i,k}$ is a piecewise polynomial of degree $<k$ which vanishes outside the interval $[t_i, t_{i+k}]$ and has possible breakpoints $t_i, \cdots, t_{i+k}$. In particular, $B_{i,k}$ is just the zero function in case $t_i = t_{i+k}$. Also, by induction, $B_{i,k}$ is positive on the open interval (t_i, t_{i+k}), since both $\omega_{i,k}$ and $1 - \omega_{i+1,k}$ are positive there. But it is a bit of a miracle that the recurrence relations produce smooth functions. This question and others are best discussed by looking at the span of *all* the $B_{i,k}$.

A spline of order k with knot sequence $\mathbf{t}$ is, by definition, a linear combination of the B-splines $B_{i,k}$ associated with that knot sequence. Let

$$(3) \qquad S_{k,t} := \left\{ \sum_{i=-\infty}^{\infty} a_i B_{i,k} : a_i \in \mathbb{R} \right\}$$

denote the collection of all such splines. We now explore this space.

We deduce from the recurrence relation that

$$(4) \qquad \sum a_i B_{i,k} = \sum (a_i \omega_{i,k} + a_{i-1}(1 - \omega_{i,k})) B_{i,k-1}.$$

On the other hand, arguing as in [2], for the special sequence

$$a_i := \psi_{i,k}(\tau) := (t_{i+1} - \tau) \cdots (t_{i+k-1} - \tau)$$

(with $\tau \in \mathbb{R}$), we find for $B_{i,k-1} \neq 0$, i.e., for $t_i < t_{i+k-1}$ that

$$a_i \omega_{i,k} + a_{i-1}(1 - \omega_{i,k}) = \psi_{i,k-1}(\tau)((t_{i+k-1} - \tau)\omega_{i,k} + (t_i - \tau)(1 - \omega_{i,k}))$$
$$= \psi_{i,k-1}(\tau)(\cdot - \tau)$$

since $f(t_{i+k-1})\omega_{i,k} + f(t_i)(1 - \omega_{i,k})$ is the straight line which agrees with f at t_{i+k-1} and t_i. This shows that

$$\sum \psi_{i,k}(\tau) B_{i,k} = (\cdot - \tau) \sum \psi_{i,k-1}(\tau) B_{i,k-1};$$

hence, by induction,

$$\sum \psi_{i,k}(\tau) B_{i,k} = (\cdot - \tau)^{k-1} \sum \psi_{i,1}(\tau) B_{i,1}.$$

This proves the following identity due to Marsden.

THEOREM 1. *For any $\tau \in \mathbb{R}$,*

$$(5) \qquad (\cdot - \tau)^{k-1} = \sum_i \psi_{i,k}(\tau) B_{i,k},$$

with $\psi_{i,k}(\tau) := (t_{i+1} - \tau) \cdots (t_{i+k-1} - \tau)$.

Since τ in (5) is arbitrary, it follows that $S_{k,t}$ contains all polynomials of degree $<k$. More than that, we can even give an explicit expression for the required coefficients, as follows.

By differentiating (5) with respect to τ, we obtain the identities

$$(6) \qquad \frac{(\cdot - \tau)^{k-\nu}}{(k - \nu)!} = \sum_i \frac{(-D)^{\nu-1}\psi_{i,k}(\tau)}{(k - 1)!} B_{i,k}, \qquad \nu > 0,$$

with Df the derivative of the function f. On using this identity in the Taylor formula

$$p = \sum_{v=1}^{k} \frac{(\cdot - \tau)^{k-v}}{(k-v)!} D^{k-v} p(\tau)$$

for a polynomial p of degree $<k$, we conclude that any such polynomial can be written in the form

$$(7) \qquad p = \sum_i \lambda_{i,k} p B_{i,k},$$

with $\lambda_{i,k}$ given by the rule

$$(8) \qquad \lambda_{i,k} f := \sum_{v=1}^{k} \frac{(-D)^{v-1} \psi_{i,k}(\tau)}{(k-1)!} D^{k-v} f(\tau).$$

Here are two special cases of particular interest. For $p = 1$, we get

$$(9) \qquad 1 = \sum_i B_{i,k}$$

since $D^{k-1}\psi_{i,k} = (-1)^{k-1}(k-1)!$, and this shows that the $B_{i,k}$ form a **partition of unity**. Further, since $D^{k-2}\psi_{i,k}$ is a linear polynomial which vanishes at $t_i^* := (t_{i+1} + \cdots + t_{i+k-1})/(k-1)$,

$$(10) \qquad l = \sum_i l(t_i^*) B_{i,k} \quad \text{for every linear polynomial } l.$$

The identity (5) also gives us various **piecewise** polynomials contained in $S_{k,t}$: Since $B_{i,k}(t_j) \neq 0$ implies $\psi_{i,k}(t_j) = 0$, the choice $\tau = t_j$ in (5) leaves only terms with support either entirely to the left or else entirely to the right of t_j. This implies that

$$(11) \qquad (\cdot - t_j)_+^{k-1} = \sum_{i=j}^{\infty} \psi_{i,k}(t_j) B_{i,k}$$

with $\alpha_+ := \max\{\alpha, 0\}$ the positive part of the number α. Since $B_{i,k}(t_j) \neq 0$ implies $D^{v-1}\psi_{i,k}(t_j) = 0$ in case $v \leq \#t_j := \{t_i : t_i = t_j\}$, the same observation applied to (6) shows that

$$(12) \qquad (\cdot - t_j)_+^{k-v} \in S_{k,t} \quad \text{for } 1 \leq v \leq \#t_j.$$

THEOREM 2. *If* $t_i < t_{i+k}$, *then the B-splines* $B_{i,k}$ *are linearly independent and the space* $S_{k,t}$ *coincides with the space* $\tilde{S}$ *of all piecewise polynomials of degree* $<k$ *with breakpoints* t_i *which are* $k - 1 - \#t_i$ *times continuously differentiable at* t_i.

Proof. It is sufficient to prove that, for any finite interval $I := [a, b]$, the restriction $\tilde{S}_{|I}$ of the space $\tilde{S}$ to the interval I coincides with the restriction of $S_{k,t}$ to that interval. The latter space is spanned by all the B-splines having some support in I, i.e., all $B_{i,k}$ with $(t_i, t_{i+k}) \cap I \neq \varnothing$. The space $\tilde{S}_{|I}$ has a basis consisting of the functions

$$(13) \quad (\cdot - a)^{k-v}, \quad v = 1, \cdots, k; \qquad (\cdot - t_i)_+^{k-v}, \quad v = 1, \cdots, \#t_i \quad \text{for } a < t_i < b.$$

This follows from the observation that a piecewise polynomial function f with a breakpoint at t_i which is $k - 1 - \#t_i$ times continuously differentiable there can be written uniquely as

$$f = p + \sum_{v=1}^{\#t_i} a_v(\cdot - t_i)_+^{k-v},$$

with p a suitable polynomial of degree $<k$ and suitable coefficients a_v. Since each of the functions in (13) lies in $S_{k,\mathbf{t}}$, by (6) and (12), we conclude that

$$(14) \qquad \tilde{S}_{|I} \subset (S_{k,\mathbf{t}})_{|I}.$$

On the other hand, the dimension of $\tilde{S}_{|I}$, i.e., the number of functions in (13), equals the number of B-splines with some support in I (since it equals $k + \sum_{a<t_i<b}\#t_i$), hence is an upper bound on the dimension of $(S_{k,\mathbf{t}})_{|I}$. This implies that equality must hold in (14), and that the set of B-splines having some support in I must be linearly independent over I. $\quad\square$

CoROLLARY 1. *All B-splines having some support on a given interval are linearly independent over that interval.*

CoROLLARY 2. *If $\hat{\mathbf{t}}$ is a refinement of the knot sequence $\mathbf{t}$, then $S_{k,\mathbf{t}} \subset S_{k,\hat{\mathbf{t}}}$.*

CoROLLARY 3. *If $t_i < t_{i+k-1}$, then the derivative of a spline in $S_{k,\mathbf{t}}$ is a spline of degree $<k-1$ with respect to the same knot sequence, i.e., $DS_{k,\mathbf{t}} \subseteq S_{k-1,\mathbf{t}}$.*

The identity (7) can be extended to all spline functions. For this, we agree, consistent with (1b), that all derivatives in (8) are to be taken as limits from the right in case τ coincides with a knot.

THEOREM 3. *If τ in definition (8) of $\lambda_{i,k}$ is chosen in the interval $[t_i, t_{i+k})$, then*

$$(15) \qquad \lambda_{i,k}\left(\sum_j a_j B_{j,k}\right) = a_i.$$

It is remarkable that τ can be chosen arbitrarily in the interval $[t_i, t_{i+k})$. The reason behind this is that $\lambda_{i,k}f$ does not depend on τ at all if f is a polynomial of degree less than k.

Proof. Assume that $\tau \in [t_l, t_{l+1}) \subset [t_i, t_{i+k})$ and let p_j be the polynomial which agrees with $B_{j,k}$ on (t_l, t_{l+1}). Then

$$\lambda_{i,k}B_{j,k} = \lambda_{i,k}p_j.$$

On the other hand,

$$p_j = \sum_{i=l+1-k}^{l} \lambda_{i,k}p_j p_i,$$

since this holds by (7) on $[t_l, t_{l+1})$, while, by Corollary 1 or directly from (7), $p_{l+1-k}, \cdots, p_l$ are linearly independent. Therefore necessarily $\lambda_{i,k}p_j$ equals 1 if $i = j$ and 0 otherwise. $\quad\square$

2. Algorithms. In this section, the basic algorithms for computing with the B-spline representation are derived. Unless otherwise stated, all algorithms refer to the spline function

$$(16) \qquad s = \sum_i a_i B_{i,k}.$$

In the following, the subscript k is omitted whenever possible, e.g., $B_i := B_{i,k}$, $\psi_i := \psi_{i,k}$, etc.

The spline s can be evaluated using the recurrence relation. By (4),

$$\sum_i a_i B_{i,k} = \sum_i (\omega_{i,k} a_i + (1 - \omega_{i,k}) a_{i-1}) B_{i,k-1} =: \sum_i a_i^1 B_{i,k-1}.$$

Iterating this identity one finally arrives at

$$s = \sum_i a_i^{k-1} B_{i,1}$$

and, by the definition of $B_{i,1}$, the right-hand side equals a_j^{k-1} on $[t_j, t_{j+1})$. This yields

ALGORITHM 1. *On the interval $[t_j, t_{j+1})$, $s = a_j^{k-1}$, with the polynomials a_i^r computed as follows*:

$$a_i^0 := a_i,$$
$$a_i^{r+1} := \omega_{i,k-r} a_i^r + (1 - \omega_{i,k-r}) a_{i-1}^r, \quad j - k + r + 1 < i \le j.$$

For the relevant range of indices, $\omega_{i,k-r}(t) \in [0, 1]$ so that $s(t)$ is computed by repeatedly forming convex combinations of the B-spline coefficients.

By Corollary 3 of Theorem 2, the derivative of s is a spline of degree $< k - 1$ with respect to the same knot sequence, i.e.,

$$(17) \qquad Ds =: \sum_i a_i' B_{i,k-1}.$$

By Theorem 3,

$$(18) \qquad a_i' = \lambda_{i,k-1}(Ds)$$

if τ is chosen in the interval (t_i, t_{i+k-1}). To relate a' to a, we express $\lambda_{i,k-1} D$ as a linear combination of the functionals $\lambda_{i,k}$, making use of the fact that $\lambda_{i,k}$ **depends linearly on** $\psi_{i,k}$ and that

$$(19) \qquad (t_{i+k-1} - t_i) \psi_{i,k-1} = \psi_{i,k} - \psi_{i-1,k}.$$

From the definition (8),

$$(\lambda_{i,k} - \lambda_{i-1,k}) f(\tau) = \sum_{v=0}^{k-1} \frac{(-D)^{k-1-v}(\psi_{i,k} - \psi_{i-1,k})(\tau)}{(k-1)!} D^v f(\tau)$$

and

$$\lambda_{i,k-1} Df(\tau) = \sum_{v=0}^{k-2} \frac{(-D)^{k-2-v}\psi_{i,k-1}(\tau)}{(k-2)!} D^{v+1} f(\tau) = \sum_{\mu=0}^{k-1} \frac{(-D)^{k-1-\mu}\psi_{i,k-1}(\tau)}{(k-2)!} D^\mu f(\tau),$$

the last equality by setting $\mu := v + 1$ and using the fact that $D^{k-1}\psi_{i,k-1} = 0$. Comparison of these two lines shows with the aid of (19) that

$$\lambda_{i,k-1} D = \frac{k-1}{t_{i+k-1} - t_i} (\lambda_{i,k} - \lambda_{i-1,k}).$$

Assuming that $B_{i,k-1} \neq 0$, i.e., that $t_i < t_{i+k-1}$, we can choose $\tau \in (t_i, t_{i+k-1}) = (t_{i-1}, t_{i+k-1}) \cap (t_i, t_{i+k})$. By Theorem 3, this yields

ALGORITHM 2. *Compute the coefficients for $\sum a_i' B_{i,k-1} := D \sum a_i B_{i,k}$ by*

$$a_i' = \frac{a_i - a_{i-1}}{(t_{i+k-1} - t_i)/(k-1)} \quad \text{if } t_i < t_{i+k-1}.$$

By Corollary 3, $S_\mathbf{t} \subset S_\hat{\mathbf{t}}$ for any refinement $\hat{\mathbf{t}}$ of the knot sequence $\mathbf{t}$, and therefore any spline $s \in S_\mathbf{t}$ can be written as a linear combination $\sum \hat{a}_i \hat{B}_i$ of the B-splines $\hat{B}_i$ which correspond to the refined knot sequence. The computation of the new coefficients $\hat{a}_i$ from the a_i constitutes the **knot insertion** algorithm used in computer aided geometric design (CAGD) [1], [4]. For this, we need to express $\hat{a}_i$ in terms of the a_i. By Theorem 3, this is equivalent to comparing the corresponding $\hat{\lambda}_i$ with λ_i. Since λ_i depends linearly on ψ_i, this requires nothing more than to express

$$\hat{\psi}_i = (\hat{t}_{i+1} - \cdot) \cdots (\hat{t}_{i+k-1} - \cdot)$$

as a linear combination of the ψ_i.

This is particularly easy when $\hat{\mathbf{t}}$ is obtained from $\mathbf{t}$ by adding just one knot, say the point $\hat{t}$. Then

$$\hat{\psi}_i = \begin{cases} \psi_i, & t_{i+k-1} \leq \hat{t}, \\ \psi_{i-1}, & \hat{t} \leq t_i, \end{cases}$$

hence there is some actual computing necessary only for $t_i < \hat{t} < t_{i+k-1}$. For this case,

$$\begin{aligned} \alpha \psi_i + \beta \psi_{i-1} &= (t_{i+1} - \cdot) \cdots (t_{i+k-2} - \cdot)[\alpha(t_{i+k-1} - \cdot) + \beta(t_i - \cdot)] \\ &= \hat{\psi}_i, \end{aligned}$$

provided $\alpha(t_{i+k-1} - \cdot) + \beta(t_i - \cdot) = (\hat{t} - \cdot)$, i.e.,

$$\alpha = \omega_i(\hat{t}) \quad \text{and} \quad \beta = (1 - \omega_i(\hat{t})).$$

Since $\hat{t}_i = t_i < \hat{t} < t_{i+k-1} = \hat{t}_{i+k}$, we can choose τ in the definition (8) in the interval $(\hat{t}_i, \hat{t}_{i+k}) = (t_{i-1}, t_{i+k-1}) \cap (t_i, t_{i+k})$. This proves

ALGORITHM 3. *If the knot sequence $\hat{\mathbf{t}}$ is obtained from the knot sequence $\mathbf{t}$ by addition of the point $\hat{t}$, then the coefficients $\hat{a}_i$ for the spline s with respect to the refined knot sequence are given by*

$$(20) \qquad \hat{a}_i = \begin{cases} a_i & \text{if } t_{i+k-1} \leq \hat{t}, \\ \omega_i(\hat{t})a_i + (1 - \omega_i(\hat{t}))a_{i-1} & \text{if } t_i < \hat{t} < t_{i+k-1}, \\ a_{i-1} & \text{if } \hat{t} \leq t_i. \end{cases}$$

Observe that $\omega(\hat{t}) \in [0, 1]$ for the relevant range of indices and thus the coefficients $\hat{a}$ are convex combinations of the coefficients a.

If $r := \#\hat{t} \leq k - 1$, then, after just $(k - 1 - r)$-fold insertion of $\hat{t}$, we obtain a knot sequence $\bar{\mathbf{t}}$ in which the number $\hat{t}$ occurs exactly $k - 1$ times. This means that there is exactly one B-spline for that knot sequence which is not zero at $\hat{t}$. Hence it must equal

1 at $\hat{t}$ and its coefficient must provide the value of s at $\hat{t}$. This makes it less surprising that the calculations in Algorithms 1 and 3 are identical.

We cannot resist the temptation to give the simple proof for the variation diminishing property of B-splines based on this fact; cf. [6]. For this, we denote by $S^-(a)$ the number of (strong) sign changes in the sequence a.

THEOREM 4. $S^-(s) \le S^-(a)$; i.e., with $x_1 < \cdots < x_r$ arbitrary,

$$S^-(s(x_1), \cdots, s(x_r)) \le S^-(a).$$

Proof.   We have just seen that $s(x_1), \cdots, s(x_r)$ is a *subsequence* of the sequence $\bar{a}$ of coefficients for s with respect to the refined knot sequence $\bar{t}$ which contains each x_i at least $k - 1$ times. Hence it is sufficient to prove that $S^-(\bar{a}) \le S^-(a)$. But this follows once we know that $S^-(\hat{a}) \le S^-(a)$, with $\hat{a}$ obtained by (20). For this simple case, though, the conclusion is immediate if we think of the construction of $\hat{a}$ from a as occurring in two steps. In the first step, we insert $\hat{a}_i$ between a_{i-1} and a_i, and this does not increase the number of sign changes since each $\hat{a}_i$ is a *convex* combination of its neighbors a_{i-1} and a_i in that new sequence. In the second step, we pull out $\hat{a}$ as a subsequence, and this may only lower the number of sign changes. $\square$

Acknowledgments.   This work was supported by the U.S. Army under contract DAAG29-80-C-0041. The work of the second author was also supported by International Business Machines Corporation and National Science Foundation grant DMS-8351187.

REFERENCES

[1]   W. BOEHM, *Inserting new knots into a B-spline curve*, Computer Aided Design, 12 (1980), pp. 199–201.
[2]   C. DE BOOR, *On calculating with B-splines*, J. Approx. Theory, 6 (1972), pp. 50–62.
[3]   ———, *A Practical Guide to Splines*, Springer-Verlag, New York, 1978.
[4]   E. COHEN, T. LYCHE AND R. RIESENFELD, *Discrete B-splines and subdivision techniques in computer-aided design and computer graphics*, Computer Graphics and Image Processing, 14 (1980), pp. 87–111.
[5]   M. G. COX. *The numerical evaluation of B-splines*, J. Inst. Math. Appl., 10 (1972), pp. 134–149.
[6]   J. LANE AND R. RIESENFELD, *A geometric proof for the variation diminishing property of B-spline approximation*, J. Approx. Theory, 37 (1983), pp. 1–4.

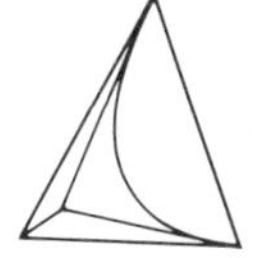

Algebraic Geometry for
Surface and Solid Modeling

THOMAS W. SEDERBERG

Abstract. Motivation is presented for incorporating some tools from algebraic geometry into computer aided geometric design. It is demonstrated that algebraic geometry provides new algorithms for problems previously unsolvable in CAGD, faster algorithms for some CAGD problems, important insights, and new approaches to solid and surface modeling.

1. Introduction. The field of computer aided geometric design has deep roots in the areas of differential geometry and approximation theory, but has not benefited from the tools of classical algebraic geometry. The purpose of this paper is to promote the use of *classical* algebraic geometry (AG) in computer aided geometric design (CAGD). The word *classical* is highlighted because we deal mainly with the constructive approaches which flourished a century ago. We[1] establish the value of AG for CAGD by demonstrating that it provides

(1)   unique insights into curves and surfaces,

(2)   faster algorithms,

(3)   algebraic solutions to problems which, to our knowledge, cannot be attacked using differential geometry or approximation theory,

(4)   new ways to approach solid modeling with free form surfaces.

Section 2 overviews some of the new insights that AG contributes to CAGD. Section 3 discusses some problems which AG addresses nicely, and § 4 presents some ideas on solid modeling with free form algebraic surface patches.

Some of the more readable books on classical algebraic geometry are listed in the references. Most of the author's limited knowledge of the field is due to Salmon [13], [14], Snyder and Sisam [23], Kurosh [6], and Sommerville [24]. These books do not

---

[1] The author is but one of several researchers who have applied AG to problems in CAGD. This paper has benefited significantly from the author's discussions with Ron Goldman, Malcolm Sabin, Dennis Arnon, Rida Farouki, Wayne Tiller and others.

presuppose a post-graduate training in mathematics. Also recommended are Abhyankar [1], Semple and Roth [22], Walker [25] and Winger [26].

Other papers which apply tools from AG to problems in CAGD which are not addressed herein include Farouki [2], Goldman [3], Kajiya [5], de Montaudouin and Tiller [9], Owen and Rockwood [10], and Sabin [12].

2. New Insights. This section presents some insights which CAGD cannot enjoy without AG. These ideas are new to the field of CAGD, but certainly not new to the mathematical world, since they were first conceived a century ago. However, these concepts fell into obscurity, partly for want of an application and partly because the algebra involved was impractically tedious for pencil and paper. Both of these obstructions have dissolved—CAGD has provided several applications, and computers now relieve the algebraic tedium—and these stunning forgotten concepts have begun to illuminate the field of CAGD.

Two of the most prominent contributions of AG are the *implicitization* and *inversion* of rational parametric curves and surfaces. Recall that curves and surfaces can be defined in basically two ways: parametrically and implicitly. A planar curve has a parametric equation of the form

$$x = \frac{x(t)}{w(t)}, \qquad y = \frac{y(t)}{w(t)},$$

and an implicit equation of the form

$$f(x, y) = 0.$$

The parametric equation of a surface is expressed as

$$x = \frac{x(s, t)}{w(s, t)}, \qquad y = \frac{y(s, t)}{w(s, t)}, \qquad z = \frac{z(s, t)}{w(s, t)},$$

and the implicit equation is of the form

$$f(x, y, z) = 0.$$

Since AG restricts itself to polynomial expressions, we restrict all of the above functions to be polynomials.

Implicitization is the process of determining the implicit equation of a parametrically defined curve or surface. One striking insight from AG is that this parametric-to-implicit conversion is *always* possible. In other words, for *any* curve or surface defined parametrically in terms of rational polynomials, there exists an implicit polynomial equation which defines precisely the same curve or surface.

The implicitization algorithm for curves (derived in Sederberg [15] and Sederberg, Anderson and Goldman [20]) proceeds as follows. Consider the parametric curve defined by

$$(1) \qquad x(t) = \frac{\sum_{i=0}^{n} a_i t^i}{\sum_{i=0}^{n} c_i t^i}, \qquad y(t) = \frac{\sum_{i=0}^{n} b_i t^i}{\sum_{i=0}^{n} c_i t^i}.$$

It is possible to obtain an equivalent implicit equation of the form

$$(2) \qquad f(x, y) = \begin{vmatrix} L_{n-1,n-1}(x, y) & \cdots & L_{0,n-1}(x, y) \\ \vdots & \vdots\vdots\vdots & \vdots \\ L_{n-1,0}(x, y) & \cdots & L_{0,0}(x, y) \end{vmatrix} = 0$$

by choosing

$$(3) \qquad \begin{aligned} L_{i,j}(x, y) &= \alpha_{i,j}x + \beta_{i,j}y + \gamma_{i,j} \\ &= \sum_{\substack{l \leq \min(i,j) \\ l+m=i+j+1}} (b_m c_l - c_m b_l)x + (a_l c_m - a_m c_l)y + (a_m b_l - a_l b_m). \end{aligned}$$

As suggested by (2), a degree n parametric curve has an implicit equation that is also of degree n. Two numerical examples will illustrate. First, the parabola

$$x = t^2 + 1, \qquad y = t^2 + 2t - 2,$$

can be processed by the implicitization algorithm to obtain

$$f(x, y) = \begin{vmatrix} 2 & x - y - 3 \\ x - y - 3 & 2x - 2 \end{vmatrix} = -x^2 + 2xy - y^2 + 10x - 6y - 13.$$

Likewise, the cubic parametric curve

$$x = \frac{2t^3 - 18t^2 + 18t + 4}{-3t^2 + 3t + 1}, \qquad y = \frac{39t^3 - 69t^2 + 33t + 1}{-3t^2 + 3t + 1}$$

can be implicitized to obtain

$$\begin{aligned} f(x, y) &= \begin{vmatrix} -117x + 69y + 564 & 117x - 6y - 636 & 39x - 2y - 154 \\ 117x - 6y - 636 & -69x - 2y + 494 & -66x + 6y + 258 \\ 39x - 2y - 154 & -66x - 2y + 258 & 30x - 6y - 114 \end{vmatrix} \\ &= -156195x^3 + 60426x^2y - 7056xy^2 + 224y^3 + 2188998x^2 \\ &\quad - 562500xy + 33168y^2 - 10175796x + 1322088y + 15631624. \end{aligned}$$

We emphasize that this conversion is a closed form process in which the coefficients of the implicit equation can be computed from the coefficients of the parametric equations using only multiplication, addition and subtraction. Thus, it is always possible to perform the conversion in exact integer arithmetic with no numerical error introduced.

Inversion is the process of computing the parameter value(s) of a point known to lie on a parametric curve or surface. Again, AG provides us with a closed form inversion equation in which the parameter(s) can be computed directly from a rational polynomial expression involving the Cartesian coordinates of the point. We refer the reader to Sederberg [15], Sederberg, Anderson and Goldman [20] or Sederberg and Anderson [19] for the particulars on this procedure, but present here without derivation that the inversion of a nonsingular point can be accomplished by solving the set of linear equations

$$f(x, y) = \begin{bmatrix} L_{0,0}(x, y) & \cdots & L_{0,n-1}(x, y) \\ \vdots & \vdots\vdots\vdots & \vdots \\ L_{n-1,0}(x, y) & \cdots & L_{n-1,n-1}(x, y) \end{bmatrix} \begin{Bmatrix} t^{n-1} \\ t^{n-2} \\ \vdots \\ t \\ 1 \end{Bmatrix} = 0.$$

For the parabola, we can obtain two inversion equations as the solution of the linear equations

$$\begin{bmatrix} 2 & x - y - 3 \\ x - y - 3 & 2x - 2 \end{bmatrix} \begin{Bmatrix} t \\ 1 \end{Bmatrix} = 0$$

from which $t = (-x + y + 3)/2$ or $t = (-2x + 2)/(x - y - 3)$. Note that, for points on the curve, either of these equations will correctly compute the parameter value. The cubic curve has an inversion equation of the form

$$t = \frac{\begin{vmatrix} 117x - 6y - 636 & 39x - 2y - 154 \\ -69x - 2y + 494 & -66x + 6y + 258 \end{vmatrix}}{\begin{vmatrix} -117x + 69y + 564 & 39x - 2y - 154 \\ 117x - 6y - 636 & -66x + 6y + 258 \end{vmatrix}}.$$

Again, inversion can be performed using only the arithmetic operations of addition, subtraction, multiplication and division on the coefficients of the parametric equations. Thus, if one is given the x,y coordinates of a point which lies exactly on a given parametric curve, the parameter value of that point can be computed exactly, in a closed form equation.

Implicitization and inversion algorithms exist for surfaces, also (see Sederberg [15] or Sederberg, Anderson and Goldman [20]). But, whereas curve implicitization yields implicit equations of the same degree as the parametric equations, surface implicitization experiences a degree explosion. A triangular surface patch, whose parametric equations are of the form

$$x = \frac{\sum_{i+j \leq n} x_{ij} s^i t^j}{\sum_{i+j \leq n} w_{ij} s^i t^j}, \quad y = \frac{\sum_{i+j \leq n} y_{ij} d^i t^j}{\sum_{i+j \leq n} w_{ij} s^i t^j}, \quad z = \frac{\sum_{i+j \leq n} z_{ij} s^i t^j}{\sum_{i+j \leq n} w_{ij} s^i t^j}, \quad i, j \geq 0,$$

generally has an implicit equation of degree n^2. A tensor product surface patch, whose parametric equations are of the form

$$x = \frac{\sum_{i=0}^{n} \sum_{j=0}^{m} x_{ij} s^i t^j}{\sum_{i=0}^{n} \sum_{j=0}^{m} w_{ij} s^i t^j}, \quad y = \frac{\sum_{i=0}^{n} \sum_{j=0}^{m} y_{ij} s^i t^j}{\sum_{i=0}^{n} \sum_{j=0}^{m} w_{ij} s^i t^j}, \quad z = \frac{\sum_{i=0}^{n} \sum_{j=0}^{m} z_{ij} s^i t^j}{\sum_{i=0}^{n} \sum_{j=0}^{m} w_{ij} s^i t^j},$$

generally has an implicit equation of degree $2mn$. Thus, a bicubic patch generally has an implicit equation $f(x, y, z) = 0$ of degree 18. Such an equation has 1330 terms!

AG also shares important theory on the nature of intersections of curves and surfaces. For example, Bezout's theorem informs us that two planar curves of degree m

and n, respectively, intersect in exactly mn points if we count complex points, multiple intersections, and points at infinity. Other insights from AG provide limits to the number of times that two space curves can intersect. For example, two twisted cubic curves can intersect in at most five points (see Goldman [3]). Bezout's theorem also applies to surfaces. Two surfaces of degree m and n, respectively, intersect in a curve of degree mn. Thus, two bicubic patches generally intersect in a curve of degree 324.

We feel that these are examples of several important contributions which AG uniquely provides to CAGD. Section 3 demonstrates that some of these concepts can be put to use solving practical problems.

3. Practical Value of AG.   Insights are inherently of *intellectual* value, but some AG insights may be so abstract that they can make little or no *practical* contribution to CAGD. Since CAGD is an applied science, the practical is to be valued more than the abstract. This section illustrates two CAGD problems which benefit concretely from AG. The first is the problem of computing the points of intersection of two planar parametric curves, and the second is the problem of detecting improperly parameterized curves and surfaces. The curve intersection problem is an example of one which AG can help to solve *faster* than existing techniques, and the improper parameterization problem is an example of a problem which, to our knowledge, cannot be solved without some ideas from AG.

3.1. Curve intersection.   The problem of computing the points of intersection of two planar parametric curves is basic to CAGD. The standard approaches have been based on subdivision (see Sederberg and Parry [21] or Lane and Riesenfeld [7]). Using implicitization, the problem can be reduced to one of finding the real roots of a univariate polynomial in a finite interval. Denote the equations of the first curve as

$$x = \frac{x_1(s)}{w_1(s)}, \quad y = \frac{y_1(s)}{w_1(s)}, \quad 0 \leq s \leq 1,$$

and the second curve as

$$x = \frac{x_2(t)}{w_2(t)}, \quad y = \frac{y_2(t)}{w_2(t)}, \quad 0 \leq t \leq 1.$$

The points of intersection can be computed by the following algorithm:

(1)   Implicitize the first curve to get $f(x, y, w) = 0$. It is convenient to express this implicit equation as a homogeneous polynomial by introducing the homogenizing variable w.

(2)   Substitute the parametric equations of the second curve into the implicit equation of the first curve to obtain $f(x_2(t), y_2(t), w_2(t)) = 0$ which is a univariate polynomial in t. Its degree is the product of the degrees of the two curves.

(3)   Compute the real roots of the univariate polynomial, in the range $0 \leq t \leq 1$.

(4)   Compute the (x, y) coordinates of those intersection points, and use the inversion equation to compute the corresponding parameter values s. Discard points for which $s < 0$ or $s > 1$.

With proper implementation, this algorithm is numerically stable and is several times faster than subdivision algorithms for curves of degree 2–4 (see Sederberg and Parry [21]).

3.2. Improper parameterization. The improperly parameterized surface problem can be motivated by noting that spheres which are modeled by rational bicubic patches are improperly parameterized in the sense that to each point on the surface there corresponds more than one set of parameters. This situation is analogous to the case of improperly parameterized curves as discussed in Sederberg [16].

Typically, a sphere modeled with bicubic patches is broken into octants with one patch per octant. It is easily seen that each bicubic patch can be expressed as a biquadratic patch. Upon conversion to power basis, the patch in the first octant of a sphere of radius r centered on the origin has the equations:

$$x = 4rstu^2, \qquad y = 2r(u^2 - s^2)tu, \qquad z = r(u^2 - t^2)(u^2 + s^2), \qquad w = (u^2 + s^2)(u^2 + t^2),$$

where we use homogeneous parameters s, t, u. Using straightforward algebraic substitution, it can be verified that the reparameterization

$$\bar{s} = 2stu, \quad \bar{t} = t(u^2 - s^2), \quad \bar{u} = u(u^2 + s^2),$$

yields the properly parameterized sphere equation

$$x = 2r\bar{s}\bar{u}, \qquad y = 2r\bar{t}\bar{u},$$
$$z = r(\bar{u}^2 - \bar{s}^2 - \bar{t}^2), \qquad w = \bar{s}^2 + \bar{t}^2 + \bar{u}^2.$$

This new equation is the equation of a Steiner patch. We know that a Steiner patch cannot easily model an octant of a sphere, because the boundary curves cannot all lie on great circles (see Sederberg and Anderson [19]). Thus, the price of having the bicubics' more flexible boundary curves is that the patch becomes improperly parameterized.

Clearly, if a surface is improperly parameterized, computations on that surface (such as intersection calculations) are much more efficiently performed on the proper parametric equations. The author is currently developing an algorithm for detecting and correcting improperly parameterized surfaces. The solution to this problem relies heavily on AG, and we doubt that it could be solved using existing CAGD tools.

4. Solid Modeling with Algebraic Volumes. Section 2 presents some rather disconcerting news about the degree of the implicit equation of popular free-form surface patches such as bicubic patches, which have implicit equations of degree 18. This raises the question, "Is it possible to model with surfaces whose implicit equation has lower degree?" In this section, we present two ways in which the answer to that question can be yes. The first approach is to constrain the parametric equation, and the second approach is to model directly with the implicit equation.

Recently, some very imaginative work has been done on the problem of blending surfaces, such as fillets which are tangent to two intersecting surfaces. These methods cleverly apply algebraic ideas, and can be found in Rockwood and Owen [11], Hoffmann and Hopcroft [4], and Middleditch and Sears [8].

4.1. Constraining the parametric equation. Recall that bilinear patches have an implicit equation of degree 2, quadratic patches have an implicit equation of degree 4, biquadratic patches have an implicit equation of degree 8, etc. It seems highly curious that there are gaps in this sequence of degrees. Are there no parametric surfaces whose implicit equation is degree 3 or 5, for example? It turns out that parametric surfaces of degree n only *generally* have implicit equations of degree n^2 and that under certain conditions that degree will decrease. To understand the nature of those conditions, we must understand why the implicit equation of a parametric surface is generally n^2. The *degree* of a surface can be thought of either as the degree of its implicit equation or as the number of times it is intersected by a line. Thus, the degree of the implicit equation of a parametric surface can be found by determining the number of times it is intersected by a line. Consider a parametric surface given by

$$x = \frac{x(s,\, t)}{w(s,\, t)}, \quad y = \frac{y(s,\, t)}{w(s,\, t)}, \quad z = \frac{z(s,\, t)}{w(s,\, t)},$$

where the polynomials are of degree n. One way we can compute the points at which a line intersects the surface is by intersecting the surface with two planes which contain the line. If one such plane is $Ax + By + Cz + Dw = 0$, its intersection with the surface is a curve of degree n in $s,\, t$ space. The second plane will also intersect the surface in a degree n curve in parameter space. The points at which these two section curves intersect will be the points at which the line intersects the surface. According to Bezout's theorem, two curves of degree n intersect in n^2 points. Thus, the surface is generally of degree n^2.

It may happen that there are values of $s,\, t$ for which $x(s,\, t) = y(s,\, t) = z(s,\, t) = w(s,\, t) = 0$. These are known as *base points* in contemporary algebraic geometry. If a base point exists, any plane section curve will contain it. Therefore, it will belong to the set of intersection points of any pair of section curves. However, since a base point maps to something that is undefined in $x,\, y,\, z$ space, it does not represent a point at which the straight line intersects the surface, and thus *the existence of a base point diminishes the degree of the implicit equation by one*. Thus, if there happen to be r base points on a degree n parametric surface, the degree of its implicit equation is $n^2 - r$.

For example, it is well known that any quadric surface can be expressed in terms of degree 2 parametric equations. However, in general, a degree 2 parametric surface has an implicit equation of degree 4, and we conclude that there must be two base points. A rigorous discussion of this example can be found in Sederberg and Anderson [19].

The problem of constraining the parametric equations to impose base points is currently under research. It appears that the constraints are reasonably flexible. For example, we have been looking at cubic parametric surfaces (implicit equation generally of degree nine) and what is required to impose six base points to force the degree of the implicit equation to be three. Expressed as a Bézier–Bernstein triangular patch, this patch has ten Bézier control points. In imposing the six base points, we are free to specify seven Bézier control points, and the remaining three are then determined.

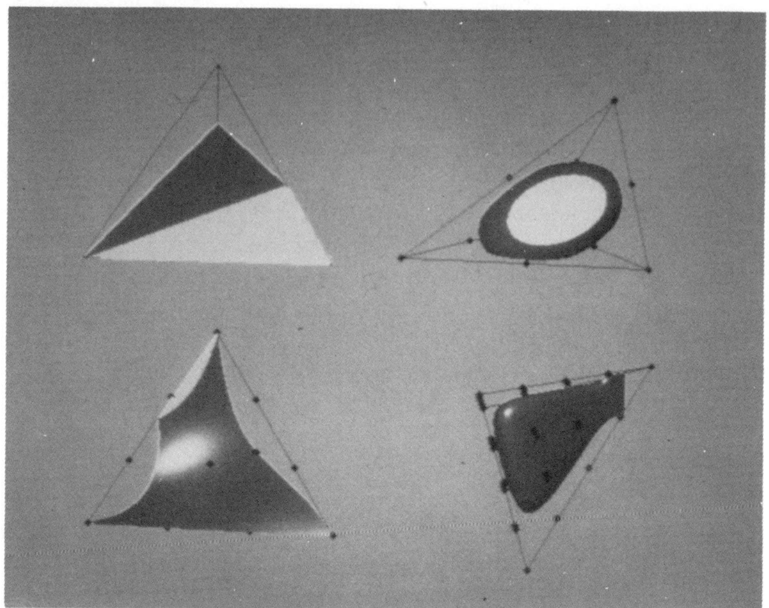

Figure 1.   Implicit surfaces of degrees one through four.

4.2. Piecewise algebraic surfaces.   Another way in which the degree of the implicit equations can be reduced is simply to work directly with the implicit equation, ignoring the parametric equation entirely. This may seem at first impression to be a foolishly radical thought, but it does appear to have some merit. A discussion of this idea, more complete than what follows, is found in Sederberg [18]. An application of these ideas to algebraic curves is found in Sederberg [17].

The main reasons that parametric surface patches have been so popular are that they can be pieced together with derivative continuity, that it is easy to compute points that lie on them, and that elegant, intuitively meaningful techniques exist for controlling their shape.

We can adapt the Bernstein–Bézier techniques in defining implicit surfaces. This provides us with a meaningful way to alter the shape of an implicit surface, and a technique to piece together implicit patches with derivative continuity. Figure 1 shows implicit surfaces of degrees one through four. Note that each surface is defined within a tetrahedral lattice of control points. If the vertices of the tetrahedron are denoted by $\mathbf{V}_{n000}$, $\mathbf{V}_{0n00}$, $\mathbf{V}_{00n0}$ and $\mathbf{V}_{000n}$, then the lattice of control points $\mathbf{V}_{ijkl}$ is defined

$$\mathbf{V}_{ijkl} = \frac{i}{n}\mathbf{V}_{n000} + \frac{j}{n}\mathbf{V}_{0n00} + \frac{k}{n}\mathbf{V}_{00n0} + \frac{l}{n}\mathbf{V}_{000n}, \qquad i, j, k, l \geq 0, \quad i + j + k + l = n.$$

The barycentric coordinates s, t, u, v of a point satisfy the linear equation

$$\mathbf{P} = s\mathbf{V}_{n000} + t\mathbf{V}_{0n00} + u\mathbf{V}_{00n0} + v\mathbf{V}_{000n}, \qquad s + t + u + v = 1.$$

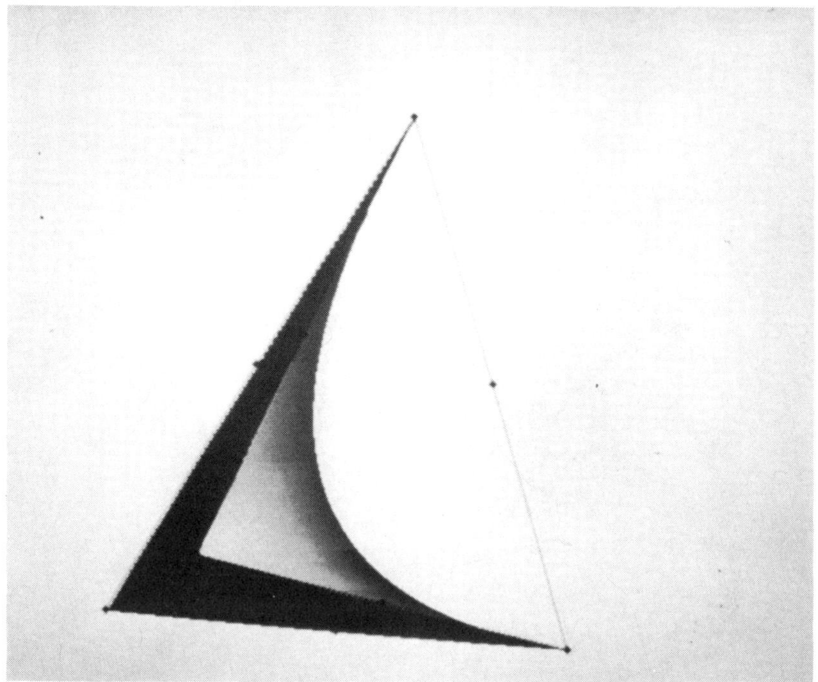

Figure 2.   Hyperbolic paraboloid as piecewise implicit surface.

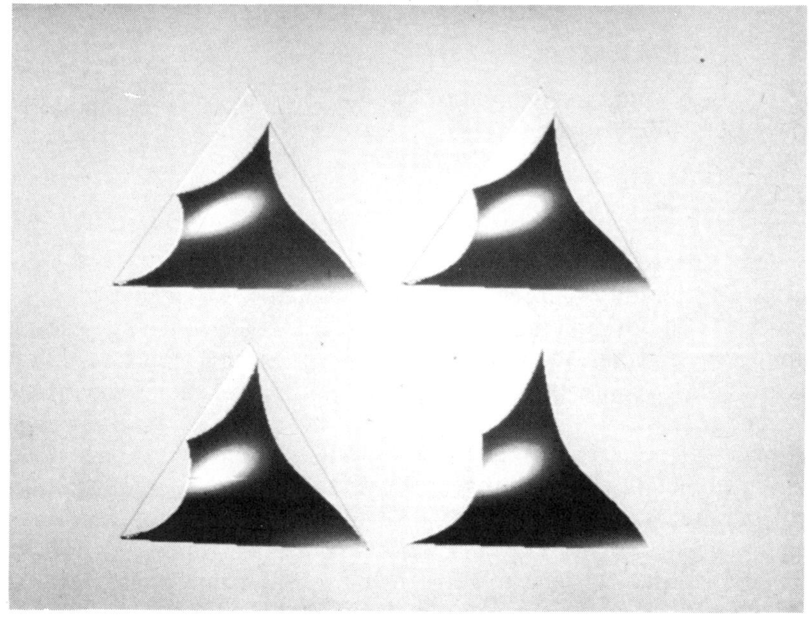

Figure 3.   Changing the weight of the top node.

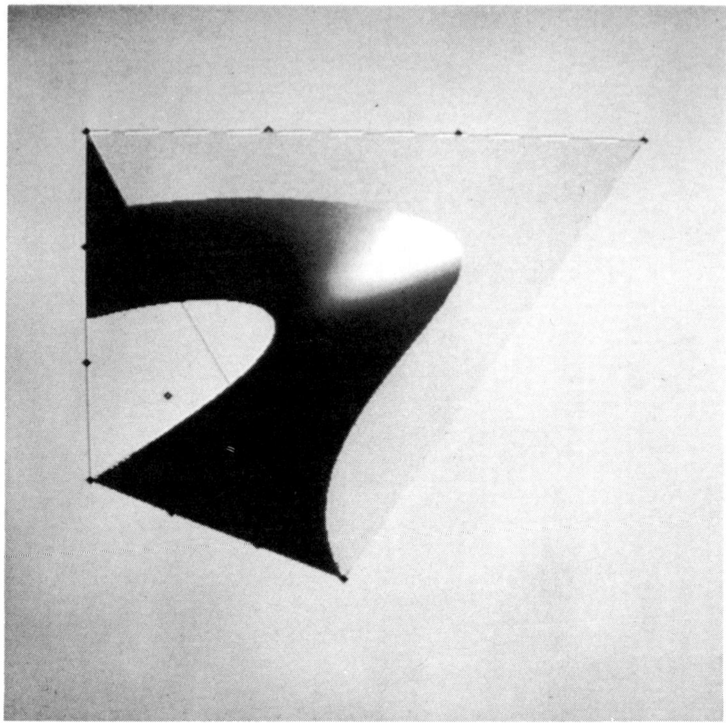

Figure 4.   Cubic implicit surface.

A weight w_{ijkl} is assigned to each control point, and the function $w(s, t, u, v)$ is defined in trivariate Bernstein polynomials as

$$w(s, t, u, v) = \sum_{i,j,k,l \geq 0} w_{ijkl} \frac{n!}{i!\,j!\,k!\,l!} s^i t^j u^k v^l, \qquad i + j + k + l = n, \quad s + t + u + v = 1.$$

The implicit surface is taken as the contour $w(s, t, u, v) = 0$, *bounded by the tetrahedron*. This bound is expressed by the inequality $s, t, u, v \geq 0$.

Note that this formulation of the function $w(s, t, u, v)$ provides a meaningful connection between the weight of a control point and the function value in the neighborhood of the control points. For example, the function value at any of the four vertices is equal to the weight of that vertex. Thus, a vertex with a weight of zero lies on the implicit surface. Also, if all control points along an edge have a weight of zero, the surface completely contains that edge. Figure 2 is a degree 2 implicit surface—a hyperbolic paraboloid—which forces the surface to contain four lines.

Figure 3 illustrates the effect of changing the weight of the top vertex from a negative value (in the top left) to zero (in the bottom right). Figures 3 through 7 are examples of cubic implicit surfaces, which appear to be quite flexible for design applications. Cubic surfaces are particularly appealing because every cubic surface can

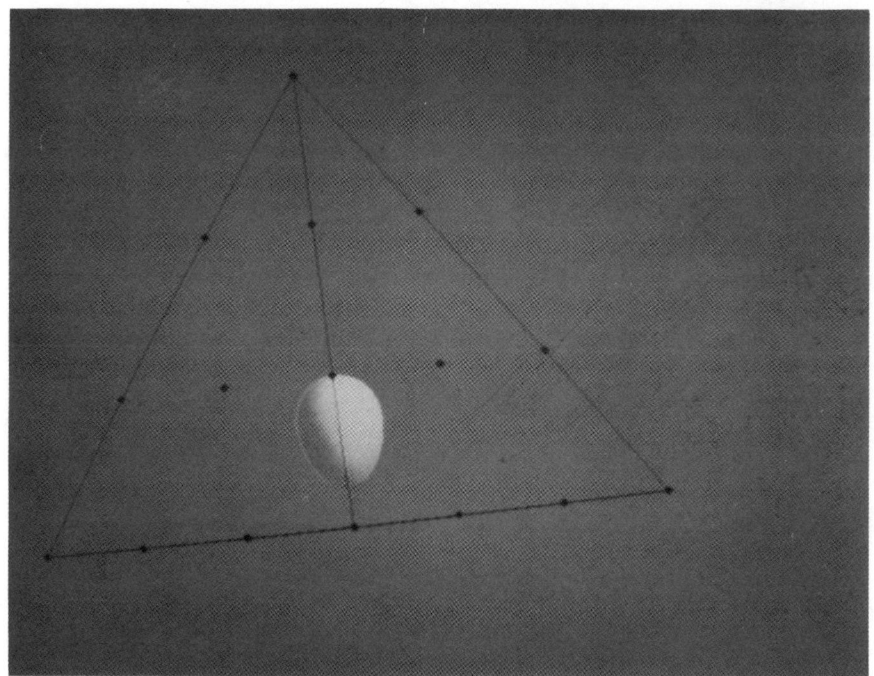

Figure 5. Control point lattice for cubic egg.

Figure 6. Egg as cubic surface.

Figure 7. C^1 cubic implicit surfaces.

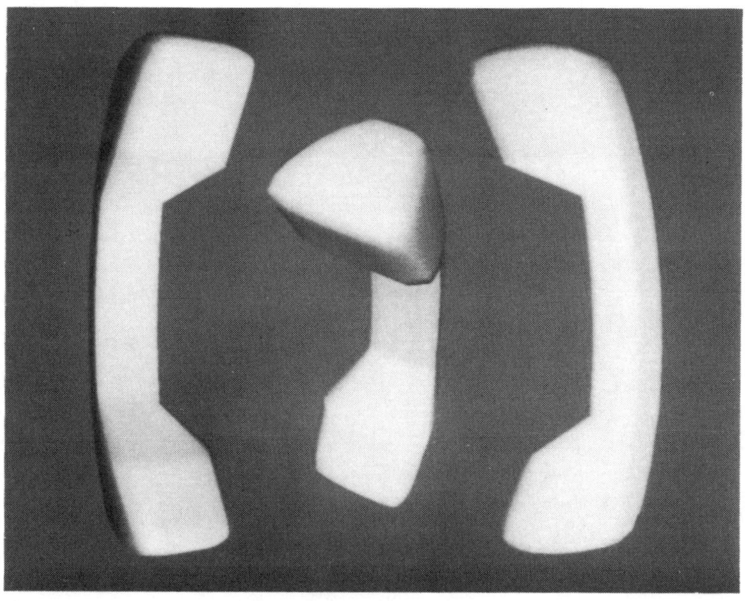

Figure 8. Telephone modeled with three implicit surfaces.

be expressed parametrically as a cubic parametric surface with six base points, or as a biquadratic surface with five base points (see Salmon [14] or Sommerville [24]).

Figures 5 and 6 show an interesting application for a cubic surface in completely modeling an egg with a single implicit equation. Note that this egg could not be modeled as a quadric, and that to model it as a surface of revolution would require a surface of at least degree six.

Figure 7 shows how two implicit surface patches can be pieced together with slope continuity. This is accomplished by forcing the Bernstein functions $w(s, t, u, v)$ of each tetrahedron to be slope continuous.

An important characteristic of implicit surfaces is that they define half spaces—$w(s, t, u, v) \geq 0$ and $w(s, t, u, v) \leq 0$. This is very useful for solid modeling applications, especially in a constructive solid geometry (CSG) context.

One final illustration, Fig. 8, shows a telephone handset which was modeled with three slope continuous implicit surfaces. This model uses tensor product Bernstein polynomials, which produces a parallelepiped lattice of control points rather than a tetrahedral lattice. The tensor product formulation creates surfaces of degree $3n$, but they are much easier to piece together. Note how naturally these implicit functions define volumes. The telephone is modeled with just three "volume primitives."

5. Conclusions.   We have presented evidence that AG teems with contributions to CAGD. This bridge provides a fresh, exciting field for research.

Acknowledgment.   Scott Parry helped create the figures.

REFERENCES

[1]   S. S. ABHYANKAR, *Historical ramblings in algebraic geometry and related algebra,* Amer. Math. Monthly, 83 (1976), pp. 409–448.

[2]   R. T. FAROUKI, *The characterization of parametric surface sections,* Computer Vision, Graphics and Image Processing, to appear.

[3]   R. N. GOLDMAN, *The method of resolvents: A technique for the implicitization, inversion, and intersection of non-planar, parametric, rational cubic curves,* Computer Aided Geometric Design, 2 (1985), pp. 237–256.

[4]   C. HOFFMANN AND J. HOPCROFT, *The potential method for blending surfaces and corners,* this Volume.

[5]   J. KAJIYA, *Ray tracing parametric patches,* Computer Graphics, 16 (1982), pp. 245–254.

[6]   A. KUROSH, *Higher Algebra,* MIR Publishers, Moscow, 1975.

[7]   J. M. LANE AND R. RIESENFELD, *A theoretical development for the computer generation and display of piecewise polynomial surfaces,* IEEE Trans., PAMI, 2 (1980), pp. 35–46.

[8]   A. E. MIDDLEDITCH AND K. H. SEARS, *Blend surfaces for set theoretic volume modelling systems,* Computer Graphics, 19 (1985), pp. 161–170.

[9]   Y. DE MONTAUDOUIN AND W. TILLER, *The Cayley method in computer aided geometric design,* Computer Aided Geometric Design, 1 (1984), pp. 309–326.

[10]   J. C. OWEN AND A. P. ROCKWOOD, *Intersection of general implicit surfaces,* this Volume, 1987.

[11]   A. ROCKWOOD AND J. OWEN, *Blending surfaces in solid modeling,* this Volume, 1987.

[12]   M. A. SABIN, *The use of potential surfaces for numerical geometry,* B.A.C. Weybridge, VTO/MS/153, 1968.

[13]   G. SALMON, *Modern Higher Algebra,* Hodges, Smith and Co., Dublin, 1866.

[14]   ———, *Analytic Geometry of Three Dimensions,* Vols. I and II, Longmans, Green and Co., London, 1912.

[15] T. W. SEDERBERG, *Implicit and parametric curves and surfaces for computer aided geometric design,* Ph.D. thesis, Purdue Univ., West Lafayette, IN, 1983.

[16] ——, *Improperly parametrized rational curves,* Computer Aided Geometric Design, 3 (1986), pp. 67–76.

[17] ——, *Piecewise algebraic curves,* Computer Aided Geometric Design, 1 (1984), pp. 241–255.

[18] ——, *Piecewise algebraic surface patches,* Computer Aided Geometric Design, 2 (1985), pp. 53–59.

[19] T. W. SEDERBERG AND D. C. ANDERSON, *Steiner surface patches,* IEEE Computer Graphics Appl., 5 (1985), pp. 23–36.

[20] T. W. SEDERBERG, D. C. ANDERSON AND R. N. GOLDMAN, *Implicit representation of parametric curves and surfaces,* Computer Vision, Graphics and Image Processing, 28 (1984), pp. 72–74.

[21] T. W. SEDERBERG AND S. R. PARRY, *A comparison of curve-curve intersection algorithms,* Computer Aided Design, 18 (1986), pp. 58–63.

[22] J. G. SEMPLE AND L. ROTH, *Introduction to Algebraic Geometry,* Clarendon Press, Oxford, 1949.

[23] V. SNYDER AND C. H. SISAM, *Analytic Geometry of Space,* Henry Holt and Company, New York, 1974.

[24] D. M. Y. SOMMERVILLE, *Analytic Geometry of Three Dimensions,* Cambridge University Press, Cambridge, 1951.

[25] R. J. WALKER, *Algebraic Curves,* Princeton University Press, Princeton, NJ, 1950.

[26] R. M. WINGER, *An Introduction to Projective Geometry,* D. C. Heath and Company, Boston, MA, 1923.

Applications of Multiple-Valued Functions

FREDERICK J. ALMGREN

Abstract. Multiple-valued functions are discussed as a computational device for minimal surfaces and other optimal geometries in the calculus of variations.

Introduction. It is common in the calculus of several variables to describe a surface S in space in several different ways. Sometimes S will be the graph of a (single-valued) function $z = f(x, y)$—in this case we say that S is represented nonparametrically. At other times S will be the set of solutions to an equation $g(x, y, z) = 0$; for example, the equation $x^2 + 2y^2 + 3z^2 - 1 = 0$ describes an ellipsoid—in this case we say that S is given implicitly. Finally, S might be given as the image of functions $x = f(u, v)$, $y = g(u, v)$, $z = h(u, v)$ defined on a parameter space of u, v coordinates—in this case S is said to be represented parametrically. Each of these methods of representation has major limitations. The nonparametric representation is perhaps the most natural and certainly is the easiest with which to work. Its disadvantage is that only surfaces which project one-to-one onto the xy plane can be so described. Recently, however, a calculus of multiple-valued functions has been developed which enables one to use many of the nice properties of nonparametric representations while describing and working with more complicated surfaces.

As a new framework in which to formulate appropriate problems in computational geometry, multiple-valued functions seem to have emerged as a practical as well as a theoretical tool in geometric analysis. A conspicuous advantage of multiple-valued techniques is that problems which are fundamentally parametric in nature such as area minimization can often be reformulated in nonparametric settings. As suggested above, such nonparametric settings can be more natural and effective computationally.

1. Minimal Surface Forms. Geometric configurations which result from constrained and unconstrained area minimization are called collectively *minimal surface* *forms* in the article by Almgren [1]. As discussed there, such forms have long been used to describe a variety of physical and biological phenomena in which isotropic surface tension forces are significant. They have also provided models for structural design in at least several distinct areas (and have been a source of inspiration for works of art). As far as I know such forms cannot be computed by present algorithms. Computation of forms such as geometries of surfaces of least area and of compound soap bubbles seems a reasonable challenge with which to test the usefulness of multiple-valued techniques. Some steps have been taken and substantial additional progress does seem accessible.

2. Area Minimizing Surfaces.   To illustrate the need for and use of multiple-valued functions we will consider surfaces which minimize area when compared to other surfaces having the same boundary. Suppose Ω is a bounded and otherwise reasonable region in the plane $\mathbf{R}^2$ and $f: \Omega \to \mathbf{R}$ is differentiable—f has Ω as its domain in the xy plane and its values are real numbers $z = f(x, y)$. The area integral

$$\text{AREA}\,(f) = \int_{\Omega} (1 + |\nabla f|^2)^{1/2}\, dx\, dy$$

gives the surface area of the graph of f. If AREA (f) has a minimum value among all functions having the same boundary values as f, then f must satisfy the minimal surface equation. The minimal surface equation states that the principal curvatures of the graph S of f must have equal magnitude but curve in opposite directions so that S is everywhere "saddle shaped". This condition of equal but opposite curvatures means that surface tension forces in the surface exactly balance each other at each point. The minimal surface equation,

$$(1 + f_y^2)f_{xx} - 2f_x f_y f_{xy} + (1 + f_x^2)f_{yy} = 0,$$

was first written down by Lagrange two and a quarter centuries ago. Any surface in space which has equal but opposite principal curvatures everywhere is called a minimal surface.

3. Cases in Which Single-Valued Functions Suffice.   (1) Suppose Ω is bounded, $\partial\Omega$ (the boundary of Ω) is smooth, and $g: \partial\Omega \to \mathbf{R}$ is smooth. If g and all its first, second, and third derivatives are *sufficiently small,* then there will be a unique function f having g as boundary values, minimizing AREA, and thus satisfying the minimal surface equation on the interior of Ω. As far as I know, how small "sufficiently small" must be is not generally known.

(2) If Ω is bounded *and convex,* $\partial\Omega$ is smooth, and $g: \Omega \to \mathbf{R}$ is smooth (but not necessarily small), then there is a unique AREA minimizing function $f: \Omega \to \mathbf{R}$ having g as boundary values. Such f is also the unique function having g as boundary values and satisfying the minimal surface equation on the interior of Ω. Solutions of the minimal surface equation are accessible computationally but, as far as I know, with somewhat less efficiency than solutions to linear partial differential equations such as Laplace's equation $f_{xx} + f_{yy} = 0$.

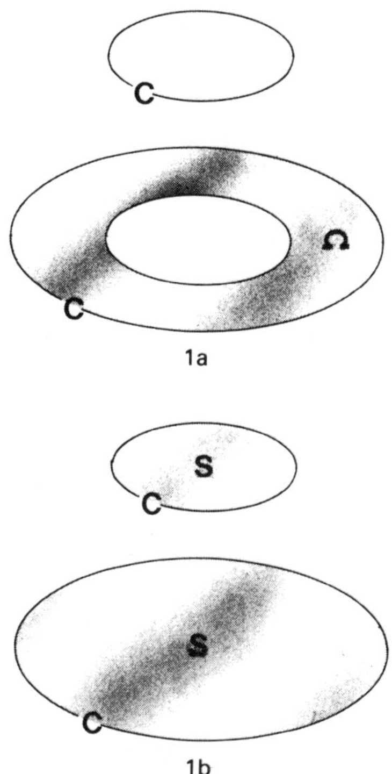

Figure 1. Figure 1a illustrates an annular domain Ω and the graph C of a prescribed boundary value function. Figure 1b illustrates the only minimal surface S spanning C; S consists of two parallel disks. Such S is the graph of a multiple-valued function defined on the convex hull of Ω.

4.   Some Cases in Which Single-Valued Functions Do Not Suffice.   Suppose Ω is bounded (but not necessarily convex), $\partial\Omega$ is smooth, and $g:\partial\Omega\to\mathbf{R}$ is smooth (but not necessarily small). What then can one say about AREA minimizing functions having g as boundary values? Also what can one do if the desired minimal surface boundary is not the graph of such a (single-valued) g? In order to see what is involved in these questions it is useful to consider several examples.

Example 1.   Suppose Ω is the annular region in the plane between concentric circles of radius 1 and 2 centred at the origin, that is $\Omega = \{(x, y) : 1 \le x^2 + y^2 \le 4\}$. Define $g:\partial\Omega\to\mathbf{R}$ by setting $g(x, y) = 0$ if $x^2 + y^2 = 4$ and $g(x, y) = 2$ if $x^2 + y^2 = 1$. Then there is no smooth function $f:\Omega\to\mathbf{R}$ having boundary values g and satisfying the minimal surface equation. Specifically, if

$$C = \{(x, y, 0) : x^2 + y^2 = 4\} \cup \{(x, y, 2) : x^2 + y^2 = 1\}$$

is the graph of g as illustrated in Fig. 1a, then general minimal surface theory guarantees that the unique area minimizing surface S spanning C consists of the

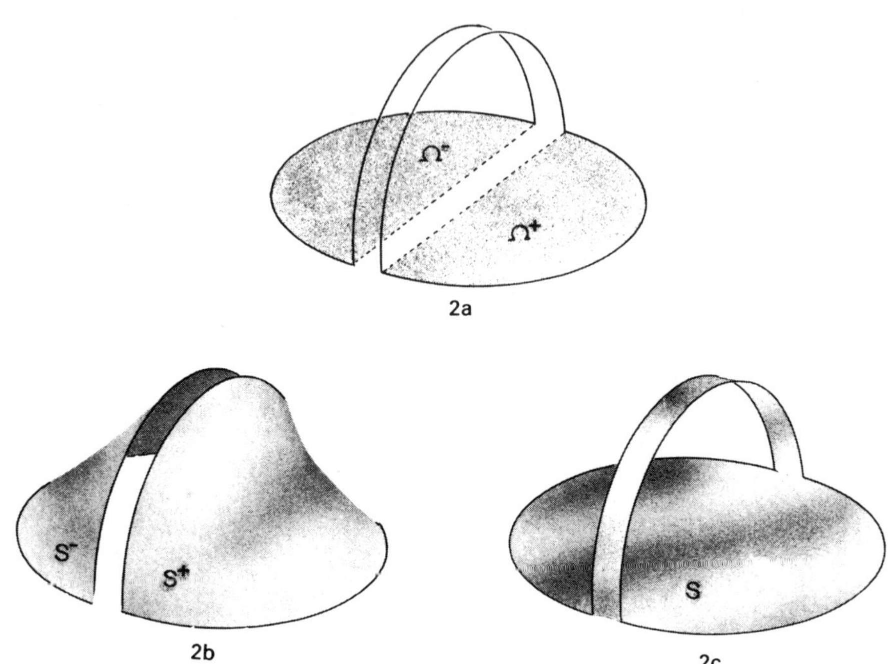

Figure 2. Figure 2a illustrates a domain Ω consisting of two half disks Ω^+ and Ω^-. The prescribed boundary values on Ω^+ and Ω^- have as their graph the set C consisting of four half circles. Figure 2b illustrates the unique minimal surface S^+ above Ω^+ and the unique minimal surface S^- above Ω^-. Figure 2c illustrates a surface S which has less area than $S^+ \cup S^-$. This surface S is the graph of a multiple-valued function defined over the convex hull of Ω.

obvious two flat disks of radius 1 and radius 2 at heights 2 and 0, respectively, i.e.,

$$S = \{(x, y, 0):x^2 + y^2 \le 4\} \cup \{(x, y, 2):x^2 + y^2 \le 1\}$$

as illustrated in Fig. 1b. S is, furthermore, the only minimal surface spanning C. Note that S is not the graph of an ordinary scalar-valued function defined in Ω. However, if we replace Ω by its convex hull $\Omega_* = \{(x, y):x^2 + y^2 \le 4\}$, then S is the graph of the multiple-valued function f_* which assumes the single value 0 if $1 < x^2 + y^2 \le 4$ and assumes the pair of values 0 and 2 if $0 \le x^2 + y^2 \le 1$.

Example 2.   Suppose ε is a small positive number (e.g., $\varepsilon = 0.001$),

$$\Omega^+ = \{(x, y):(x - \varepsilon)^2 + y^2 \le 1 \text{ with } x \ge \varepsilon\},$$

$$\Omega^- = \{(x, y):(x + \varepsilon)^2 + y^2 \le 1 \text{ with } x \le -\varepsilon\},$$

and $\Omega = \Omega^+ \cup \Omega^-$. Suppose also $g:\partial\Omega \to \mathbf{R}$ is given by requiring $g(x, y) = 0$ if $|x| > \varepsilon$ and $g(-\varepsilon, y) = g(\varepsilon, y) = (1 + y^2)^{1/2}$ if $|x| = \varepsilon$. The graph C of g thus consists of four half circles (two half circles in the xy plane and two in planes parallel to and close to the yz plane) as illustrated in Fig. 2a. Then, as indicated in (2) above, there are unique AREA minimizing functions $f^+:\Omega^+ \to \mathbf{R}$ and $f^-:\Omega^- \to \mathbf{R}$ having boundary values

$g \mid \partial\Omega^{+}$ and $g \mid \partial\Omega^{-}$, respectively. Furthermore, the graphs S^{+} and S^{-} of f^{+} and f^{-} as illustrated in Fig. 2b are the only minimal surfaces (in any reasonable sense) spanning $C \cap (\Omega^{+} \times \mathbf{R})$ and $C \cap (\Omega^{-} \times \mathbf{R})$, respectively (and individually). However, there is a minimal surface S having boundary C and having substantially less area than the area of $S^{+} \cup S^{-}$. As illustrated in Fig. 2c, S consists of a nearly flat elongated disk with a thin semicircular ribbon of surface arching over it. If we replace Ω by, say, its convex hull Ω_{*}, then S can be realized as the graph of a multiple-valued function f_{*} defined on Ω_{*}. Such f_{*} takes single values which are close to 0 in the interior of Ω and takes double values (one of which is close to 0) in the interior of $\Omega_{*} \sim \Omega$ except very near to the two tiny line segments of $\partial\Omega_{*} \sim \partial\Omega$ where it takes no values. Nothing of substance is changed in this example if the four corners of Ω and of C are rounded off slightly so that everything is smooth.

Example 3.   Suppose Ω is the two-dimensional disk $\{(x, y);\ x^2 + y^2 \le 4\}$ and the multiple-valued boundary value function g takes the triple of values $-1, 0, +1$ for each (x, y) in $\partial\Omega$. The graph C of g is thus the three coaxial circles of radius 2 at heights -1, 0, and $+1$—that is,

$$C = \{(x, y, z): x^2 + y^2 = 4 \text{ with } z = -1, 0, +1\}$$

as illustrated in Fig. 3a. Depending on the notion of spanning specified for the problem, surfaces of least area "spanning" a boundary such as C can resemble each of the five surfaces illustrated in Figs. 3b, 3c, 3d, 3e, and 3f, among various other possibilities. Each of these admits a reasonable representation as the graph of a multiple-valued area-minimizing function.

5.   Weierstrass's Parametric Representation of Minimal Surfaces.   Minimal surface theory is one of the classical areas of mathematics and has had a close association with complex function theory. The nineteenth century mathematician Weierstrass discovered that any classical minimal surface has a standard parametric representation determined explicitly by a pair of complex functions, and that any suitable pair of such functions gives a minimal surface. Since it is almost impossible to find these functions for a minimal surface required to span a prescribed boundary, Weierstrass's representation has not been used extensively in the study of area-minimizing surfaces. On the other hand, D. Hoffman and W. Meeks recently have used such parametric representations to discover previously unknown types of minimal surfaces (which extend to infinity and are area minimizing only in the small). Their computer drawn color pictures are spectacular (see Peterson [6]).

6.   Multiple-Valued Functions with Multiplicities and Orientations.   By multiple-valued functions in this article we mean a function f having as its domain a suitable region in the plane $\mathbf{R}^2$ (e.g., a disk or all of $\mathbf{R}^2$) and taking at each point of the domain not necessarily a single scalar value but, typically, several scalar values (or none), each representing a separate sheet of surface or several coinciding sheets of surface. In many cases we wish each of these scalar values z carries as part of its structure a positive integer multiplicity (representing the number of sheets of surface at that height above (x, y)) and also a $(+)$ or $(-)$ orientation. In particular, as a possible range value we

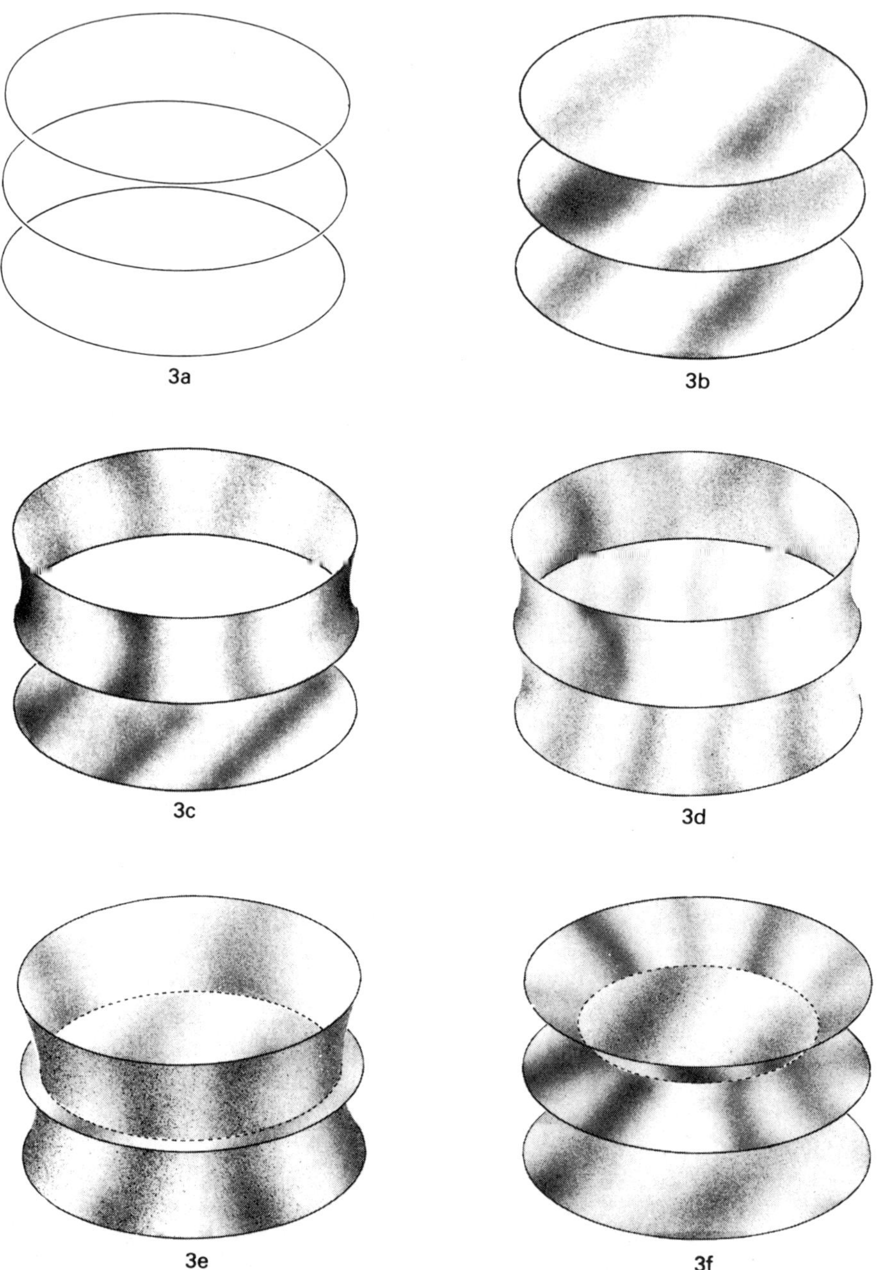

Figure 3. The three parallel circles illustrated in Fig. 3a are the graph C of a multiple-valued function g defined on the boundary of a disk. Figures 3b, 3c, 3d, 3e and 3f illustrate different area minimizing surface geometries corresponding to different definitions of "spanning" a boundary such as C and different orientations and multiplicities for the circles in C.

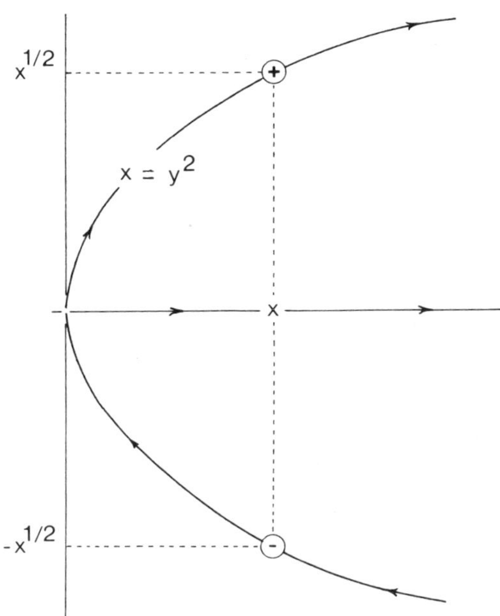

Figure 4. Figure 4 illustrates the parabola $x = y^2$ expressed as the graph of a multiple-valued function which is defined for all real numbers. For $x > 0$, $f(x) = [\![x^{1/2}]\!] - [\![-x^{1/2}]\!]$. For $x \leq 0$, $f(x)$ contains no heights.

permit no heights at all—if one copy of height $z \in \mathbf{R}$ with $(+)$ orientation happens to coincide with another copy of z with $(-)$ orientation above the same point (x, y), then they cancel each other and leave no height there. (If one copy of height z with, say, $(-)$ orientation happens to coincide with another copy of z also with $(-)$ orientation then they add to form height z with multiplicity 2 and $(-)$ orientation.) It is customary to denote by $[\![z]\!]$ the height z with multiplicity one and $(+)$ orientation, by $-[\![w]\!]$ the height w with multiplicity one and $(-)$ orientation, and by 0 the range value corresponding to no heights at all; $-7[\![w]\!]$ thus refers to the height w with $(-)$ orientation and multiplicity 7 (or seven copies of the same height all on top of each other). A useful example is the following. Consider the function f defined on all the real numbers by setting

$$f(x) = [\![x^{1/2}]\!] - [\![-x^{1/2}]\!] \quad \text{if } x > 0,$$
$$f(x) = 0 \quad \text{if } x \leq 0.$$

The "graph" of f is the oriented parabola $x = y^2$, oriented like the x axis for $y > 0$ and opposite to the x axis for $y < 0$ (see Fig. 4). Similarly, a piece of surface represented by a sheet of positively oriented points is oriented so that it projects positively onto the xy plane with its usual orientation while a piece of surface represented by a sheet of negatively oriented points projects negatively onto the xy plane.

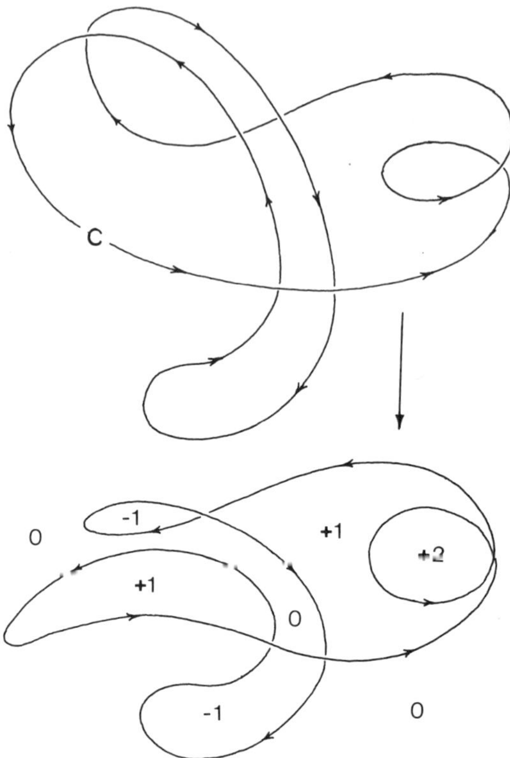

Figure 5.   In the representation of an oriented surface S as the graph of a multiple-valued function f, the number of $(+)$ heights less the number of $(-)$ heights in $f(x, y)$ is determined by counting (with appropriate signs) the number of times a line connecting (x, y) to ∞ must cross the projection $((x, y, z) \to (x, y))$ of the boundary C of S. These are the numbers indicated in the various domain regions.

Suppose S is an oriented surface in space spanning an oriented boundary curve C. If S is written as the graph of a multiple-valued function f defined in $\mathbf{R}^2$, then the difference between the total number of $(+)$ heights counted with multiplicity and the total number of $(-)$ heights counted with multiplicity in $f(x, y)$ can change only across a curve segment in the projected image of C and not inside any of the regions delineated by this projected image (see Fig. 5).

7.   Super's Computational Scheme for Area Minimizing Surfaces.   In his algorithm for computing area minimizing surfaces, Super [8] started with a triangulated approximation to the domain $\Omega = \{(x, y): x^2 + y^2 \leq 1\}$ and considered boundary value mappings g for which total $(+)$ multiplicity equaled total $(-)$ multiplicity. As a way of getting started, prescribed boundary curves were incorporated into the problem as sums of graphs of single-valued functions $g_1 \leq g_2 \leq \cdots \leq g_{2N}$ with alternating orientations. Figure 6 illustrates one of the less obvious possibilities. Each g_i was extended to f_i defined within Ω by choosing vertex heights randomly within suitable ranges subject to

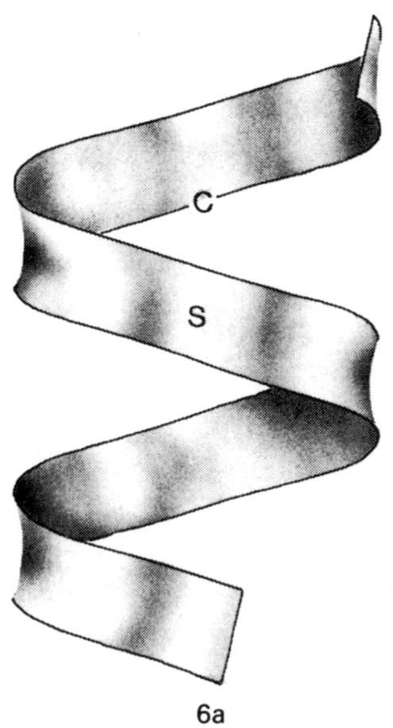

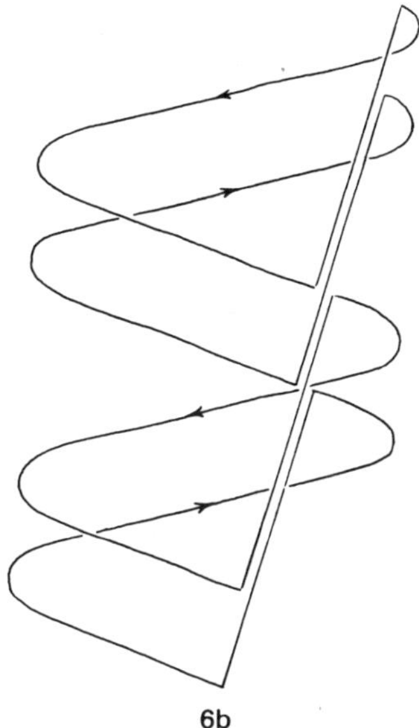

Figure 6.   The oriented minimal surface S illustrated in Fig. 6a has oriented boundary curve C contained in a vertical cylinder. In the context of multiple-valued functions, such C is equivalent to four single-valued graphs. Figure 6b illustrates possible boundary curves for such single-valued functions; the graphs of these functions can add up to C because of cancellation of oriented heights by the same heights with opposite orientation. Such representation of complicated curves as sums of single-valued graphs has been useful in computations.

the constraint $f_1 \leq f_2 \leq \cdots \leq f_{2N}$. The sum of these graphs was the "seed surface" S with which to begin the area minimization routine. Topological holes and/or handles in S occur since positive orientations cancel with negative orientations on sheets of opposite orientations when they coincide (see Fig. 7). Formally, the cancelled vertices and triangles remained in computer memory but the area of cancelled triangles was set to zero in the area minimization routines and they were not drawn in the graphics routine. Super's scheme and its results are discussed somewhat and illustrated in Almgren and Super [4], while the complete program is contained in Super [8].

8.   The Abstract Theory Behind Multiple-Valued Functions.   It is the suggestion of this article that multiple-valued function techniques merit serious consideration as a computational framework for many geometric problems including, in particular, problems of least area. For those interested in abstract foundations, the theoretical strengths of such an approach include the following facts.

(1) The surface S of virtually any solid which one might wish to consider in abstract problems can be strongly approximated as the image (typically as the "graph") of a *Lipschitz* multiple-valued function f. To say that f is Lipschitz means that the distance between $f(x, y)$ and $f(x', y')$ is not more than a fixed constant multiple (called the Lipschitz constant) of the distance between (x, y) and (x', y'). For ordinary

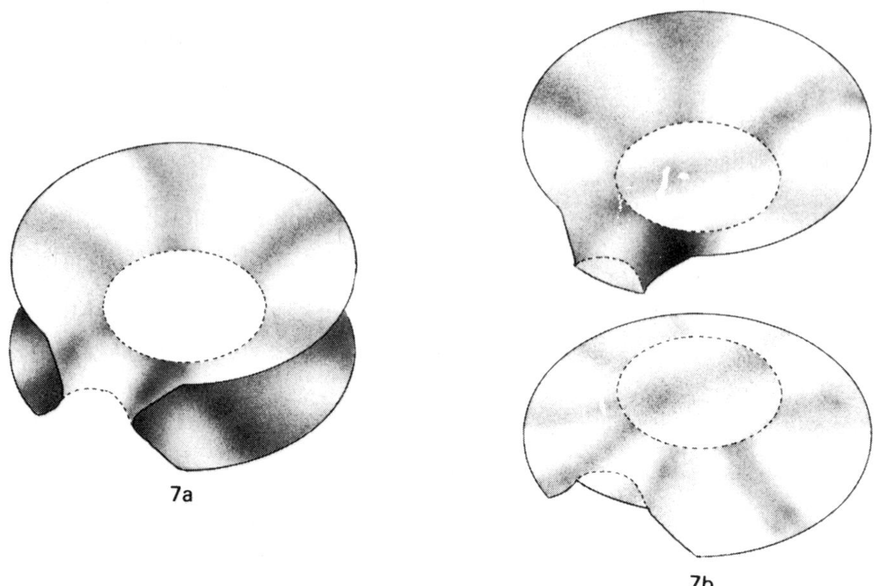

Figure 7. Because of orientation cancellations of multiple-valued functions, the surface illustrated in Fig. 7a is equivalent to the sum of the two graphs illustrated in Fig. 7b (in an exploded view). Cancelled regions remain in the computer memory but contribute zero area and usually are not drawn.

differentiable functions, the Lipschitz constant is the maximum slope. For multiple-valued functions, computation of the distance between two range values is illustrated in the following example:

$$f(x, y) = 3[\![z]\!] + 4[\![z_*]\!] - 7[\![w]\!], \qquad f(x', y') = 5[\![z']\!] - 5[\![w']\!],$$
$$\text{dist}\,(f(x, y), f(x', y')) = |3z + 4z_* - 7w - 5z' + 5w'|.$$

The multiplicity and the Lipschitz constant of an approximating multiple-valued function can be estimated *explicitly* and a priori in terms of the degree of accuracy desired in the approximation; errors of approximation are measured in terms of the size of their orthogonal projections. For example, if ε denotes the amount of area of orthogonal projection in the xy plane which is acceptable as error, then the required multiplicity and required Lipschitz constant need not exceed 26 times the area of S divided by ε. We emphasize that these multiplicity and Lipschitz constant estimates and error estimates are *independent* of the shape and number of handles (and how they are intertwined) of the surface being represented. These assertions are proved in Theorems 3.16, 3.17, and 4.1 of Almgren [3]. The effect of these estimates is that for given acceptable error one can search for optimal surfaces in tightly controlled spaces of functions. The domains of the functions are fixed to be either all of the plane or, perhaps, a closed disk or rectangle in the plane and the ranges of the functions are collections of oriented points with multiplicities with zero net orientation. Neither the domain of the functions nor the range changes in searching for optimal geometries although the specific shape and number and configurations of the handles being represented can vary wildly. In particular, the entire topology of optimal geometries computed is part of the output and not part of the input of the problem.

(2) If one knows ahead of time that the desired optimal surface will be smooth with nonpositive Gauss curvatures at all points and with bounds on its fourth derivatives (as is the case for area minimizing surfaces) then Nance [5] has shown how to estimate (in terms of variable orthonormal coordinate systems) both the multiplicity and the Hölder constant (with exponent $\frac{1}{3}$) for *exact* multiple-valued representation.

(3) Although functions taking multiple values are a bit more complicated to manipulate than functions taking single values in vectorspaces, the combinatorial interpolation schemes set forth in §§ 3.6–3.10 of Almgren [3] are readily implemented computationally. The best methods of computation for achieving optimal designs using multiple-valued functions are a long way from being understood; it is not clear, for example, whether any extension of the Galerkin method has a chance of working or not. My best guess is that realistic solutions to general optimal geometry problems of the types under consideration will involve a composite of techniques, likely varying with the type of problem to be solved. In the case of the least area computations of Super, the initial geometry was determined randomly in the multiple-valued function context and area minimization was achieved by sequentially moving nonboundary vertices to reduce area (typically 20 full passes sufficed). As became clear from experience with these computations, it was necessary to delete vertices that bunched together and to add vertices when triangles became too thin. Although in practice everything seemed to work nicely, it is certainly possible that additional experience will uncover other problems and lead to further modification and improvement.

(4) Multiple-valued functions at the present time are a substantial theory in pure mathematics. They were developed during a period of more than a decade as a device to study the singularities of oriented area minimizing surfaces in general dimensions and codimensions (e.g., 27 dimensional surfaces in 291 dimensional spaces) (see Almgren [2]). The study of such singularities is one of the central problems of the calculus of variations in the context of geometric measure theory. (Surfaces are called regular at those points around which they are smoothly curving and other points are called singular, i.e., where the surface has a cusp, a crease, a junction, or other complicated structure.) As it turns out there is a natural extension of harmonic functions (solutions to Laplace's equation) to the multiple-valued function context which leads to a beautiful theory in its own right. In contrast to the single-valued case, such Dirichlet integral minimizing multiple-valued functions need not be unique and can have small singularities. These functions, however, do closely approximate nearly flat multiple-valued area minimizing functions. Multiple-valued function techniques have also provided new compactness theorems for spaces of surfaces and new proofs of earlier compactness theorems as shown in Almgren [3] and Solomon [7]; it is these theorems which typically guarantee the abstract existence of area minimizing surfaces and other optimal geometries.

Acknowledgment.   I am happy to acknowledge the generous help of J. E. Taylor in writing this article. This work was supported in part by grants from the National Science Foundation.

REFERENCES

[1] F. ALMGREN, *Minimal surface forms,* Math. Intelligencer, 4 (1982), pp. 164–172.

[2] ———, *Q valued functions minimizing Dirichlet's integral and the regularity of area minimizing rectifiable currents up to codimension two,* Bull. Amer. Math. Soc., 8 (1983), pp. 327–328.

[3] ———, *Deformations and multiple valued functions,* Proc. Sympos. Pure Math., 44 (1986), pp. 29–130.

[4] F. ALMGREN AND B. SUPER, *Multiple valued functions in the geometric calculus of variations,* Astérisque, 118 (1984), pp. 13–32.

[5] D. NANCE, *A priori integral geometric estimates for nonpositively curved surfaces,* Ph.D. thesis, Princeton Univ., Princeton, NJ, 1983.

[6] I. PETERSON, *Three bites in a doughnut. Computer generated pictures contribute to the discovery of a new minimal surface,* Science News, Vol. 127, No. 11, March 16, 1985, pp. 168–169.

[7] B. SOLOMON, *A new proof of the closure theorem for integral currents,* Indiana J. Math., 33 (1984), pp. 393–418.

[8] B. SUPER, *Computational algorithms for generating minimal surfaces,* Senior thesis, Princeton Univ., Princeton, NJ, 1983.

Subdividing Bézier Curves
and Surfaces

ART J. SCHWARTZ

Abstract. This article develops direct (i.e., nonrecursive) formulas for subdividing Bézier curves and surfaces by applying elements of the commutative algebra of shift operators. The technique is motivated by an example using the so-called Umbral or Blissard calculus. Applications to de Casteljau's algorithm and degree elevation are also given.

1. Introduction. The aim of this paper is to derive subdivision formulas for Bézier curves and surfaces. For this purpose we define the Bernstein polynomials,

$$(1.1) \qquad B_{nk}(u) = \binom{n}{k} u^k (1-u)^{n-k}$$

where

$$(1.2) \qquad \binom{n}{k} = \begin{cases} \dfrac{n!}{(n-k)!\,k!} & \text{if } 0 \le k \le n, \\ 0 & \text{if } k < 0 \text{ or } k > n; \end{cases}$$

and we assume that we are given a list, $\mathbf{X}$, of $n+1$ vectors

$$\mathbf{X} = (\mathbf{x}_0, \mathbf{x}_1, \cdots, \mathbf{x}_n), \qquad \mathbf{x}_i \text{ in } R^3.$$

We then define the Bézier curve, $\mathbf{x}(t)$, with control points $\mathbf{X}$, as follows:

$$(1.3) \qquad \mathbf{x}(t) = \sum_k B_{nk}(t)\mathbf{x}_k.$$

Note that since $\binom{n}{k} = 0$ for $k < 0$ or $k > n$ we can often dispense with limits on our summation signs. We sometimes write $[\mathbf{X}](t)$ instead of $\mathbf{x}(t)$ to emphasize the dependence on $\mathbf{X}$.

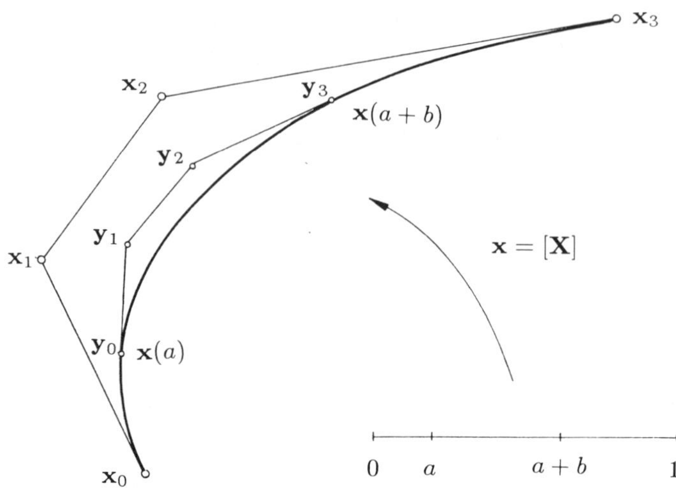

Figure 1.

Now suppose, for fixed numbers a and b, we reparameterize this curve by the substitution $t = a + bu$ and seek a new list of control points, $\mathbf{Y}$, such that

$$(1.4) \qquad \mathbf{y}(u) = \mathbf{x}(a + bu).$$

Since

$$\{\mathbf{y}(u) \mid 0 \leq u \leq 1\} = \{\mathbf{x}(t) \mid a \leq t \leq a + b\},$$

and since $\{\mathbf{y}(u) \mid 0 \leq u \leq 1\}$ is contained in the convex hull of $\mathbf{Y}$ by virtue of the fact that

$$\sum_k B_{nk}(u) = (u + (1 - u))^n = 1,$$

and

$$B_{nk}(u) \geq 0 \quad \text{when } 0 \leq u \leq 1,$$

we see that by determining $\mathbf{Y}$ we can find an estimate of the extent of $\{\mathbf{x}(t) \mid a \leq t \leq a + b\}$, a subdivision of $\mathbf{x}(t)$. See Fig. 1.

As an example of the formulas we are going to derive, in the case where $a = 0$, we shall find that

$$(1.5) \qquad \mathbf{y}_i = \sum_j B_{ij}(b)\mathbf{x}_j.$$

Turning to the investigation of surfaces, we consider an array of vectors

$$\mathbf{W} = \begin{bmatrix} \mathbf{w}_{00} & \cdots & \mathbf{w}_{0n} \\ \vdots & & \vdots \\ \mathbf{w}_{m0} & \cdots & \mathbf{w}_{mn} \end{bmatrix}, \qquad \mathbf{w}_{ij} \text{ in } \mathbf{R}^3,$$

and we define

$$(1.6) \qquad \mathbf{w}(u, v) = \sum_k \sum_l B_{mk}(u) B_{nl}(v) \mathbf{w}_{kl},$$

which parameterizes a Bézier surface. Our problem, now, is to find an array, Z, such that

$$(1.7) \qquad \mathbf{z}(r, s) = \mathbf{w}(a + br, c + ds)$$

for given numbers a, b, c and d. In the case $a = c = 0$, we shall find that

$$(1.8) \qquad \mathbf{z}_{jk} = \sum_{p,q} B_{jp}(b) B_{kq}(d) \mathbf{w}_{pq}.$$

The algorithms we derive here are direct rather than recursive. For an interesting development of a recursive algorithm see Goldman [7].

2. Heuristic Development. Suppose we wish to calculate $(d^2/dt^2)\mathbf{x}(t)$. We observe that

$$x(t) = \sum_k \binom{n}{k} t^k (1 - t)^{n-k} \mathbf{x}_k$$

is intimately related to a binomial expansion. We exploit this by making the symbolic substitution of "$\mathbf{x}^k$" for "$\mathbf{x}_k$". Then

$$\mathbf{x}(t) \text{ "="} (t\mathbf{x} + (1 - t))^n,$$

and

$$\frac{d^2}{dt^2} \mathbf{x}(t) \text{ "="} n(n - 1)(t\mathbf{x} + (1 - t))^{n-2}(\mathbf{x} - 1)^2$$

$$\text{"="} n(n - 1) \sum_k B_{n-2,k}(t)(\mathbf{x}^{k+2} - 2\mathbf{x}^{k+1} + \mathbf{x}^k)$$

$$\text{"="} n(n - 1) \sum_k B_{n-2,k}(t)(\mathbf{x}_{k+2} - 2\mathbf{x}_{k+1} + \mathbf{x}_k).$$

Thus it seems that

$$\frac{d^2}{dt^2} \mathbf{x}(t) = n(n - 1)\mathbf{x}^{\langle 2 \rangle}(t)$$

where

$$\mathbf{x}_j^{\langle 2 \rangle} = \mathbf{x}_{j+2} - 2\mathbf{x}_{j+1} + \mathbf{x}_j.$$

We have employed what is known as the Umbral or Blissard calculus (see Bell [1] or Rota et al. [9]). In the next section we shall develop a rigorous formalism that will substantiate the above, somewhat shady, approach.

3. The Shift Operator Calculus. We shall make heavy use of the shift operator, L, defined by

$$(3.1) \qquad L(\mathbf{X}) = (\mathbf{x}_1, \mathbf{x}_2, \cdots, \mathbf{x}_n, \mathbf{0}).$$

The k-fold composition of L with itself, L^k, thus has the effect

$$(3.2) \qquad L^k(\mathbf{X}) = (\mathbf{x}_k, \cdots, \mathbf{x}_n, \mathbf{0}, \cdots, \mathbf{0}),$$

which indicates why $\mathbf{x}_k$ may sometimes be treated as a kth power. As is conventional we set

$$(3.3) \qquad L^0(\mathbf{X}) = I(\mathbf{X}) = \mathbf{X}.$$

The sum and scalar product of these operators are defined in the usual fashion

$$(3.4) \qquad (aL^k + bL^j)(\mathbf{X}) = aL^k(\mathbf{X}) + bL^j(\mathbf{X}),$$

so that the set of linear combinations of powers of L forms a commutative algebra.
We further define the projection, P, by

$$(3.5) \qquad P(\mathbf{X}) = \mathbf{x}_0,$$

so that

$$(3.6) \qquad PL^k(\mathbf{X}) = \mathbf{x}_k.$$

Using (3.1)–(3.6) we obtain

$$(3.7) \qquad P(tL + (1 - t)I)^n(\mathbf{X}) = \sum_k B_{nk}(t)PL^k\mathbf{X}$$
$$= \mathbf{x}(t).$$

Also, by setting $A = (t + h)L + (1 - t - h)I$, $B = tL + (1 - t)I$, and noting that $A - B = h(L - I)$ we find that

$$\mathbf{x}'(t) = \lim_{h \to 0} P(A^n - B^n)/h\mathbf{X}$$
$$(3.8) \qquad = \lim_{h \to 0} P(A - B)(A^{n-1} + A^{n-2}B + \cdots + B)^{n-1}/h\mathbf{X}$$
$$= nPB^{n-1}(L - I)\mathbf{X}$$
$$= n[(L - I)\mathbf{X}](t).$$

In other words, the control point list for $\mathbf{x}'(t)$ is $n(L - I)\mathbf{X}$. Note that we use the convention that

$$(L - I)\mathbf{X} = (\mathbf{x}_1 - \mathbf{x}_0, \mathbf{x}_2 - \mathbf{x}_1, \cdots, \mathbf{x}_n - \mathbf{x}_{n-1}),$$

a string of length n, is the argument for [ ]. We rely on the context to indicate the length parameter so as to avoid cluttering the text.
The key formulas that we shall use are

$$(3.9) \qquad \mathbf{x}^{(k)}(t) = (n)_k[(L - I)^k\mathbf{X}](t),$$

and

$$(3.10) \qquad \mathbf{x}^{(k)}(0) = (n)_k P(tL + (1 - t)I)^{n-k}(L - I)^k\mathbf{X}|_{t=0}$$
$$= (n)_k P(L - I)^k\mathbf{X}.$$

Here $f^{(k)}(t)$ denotes the kth derivative, $(d^k/dt^k)f(t)$ and $(n)_k$ denote the falling factorial, $n(n-1)\cdots(n-k+1)$.

4. The Work of Hosaka and Kimura. The use of the shift operator was introduced by Hosaka and Kimura [8] and has been discussed by Chang [4]. They use notation which is compact but which may be misleading and may even appear to lack precision. Their basic definition is

$$E\mathbf{P}^i = \mathbf{P}^{i+1}$$

instead of

$$L\mathbf{X} = (\mathbf{x}_1, \cdots, \mathbf{x}_n, \mathbf{0}),$$

and

$$P\mathbf{X} = \mathbf{x}_0.$$

Such abbreviated expressions are common in mathematics dealing with sequence manipulation but they can lead to confusion: $\mathbf{P}^i = \mathbf{P}^j$ does not imply that $E\mathbf{P}^i = E\mathbf{P}^j$; the operator E more or less corresponds to PL but a polynomial in E,

$$f(E) = a_r E^r + \cdots + a_0 I$$

corresponds to $Pf(L)$. Moreover, the arguments of these operators are lists of vectors, not individual vectors.

By using a more explicitly precise notation, it is easier to derive and simplify formulas. For example, Hosaka and Kimura do not simplify the formula for degree elevation ([8, Formula 2.16], see § 7 of this article), nor do they derive explicit formulas for curve and surface subdivision ([8, relations 2.18], see §§ 6 and 9 of this article). The approach here is not merely to use the shift operator calculus to set up various relationships but also to use it to manipulate those relationships.

5. The de Casteljau Algorithm. The formula $\mathbf{x}(t) = P(tL + (1-t)I)^n\mathbf{X}$ will be used in this article primarily as a device for deriving subdivision formulas; however, we may also use it to calculate $\mathbf{x}(t)$ for a given t by using the following recursive approach. Set

(5.1) $$\mathbf{x}_{i,j} = P(tL + (1-t)I)^i L^j\mathbf{X},$$

and note that

(5.2) $$\mathbf{x}_{0,j} = \mathbf{x}_j \quad \text{and} \quad \mathbf{x}_{n,0} = \mathbf{x}(t).$$

We now obtain a recursive relation

(5.3)
$$\begin{aligned}
\mathbf{x}_{i+1,j} &= P(tL + (1-t)I)(tL + (1-t)I)^i L^j\mathbf{X} \\
&= tP(tL + (1-t)I)^i L^{j+1}\mathbf{X} \\
&\quad + (1-t)P(tL + (1-t)I)^i L^j\mathbf{X} \\
&= t\mathbf{x}_{i,j+1} + (1-t)\mathbf{x}_{i,j}.
\end{aligned}$$

Thus we may successively compute

$$(5.4)\qquad \begin{matrix} \mathbf{x}_{0,0} & \cdots & \mathbf{x}_{0,n-1}, & \mathbf{x}_{0,n} \\ \mathbf{x}_{1,0} & \cdots & \mathbf{x}_{1,n-1} \\ \vdots & & \\ \mathbf{x}_{n,0} & & \end{matrix}$$

This method is the de Casteljau algorithm as described in Boehm et al. [2].

6. Derivation of the Subdivision Formula. Observe that there are two lists which determine $\mathbf{x}(t)$; namely, the control point list $\mathbf{X}$ as well as a list of normalized derivatives $\mathbf{Q}$ defined by

$$(6.1)\qquad \begin{aligned} \mathbf{q}_k &= \frac{\mathbf{x}^{(k)}(0)}{(n)_k} \\ &= P(L - I)^k \mathbf{X}. \end{aligned}$$

It is the relationship between $\mathbf{X}$ and $\mathbf{Q}$ that is critical to our derivation. We can express $\mathbf{Q}$ directly in terms of $\mathbf{X}$ as follows:

$$(6.2)\qquad \begin{aligned} \mathbf{q}_k &= P \sum_s \binom{k}{s} (-1)^{k-s} L^s \mathbf{X} \\ &= \sum_s \binom{k}{s} (-1)^{k+s} P L^s \mathbf{X} = \sum_s \binom{k}{s} (-1)^{k+s} \mathbf{x}_s. \end{aligned}$$

By treating a list as a vector with vector components we may express this as

$$(6.3)\qquad \mathbf{Q} = \mathbf{MX}$$

where the matrix $\mathbf{M}$, defined by

$$(6.4)\qquad m_{ij} = (-1)^{i+j} \binom{i}{j},$$

is a lower triangular matrix since $\binom{i}{j} = 0$ if $j > i$.

The inverse relationship may be obtained as follows:

$$(6.5)\qquad \begin{aligned} \mathbf{x}_k &= P L^k \mathbf{X} \\ &= P(I + (L - I))^k \mathbf{X} \\ &= P \sum_j \binom{k}{j} (L - I)^j \mathbf{X} \\ &= \sum_j \binom{k}{j} \mathbf{q}_j. \end{aligned}$$

Thus we see that $\mathbf{M}^{-1}$ is defined by

$$(\mathbf{M}^{-1})_{kj} = \binom{k}{j},$$

that is

$$(6.6) \qquad \sum_r \binom{k}{r}\binom{r}{s}(-1)^{r+s} = \delta_{ks}.$$

It is important to note that since $\mathbf{M}$ is nonsingular, from $\mathbf{x}(t) = \mathbf{y}(t)$ it follows that $\mathbf{MX} = \mathbf{MY}$ and thus $\mathbf{X} = \mathbf{Y}$.

Now given any polynomial curve, $\mathbf{c}(t)$, of degree at most n, we may compute

$$(6.7) \qquad \mathbf{Q}_c = \left(\mathbf{c}(0), \frac{\mathbf{c}'(0)}{n}, \cdots, \frac{\mathbf{c}^{(n)}(0)}{(n)_n} \right)$$

so that if we set

$$(6.8) \qquad \mathbf{X} = \mathbf{M}^{-1}\mathbf{Q}_c,$$

it will follow that

$$\mathbf{x}(t) = \mathbf{c}(t),$$

since

$$(6.9) \qquad \mathbf{Q_X} = \mathbf{MX} = \mathbf{MM}^{-1}\mathbf{Q_C} = \mathbf{Q_C},$$

and $\mathbf{c}(t)$ and $\mathbf{x}(t)$ are polynomial curves of degree at most n.

Thus every polynomial curve of degree at most n has a unique Bézier representation of degree n and its control points may be determined from its derivatives at $t = 0$.

We are ready to subdivide $\mathbf{x}(t)$ in the case where $a = 0$; that is we shall compute $\mathbf{X}^b$ such that

$$(6.10) \qquad \mathbf{x}^b(t) = \mathbf{x}(bt).$$

By virtue of the chain rule we have, differentiating s times,

$$(6.11) \qquad \mathbf{x}^{b(s)}(t) = b^s \mathbf{x}^{(s)}(bt),$$

so that

$$(6.12) \qquad \mathbf{q}_k^b = b^k \mathbf{q}_k;$$

thus using (6.5) and (6.1) we obtain

$$(6.13) \qquad \begin{aligned}
\mathbf{x}_k^b &= \sum_j \binom{k}{j} b^j \mathbf{q}_j \\
&= P \sum_j \binom{k}{j} (b(L - I))^j \mathbf{X} \\
&= P(I + b(L - I))^k \mathbf{X} \\
&= P(bL + (1 - b)I)^k \mathbf{X}.
\end{aligned}$$

From this we obtain

$$(6.14) \qquad \mathbf{x}_k^b = \sum_j \binom{k}{j} b^j (1-b)^{k-j} \mathbf{x}_j$$

$$= \sum_j B_{kj}(b) \mathbf{x}_j,$$

so that

$$(6.15) \qquad \mathbf{X}^b = \mathbf{T}\mathbf{X}$$

where

$$(6.16) \qquad t_{ij} = \binom{i}{j} b^j (1-b)^{i-j}.$$

We may also interpret (6.12) as follows:

$$(6.17) \qquad \mathbf{x}_k^b = \mathbf{x}_{k,0}$$

where we have set $t = b$ in the de Casteljau approach described by (5.1) and (5.3).

We complete the discussion of subdivided curves by considering the case $a \neq 0$. We now seek $\mathbf{X}^{a,b}$ such that

$$\mathbf{x}^{a,b}(t) = [\mathbf{X}^{a,b}](t) = [\mathbf{X}](a + bt).$$

For any given t we have

$$[\mathbf{X}](a + bt) = [\mathbf{X}]\left(a \left(1 - \left(\frac{-b}{a} \right) t \right) \right)$$

$$= [\mathbf{X}^a]\left(1 - \left(\frac{-b}{a} \right) t \right)$$

$$= [(\mathbf{X}^a)^*]\left(\left(\frac{-b}{a} \right) t \right)$$

$$= [((\mathbf{X}^a)^*)^{-b/a}](t)$$

where Y^* designates the reversed list, $(\mathbf{y}_n, \cdots, \mathbf{y}_0)$. Thus

$$(6.18) \qquad \mathbf{X}^{a,b} = ((\mathbf{X}^a)^*)^{-b/a}.$$

7. Degree Elevation. Given $\mathbf{X} = (\mathbf{x}_0, \cdots, \mathbf{x}_n)$, it is sometimes useful to find $\mathbf{X}^+ = (\mathbf{x}_0^+, \cdots, \mathbf{x}_{n+1}^+)$ such that $\mathbf{x}(t) = \mathbf{x}^+(t)$, that is to say, to find a higher order representation of the *same* curve. By adding points we increase the "flexibility" of the representation (Farin [6]), and if we add sufficiently many points, the control points approach $\mathbf{x}(t)$ (Farin [5], as quoted in Boehm et al. [2]). We can use the shift operator calculus to find $\mathbf{X}^+$ as follows:

For $0 \leq k \leq n + 1$ we have (noting that $(n)_{n+1} = 0$),

$$(n + 1)_k \mathbf{q}_k^+ = (n)_k \mathbf{q}_k,$$

so

$$\mathbf{q}_j^+ = \left(1 - \frac{j}{n+1}\right)\mathbf{q}_j.$$

Thus by using (6.5), the identity

$$j\binom{k}{j} = k\binom{k-1}{j-1}$$

and (6.1), we obtain

$$
\begin{aligned}
\mathbf{x}_k^+ &= \sum_j \binom{k}{j}\left(1 - \frac{j}{n+1}\right)\mathbf{q}_j \\
&= \mathbf{x}_k - \frac{k}{n+1}\sum_j \binom{k-1}{j-1} P(L-I)^{j-1}(L-I)\mathbf{X} \\
&= \mathbf{x}_k - \frac{k}{n+1}(\mathbf{x}_k - \mathbf{x}_{k-1}) \\
&= \left(1 - \frac{k}{n+1}\right)\mathbf{x}_k - \frac{k}{n+1}\mathbf{x}_{k-1}.
\end{aligned}
$$

(7.1)

8. Bézier Surfaces.   In the remaining sections of this article we take the methods developed above for curves and extend them in a straightforward way to tensor product surfaces. Tensor product interpolation is discussed by de Boor [3]. One could treat the Bézier curves as interpolants of the $\mathbf{Q}$ data and then use the methods described by de Boor to extend our results to tensor product surfaces. It is, however, quite easy to simply generalize the shift operators and mimic our previous techniques. We shall skip many details.

We are now given a two–dimensional array of control vectors

$$\mathbf{W} = \begin{bmatrix} \mathbf{w}_{00} & \cdots & \mathbf{w}_{0n} \\ \vdots & & \vdots \\ \mathbf{w}_{m0} & \cdots & \mathbf{w}_{mn} \end{bmatrix},$$

from which we construct a surface with parameteric representation

(8.1) $$\mathbf{w}(u, v) = \sum_j \sum_k B_{mj}(u)B_{nk}(v)\mathbf{w}_{jk}.$$

We define two shift operators, S and T:

$$(S\mathbf{W})_{jk} = \begin{cases} \mathbf{w}_{j+1,k} & \text{if } j < m, \\ 0 & \text{if } j = m, \end{cases}$$

and

(8.2) $$(T\mathbf{W})_{jk} = \begin{cases} \mathbf{w}_{j,k+1} & \text{if } k < n, \\ 0 & \text{if } k = n. \end{cases}$$

In other words, S shifts the components of the array to the left, padding the rightmost column with zeros; T shifts the components upward, padding the bottom row with zeros. We further define

$$(8.3) \qquad\qquad PW = \mathbf{w}_{00}.$$

In the spirit of the preceding sections we obtain

$$(8.4) \qquad P(uS + (1-u)I)^m(vT + (1-v)I)^n \mathbf{W} = \sum_j \sum_k B_{mj}(u)B_{nk}(v)\mathbf{w}_{jk}$$
$$= \mathbf{w}(u, v).$$

Furthermore,

$$\frac{\partial}{\partial u}\mathbf{w}(u, v) = mP[(S - I)\mathbf{W}](u, v),$$

$$\frac{\partial}{\partial v}\mathbf{w}(u, v) = nP[(T - I)\mathbf{W}](u, v),$$

and

$$(8.5) \qquad \frac{\partial p + q}{\partial u^p v^q}\mathbf{w}(u, v) = (m)_p(n)_q P[(S - I)^p(T - I)^q \mathbf{W}](u, v),$$

so that

$$(8.6) \qquad \frac{\partial p + q}{\partial u^p \partial v^q}\mathbf{w}(0, 0) = (m)_p(n)_q P(S - I)^p(T - I)^q \mathbf{W}.$$

9. Subdividing Surfaces. We can also extend the idea of the list, $\mathbf{Q}$, to that of the array, $\mathbf{R}$, defined as follows:

$$\mathbf{R}_{jk} = \frac{\partial j + k}{\partial u^j \partial v^k}\mathbf{w}(0, 0)/(m)_j(n)_k$$
$$(9.1) \qquad\qquad = P(S - I)^j(P - I)^k \mathbf{W}$$
$$= \sum_h \sum_i \binom{j}{h}\binom{k}{i}(-1)^{j+k-h-i}\mathbf{w}_{hi}.$$

Thus the normalized derivatives, $\mathbf{R}_{jk}$, can be computed in terms of the control point array, $\mathbf{W}$. To get the inverse relationship we merely exploit our previous work as follows: for fixed s and t we find using (9.1) and (6.6) that

$$(9.2) \qquad \sum_j \sum_k \binom{s}{j}\binom{t}{k}\mathbf{r}_{jk} = \mathbf{w}_{st}.$$

We are now ready to derive a subdivision formula for surfaces. Given a and b we seek an array of control points $\mathbf{W}^{a,b}$ such that

$$\mathbf{w}^{a,b}(u, v) = \mathbf{w}(au, bv).$$

Using the chain rule, we find

$$\mathbf{r}_{jk}^{a,b} = a^j b^k \mathbf{r}_{jk}.$$

It follows from (9.2) that

(9.3)
$$\mathbf{w}_{st}^{a,b} = \sum_p \sum_q B_{sp}(a) B_{tq}(b) \mathbf{w}_{pq}.$$

Note that the s, t coefficient of $\mathbf{W}^{a,b}$ is given by

(9.4)
$$\mathbf{w}_{st}^{a,b} = \mathbf{w}_{\langle s,t\rangle}(a, b)$$

where

(9.5)
$$\mathbf{W}_{\langle s,t\rangle} = \begin{bmatrix} \mathbf{w}_{00} & \cdots & \mathbf{w}_{0t} \\ \vdots & & \vdots \\ \mathbf{w}_{s0} & \cdots & \mathbf{w}_{st} \end{bmatrix}.$$

Finally, we obtain the more general formula for $\mathbf{W}^{\#}$ such that

$$\mathbf{w}^{\#}(u, v) = \mathbf{w}(a + bu, c + dv)$$

(assuming $a \neq 0$ and $c \neq 0$), as follows:

(9.6)
$$\mathbf{w}^{\#}(u, v) = \mathbf{w}\left(a\left(1 - \left(\frac{-b}{a}\right)u\right), c\left(1 - \left(\frac{-d}{c}\right)\right)\right)$$
$$= [((\mathbf{W}^{a,c})^*)^{-b/a,\,-d/c}](u, v)$$

where we define $\mathbf{B}^*$ as the row and column reversal of $\mathbf{B}$,

$$\mathbf{B}^* = \begin{bmatrix} \mathbf{b}_{mn} & \cdots & \mathbf{b}_{m0} \\ \vdots & & \vdots \\ \mathbf{b}_{0n} & \cdots & \mathbf{b}_{00} \end{bmatrix}.$$

Acknowledgments. Research for this article was supported by the Computer Aided Engineering Systems Department of the Ford Motor Company.

REFERENCES

[1] E. BELL, *The history of Blissard's symbolic method with sketch of its inventor's life*, Amer. Math. Monthly, 45 (1938), pp. 414–419.

[2] W. BOEHM, G. FARIN AND J. KAHMANN, *A survey of curve and surface methods in CAGD*, Computer Aided Geometric Design, 1 (1984), pp. 1–60.

[3] C. DE BOOR, *A Practical Guide to Splines*, Springer-Verlag, New York, 1978.

[4] G-Z. CHANG, *Bernstein polynomials via the shifting operator*, Amer. Math. Monthly, 91 (1984), pp. 634–638.

[5] G. FARIN, *Konstruktion und Eigenschaften von Bézier–Kurven und Bézier–Flächen*, Diplom-Arbeit, TU Braunschweig, 1977.

[6] ———, *Algorithms for rational Bézier curves*, Computer Aided Design, 15 (1983), pp. 73–77.

[7] R. GOLDMAN, *Using degenerate Bézier triangles and tetrahedra to subdivide Bézier curves,* Computer Aided Design, 14 (1982), pp. 307–311.

[8] M. HOSAKA AND F. KIMURA, *Synthesis methods of curves and surfaces in interactive* CAD, Proc. Interactive Techniques in CAD, Bologna (1978), pp. 151–156.

[9] G. C. ROTA, D. KAHANER AND A. ODLYZKO, *Finite operator calculus,* J. Math. Anal. Appl., 42 (1973) pp. 684–760.

Part II:
Solid Modeling and Multivariate Surfaces

As in the previous section, most of the articles in this section are surveys in nature.

R. Goldman gives an account of the history of a special aspect in solid modeling, namely, of the evolution of surfaces in this discipline. The paper also includes a discussion of several philosophies of solid modeling and their bearing on the addressed topics.

D. Field describes general mathematical problems, several of which, as it turns out, relate to the proper handling of surfaces in a solid modeling context, so there is a close link between this paper and the preceding one.

Since many solid modelers cannot handle full free-form surfaces, one way to feed such surfaces into the modeler is through suitable preprocessing. The paper by L. Lichten and M. Samek describes one such method—it approximates the given sculptured surface by a piecewise linear, polygonal surface.

Solid models are only one way in which one can generalize surfaces; another possibility is given by multivariate surfaces. A good impression of the problems (and solutions) in that field is given in the article by R. Barnhill, G. Makatura and S. Stead. One of the main tasks when dealing with such surfaces is of course their visualization—this problem is also addressed by Peterson, Piper and Worsey in the "Intersection Algorithms" section.

While the paper by Barnhill et al. is application-oriented, the two following papers deal with some underlying concepts. In the theory of piecewise polynomials, Bernstein–Bézier methods have become an accepted standard. The paper by C. de Boor gives a comprehensive treatment of Bernstein–Bézier triangular patches for the general case of n dimensions.

The theory of multivariate piecewise polynomials contains surprisingly many unsolved questions, even for the "simple" case $n = 2$. One of the most fundamental of these is the problem of finding the dimensions of piecewise polynomial function spaces. This problem is unsolved, and P. Alfeld gives several indications why this is indeed a "hard" problem.

The Role of Surfaces
in Solid Modeling

RONALD N. GOLDMAN

Abstract. A brief survey of geometric modeling schemes is presented. Solid modeling emerges as the technique with the most potential advantages and the fewest intrinsic limitations. Nevertheless to satisfy the needs of a large variety of applications, solid modeling systems must include many different surface types. The nature of these surfaces, the applications for which they are appropriate, and the ways in which they fit into current solid modeling systems are discussed. Several strategies for incorporating freeform design into solid modeling are also analyzed.

1. Introduction. Geometric modeling is rapidly becoming an important tool in modern industrial design and manufacture. Computer models are replacing physical models. They are cheaper to construct, easier to change, and simpler to analyze. Computer simulations save both time and money, and computer analyses of geometric models lead to better and cheaper products. Stress analysis, interference checking, process planning, N.C. verification, mass properties calculations, and a host of other applications either are already being performed today or will be performed in the very near future directly from computer models. Architectural engineering, electrical engineering, industrial engineering, and mechanical engineering will all be re-volutionized by this new technology.

Underlying this new technology are the mathematical models themselves. In this paper our plan is to focus attention on these computer models rather than on simulations, analyses, and applications. We begin in § 2 with a general overview of geometric modeling. Here we shall try to elucidate the advantages and disadvantages of the various types of mathematical models. Of these modeling schemes solid modeling appears to have the most potential strengths and the fewest inherent weaknesses, but to achieve widespread application a large variety of surface types must be included in solid modeling systems. In § 3 we take a closer look at these surfaces and try to see not only how they fit in with various solid modeling schemes, but also the

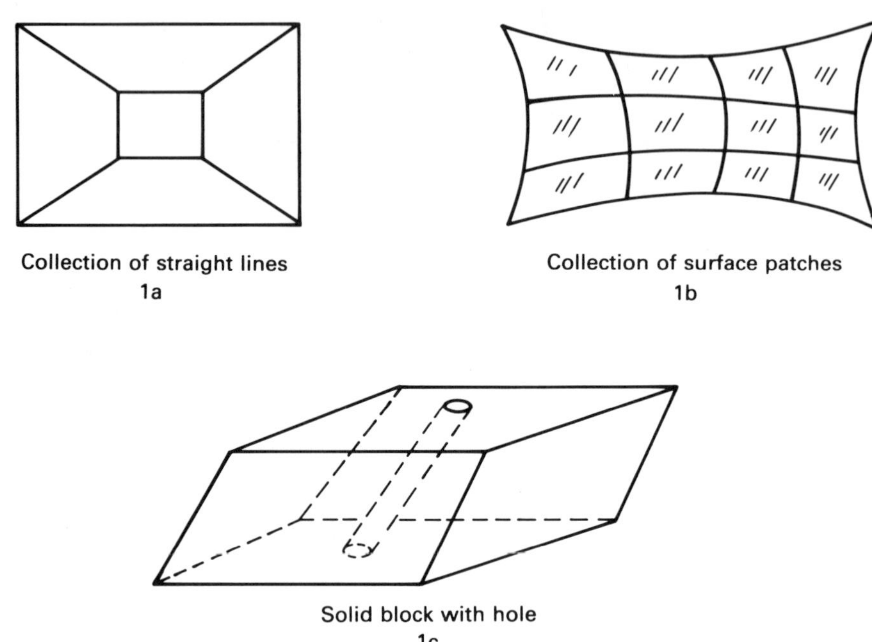

Figure 1. Examples. 1a wire frame; 1b surface; 1c solid.

applications for which they are really appropriate. We go on in § 4 to discuss some strategies for incorporating complex freeform surfaces into solid modeling systems. Our conclusions are summarized in § 5.

2. An Overview of Geometric Modeling. Three kinds of computer aided geometric modeling systems have been developed for use by industry: one based on curves, another on surfaces, and a third on solids (see Fig. 1). We begin by exploring the strengths and weaknesses of each approach. In doing so, we shall attempt to show why solid modeling is now emerging as the system of choice for sophisticated users.

2.1. Wire frame models. The first computer aided geometric modeling systems were simple two-dimensional wire frame systems. Used to display projections of mechanical parts, these were essentially computer aided drafting systems. The data stored and displayed by these systems generally consisted of just straight lines and circular arcs, although other conics and even some simple splines might be included as well. Though these curves are supposed to represent the boundaries of surfaces surrounding real, manufacturable, solid objects, unfortunately, with two-dimensional wire frame systems, it is all too easy to construct models which are physically absurd. That is, two-dimensional pictures can readily be drawn for which there are no corresponding physical three-dimensional objects (see Fig. 2). Since in a two-dimensional wire frame system there is no way to insure the integrity of the part,

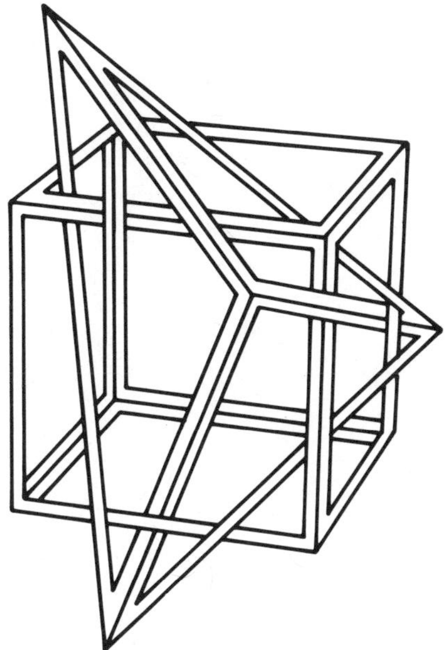

Figure 2.   Impossible wire frame.

nonsense designs may simply be passed on to manufacturing. Thus a good deal of time and money may be wasted before mistakes are corrected. Of course, the situation is no worse than if the parts were designed by hand, but one would have hoped that the computer could be used to detect and avoid such flagrant human errors.

True three-dimensional wire frame models do avoid some of these optical illusions. For example, since the data is now really three-dimensional the situation depicted in Fig. 2 cannot occur because it is not possible for the end points of one line to be in front of a second line while the first line itself passes behind the second line. Nevertheless it is possible to design ludicrous components even in three dimensions (see Fig. 3). Moreover even if the three-dimensional wire frame geometry is not absurd it may still be ambiguous (see Fig. 4). In industry ambiguity may be worse than nonsense because it can be more costly. Indeed, now the wrong component can actually be manufactured before the error is detected.

Wire frame modeling also has other problems. Since each edge, each arc, and each spline must be entered individually, design is a very tedious and error prone process. Thus many wire frame models are simply incomplete because one or several edges have been mistakenly omitted by the designer. Again this problem can raise havoc when designs are passed on to manufacturing.

To sum up, though wire frame models are the simplest models used by industry, they have many drawbacks. Some are impossible; many are ambiguous; most are incomplete; and all are tedious to construct. Also due to the lack of explicit surfaces, holes are difficult to locate and complex shapes are hard to describe. For all these

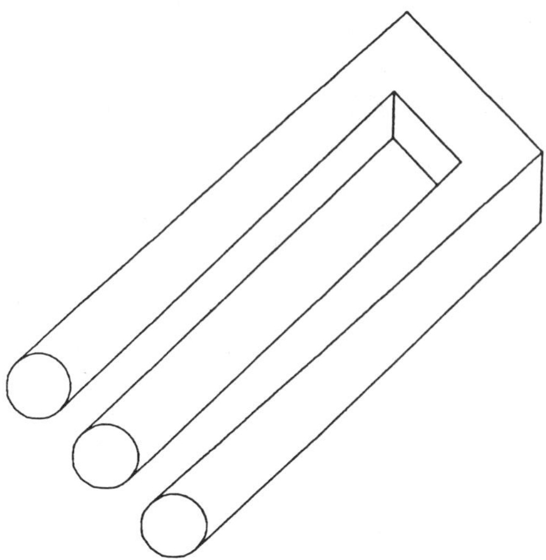

Figure 3. Impossible geometry.

reasons wire frame modeling has generally given way to other more sophisticated techniques of geometric modeling.

2.2. Surface models. To overcome the deficiencies of wire frames, people began to model with surfaces. Two basic types of surface modeling schemes are currently employed: transfinite interpolation and discrete approximation.

In transfinite interpolation a surface must be constructed to pass through a given collection of curves; thus the surface must interpolate a transfinite number of points.

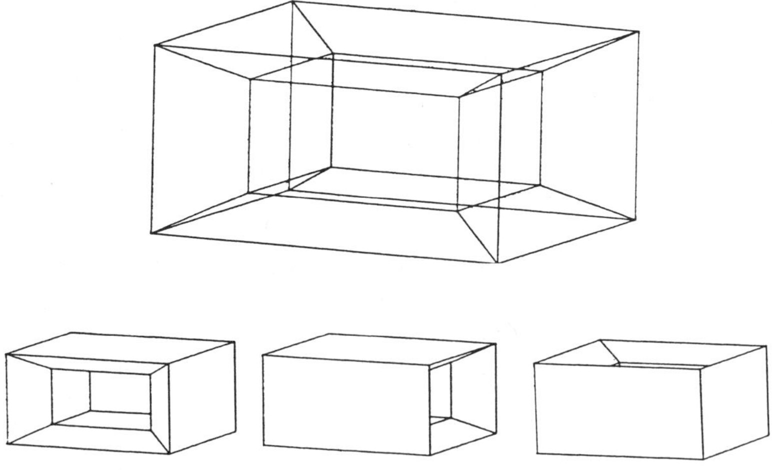

Figure 4. Ambiguous geometry.

The curves themselves may represent cross-sections, or contour lines, or other data, and they can be designed *ab initio,* or defined from digitized data, or derived in some other manner. In any event, the curve data is regarded as sacred and what is wanted is a surface which interpolates these curves and behaves otherwise in a reasonable fashion—that is, neither too wiggly, nor too flat, nor too curved.

Two basic methods are now being used to perform transfinite interpolation: one is local, the other global. Coons [11] developed a local four-sided method, the Coons patch, which is both defined by and interpolates to its four boundary curves. By piecing together many Coons patches, one can build large well behaved surfaces. Gordon [18], on the other hand, developed a global method, the Gordon surface, by using cubic cardinal splines to interpolate any arbitrary rectangular mesh of curves. The American automobile industry currently employs both techniques: Ford uses Coons patches; General Motors uses Gordon surfaces. This is not the place to go into the relative merits of each approach, though clearly both are viable. What is apparent is that both techniques are really methods for fitting a surface skin on top of a wire frame skeleton. Since the curves determine the surfaces and since curves, not surfaces, are considered inviolable, both of these methods are dominated by wire frame techniques.

Not so discrete approximation. Here the users specify only a relatively small, finite array of data points called control points. The system then generates a surface to approximate the shape described by these points. In general, this surface is not required to interpolate the control points. Now the surface itself is considered primary and the control points which specify the surface can be changed if the shape of the surface does not conform to the designer's intent.

Like transfinite interpolation, discrete approximation also allows both local and global methods. The most popular global method is due to Bézier [3]. Here Bernstein polynomials are used to construct a single polynomial surface, called a Bézier patch, to approximate all the given data (Forrest [15]). There are two fundamental problems with this approach. First, moving any control point affects the entire surface; thus there is no local control over the shape of the patch. Second, if there is a lot of data, high degree polynomials are required. Some of these problems can be alleviated by resorting to piecewise Bézier patches, but then special techniques must be invoked to guarantee smoothness between adjacent patches. The most popular local method, introduced into computer aided geometric design by Riesenfeld [32], uses B-spline blending functions to construct piecewise polynomial surfaces. Since B-splines have local support, the affect of any control point is local. Thus moving a single control point changes the shape of only a small part of the surface. Also, since the surfaces are piecewise polynomials, additional data does not require any increase in polynomial degree. Moreover, the use of B-splines guarantees smoothness between adjacent patches; no additional special tricks are required.

Surface models have many advantages over wire frame models. When surfaces are explicitly included in the computer model, fewer impossible objects can be designed. For example, the impossible objects in Figs. 2 and 3 cannot be constructed in surface modeling systems because it is impossible to flesh out these pictures with real surfaces. The inclusion of surfaces also removes the ambiguity from many wire frame designs. In particular, once surfaces are included, the object in Fig. 4 is no longer ambiguous.

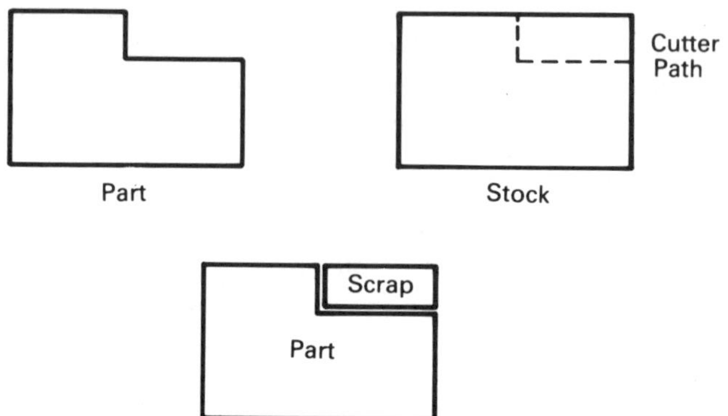

Figure 5.   Did we correctly cut the part from the stock?

$$\text{PART} + \text{SCRAP} \neq \text{PART}$$

Without some knowledge of connectedness, the computer would answer "NO". To correctly answer this question, the computer must be able to recognize scrap; that is, it needs to recognize distinct connected components.

Holes can now be located easily and unambiguously. Computer aided geometric design with freeform surfaces is also an excellent tool for describing complex shapes. Car bodies, ship hulls, airplane wings, turbine blades, and all kinds of ducts, blends, chamfers, and fillets are readily described in surface modeling systems.

Nevertheless computer modeling with surfaces has disadvantages too. Not all objects which can be modeled with surfaces can be manufactured on the shop floor. For instance, there is nothing to prevent a designer from modeling a Klein bottle although there is no way to manufacture one in three dimensions. Also, modeling with surfaces remains tedious because there are no higher level primitives. Thus each individual surface must be modeled explicitly. Moreover systems employing transfinite interpolation must rely on an already tedious and error prone wire frame design.

But beyond these problems, surface modeling systems are inherently inadequate for two additional reasons. First, the surfaces are not guaranteed to bound a solid object. Thus, like wire frame models, surface models are often incomplete. An incomplete model may not be so bad if the design is passed on to people who can provide the missing information from their own experience, but it may be fatal if the incomplete computer model is passed on to other programs for further analysis. For example, mass properties cannot be computed automatically from surface models precisely because too much information is missing from these models.

The second major flaw in surface modeling systems is that generally they lack connectivity information. That is, even though the individual surfaces are completely described mathematically, there is no way to determine directly from the data base which surface is adjacent to a particular patch along a specified boundary. The absence from the data base of this kind of explicit topological information makes it all but impossible to solve certain problems automatically, Consider, for example, the following N.C. comparison problem suggested by Professor Voelcker of the University

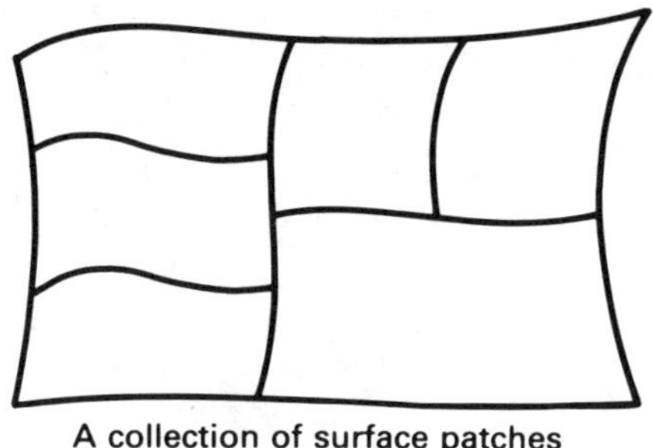

A collection of surface patches

Figure 6. A designer wants to change the shape of one of the surface panels. If the surface boundary is affected in any way, then adjacent panels must also be changed: otherwise these panels will not join evenly. Without some topological knowledge of connectedness and juxtaposition, the system cannot automatically update these surfaces or even cue the user to the problem. Thus we must rely totally on the user. This is a tedious and error prone procedure.

of Rochester. Given a computer model of a part, a stock, and an N.C. cutter path (see Fig. 5), can a computer determine if the part was correctly cut from the stock? Unless the computer model contains some topological information, the computer will have no notion of connected components and without looking at individual components the computer will not be able to distinguish the part from the scrap. Automatic verification of N.C. cutter paths is potentially an important industrial application of computer aided geometric modeling because it can save both time and money (Voelcker and Hunt [57]). Unfortunately surface modeling systems do not lend themselves to this application because they lack the necessary topological information. Another type of problem that requires topology is epitomized by Ford's surface panel problem (see Fig. 6). Here a hood, or a fender, or a door panel is described by a collection of surface patches. A designer then decides to alter the shape of one of these patches. If a boundary is affected, then adjacent patches must also be changed; otherwise these panels will not join evenly and they will separate and tear. The problem for the computer is to fix these adjacent patches automatically or at the very least to cue the designer to the location of the problem patches. Again unless some topology is built into the geometric model, the computer will lack the ability to determine connectedness and juxtaposition and therefore it will be unable to solve the problem. In fact, this problem has never been solved by computer at Ford. Designers proceed at their own risk, and they must fix problems as they arise without computer assistance. Needless to say, this is a tedious, time consuming, and error prone procedure.

Surface modeling systems are widespread in the automobile, aerospace, and ship building industries, and these modeling systems have many important and useful applications. Nevertheless the inherent limitations of these design systems have led people to search for more robust techniques of computer aided geometric modeling.

2.3. Solid models. The motivation for modeling with solids is to provide complete, unambiguous descriptions of real, manufacturable parts. A solid is, by

Figure 7.   ICEM solid model.

definition, a three-dimensional object with a well defined inside and outside separated by a two-dimensional boundary. Thus for each point in space it is possible, in principle, to determine whether the point is in, out, or on the boundary of the solid. Many techniques have been developed for generating and storing computer models of solid objects. These include constructive solid geometry (CSG), boundary representations (B-Rep), octrees, primitive instancing, and others (Requicha [30]). Here we shall concentrate on the two most important and popular of these computer models: constructive solid geometry and boundary representations.

In constructive solid geometry complex objects are built by performing Boolean operations on a fixed collection of simple primitives. The primitives usually include several solids with simple surfaces such as boxes, cylinders, spheres, cones, and tori, but primitives with more complicated surfaces can also be incorporated into these systems (see Fig. 7). By performing the Boolean operations of union, difference, and intersection an unlimited variety of parts can be constructed from a relatively small collection of primitive solids. In the data base a solid is then stored as a binary tree (CSG tree) where the intermediate nodes are Boolean operations and the terminal nodes are solid primitives (Requicha and Voelcker [29]). This representation scheme is very powerful for several reasons. First, it is very concise. Second, each binary tree is

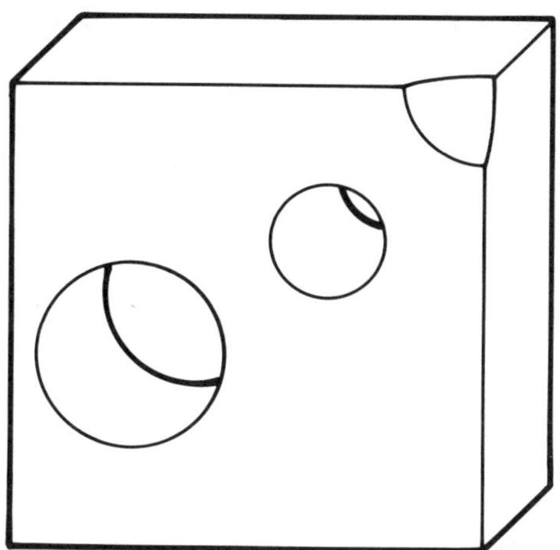

Figure 8.   Solid model.

guaranteed to represent a unique, physical solid provided that the primitives themselves have complete, unambiguous descriptions. That is, once we have insured that the primitives are well defined, impossible or ambiguous objects can never occur. Third, Boolean operations can be used to mirror manufacturing processes. For example, drilling a hole is mathematically equivalent to subtracting a cylinder; welding two parts is equivalent to taking their union. Thus if done correctly a CSG tree will contain not only a mathematical description of part geometry, but also a physical prescription for part manufacture. Therefore ideally a CSG description will lead directly to a process plan. An example of a simple part and the solid primitives used to represent it is given in Figs. 8 and 9. Notice that although this scheme provides a complete description of part geometry, topological information is still lacking.

A boundary representation of a solid is more akin to a surface model than a CSG tree. Nevertheless only complete solids are represented and topological information is always stored. In a boundary representation a solid is described by storing a representation of each of its bounding surfaces (geometry) along with specifications for how these surfaces are joined together (topology) (Braid [6]). A boundary representation of the solid in Fig. 8 is given in Fig. 10. Offhand there is nothing to guarantee that a boundary representation will be any more complete or manufacturable than any standard surface model. However if a CSG tree is used as input to a B-Rep system, then complete, physical models are guaranteed. Thus some solid modelers use CSG in the front end as input to guarantee physical integrity and B-Rep in the data base to provide explicit geometry and topology (Hillyard [20]), (Boyse and Gilchrist [5]). A nice feature of these systems is that they allow the designer to work with high level solid primitives instead of individual curves and surfaces. By shifting the burden of computing the individual faces and edges to the machine instead of the human, these systems make design much less tedious and error prone. Unfortunately boundary representations are expensive to compute and store because they contain so much

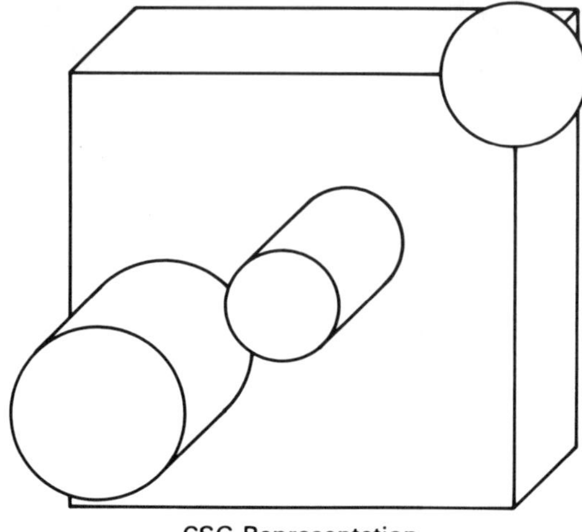

Figure 9. Solid model.

explicit information. Since CSG trees are so simple and concise, some systems like SynthaVision store just CSG trees and compute B-Reps only when they are required by an application. The fact that B-Reps can always be computed from CSG trees says that topological information is always present at least implicitly even in CSG trees.

Though wire frame and surface modeling systems will always have their place for very simple or highly specialized applications, knowledgeable users will generally prefer solid models because they contain complete, unambiguous descriptions of real,

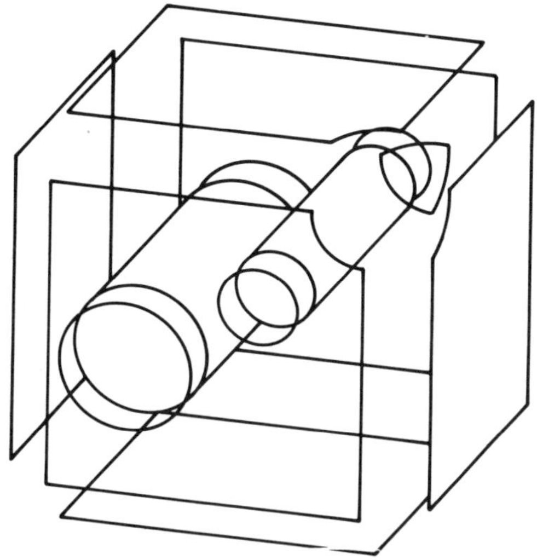

Figure 10. Solid model.

manufacturable parts. Nevertheless, solid modeling still has many drawbacks. In the past decade much work has gone into studying the various technical aspects of solid modeling, but applications which run directly off of solid models are still few and far between. Image rendering, interference checking, and mass properties calculations are currently the main industrial applications of solid modeling (Requicha and Voelcker [31]). To make their mark in industry, solid modeling systems must also be used to perform N.C., FEM, process planning, kinematics, robotics, and a host of other real applications. Research is now progressing in all of these areas and the future, at least, looks promising. Another problem is that due to the complexity of their software, solid modeling systems have been slow to incorporate complicated freeform surfaces. This makes it difficult to model parts with complex shapes or nontrivial blends. Fortunately this too has recently begun to change, and complicated freeform surfaces are now making their way into several solid modeling systems (Chiyokura and Fumihiko [7], Sarraga and Waters [40], Riesenfeld [35], Tiller [51], and Veenman [54]). While other modeling techniques have built-in flaws and limitations, solid modeling systems seem to have none of these inherent problems. It is for this reason that solid modeling will eventually become the system of choice for sophisticated users.

3. The Role of Surfaces in Solid Modeling.   Which types of surfaces need to be incorporated into solid modeling systems? The answer to this question is crucial for the people who design, implement, and use these systems. Yet even with only a few very simple surfaces, the software needed to perform solid modeling is already large and complex. Until recently the people who design and implement this software have preferred not to compound their problems any further by including complicated surface types. Yet without these surface types many applications remain beyond the scope of these systems. Clearly then *the surfaces included in a solid modeling system must depend on the applications for which the system is ultimately intended.* We shall now examine the advantages and disadvantages of various surface types keeping a critical eye open for the applications to which they can be applied.

3.1. Planes.   Planes are the simplest possible surfaces, so they were the first to be incorporated into solid modeling systems (Baumgart [2], Braid [6], Borkin et al. [4], Voelcker et al. [56]). Since two planes always intersect in a straight line, solid modeling systems can be built with only one curve type and one surface type (Turner [52]). These polyhedral solids are easy to render because the surface normal along any face does not vary. Also because they have only linear edges and planar faces, hidden line and hidden surface algorithms are particularly simple. Mass properties of polyhedral solids are easy to compute either exactly by calculating the integrals in the divergence theorem or approximately by using standard ray firing techniques (Roth [38]). Another justification for building polyhedral systems is that they can be used to test the effectiveness of topological data structures without resorting to complicated mathematics. If these data structures are adequate for planar models, they will also work for more complicated geometry since topology is essentially independent of geometry.

Polyhedral systems are adequate for many architectural applications and they can also be used to model some simple mechanical parts. However they are not sufficient for modeling even very commonplace features such as round holes or circular fillets. Certainly freeform aesthetic design and complex blends are out of the question with planar models. Thus outside of architecture there seem to be no applications for which planar models are satisfactory.

3.2. Quadrics. The attempt to build solid modeling systems which lend themselves to more practical applications while at the same time keeping the mathematics as simple as possible led to the introduction of quadric surfaces (Braid [6], Levin [26], Voelcker and Requicha [55]). These second degree surfaces were well studied in the 19th century, and a fully developed, self contained, ready to implement mathematical theory was already well established (Dresdin [14]).

To convert from a CSG tree to a B-Rep, one must be able to compute and represent the intersection of any two surfaces. The intersection of a plane and a quadric surface is always a conic section so calculating cross sections is no problem. Moreover, Levin [27] showed that the intersection of any two quadric surfaces can be represented analytically and he gave an algorithm for finding these intersection curves (see also Sarraga [39]). Thus, at least theoretically, converting from a CSG tree to a B-Rep is not an insurmountable problem. Also, since quadrics are second order surfaces, ray firing at quadrics requires solving only a quadratic equation. Therefore quadric models are simple to render and their mass properties are easy to analyze by standard ray firing techniques.

Still there are problems. Not all quadric surfaces are needed in design. For example, hyperbolic paraboloids are hardly ever used; yet overhead is paid for these surfaces if the full quadric model is implemented. Therefore some systems implement only special subsets of the quadrics such as the natural quadrics (Hakala et al. [19]) or the quadrics of revolution (Goldman [16]). Even so there are still problems for while a rigorous mathematical theory of these surfaces exists on paper, some of these algorithms are not robust for numerical reasons when implemented on a digital machine (Goldman [17]). People have been fighting these numerical problems for many years, but even today they are still not fully solved.

Solid modeling systems based on quadrics have many more potential applications than simple polyhedral systems. Even architecture requires cylinders once in a while, and spheres and cones are sometimes useful too. With the introduction of the cylinder, piping design becomes possible. Also, simple, but realistic, mechanical parts can now be modeled, and N.C. tool path verification begins to be practical.

But quadric surfaces are still quite inflexible structures. They are very good for modeling rigid objects such as pipes, engine blocks, ball bearings, and even tool paths, but not for freeform entities such as airplane wings, car bodies, ship hulls, and turbine blades. Also, it is not clear how to blend two quadrics using only other quadrics. Even tori are not present for this purpose so pipes, for example, cannot be made to turn corners. The introduction of quadric surfaces just begins to make solid modeling useful for real world applications, but these systems still leave much to be desired.

3.3. Tori. Once quadric surfaces are implemented, the need for the torus in solid modeling quickly becomes apparent. Since quadrics are only second order surfaces, they cannot turn corners. Quadrics are good for modeling objects which generically are like rigid blocks with pipes and holes, but they are inadequate for modeling objects which have bends like ducts and hoses. Even piping requires surfaces other than quadrics simply to turn corners. Thus the torus is needed to bend around corners and to connect quadrics to quadrics.

The torus is a quartic surface; this is both good and bad for solid modeling. The good news is that ray firing at tori requires solving only a fourth degree equation. Thus solid models which include both quadrics and tori are still relatively simple to render and their mass properties easy to calculate by standard ray firing techniques. The bad news is that while the intersection of two quadrics is an algebraic curve of degree four, the intersection of two tori is an algebraic curve of degree sixteen. Thus, unlike quadrics, it is no longer practical to compute and store an exact analytic representation for the intersection of two tori. Rather, the intersection curve must be approximated by some numerical technique. Not only are numerical procedures relatively slow compared to analytic methods, but also they introduce additional approximations into the model. These numerical approximations make it even harder to convert without error from a CSG tree to a B-Rep because the approximations are not always geometrically and topologically consistent.

The torus is required to round out and complete the quadric surface system, but it does not remove all the inherent inadequacies of this modeling scheme. Quadrics and tori together are sufficient for modeling most rigid mechanical parts, but they are still inadequate for complex blends and freeform design.

3.4. Sweeps. Many mechanical parts are designed in two dimensions first and then extruded into three dimensions along either a straight line or a circular arc. This rather standard two and one-half-dimensional technique is a design paradigm that can be captured readily by solid modeling systems (Shirma et al. [47]). In CSG systems all that is required are two new solid primitives: a solid slab to model straight line extrusions, and a solid of revolution to model circular sweeps. Boundary file systems are a bit more complicated. Here one needs to add new surface types for each curve that can be extruded as well as algorithms for computing all the possible novel surface-surface intersection curves. Even if the design curves are restricted to just straight lines and conic sections, this can lead to many new surface types. Indeed if a conic is extruded along a line which is not perpendicular to its plane of definition or if it is swept around an axis which is not one of its own axes of symmetry, then the resulting surface will not be a quadric.

Nevertheless solids containing sweeps are not hard to analyze. Ray firing, for example, can be reduced to a two-dimensional problem (Kajiya [24], van Wijk [53]). That is, for swept surfaces the three-dimensional ray-surface intersection problem can always be reduced to a two-dimensional curve-curve intersection problem and this two-dimensional problem can generally be solved quite easily especially if the generating curves are of low degree (Sederberg [41], Sederberg et al. [43]). Thus swept solids are easy to render and their mass properties are simple to analyze by standard ray firing techniques. Although it is not practical to calculate intersections analytically for two general swept surfaces, good approximations can be obtained simply by intersecting one surface with a series of generating lines or circles on the other surface. This approach reduces a hard surface-surface intersection problem to a series of easy curve-surface intersection problems. Thus the surface-surface intersection problem is greatly simplified but at the cost of introducing numerical approximations into the answer.

Sweeps permit us to model essentially all two and one-half-dimensional mechanical parts. Solid modeling systems which include linear and circular sweeps generalize the power of quadric-torus modeling schemes by allowing more general design surfaces and more extensive ducts and blends. Still the kinds of objects that can be modeled are much the same—that is, rigid mechanical parts. For complex blends and true freeform design we must look elsewhere.

3.5. Quadratics and biquadratics. With parametric quadratic and biquadratic patches we can begin to achieve true freeform design. The reason for choosing to use parametric quadratics and biquadratics is simply that we need to go at least to degree two in the parameters in order to find surfaces with any kind of flexibility. Indeed, linear surfaces are planes and bilinear surfaces are just hyperbolic paraboloids.

A straight line can intersect a quadratic patch in at most four points and a biquadratic patch in at most eight points. Thus ray firing at these surfaces requires solving, at worst, a degree four or a degree eight polynomial equation. Therefore solids containing these surface patches are still relatively easy to render and their mass properties not difficult to analyze by standard ray firing algorithms (Sederberg and Anderson [42]). Indeed since quadratics are really degree four surfaces, they are intrinsically no more complex than the torus. Moreover the rational quadratics include the quadrics and the torus as special cases so it is possible to build a modeling system based solely on these patches that includes the quadric-torus system.

A lot more blending and much more freeform design is possible using second degree parametric patches. Doo and Sabin [12], [13] have developed an algorithm for generating smooth freeform surfaces from polyhedra by repeatedly chamfering edges and vertices, and Riesenfeld [33] has proved that, in general, this recursive subdivision technique converges to a collection of piecewise biquadratic patches. Sederberg and Anderson [44] have shown how to use quadratic triangular patches to round the corners on a cube and to blend the intersection of three rectangular blocks, and Steinberg [49] has used biquadratic rectangular patches to model such diverse objects as shoes, airfoils, dolls, and electric shavers.

However second degree parametric patches also have their problems. The intersection of two quadratic patches is a curve of degree sixteen and the intersection of two biquadratic patches is a curve of degree sixty-four. Thus while higher degree surfaces add flexibility to our modeling scheme, they also increase the burden on our algorithms. Clearly exact intersection algorithms are out of the question; the challenge is to construct numerical algorithms sufficiently robust to use in B-Rep systems. This is a difficult and challenging technical problem which, even today, has not been totally solved. And yet while from this perspective the degree of these surface patches is already too high, from other perspectives it is not quite high enough. Piecewise quadratic and biquadratic patches can only be pieced together with C^1-continuity, not with C^2-continuity. Yet C^2-continuity is often required for many applications; hence the popularity of cubic and bicubic patches in many surface modeling systems. It may be possible to circumvent this problem by stitching together quadratic and biquadratic patches in a curvature (V^2 or G^2) continuous manner (Kahmann [22]). Unfortunately piecewise quadratic and biquadratic patches are not as flexible for freeform design as one might initially expect. Sederberg and Anderson [44] experimented with piecewise, quadratic, triangular, Bézier patches and found that they very soon ran out of freedom in the placement of the control points if they insisted on just C^1-continuity across the boundaries.

Parametric surfaces which are second degree in their parameters only just begin to give us the ability to perform complex blends and true freeform design. Unfortunately these blends are not always sufficiently smooth and the amount of freeform design is still severely restricted. Thus while quadratics and biquadratics have many nice properties, they are far from the last word on surfaces in solid modeling.

3.6. Freeform surfaces.   Up to now two somewhat different types of polynomial surfaces have been competing for our attention: algebraic surfaces such as planes, quadrics, and tori, and parametric patches such as sweeps, quadratics, and biquadratics. For the surfaces we have encountered so far it is always possible to convert from one form to the other. For example, every quadric surface has a rational quadratic parameterization, and every quadratic patch can be represented by an implicit, degree four, polynomial equation. However, there do exist quartic algebraic surfaces which have no rational polynomial parameterizations (Walker [58]). On the other hand, all rational polynomial patches can be represented as algebraic surfaces (Sederberg [41]). To explore possible paradigms for freeform design, we shall need to look at both types of surfaces. We begin with parametric patches.

If we allow B-Rep solid modelers to include parametric polynomial patches up to some arbitrarily high degree, we will certainly have enough power to construct complex blends and to perform freeform design because we will have all the power of standard surface modeling systems (Tiller [51]). Since rational parametric patches include all of the surface types we have looked at so far, a solid modeling system which incorporates these patches will necessarily have the capacity to accommodate all the applications of previous systems. To simplify the mathematical algorithms and computations, we can choose to work with whatever polynomial basis is most convenient. Riesenfeld [35] has constructed a solid modeling system called Alpha-1 based totally on tensor product rational B-splines. He chose B-splines in order to take advantage of their many special properties such as local control, subdivision, and knot refinement. However other choices such as Bernstein polynomials, beta-splines, and even multivariate B-splines are clearly possible.

Once we permit parametric polynomial patches of even moderately high degree, analysis becomes a real problem. A straight line can intersect a triangular patch of degree N in N^2 points and it can intersect a rectangular patch of bidegree N in $2N^2$ points. Ray firing at such patches by solving polynomial equations soon becomes prohibitively expensive even for reasonably tame values of N (Kajiya [23]). But if ray firing at these patches by analytic methods is prohibitively expensive, intersecting two such patches analytically is astronomically costly. Two triangular degree N patches intersect in a curve of degree N^4 and two rectangular bidegree N patches intersect in a curve of degree $4N^4$. In place of analytic methods recursive subdivision is currently employed to render these solids, analyze their mass properties, and intersect their bounding curves and surfaces while performing Boolean operations (Cohen et al. [10], Lane and Riesenfeld [25], Riesenfeld et al. [34], Thomas [50]). Unfortunately, subdivision is an iterative, relatively slow, nonanalytic technique (Sederberg [46]). Nevertheless this seems to be the price we must pay for using what are really very high degree—N^2, $2N^2$—algebraic surfaces in solid modeling systems.

There is another way. Between the quadric surfaces of most standard solid modeling systems and the bicubic patches of most typical surface design systems lie many untapped degrees of freedom. Quadrics are, after all, algebraic surfaces of degree two and bicubic patches are, in reality, algebraic surfaces of degree eighteen; between two and eighteen there is lots of room for a whole host of useful surface types. Some of these we have already encountered such as tori (degree four), quadratics (degree four), biquadratics (degree eight), and even cubics (degree nine), but what about trying algebraic surfaces of degree three, the next step up after quadrics? Also, why limit ourselves only to those algebraic surfaces which have rational polynomial parameterizations (genus zero); why not try modeling with arbitrary algebraic patches?

Many benefits could accrue from such an approach. Keeping the degree of the surfaces low would permit ray firing to be performed analytically by solving low degree polynomial equations. Therefore solids bounded by these surfaces would be easy to render and their mass properties simple to compute by standard ray firing techniques. Also with algebraic surfaces of low degree classical analytic techniques from algebraic geometry become practical for computing the surface-surface intersections required by B-Rep systems. Unlike parametric patches, algebraic surfaces are closed under offset. Thus offsets could be modeled exactly instead of being approximated as they are now in systems based on parametric patches. Nice blends are possible using algebraic patches. Both Hoffman and Hopcroft [21] and Middleditch and Sears [28] have shown how to blend any two quadric surfaces smoothly and naturally with algebraic surfaces of degree four, and their methods can be used to blend any two arbitrary algebraic surfaces without an explosion of degree. Thus complex blends, ducts, fillets, and chamfers can all be done quite easily and naturally with algebraic surfaces.

Since all parametric polynomial patches are algebraic surfaces, freeform design is certainly possible with algebraic surfaces; what then prevents algebraic patches from being used more extensively in geometric modeling today? Unlike parametric patches, there are no well established, intuitive techniques for defining and controlling the shape of algebraic surfaces. Changing a coefficient in the implicit equation of an algebraic surface is not at all like moving a control point in a Bézier patch, for although changing a coefficient will alter the shape of the surface, it will not do so in a simple intuitive way. Sederberg [45] has begun to experiment with techniques for defining algebraic surfaces with control points, but much work still remains to be done along these lines. Transfinite interpolation also needs to be accommodated with algebraic surfaces so that lofted solids can be constructed, but, as yet, it is unclear how this is to be accomplished. Finally, if we allow algebraic surfaces with nonzero genus, then by the Clebsch Uniformization Theorem (Clebsch [8]) we will be unable to construct rational polynomial parameterizations for these patches. Since currently parameterizations are required by many mathematical algorithms, we shall need either to construct useful nonpolynomial parameterizations or to develop new algorithms which work directly from the implicit polynomial equations. As of now neither approach has been seriously attempted.

3.7. Other possibilities.   Other possibilities exist for modeling both surfaces and solids. Implicit equations need not be restricted to polynomials. Rockwood [36] has used implicit nonpolynomial surface equations successfully as blends for solid models. There have also been experiments with procedurally defined surfaces and solids. Veenman [54] has constructed freeform solids by applying the Sabin–Doo recursive subdivision algorithm. Here the sculpting paradigm is pursued by repeatedly chopping off the corners of a polyhedral solid until a reasonably smooth shape appears. Rossignac and Requicha [37] have introduced constant radius blends into CSG modelers by using offset solids to define blending primitives. Barr [1] has suggested that useful surfaces and solids can be constructed by applying smooth deformations to simple primitives, and Cobb [9] has applied these operations to rational B-spline surfaces to produce many interesting and practical objects. Researchers at Ford, General Electric, PDA, and several other companies have been experimenting with trivariate methods for defining solids. These techniques seem to be especially appropriate for solids composed of heterogeneous materials and for finite element applications (Stanton et al. [48]). Finally from algebraic geometry comes the idea of a more general parametric polynomial surface; that is, a surface defined not by two parameters but by several parameters related by algebraic equations. All of these possibilities seem promising and are worth pursuing in future investigations of solid modeling, but, for now, we choose to rest this discussion here and go on to other more pressing matters.

4. Some Strategies for Introducing Freeform Surfaces into Solid Modeling Systems.   The generation and analysis of solid models necessitates the computation of intersections. Ray firing calls for a line-surface intersection routine for each new surface type. B-Rep systems also require surface-surface intersection routines in order to form the edges of their bounding faces. Now if there are N distinct surface types, this will lead either to a nightmare of N^2 surface-surface intersection routines or to one very general, but dreadfully slow algorithm. To model more objects and to cover more applications, we are drawn to use more and more complicated surfaces. But the more complex the surfaces, the harder it is to compute fast, accurate intersections. This dilemma must be faced sooner or later by all solid modeling systems which wish to incorporate freeform design. Currently four strategies for attacking this problem are being pursued. We shall look at each in turn.

4.1. Ray firing at freeform CSG primitives.   Freeform surfaces can be introduced into a CSG system by inserting primitive solids bounded by Bézier, B-spline, Coons, Gordon, or any other surface type. Ray firing can then be used to render and analyze these solids. Thus for each additional surface type only one new, relatively simple, subroutine is required, a line-surface intersection algorithm. This approach simply eschews B-Reps altogether and attempts to do all design and analysis by firing rays at CSG trees. These tactics completely circumvent the problem of computing surface-surface intersections.

For complicated surfaces even line-surface intersection algorithms may be very slow, but ray firing can be speeded up by using coherence, or parallel processing, or specialized hardware. This strategy is simple and can be used to extend the capabilities of existing CSG systems such as PADL or SynthaVision. Still it remains to be seen whether all the required applications can really be run from a strict CSG representation. The lack of explicit topology in the data base of the CSG representation makes this claim extremely dubious.

4.2. Homogeneous B-Rep systems. Adding a new surface type to a B-Rep system is not so easy. Either a new surface-surface intersection routine must be written for each surface type in the system or a very general, very slow, intersection algorithm must be invoked. Neither alternative is very appealing.

One way to avoid this predicament is to find a single canonical surface type which is sufficiently general to handle most applications. Then represent all surfaces in this homogeneous form, and develop fast algorithms, and if need be specialized hardware, to intersect these canonical patches. These tactics are employed by Riesenfeld in the Alpha-1 system where all surfaces are represented as tensor product rational B-splines, and subdivision, knot refinement, and the Oslo Algorithm are used to intersect two B-spline patches.

Though attractively simple, there are some obvious drawbacks to this approach. First, easy things are made difficult. For example, even simple surfaces like planes and spheres are going to be given complex representations as, say, rational B-splines. This complicates the data base unnecessarily. Moreover, finding the intersection of two planes as the intersection of two rational B-splines is certainly excessively complicated and will lead to unnecessary approximations to a simple straight line. Nevertheless, this is the price one pays for generality. Second, the system is not extendible. Each time new, more general, more flexible surface types are invented (e.g., beta-splines, multivariate B-splines, algebraic patches, etc.), it may be necessary to scrap the old system and start over. This debacle has already occurred to systems based entirely on quadric surfaces. Thus one is really betting everything on a particular surface type. Still the strategy of homogeneous surface representations does solve the problem of how to add freeform surfaces to B-Rep systems and it can be implemented with today's technology.

4.3. Faceting and other approximation schemes. Instead of using complicated patches to represent simple surfaces, why not use simple surfaces to approximate complicated patches? There are many alternatives to this approach since there is an obvious trade-off between the simplicity of the approximating surfaces and the number of surfaces required to generate a reasonable approximation. Planes, quadrics, quadratics, and biquadratics have all been suggested as possible approximating surfaces.

Planar, or faceted, models are appealing because of their obvious simplicity.

Planes are certainly easy to intersect, and specialized hardware already exists for rendering polygonal objects. The main drawback of faceted modelers is the huge amounts of storage required to represent accurately complicated curved shapes. Faceted models are useful for generating quick, gross approximations and fast, crude pictures, but not for accurate modeling of complex freeform shapes.

Other approximation schemes run into similar problems. Even when quadrics are used just to approximate a torus the results are not particularly impressive. Quadratics and biquadratics seem to hold the most promise since they are the lowest degree freeform parametric patches. Unlike planes and quadrics, quadratics and biquadratics are already adequate for some freeform design, and their low degree helps to simplify the intersection problem. Still it is not clear exactly how one is to approximate higher order surfaces by quadratics and biquadratics. Indeed this looks like a very challenging, cumbersome, and difficult numerical problem.

4.4. Low degree algebraic surfaces.   Another alternative is simply to abandon parametric patches altogether and attempt to model everything with low degree algebraic surfaces. Algebraic surfaces have much to recommend them. They are more general than parametric patches, and lots of interesting freeform shapes and complex blends are possible with algebraic surfaces of quite moderate degree. Keeping the degree low is essential because we will still need to construct fast, accurate intersection algorithms.

However before this approach can be effective, much work remains to be done. Blending techniques seem under control, but reasonable methods for transfinite interpolation and discrete approximation have yet to be established. That is, design paradigms for algebraic surfaces need to be developed. Also, algorithms which run directly from implicit rather than parametric equations need to be investigated. This general strategy seems very promising, but a lot of research remains to be done to bring it to fruition.

5. Conclusions.   Solid modeling is a powerful tool for the design and analysis of structures and assemblies. Unlike other modeling schemes, solid modeling provides complete, unambiguous descriptions of real, manufacturable objects. Solid models also contain topological information essential for solving problems involving connectedness and juxtaposition. In addition, solid modeling encourages the use of high level primitives making design easier, less tedious, and less error prone. For all these reasons solid modeling is the technique with the most potential for the future of computer aided design.

Nevertheless many problems still remain. Currently only a very few applications can actually be run directly from solid models. To fulfill its promise, solid modeling must embrace a whole host of new applications. A good deal of research is now being devoted to expanding the applications of solid modeling systems, but a major impediment to progress is that freeform surfaces have yet to be fully integrated with solids. Several approaches for correcting this deficiency are being pursued, but they all have distinct drawbacks and limitations. How this problem will ultimately be solved is an important subject for future investigations.

REFERENCES

[1] A. L. BARR, *Global and local deformations of solid primitives*, Computer Graphics, 18 (1984), pp. 21–30.

[2] B. G. BAUMGART, *Geometric modeling for computer vision*, Tech Rept. STAN-CS-74-463, Stanford Artificial Intelligence Laboratory, Stanford Univ., Stanford, CA, 1974.

[3] P. BÉZIER, *Procédé de définition numérique des courbes et surfaces non mathematiques*, Systeme UNISURF, Automatisme 13, 1968.

[4] H. J. BORKIN, J. F. MCINTOSH AND J. A. TURNER, *The development of three-dimensional spatial modeling techniques for the construction planning of nuclear power plants*, Computer Graphics, 12 (1978), pp. 341–347.

[5] J. W. BOYSE AND J. E. GILCHRIST, GMSOLID: *interactive modeling for design and analysis of solids*, IEEE Computer Graphics Appl., 2 (1982), pp. 27–40.

[6] I. C. BRAID, *Designing with Volumes*, Cantab Press, Cambridge, England, 1974.

[7] H. CHIYOKURA AND K. FUMIHIKO, *Design of solids with free-form surfaces*, Computer Graphics, 17 (1983), pp. 289–298.

[8] A. CLEBSCH, J. fur Math., 64 (1865), pp. 43–65.

[9] E. COBB, *Design of sculpted surfaces using the B-spline representation*, Ph.D. thesis, Dept. Computer Science, Univ. Utah, Salt Lake City, UT, 1984.

[10] E. COHEN, T. LYCHE AND R. RIESENFELD, *Discrete B-splines and subdivision techniques in computer-aided geometric design and computer graphics*, Computer Graphics and Image Processing, 14 (1980), pp. 87–111.

[11] S. A. COONS, *Surfaces for computer-aided design of space forms*, MIT MAC-TR-41, Cambridge, MA, 1967.

[12] D. V. H. DOO, *A recursive subdivision algorithm for fitting quadratic surfaces to irregular polyhedrons*, Ph.D. thesis, Brunel Univ., 1978.

[13] D. V. H. DOO AND M. SABIN, *Behavior of recursive division surfaces near extraordinary points*, Computer Aided Design, 10 (1978), pp. 356–360.

[14] A. DRESDIN, *Solid Analytic Geometry and Determinants*, Dover, New York, 1964.

[15] A. R. FORREST, *Interactive interpolation and approximation by Bézier polynomials*, Computer J., 15 (1972), pp. 71–79.

[16] R. N. GOLDMAN, *Quadrics of revolution*, IEEE Computer Graphics Appl., 3 (1983), pp. 68–76.

[17] ———, *Two approaches to a computer model for quadric surfaces*, IEEE Computer Graphics Appl., 3 (1983), pp. 21–24.

[18] W. J. GORDON, *Spline-blended surface interpolation through curve networks*, J. Math. Mech., 18 (1969), pp. 931–952.

[19] D. G. HAKALA, R. C. HILLYARD, P. J. MALRAISON AND B. E. NOURSE, *Natural quadrics in mechanical design*, Proc. CAD/CAM VIII, pp. 363–378.

[20] R. C. HILLYARD, *The BUILD group of solid modelers*, IEEE Computer Graphics Appl., 2 (1982), pp. 43–52.

[21] C. HOFFMANN AND J. HOPCROFT, *Automatic surface generation in computer aided design*, The Visual Computer, 1 (1985), pp. 92–100.

[22] J. KAHMANN, *Continuity of curvature between adjacent Bézier patches*, in Surfaces in Computer Aided Geometric Design, R. Barnhill and W. Boehm, eds., North-Holland, Amsterdam, pp. 65–75.

[23] J. T. KAJIYA, *Ray tracing parametric patches*, Computer Graphics, Proc. Siggraph '82, 16, pp. 245–254.

[24] ———, *New techniques for ray tracing procedurally defined objects*, Computer Graphics, Proc. Siggraph '83, 17, pp. 91–102.

[25] J. M. LANE AND R. F. RIESENFELD, *A theoretical development for the computer generation and display of piecewise polynomial surfaces*, IEEE Trans. on Pattern Analysis and Machine Intelligence, PAMI-2, pp. 35–46.

[26] J. Z. LEVIN, *A parametric algorithm for drawing pictures of solid objects composed of quadric surfaces*, Comm. ACM, 19 (1976), pp. 555–563.

[27] J. Z. Levin, *Mathematical models for determining the intersections of quadric surfaces*, Computer Graphics and Image Processing, 11 (1979), pp. 73–87.

[28] A. E. Middleditch and K. H. Sears, *Blend surfaces for set theoretic volume modelling systems*, Computer Graphics, Proc. Siggraph '85, 19, pp. 161–170.

[29] A. A. G. Requicha and H. B. Voelcker, *Constructive Solid Geometry*, Tech. Memo. No. 25, Production Automation Project, Univ. Rochester, Rochester, NY, 1977.

[30] A. A. G. Requicha, *Representations for rigid solids: Theory, methods, and systems*, Computing Surveys, 12 (1980), 437–464.

[31] A. A. G. Requicha, and H. B. Voelcker, *Solid modeling: A historical summary and contemporary assessment*, IEEE Computer Graphics Appl., 2 (1982), pp. 9–24.

[32] R. F. Riesenfeld, *Applications of B-spline approximation to geometric problems of computer aided design*, Ph.D. thesis, Dept. Systems and Information Science, Syracuse Univ., Syracuse, NY, 1973.

[33] ———, *On Chaikin's algorithm*, Computer Graphics Image Process., 4 (1975), pp. 304–310.

[34] R. F. Riesenfeld, E. Cohen, R. Fish, S. Thomas, E. Cobb, B. Barsky, D. Schweitzer and J. Lane, *Using the Oslo Algorithm as a basis for CAD/CAM geometric modeling*, Proc. NCGA, 1981.

[35] R. F. Riesenfeld, *Summary of the Concepts of the Alpha-1 CAGD system*, Detroit Engineer, (1982), pp. 8–11.

[36] A. P. Rockwood, *Introducing sculpted surfaces into a geometric modeler*, in Solid Modeling by Computers: From Theory to Applications, Plenum Press, New York, 1984, pp. 237–258.

[37] J. R. Rossignac and A. A. G. Requicha, *Constant-radius blending in solid modeling*, Comput. Engrg., (1985), pp. 65–73.

[38] S. D. Roth, *Ray casting for solid models*, Comput. Graphics Image Process., 18 (1982), pp. 109–144.

[39] R. F. Sarraga, *Algebraic methods for intersections of quadric surfaces in GMSOLID*, Computer Vision, Graphics, and Image Processing, 22 (1983), pp. 222–238.

[40] R. F. Sarraga and W. C. Waters, *Free-form surfaces in GMSOLID: Goals and issues*, in Solid Modeling by Computers: From Theory to Applications, Plenum Press, New York, 1984, pp. 187–209.

[41] T. W. Sederberg, *Implicit and parametric curves and surfaces for computer aided geometric design*, Ph.D. thesis, Dept. Mech. Engrg., Purdue Univ., West Lafayette, IN, 1983.

[42] T. W. Sederberg and D. C. Anderson, *Ray tracing of Steiner patches*, Computer Graphics, Proc. Siggraph '84, 18 (1984), pp. 159–164.

[43] T. W. Sederberg, D. C. Anderson and R. N. Goldman, *Implicit representation of parametric curves and surfaces*, Computer Vision, Graphics, Image Process., 28 (1984), pp. 72–84.

[44] T. W. Sederberg and D. C. Anderson, *Steiner surfaces patches*, IEEE Computer Graphics Appl., 5 (1985), pp. 23–36.

[45] T. W. Sederberg, *Piecewise algebraic surface patches*, Computer Aided Geometric Design, (1985), pp. 53–59.

[46] ———, *A comparison of three curve intersection algorithms*, Computer Aided Design, 18 (1986), pp. 58–63.

[47] Y. Shirma, N. Okino and Y. Kakuzu, *Research on 3-D geometric modeling by sweep primitives*, Proc. Computer Aided Design, 1983, pp. 671–680.

[48] E. L. Stanton, L. M. Crain and T. F. Neu, *A parametric cubic modelling system for general solids of composite material*, Internat. J. Numer. Meth. Engrg., 11 (1977), pp. 653–670.

[49] H. A. Steinberg, *A smooth surface based on biquadratic patches*, IEEE Computer Graphics Appl., 4 (1984), pp. 20–23.

[50] S. W. Thomas, *Modeling volumes bounded by B-spline surfaces*, Ph.D. thesis, Dept. Computer Science, Univ. Utah, Salt Lake City, UT, 1984.

[51] W. Tiller, *Rational B-splines for curve and surface representation*, IEEE Computer Graphics Appl., 3 (1983), pp. 61–69.

[52] J. A. Turner, *A set-operation algorithm for two and three dimensional geometric objects*, Proc. Fourth Annual Pacific Northwest Computer Graphics Conference, 1985.

[53] J. J. van Wijk, *Ray tracing objects defined by sweeping planar cubic splines*, Trans. Graph., 3 (1984), pp. 223–237.

[54] P. Veenman, *The design of sculpted surfaces using recursive subdivision*, Proc. of the Conference on CAD/CAM Technology in Mechanical Engineering, MIT, Cambridge, MA, 1982, pp. 54–64.

[55] H. B. VOELCKER AND A. A. G. REQUICHA, *Geometric modeling of mechanical parts and processes,* Computer, 10 (1977), pp. 48–57.

[56] H. VOELCKER, A. REQUICHA, A. HARTQUIST, E. FISHER, W. METZGER, R. TILOVE, N. BIRRELL, W. HUNT, G. ARMSTRONG, T. CHECK, R. MOOTE AND J. MCSWEENY, *The* PADL-1.0/2 *system for defining and displaying solid objects,* Computer Graphics, Proc. Siggraph '78, 12 (1978), pp. 257–263.

[57] H. B. VOELCKER AND W. A. HUNT, *The role of solid modelling in machining-process modelling and* NC *verification,* SAE Technical Paper Series 810195, 1981.

[58] R. J. WALKER, *Algebraic Curves,* Princeton University Press, Princeton, NJ, 1950.

Mathematical Problems in Solid
Modeling: A Brief Survey

DAVID A. FIELD

Abstract.   Intrinsic mathematical issues in three-dimensional solid modeling are surveyed by referencing an early version of GMSOLID, a solid modeler which incorporated both a CSG tree and boundary representations, keystones in today's solid modeling systems. Mathematical problems that are generally applicable to all solid modelers are presented. Discussions of these problems include constraints imposed by computer systems on solutions incorporated into initial solid modeling efforts. Mathematical aspects of solid modeling based applications and precision of numerical calculation are also discussed.

1.   Introduction.   The adoption of CAD technology has been viewed as essential to competitiveness in the auto industry. Since the early 1960s General Motors has used proprietary CAD systems to design sheet metal panels and, more recently, solid modeling has been introduced to design and analyze solid automotive components. The technology of solid modeling is so sufficiently advanced that in 1982 many commercial solid modelers were already available [16].

Characterized as highly interactive with extremely large data bases, today's solid modelers require the implementation of computationally intensive and sophisticated mathematical software. Although the underlying concepts of solid modeling involve simple Boolean operations of unions, intersections, and differences of three-dimensional sets, these Boolean operations quickly become computationally expensive as the solid model becomes more complex. Boolean operations with the primitive sets shown in Fig. 1.1 are not always sufficient and objects with traditional sculptured surfaces found in wire frame design systems have made calculations and representations of intersections, curves and surfaces a complex area of mathematical research. Solid modelers must also communicate through computer graphic displays which often require time-consuming calculations. In spite of heavy computational burdens and managing a large data base necessary for design integrity and applications, the solid modeler must also be responsive to users with a variety of accurate and

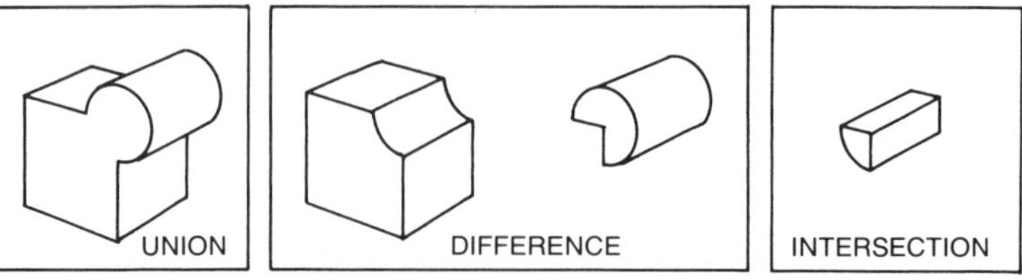

Figure 1.1. Boolean operations with block and cylinder.

helpful design tools. Accuracy, user response and managing data bases often impose conflicting demands.

It is interesting to note that well-known mathematicians among the developers of CADANCE (a General Motors proprietary wire frame system) included W. J. Gordon and J. R. Rice (see Myers [23] for a list of General Motors Research Laboratories Mathematics Department Publications). An enhanced CADANCE system is still extensively utilized in designing automotive body exteriors. By defining a thickness to surfaces, applications such as calculating mass properties (volumes, moments of inertia, etc.) are supported for these essentially two-dimensional geometries. However, it was the awkward and difficult task of computing interference checks with CADANCE which launched General Motors to develop its own solid modeling system (Boyse and Gilchrist [1]). Compared to the parametric surfaces developed for CADANCE, the primitive solids used by GM's first solid modelers were simple. In this paper, an early but well developed version of GMSOLID, GM's in-house solid modeler (Boyse and Gilchrist [1]) is discussed.

Presenting GMSOLID allows a focus on basic mathematical issues in three-dimensional solid modeling. For example, this early version of GMSOLID also incorporates both a CSG tree and a boundary representation, keystones in today's solid modeling systems. Although specific aspects of GMSOLID are referenced in the next section, the third section draws upon these aspects to present mathematical problems that are generally applicable to all solid modelers. The fourth section is a brief introduction to mathematical aspects of solid modeling based applications. Based upon the examples of the previous sections, the final section of the paper discusses precision of mathematical calculations in solid modeling.

2. Mathematical Representations of Solids. During the late 1970s GMSOLID was developed through collaboration of the General Motors Research Laboratories Computer Science Department and the Production Automation Project (PAP) at the University of Rochester (Brown [3]). GMSOLID initially followed concepts from the Part and Assembly Language, PADL, developed by PAP and based its Boolean set operations on primitive solids having quadric surfaces. But the need to interface with CADANCE was also a requirement. Since an extensive interactive graphics system (CADANCE) and associative data base managers were already functioning, certain data representations and graphics interfaces had early influences.

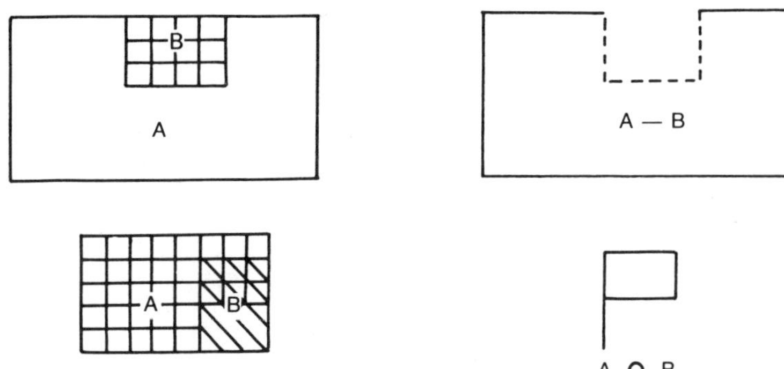

Figure 2.1. Two-dimensional examples of ordinary Boolean operations that do not produce physically realizable objects.

Among the chief reasons for patterning GMSOLID after PADL was the potential that quadric surface intersections would provide computational accuracy. Dealing with solids bounded by quadric surfaces also simplified the Boolean operators of union, intersection and differencing, especially since the original primitives were restricted to solid spheres, blocks, cylinders and cones. Note that the difference of set A with set B, $A - B = A \cap B'$ where $'$ denotes complement, will not always define a closed set or even a three-dimensional object (Fig. 2.1). To create Boolean operators mathematically consistent with physical objects, closure is used to "regularize" Boolean operators. Requicha and Tilove [28] present a thorough mathematical treatment of regularized Boolean operations used in solid modeling.

To record the design of a solid through Boolean operations, GMSOLID uses a constructive solid geometry tree (see Fig. 2.2). The leaves of the CSG tree represent solid primitives and the nodes represent Boolean operations. The top of the CSG tree represents a true solid which is mathematically well defined. The unambiguous definition of a solid is not unique however (see Fig. 2.3).

Determining when two solids are identical but have different CSG trees is still open for research. Determination can be laboriously carried out by comparing each object's boundary representation. On the other hand, the determination can be succinctly stated mathematically as object A equals object B if and only if $(A - {}^{*}B) \cap (B - {}^{*}A)$ is the null object (* denotes regularized Boolean operations). Detecting null objects have certain mathematical advantages which are exploited by Tilove [34]. However, even with Tilove's insights, the computational cost for such a simply stated problem is heavy. A closely related and important research direction is to ascertain which sequences of Boolean operations are most efficient in defining given objects.

CSG tree nodes record rigid translations and rotations imposed on Boolean operation arguments. They also contain physical dimensions of primitives and pointers to previous operators. Representation of this information is crucial because fundamental geometric data may be transformed into equivalent representations several times before the results of Boolean operations are calculated. Thus, managing information through the CSG tree can be a significant source of computational errors and inefficiencies.

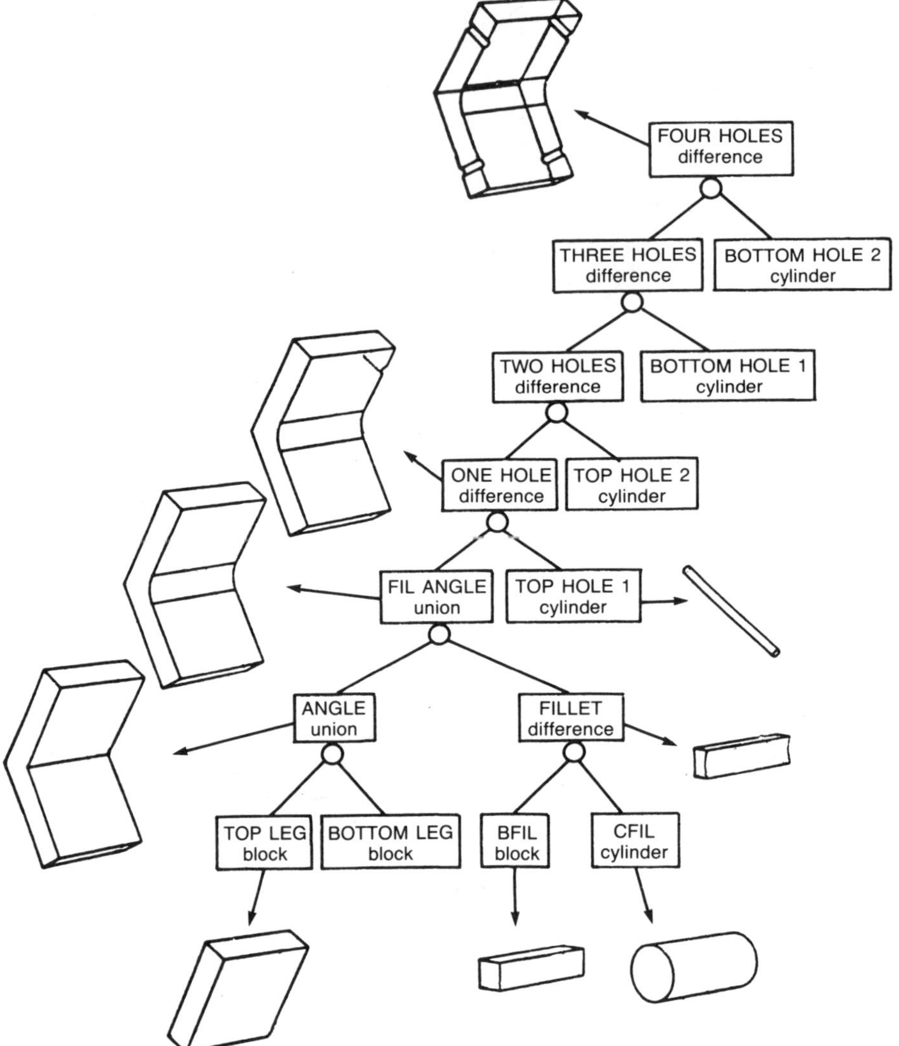

Figure 2.2.   A constructive solid geometry tree.

GMSOLID also supports a boundary representation which records a solid's current stage of construction with definitions of every vertex, edge and face on its surface along with their relationships to one another. Since each Boolean operation creates new vertices, edges and faces, these entities and the old boundary representation must be reclassified as inside, on, or outside of the new solid and new relationships must be established (Figs. 2.4 and 2.5).

An important consideration in using boundary representations is to only allow relationships among vertices, edges and faces which unambiguously define not only

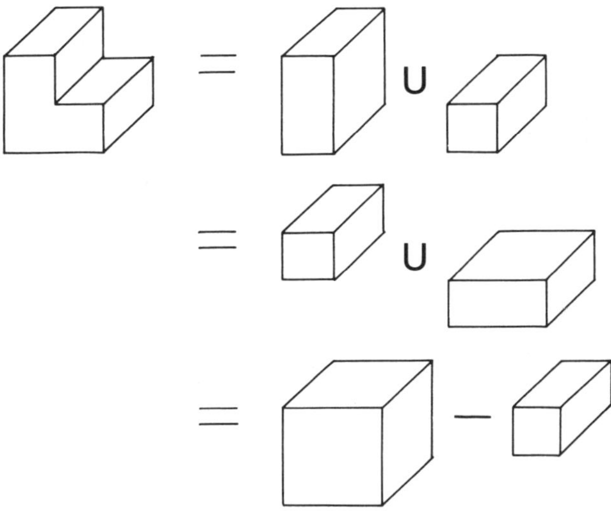

Figure 2.3.   Three different Boolean operations defining the same solid.

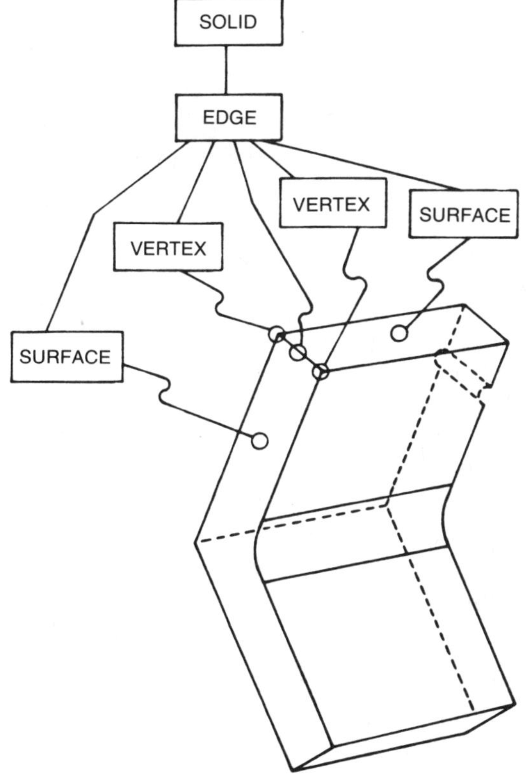

Figure 2.4.   Elements and relations is the boundary representation data structure.

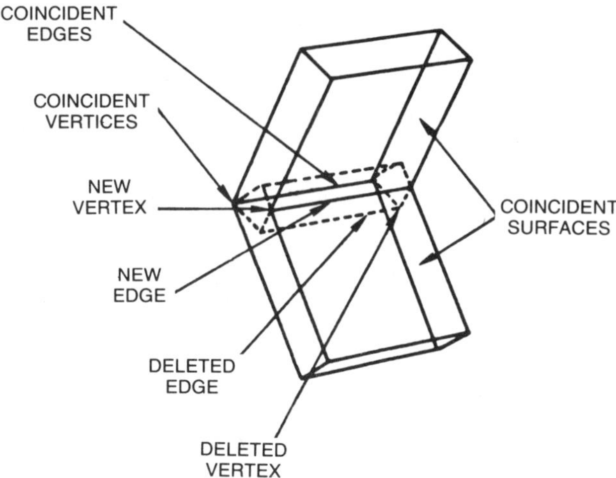

Figure 2.5.   A portion of the reclassification of boundary elements created by the union of two blocks.

valid objects but the object the designer has in mind. Rules derived from algebraic topology have provided necessary conditions for classification systems and algorithms to update and test the validity of reclassifying boundary entities and their relationships. A notable example are Euler operators, a calculus of rules to ensure that certain incremental changes to a true solid maintain topologically valid boundary representations (Braid, Hillyard and Stroud [2]).

Boundary representations generate orders of magnitude more data than CSG trees. Managing and updating boundary representations is time consuming and often involves many nontrivial mathematical calculations. In contrast to CSG trees, boundary representations carry no record of the Boolean operations and primitives used to construct solids. By executing CSG tree records of Boolean operations with varying sizes and locations of primitives, families of solids can be easily produced. More substantial changes in a solid model can be traced to particular nodes in its CSG tree so that initial Boolean operations can be automatically re-executed. On the other hand, computer graphics applications have generally been much easier and faster with boundary representations. Real time applications are very important since engineers interact with solid modelers through extensive computer graphics displays. Although there are other representations of solids, most industrial solid modeling systems support both CSG and boundary representations (Requicha [27]). The next two sections consider mathematical problems encountered in the design and application of solid modeling systems.

3. Mathematical Problems in the Design Process. A major mathematical problem that has consumed substantial efforts in the development of all solid modelers is the calculation and displaying of intersection and feature curves created by Boolean

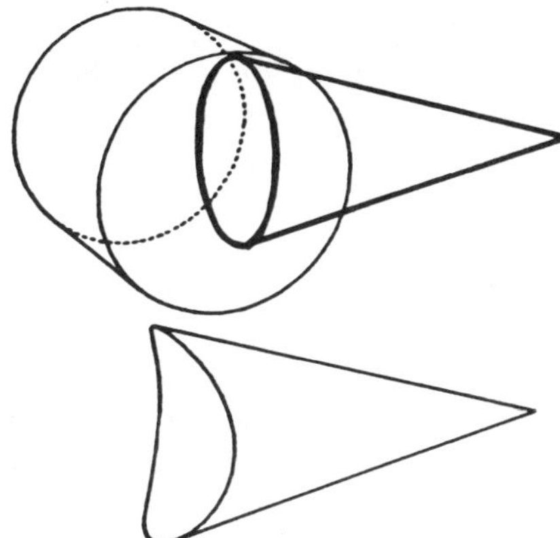

Figure 3.1. Intersection curve for cylinder and cone.

operations. Decisions to include quadric surface primitives in early modeling systems were significantly influenced by the need to accurately calculate intersection curves. Although parametric surfaces are now commonplace in solid modelers, intersection problems with simpler primitives still provide a good introduction into significant aspects of this major problem. Parametric surface intersection problems are considerably more difficult and require a more detailed explanation than this survey will allow. These important problems are addressed by Sederberg in this book and in Sederberg, Anderson and Goldman [31].

Accurate representations of intersection curves are vital to solid modeling systems. Intersection curves are not only used to graphically display solids, they are often used as references in precise technical drawings, manufacturing operations, process planning and many other solid modeling applications. A more crucial and immediate issue is that tangents to curves and surfaces are frequently used to create solid models. Algorithms for quadric surface computations devote special attention to tangencies because general algorithms often fail whereas a known tangency can greatly simplify calculations. Special case algorithms have been developed where computational errors have been investigated and have directed algorithm development (Pfeifer [26]).

In the first solid modelers, calculations and representations of intersection curves were often influenced by data management and computer graphics considerations. To attain fast computer graphics response, points along intersection curves were often stored and their calculation had to be simple and fast. For example, the intersection curve for the cylinder and cone in Fig. 3.1 was computed using the intersections of a ruling of lines on the cylinder's surface with the cone. Even though each line intersection computation involves evaluating the quadratic formula, special care was taken when lines were nearly tangent to the cone's surface. For more complex intersections two real roots from the quadratic formula were carefully sorted to distinguish self-intersecting curves from nearly self-intersecting curves. Unless the self-intersection point was a calculated root there was no simple way to distinguish between a self intersection and a near self intersection. Thus a fast computer graphic solution for determining intersection curves lacked the geometric content to make subsequent geometric decisions during the design of the solid model.

An additional disadvantage of representing intersection curves by ordering large numbers of intersection points is that data management systems are severely burdened. Fortunately analytic methods for parameterizing quadric intersection curves can reduce data management problems and also take advantage of improvements in computer graphics software and hardware. These analytic methods also contain relevant geometric information to identify self intersections of curves, solution branches, and control the ordering and density of plotting points.

A powerful method for determining intersection curves of quadric surfaces involves classical algebraic and projective geometry (Ocken and Schwartz [24]). A crucial aspect is that any quadric surface

$$Ax^2 + By^2 + Cz^2 + Dxy + Exz + Fyz + Gx + Hy + Iz + J + 0,$$

where $|ABC| + |AB| + |AC| + |BC| \neq 0$, can be transformed into one of three common forms in (x, y, z, w) projective space:

$$x^2 + y^2 + z^2 - w^2 = 0, \qquad \text{sphere,}$$
$$x^2 + y^2 - z^2 - w^2 = 0, \qquad \text{single sheet hyberboloid,}$$
$$y^2 - z^2 - w^2 = 0, \qquad \text{cylinder.}$$

In projective space, further transformations based on the use of pencils (Hodge and Pedoe [13]), are invoked to obtain one-dimensional rational algebraic parameterizations to intersection curves in projective space coordinates. Points along the parameterized curve are then transformed from projective space to Euclidean space.

Parameterizing quadric surface intersection curves directly in Euclidean space is possible for very general classes of surface intersections (Levin [19] and Meyer [22]). Sarraga [30] discusses examples of current parameterization methods of quadric surface intersections involving cylinders and cones.

Every solid modeler must efficiently and accurately determine when points, edges and faces lie inside, on or outside a solid model. It is this classification trichotomy of vertices, edges and surfaces that makes boundary representations of solid models an intrinsic three-dimensional problem. These trichotomous tests are often the most frequently invoked calculations during the creation of solid models.

Implicit representations of quadric surfaces can simplify trichotomous classification of points. Namely, substitution of coordinates into normalized implicit representations yields negative values for points outside primitive solid and positive values for interior points. By referencing the Boolean operations in the CSG tree, points can be quickly classified. However referencing the CSG tree is not as easy when surfaces are given in parametric forms which are not readily transformed to implicit representations.

A standard classification method in solid modeling is the three-dimensional version of the point-in-polygon problem, Fig. 3.2. A ray is cast toward a point P from a point V known to be outside the solid. The ray is parameterized as $W = V + t*U$, where U is a unit vector in the direction P from V. Intersections of the ray with surfaces in the

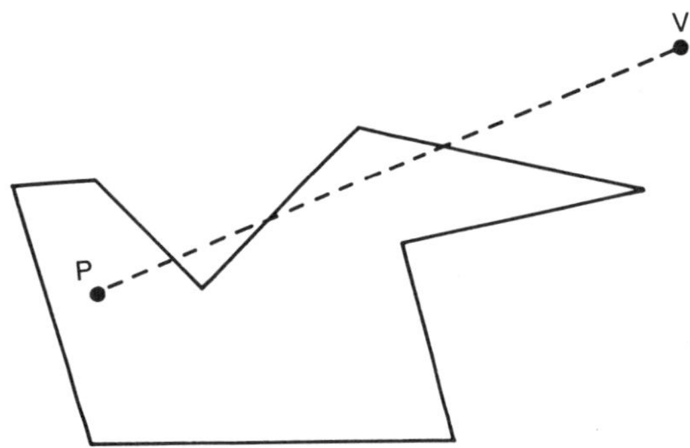

Figure 3.2.   Special cases must be considered when a vertex of the polygon lies on the line joining V and P.

boundary representation give ordered values $t_1, t_2, \cdots, t_n$ along the positive real axis. Intervals $[t_i, t_{i+1}]$ are then alternately classified as outside and inside the solid so that an open interval containing P uniquely classifies P. Since a single ray may produce special cases such as tangency with a bounding surface, more than one ray is cast to correctly classify P. For primitive solids with quadric surfaces, determining values for t involves the quadratic formula. For primitives with parameterized surfaces, values for t can involve solving systems of polynomial equations.

An intersection curve is classified according to the trichotomous classification of an ordered sequence of key points along its path. When solid models are modified through Boolean operations, boundary edges become temporary edges and points along these edges must have their *neighborhood* reclassified. The neighborhood is a two-dimensional disk of epsilon radius and perpendicular to the intersection curve passing through its center. As shown in Fig. 3.3, shaded portions of disks indicate how much of an epsilon disk is contained in the solid primitives. A disk is then classified with respect to a new primitive by applying the new Boolean operation to the disk (Fig. 3.3) where the union of two partially shaded disks is now totally shaded to indicate that a boundary edge has now become an interior edge.

Classifying epsilon disks of points is made efficient by utilizing the fact that each geometric entity comprising a solid model is given its own bounding box, a box enclosing the entity and oriented with its edges parallel to the x, y and z axes. Since an intersection of two bounding boxes is simply a test on diagonally opposite vertices, many calculations such as updating epsilon disks can be quickly resolved. On the other hand, the orientation of bounding boxes is constrained but their enclosed objects are arbitrarily oriented, so that boxes tend to be almost empty. More stringent quick tests to indicate that further calculations are required when bounding boxes do intersect can be found in Field and Morgan [8].

This use of bounding boxes in solid modeling is not exclusive to GMSOLID. Bounding boxes are also practical alternatives to more elaborate and computationally expensive algorithms (Tilove [34]). They significantly reduce unnecessary calculations and are extensively used in calculating parametric surface intersection curves and initializing recursive subdivision methods.

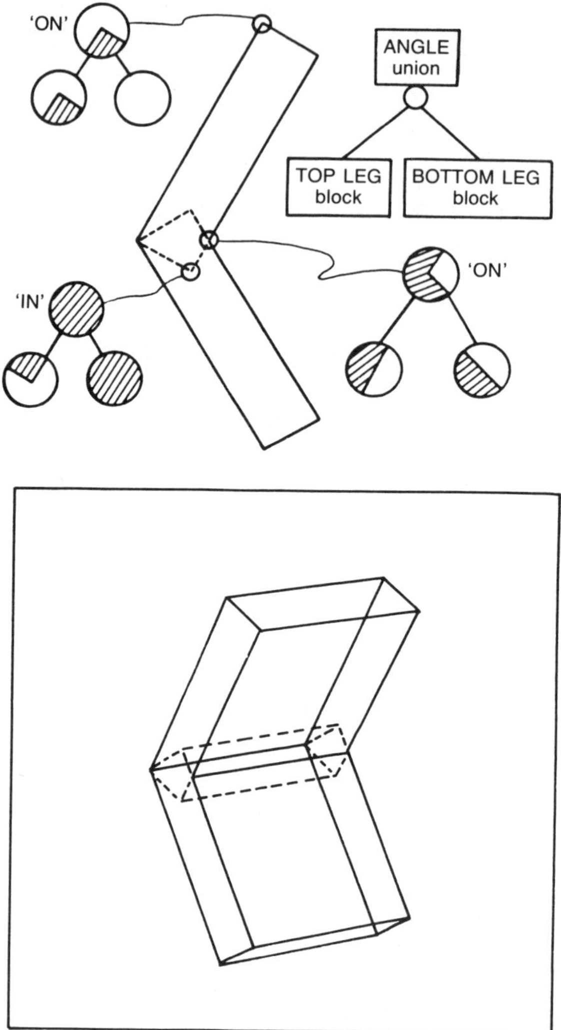

Figure 3.3.   Boolean operation on neighborhoods for edge classification.

Faces in solid modelers are usually represented by edges. Redundant information such as outward normals and adjacency relations are stored as well. Classification of a face is done essentially through classifying its edges and drawing logical conclusions from the redundancy of stored information. Redundancies of information are also very useful in reducing searches and are convenient for graphic displays.

4. Mathematical Problems in Solid Model Applications. A partial list of applications intended for early solid modelers included part, tool and die designs, calculating mass properties of solids, packaging and interference checking, kinematics, finite element analysis, design drawings, technical illustrations, process planning and numerically controlled machining. Some of these applications require sophisticated computer graphic displays based upon information extracted from the solid modeler's comprehensive data base. Other applications require extensions of the solid modeling system. For example, to provide rounding, filleting and offsetting (expansion and shrinking) of solids, Rossignac and Requicha [29] investigated the mathematical consistency of these new operations with existing Boolean operations and incorporated these new operations into a solid modeler.

This section briefly discusses two applications. With these examples, the contributions of mathematics and solid modelers are emphasized. Important points are that solid modelers can be a resource and extensive enhancements of solid modelers are not always necessary. Another point illustrated especially by the second example is that mathematics stimulated by solid modelers can develop separately from the solid modeler.

4.1. Volumes. This relatively simple mass property is generally solved by the classical Greek method of exhaustion (see Fig. 4.1) which displays typical decompositions that solid modelers can support. For example, in Figs. 4.1a and b, the decompositions are defined by the union of intersections of a solid model with solid blocks of fixed cross section. However, in Fig. 4.1b a time consuming and error prone boundary evaluation is required to obtain intersection curves and surface patches for each column. The decomposition in Fig. 4.1c is particularly suitable for Octree-based solid modeling systems (Meagher [21]). Whereas these decompositions clearly rely on quickly determining when corners of solid blocks lie inside, on or outside the solid model, other integration methods including using the divergence theorem have been implemented with limited success on curved surfaces (Lee and Requicha [17], [18]).

An important feature of this example is that applications often have a variety of solutions. Although mathematics provides a rich variety of solutions, it is accuracy, cost and especially a solid modeler's capabilities which direct the final choice of

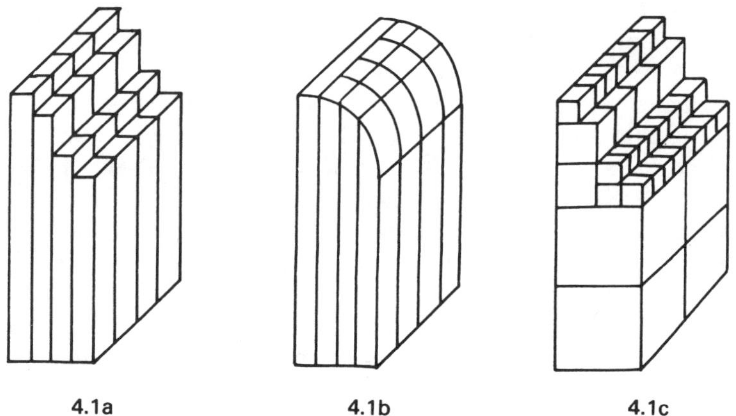

4.1a   4.1b   4.1c

Figure 4.1. Typical decompositions used for volume approximations.

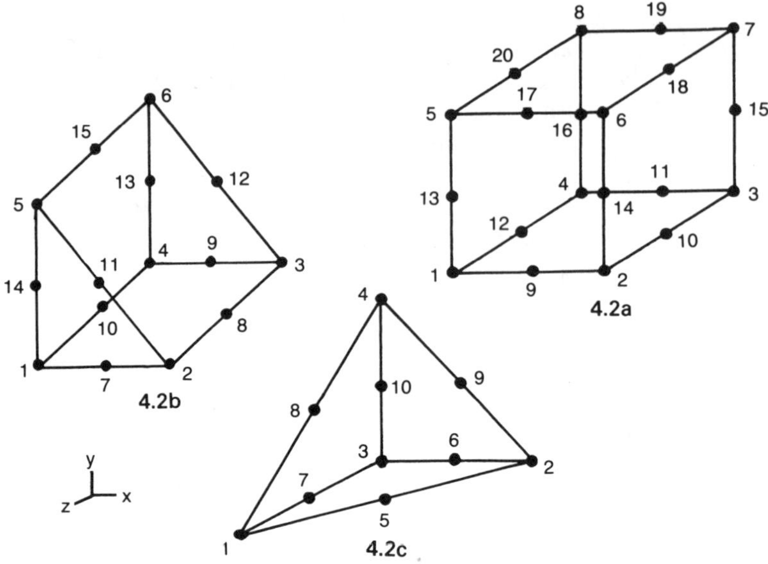

Figure 4.2. The quadratic isoparametric elements. 4.2a hexahedron; 4.2b pentahedron; 4.2c tetrahedron.

numerical integration technique. In this case it is important that the modeler is not enhanced or radically slowed to accommodate a mathematically elegant solution. In addition to mathematical accuracy, criteria for selecting an appropriate solution are chiefly determined by the ease of implementing an algorithm in a solid modeler. The decompositions in Figs. 4.1a and b are methods commonly adopted.

4.2. Mesh generation. Solid modeling and analysis of finite element solutions are obviously tightly connected in the design-redesign cycle. This application will focus on linking two extremely large and independently developed codes, a solid modeling system and a commercial finite element analysis code. The enormous size and sophistication of such software requires an independence even though both are intimately connected through the geometry of solid models.

The finite element method is an algorithm for numerically solving partial differential equations such as those arising in structural analysis of solid automotive components. A finite element solution is a linear combination of piecewise polynomials where each polynomial is defined on its own distinct subregion (called a finite element) of a solid's geometry. A particular difficulty with the finite element method is that solids must be decomposed into a union (mesh) of relatively small finite elements. Three-dimensional geometries are usually decomposed into three element types, tetrahedra, pentahedra and hexahedra where each element is defined by assigning three-dimensional coordinates to its nodes (Fig. 4.2). Decompositions of three-dimensional solids into finite elements are generally constrained so that the edge of an element cannot lie on the interior of a face of another element (Fig. 4.3).

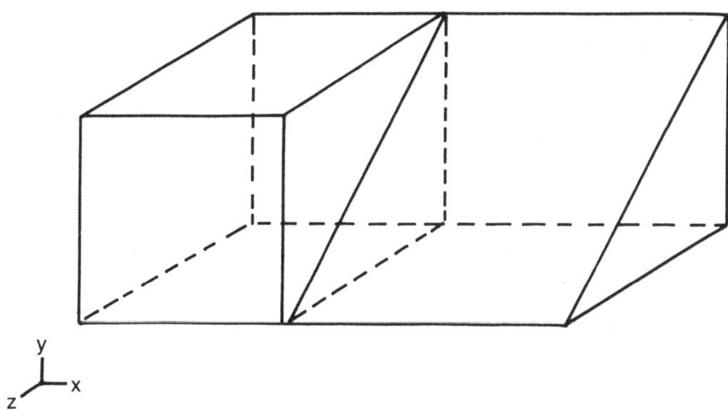

Figure 4.3.   This union of a hexahedron and a pentahedron is not allowed.

Hexahedra, often used for their mathematical properties and for their relative ease in decomposing many geometries, cannot adequately decompose geometries produced by sophisticated solid modelers such as GMSOLID. Graphic displays traditionally used to guide the construction of meshes are often unintelligible because of the concentration of element edges. This is especially true of tetrahedral meshes. Difficulty in visualizing meshes during their construction is a major reason why automation of mesh generation is a serious concern in three-dimensional structural analysis.

At this point it is worth commenting on computerized mesh generators that have appeared in research journals or are commercially available (Krouse [16]). The vast majority of three-dimensional mesh generators rely upon piecewise polynomial interpolants (DeSalvo [5], Grieger and Kamel [12], McCormick [20], PDA [25]). These generators require solids to be partitioned into simpler blocks having curvilinear boundaries which can be described with networks of curves. Through transfinite interpolation functions, a natural coordinization compatible with hexahedral finite element decompositions can be induced. Although Boolean operations are sometimes allowed, decomposition of many solids into convenient blocks is difficult and practically impossible with these mesh generators. Indeed, solid modeling systems such as GMSOLID were developed because of this inadequacy.

A successful approach to mesh generation uses a solid modeling data base and extends two-dimensional triangulation algorithms to construct three-dimensional tetrahedra (Cavendish, Field and Frey [4], Shephard and Yerry [32], Wordenweber [37], and Fig. 4.4). That tetrahedra are used is largely due to the complicated three-dimensional geometries solid modelers can create. Since solid modelers have adequate data bases and means to provide triangulation points on and inside solid models, once a list of triangulation points is created, the mathematical problem lies in designing robust three-dimensional triangulation algorithms. Although the mathematics used involves combinatorial and computational geometry, knowledge of finite element analysis can be very helpful and can influence how triangulation points are generated (Field and Frey [7]).

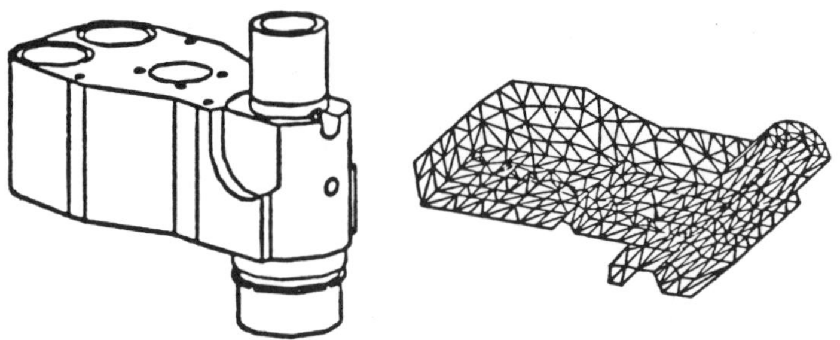

Figure 4.4.   A solid model and its tetrahedral finite element mesh.

The strategy first used by Cavendish, Field and Frey [4] is to create Delaunay triangulations which are characterized by producing a convex hull of tetrahedra where the four vertices of each tetrahedron defines a sphere which contains none of the remaining triangulation points. This latter condition is the basis for the triangulation algorithm originally proposed by Watson [36]. However, a convex hull of tetrahedra also fills concave surfaces and cavities. A simple solution to this problem is to create a list of each tetrahedron's centroid and let the modeler classify each tetrahedron as inside or outside the solid model according to whether its centroid is inside or outside the solid model.

The finite element mesh generator satisfies two very important criteria. First, it requires a simple communication with the solid modeler, namely sequential lists of points are exchanged. Secondly, the mesh generator can be developed independently from a solid modeler whose code is already unwieldy. A significant consequence of these criteria is that the mesh generator can communicate with any solid modeler. It remains to be seen how much additional interaction among solid modeler, mesh generator and finite element analysis will be required for more sophisticated mesh generation ideas such as adaptive mesh refinement.

5.   Observations on Computational Issues.   Data bases for solid modelers are usually organized to quickly access boundary information for computer graphic displays. Thus prompted by computer graphics, it was seen in § 3 that points along intersection curves were stored for quick retrieval and points were determined from line and quadric surface intersections. Although Forsythe [9], Forsythe, Malcolm and Moler [10], and Kahan [15] have shown accurate usage of the quadratic formula and other algorithms, the solid modeling process itself introduces inherent errors that cannot be compensated for by accurate calculations of intersection points. Control of numerical errors during the creation of solid models is critical to prevent solid modelers from creating geometrically inconsistent solids.

An example of these errors is recorded in the CSG tree. Each node of the CSG tree records rigid body motions given to each Boolean operand. Whenever an operand is a new primitive, the operand has undergone at most only a rotation and a translation. On the other hand, when an operand consists of subsets of many primitives, each primitive has undergone a rigid body motion nearly every time it was part of a Boolean operand. Solid models often contain subsets of a primitive where the boundaries of each subset contain intersection curves that have been calculated with different orientations of the primitive. Moreover, since each orientation can be the composition of many translations and rotations, the accumulated errors of calculating many rigid body motions has produced a distorted primitive solid, i.e., an ellipsoid rather than a sphere. Situations have occurred where two neighboring intersection curves have been calculated when in reality there was only one intersection curve and the two curves could be distinguished on a graphics terminal. Ordinarily the problem of intersection curves calculated with different orientations of the same primitive will cause no difficulties. Difficulties tend to occur when tangencies are involved and when there are double and triple roots of polynomial equations. Generally, data bases which represent solids as closely as possible to their geometry help numerical algorithms produce more accurate calculations.

Some numerical errors due to composition of functions are unavoidable. It is not surprising then that multiplying orthogonal matrices creates a matrix which is no longer orthogonal. Matrix representations of rotations imply that each rotation requires storing nine numbers whose accuracy must be monitored. Matrices deviating from orthogonality can be reorthogonalized but a more efficient representation of rotations to reduce the number of arithmetic operations and storage would be more effective. Quaternions provided such a representation for Taylor to represent rotations about an arbitrary vector on the unit sphere (Taylor [33]). Quaternions can be represented by four numbers and a composition of two rotations require 16 multiplications and 12 additions compared to 27 multiplications and 18 additions for matrix multiplications. Moreover, orthogonalization of quaternions is given by $q/|q|$ where $|q| = (q \cdot q)$ is the ordinary dot product.

The significance of applying quaternions during the design of solid models is not diminished by the following observation. When a solid model is completed, each primitive had reached its final orientation. Through its CSG tree the solid can be reconstructed with each primitive in its final orientation. Therefore each intersection curve can be recalculated with primitive solids which have undergone only one rigid body motion. The new solid may be different from the original solid but both solids should be topologically and geometrically equivalent.

The effect of using double precision arithmetic is mixed. Double precision calculation can unreasonably slow down the performance of solid modelers. On the other hand, there are certain calculations where using a double precision version of an algorithm just does not provide enough accuracy. These calculations require special purpose algorithms.

Algebraic methods for intersections of quadric surfaces can require determining eigenvalues of matrices (Sarraga [30]). Rather than use characteristic polynomials of degree three or four, GMSOLID uses codes from the LINPACK (1985) and IMSL libraries [14] to solve eigenvalue and generalized eigenvalue problems. Although there are robust algorithms for solving polynomial equations, characteristic polynomials are avoided because their coefficients involve determinants whose entries already contain numerical errors.

Acknowledgments. The author appreciates many discussions with Ramon Sarraga and William Frey during the preparation of this paper.

REFERENCES

[1] J. W. BOYSE AND J. E. GILCHRIST, GMSOLID-*interactive modeling for design and analysis of solids*, IEEE Computer Graphics Appl., 2 (1982), pp. 27–40.

[2] I. C. BRAID, R. C. HILLYARD AND I. A. STROUD, *Stepwise construction of polyhedra in geometric modeling*, in Mathematical Methods in Computer Graphics and Design, K. W. Brodlie, ed., Academic Press, New York, 1980.

[3] C. M. BROWN, PADL-2: *A technical summary*, IEEE Computer Graphics Appl., 2 (1982), pp. 69–84.

[4] J. C. CAVENDISH, D. A. FIELD AND W. H. FREY, *An approach to automatic three-dimensional finite element mesh generation*, Internat. J. Numer. Methods Engrg., 21 (1985), pp. 329–347.

[5] G. DESALVO, PREP7: *Preprocessor for finite element program* ANSYS, in ANSYS User's Manual 1, Swanson Analysis Systems, Inc., Houston, PA, 1985.

[6] J. J. DONGARA, C. B. MOLER, J. R. BUNCH AND G. W. STEWART, LINPACK *User's Guide*, Society for Industrial and Applied Mathematics, Philadelphia, PA, 1979.

[7] D. A. FIELD AND W. H. FREY, *Automation of Tetrahedral Mesh Generation*, General Motors Research Publication GMR-4967, 1985.

[8] D. A. FIELD AND A. P. MORGAN, *A quick method for determining whether a second degree polynomial has a solution in a given box*, IEEE Computer Graphics Appl., 2 (1982), pp. 65–68.

[9] G. E. FORSYTHE, *Pitfalls in computation or why a math book isn't enough*, Amer. Math. Monthly, 77 (1970), pp. 931–956.

[10] G. E. FORSYTHE, M. A. MALCOLM AND C. B. MOLER, *Computer Methods for Mathematical Computations*, Prentice-Hall, Englewood Cliffs, NJ, 1967.

[11] W. J. GORDON, *Spline-blended surface interpolation through curve networks*, J. Math. Mech., 10 (1968), pp. 931–952.

[12] I. GRIEGER AND H. A. KAMEL, *Application of interactive graphics and other programming aids, state of the art surveys on finite element technology*, American Society of Mechanical Engineers Publication, 1983, pp. 341–361.

[13] W. V. D. HODGE AND D. PEDOE, *Methods of Algebraic Geometry*, 3 vols., Cambridge University Press, Cambridge, England, 1952.

[14] IMSL, IMSL *Library User's Manual*, Houston, TX, 1985.

[15] W. KAHAN, *Implementation of Algorithms, Parts* I *and* II, National Technical Information Service AD-769124, Springfield, VA, 1973.

[16] J. K. KROUSE, *Industry gets serious about solid modeling*, Computer Aided Engineering, Nov.–Dec. (1982), pp. 22–26.

[17] Y. T. LEE AND A. A. G. REQUICHA, *Algorithms for computing the volume and other integral properties of solids. I. Known methods and open issues*, Commun. ACM, 25 (1982), pp. 635–641.

[18] ———, *Algorithms for computing the volume and other integral properties of solids. II. A family of algorithms based on representation conversion and cellular approximation*, Commun. ACM, 25 (1982), pp. 643–650.

[19] J. Z. LEVIN, *Mathematical models for determining the intersections of quadric surfaces*, Computer Graphics and Image Processing, 11 (1979), pp. 73–87.

[20] C. W. MCCORMICK, MSC/NASTRAN *User's Manual* 1, The MacNeal–Schwindler Corporation, Los Angeles, CA, 1983.

[21] D. H. MEAGHER, *Octree Encoding*: A New Technique for the Representation, Manipulation, and Display of Arbitrary Three Dimensional Objects by Computer IPL-TR-80-111, Image Processing Laboratory, Rensselaer Polytechnic Institute, Troy, NY, 1980.

[22] W. H. MEYER, personal communication, 1984.

[23] J. A. MYERS, *General Motors Research Laboratories Mathematics Department Publications* GMR-4432, General Motors Research Laboratories, Warren, MI, 1985.

[24] S. OCKEN AND J. T. SCHWARTZ, *Precise Implementation of* CAD *Primitives Using Rational Parametrizations of Standard Surfaces*, in Solid Modeling by Computers, Pickett and Boyse, eds., Plenum Press, New York, 1984, pp. 259–269.

[25] PDA, PATRAN *User's Guide*, PDA Engineering, Santa Anna, CA, 1984.

[26] H-U. Pfeifer, *Methods used for intersecting geometric entities in the* GPM *module for volume geometry*, Computer Aided Design, 17 (1985), pp. 311–318.

[27] A. A. G. Requicha, *Representation for rigid solids: theory, methods, and systems*, Comput. Surveys, 12 (1980), pp. 437–464.

[28] A. A. G. Requicha and R. B. Tilove, *Mathematical Foundations of Constructive Solid Geometry: General Topology of Closed Regular Sets*, TM-27a, Production Automation Project, Univ. Rochester, Rochester, NY, 1978.

[29] J. R. Rossignac and A. A. G. Requicha, *Constant-radius blending in solid modeling*, Comput. Mechan. Engrg., 3 (1984), pp. 65–72.

[30] R. F. Sarraga, *Algebraic methods for intersections of quadric surfaces in* GMSOLID, Computer Vision, Graphics Design Processing, 22 (1982), pp. 222–238.

[31] T. W. Sederberg, D. C. Anderson and R. N. Goldman, *Implicitization, inversion, and intersection of planar rational cubic curves*, Computer Vision, Graphics, Image Processing, 31 (1985), pp. 89–102.

[32] M. S. Shephard and M. A. Yerry, *Finite-element mesh generation for use with solid modeling and adaptive analysis*, in Solid Modeling by Computers, Pickett and Boyse, eds., Plenum Press, New York, 1985, pp. 53–77.

[33] R. H. Taylor, *Planning and execution of straight line manipulators and trajectories*, IBM J. Res. Develop., 23 (1979), pp. 429–436.

[34] R. B. Tilove, *Set membership classification: A unified approach to geometric intersection problems*, IEEE Trans. Comput., C-29 (1980), pp. 874–883.

[35] ——, *A null-object detection algorithm for constructive solid geometry*, Commun. ACM, 27 (1984), pp. 684–694.

[36] D. F. Watson, *Computing the n-dimensional Delaunay tesselation with applications to Voronoi polytopes*, Comput. J., 24 (1981), pp. 167–172.

[37] B. Wordenweber, *Finite-element mesh generation from geometric models*, COMPEL, Internat. J. Comput. Math. Electric. Electron. Engrg., 1 (1983), pp. 23–33.

Integrating Sculptured Surfaces into a Polyhedral Solid Modeling System

LARRY LICHTEN AND MARCEL SAMEK

Abstract. Sculptured surface definition and manipulation facilities have been integrated into a polyhedral-based solid modeling computer aided design system. Interpolated and approximated curves and surfaces are approximated within a faceting tolerance. Wireframe models or "very thin" polyhedra result; subsequently, CSG Boolean operations can be performed on these surfaces and on other polyhedra. Direct applications include stamping dies and mold design, areas not generally addressed by CSG based solid modelers. Methods used to define and fit surfaces, approximating and faceting processes, model size, and loss of accuracy are discussed.

1. Introduction. GDP (Geometric Design Processor) (Fitzgerald [5]) is an experimental, constructive solid geometry (CSG) based, computer aided design system. Using a small number of primitive solids as construction blocks, GDP allows modeling of complex three-dimensional solids as planar-faceted polyhedra in a boundary representation. Although such modeling approaches are useful for many applications, as well as very simple to use, they generally do not lend themselves to the design of sculptured surfaces.[1]

A constructive solid modeler supports design of objects based on simple geometric volumes. In GDP, these include cuboids, cones, hemispheres, cylinders, extruded volumes (swept polygons), and revolute objects (rotated polygons). Although GDP retains the original definitions of these volumes, the fundamental stored entity is a polyhedral volume bounded by a finite set of planar faces that approximates a primitive volume. Models are constructed by "adding" or "subtracting" such approximations to form more complex polyhedra. Surfaces cannot be modeled, however, since they do not necessarily bound any specific volume.

---

[1] A "sculptured" or "free form" surface is generally defined as having double curvature; sculptured surfaces are often represented by bicubic or higher degree polynomials in CAD systems.

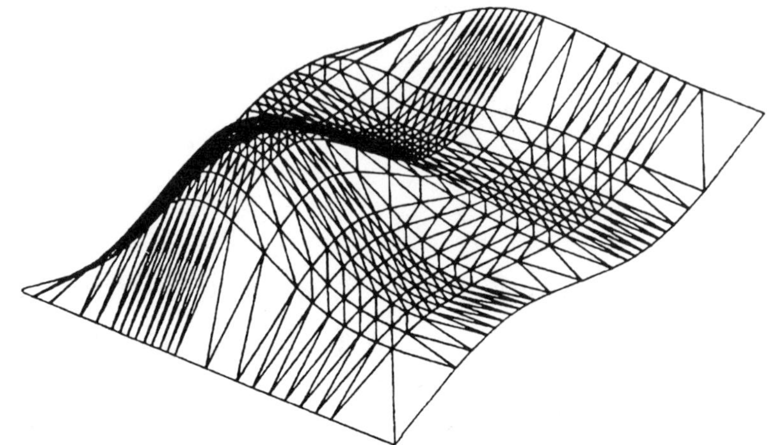

Figure 1. GDP approximation of a Ferguson–Coons surface.

While GDP is a powerful design system with flexible data structuring and efficient numerical and display algorithms, it currently cannot generate or manipulate such surfaces. Introduction of surface capabilities into GDP could significantly increase its power and enhance its appeal as a general purpose design system.

NGS (Numerical Geometry System) (Mayfield [7]) is a surface-oriented CAD system designed in the mid 1970s by IBM and Teledyne Ryan Aeronautical. NGS, like GDP, allows creation of three-dimensional models. NGS, however, relies primarily on analytically represented surfaces to define topological boundaries of models. Several curve and surface fitting techniques are supported by NGS. Solid modeling systems often sacrifice accuracy by using approximations rather than analytic representations in order to achieve reasonable interactive response times. It is therefore highly desirable to integrate some NGS-like analytic curve and surface capabilities into an approximating CSG system like GDP. This paper describes an extended version of GDP that supports definition and approximation of Ferguson–Coons, Bézier, and B-spline curves and surfaces. Existing GDP algorithms can operate on such surfaces as if they were native objects.

An example of surface approximation is shown in Fig. 1; a primitive cuboid and another surface are shown in Fig. 2. Our general approach to define a surface in the solid modeling system is outlined as follows:

Definition: Support interactive entry of a set of control points or curves and other parameters necessary to define a surface using a particular surface fitting algorithm.

Interpolation: Using algorithms and data structures similar to those of NGS, calculate an analytic representation of the surface.

Refinement: Convert the analytically represented surface (e.g., by a set of control point coordinate and tangent values) to a format (independent of the surface-fitting technique) convenient for faceted approximation. By recursive subdivision, refine the surface into a set of patches until a predefined tolerance is achieved.

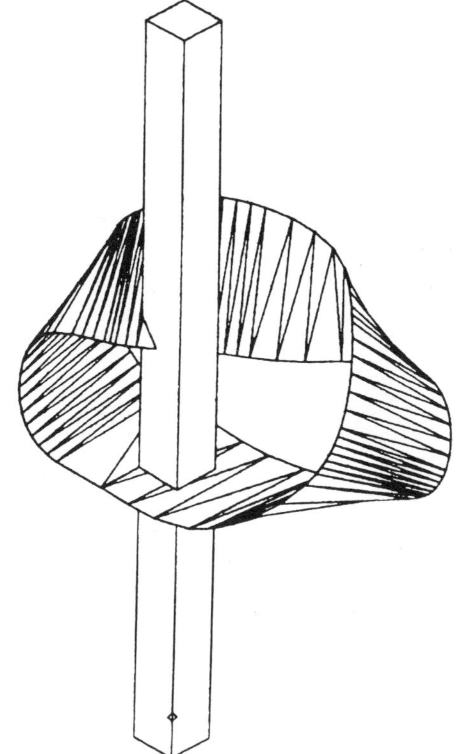

Figure 2.   Primitive cuboid and Mobius band.

Triangulation: Approximate each patch of the refined surface by a pair of planar triangular facets.

Polyhedral approximation: Convert the planar-faceted approximation of the surface into a polyhedron by giving the surface a "back" side, thus creating a "very thin" object. A volumetric model, as required by GDP, is thereby created.

Conversion: Create a GDP polyhedron from the approximated "surface" by extracting vertex, edge, face, and loop information from the "polyhedral surface."

Subsequent sections of this paper briefly overview the mathematical bases and data structures of NGS and GDP's polyhedral structure. Approximation and refinement algorithms used for converting analytic surfaces into polyhedra and alternative approaches are then discussed, and several examples are provided.

2. Mathematical Background. Ferguson–Coons, Bézier and B-spline curves and surfaces were chosen for our research since they are common design and fitting techniques. Initially, Ferguson–Coons surfaces were incorporated into GDP; we therefore provide overviews of the mathematical bases of these three fitting techniques, while subsequent sections focus primarily on Ferguson–Coons interpolation.

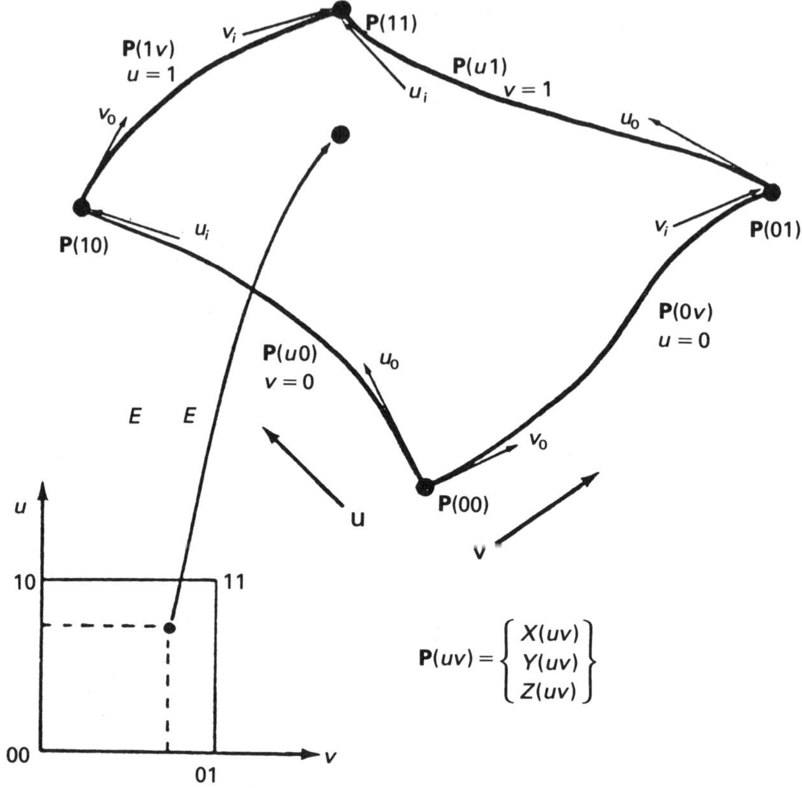

Figure 3.   Ferguson–Coons bicubic patch.

2.1. Ferguson–Coons networks. A Ferguson–Coons surface or network, known in NGS as an FCNet, is a surface interpolated through a rectangular grid of control points.[2] The control points define a set of Ferguson–Coons bicubic parametric patches which are blended to satisfy continuity constraints between adjacent patches.

In (u, v) parametric space, each Ferguson–Coons patch is normalized over the intervals $u, v \in [0, 1]$ and is defined as

$$\mathbf{P}(u, v) = [u^3 u^2 u 1] N \mathbf{B} N^T [v^3 v^2 v 1]^T$$

where

$$N = \begin{bmatrix} 2 & -2 & 1 & 1 \\ -3 & 3 & -2 & -1 \\ 0 & 0 & 1 & 0 \\ 1 & 0 & 0 & 0 \end{bmatrix}$$

and B is a 4-by-4 matrix of coordinate and tangent information at control points.

---

[2] S. A. Coons originally formulated this surface as blended individual patches, while J. C. Ferguson viewed the surface as bidirectional spline fitting.

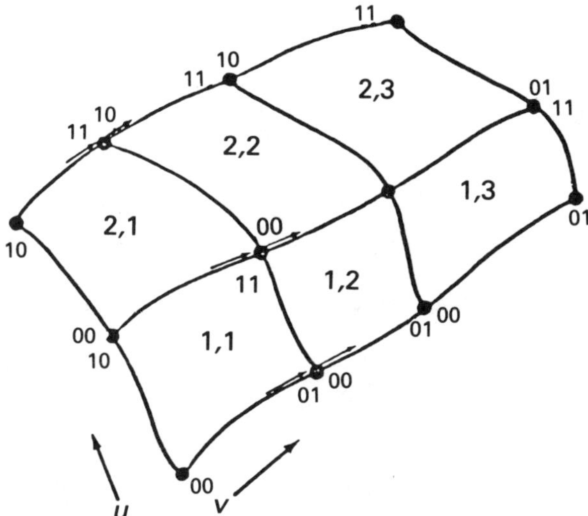

Figure 4. Ferguson–Coons network.

Figure 3 shows a Ferguson–Coons patch and its defining information. An FCNet is a continuous-in-first-derivative rectangular array of such patches; a six-patch FCNet is shown in Fig. 4. Each control point of an FCNet is characterized by the following information:

(1) point location: the absolute location of the control point in Cartesian world coordinates.

(2) u-tangent vector: the end point of the vector form of the partial derivative in the U direction of the patch, at the control point. Since there are two adjacent patches in the U direction, there are also two u-tangents, referred to as "incoming" and "outgoing" tangents with respect to the direction away from $(u, v) = (0, 0)$.

(3) v-tangent vector: the end point of the vector form of the partial derivative in the V direction of the patch, at the control point. Since there are two adjacent patches in the V direction, there are both incoming and outgoing tangents.

(4) uv-tangent vector point: the end point of the cross tangent (twist) vector at the control point.

Thus, an FCNet can be represented as an array of point and tangent information about the original control points that can be stored very compactly. Ferguson [4], Coons [1] and Silverman [12] provide in-depth discussions of Ferguson–Coons curve and surface fitting; Faux and Pratt [3] and Rogers [11] contain excellent background on these and other techniques.

2.2. Bézier surfaces. Bézier surfaces are also defined by a rectangular grid of points. In (u, v) parametric space, a Bézier surface is normalized over the intervals $u, v \in [0, 1]$ and is defined as

$$\mathbf{Q}(u, v) = \sum_{i=0}^{n} \sum_{j=0}^{m} \mathbf{B}_{i+1, j+1} J_{n,i}(u) K_{mj}(v),$$

where

$$\mathbf{B} \text{ is a control point,}$$

$$J_{n,i} = \binom{n}{i} u^i (1-u)^{n-i},$$

$$K_{m,j} = \binom{m}{j} v^j (1-v)^{m-j}.$$

2.3. B-spline surfaces. A B-spline surface may be defined as

$$\mathbf{Q}(u, v) = \sum_{i=0}^{n} \sum_{j=0}^{m} \mathbf{B}_{i+1,j+1} N_{i,k}(u) M_{j,l}(v)$$

where

$$N_{i,1}(u) = \begin{cases} 1 & \text{if } x_i \leq u < x_{i+1}, \\ 0 & \text{otherwise,} \end{cases}$$

$$N_{i,k}(u) = \frac{(u - x_i)N_{i,k-1}(u)}{x_{i+k-1} - x_i} + \frac{(x_{i+k} - u)N_{i+1,k-1}(u)}{x_{i+k} - x_{i+1}},$$

$$M_{j,1}(v) = \begin{cases} 1 & \text{if } y_j \leq v \leq y_{j+1}, \\ 0 & \text{otherwise,} \end{cases}$$

$$M_{j,l}(v) = \frac{(v - y_j)M_{j,l-1}(v)}{y_{j+l-1} - y_j} + \frac{(y_{j+1} - v)M_{j+1,l-1}(v)}{y_{j+l} - y_{j+1}},$$

x_i is the ith value in the knot vector in the u direction, y_i is the ith value in the knot vector in the v direction and $\mathbf{B}$ is a control point.

The knot vector specifies values of the parameter at successive spans on the B-spline mesh. Fuhr [6] and Mortenson [8] discuss B-spline techniques in depth.

3. Representation of GDP Polyhedra. GDP data structures typify polyhedral approximating boundary representation schemes. GDP polyhedra are uniformly structured, consisting of lists of vertices, edges, loops of edges, and faces, stored in tables. Indices indicate relationships between entries, and information is maintained on loops that bound a face and on edges that are on the intersection of particular faces. GDP tables are described below.

3.1. Vertex table. A GDP vertex table is a list of all vertices (in absolute Cartesian world coordinates) of a polyhedron. No vertex entry is duplicated, regardless of how many edges contain it.

3.2. Edge table. An edge table contains information including the indices in the vertex table of the "head" and "tail" vertices of each edge and the indices in the face table of faces containing the edge. An edge must be contained in exactly two faces, and, as in the case of the vertex table, no entries are duplicated.

Face Table

Face	Normal	Dist	Begin	End
1	0, −1, 0	0	1	5
2	1, 0, 0	1	6	10
3	0, 1, 0	1	11	15
4	0, 0, −1	0	16	20
5	0, 0, 1	1	21	25
6	−1, 0, 0	0	26	30

Loop Table

Index	Dir	Edge	Index	Dir	Edge
1	1	1	11	0	6
2	1	2	12	1	8
3	1	3	13	1	9
4	1	4	14	1	10
5	0	0	15	0	0
6	1	5	16	0	5
7	1	6	17	0	1
8	1	7	18	1	11
9	0	2	19	0	8
10	0	0			

Edge Table

Edge	Tail	Head	Face	Face
1	1	4	1	4
2	4	3	1	2
3	3	2	1	5
4	2	1	1	6
5	4	7	2	4
6	7	6	2	3
7	6	3	2	5
8	7	8	3	4
9	8	5	3	6
10	5	6	3	5
11	1	8	4	6
12	5	2	5	6

Vertex Table

Vertex	Coords
1	0, 0, 0
2	0, 0, 1
3	1, 0, 1
4	1, 0, 0
5	0, 1, 1
6	1, 1, 1
7	1, 1, 0
8	0, 1, 0

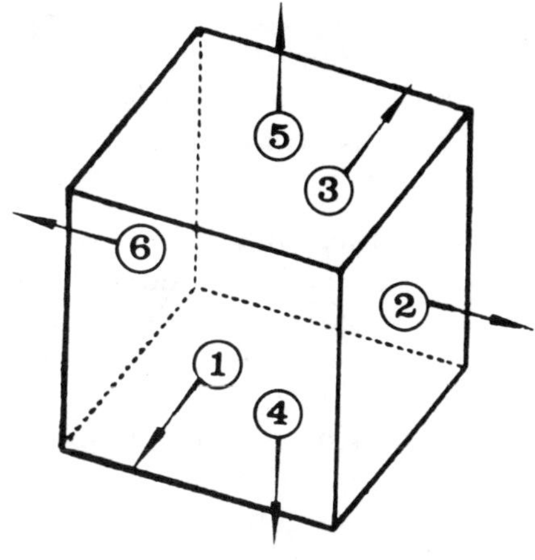
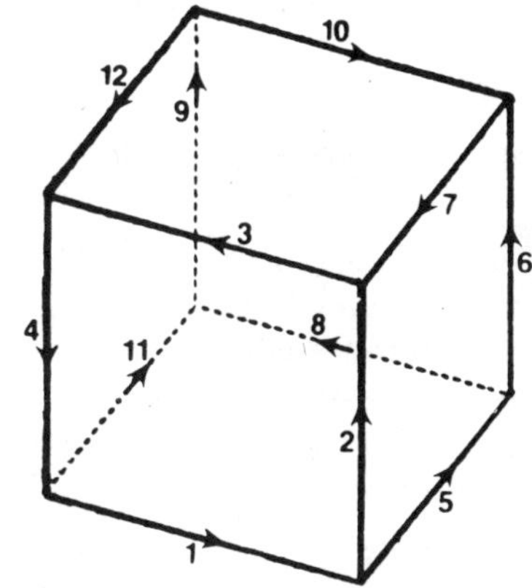
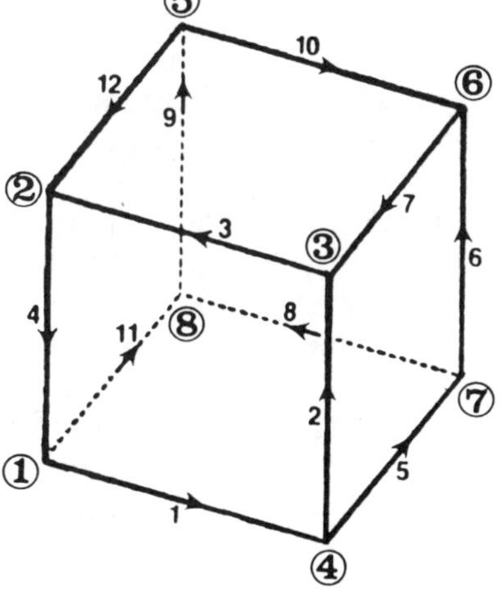

Figure 5. GDP vertex, edge, loop, and face tables.

3.3. Loop table. Each loop table entry holds the index of an edge in the edge table and information on whether the edge is pointing in a clockwise or counterclockwise direction. The latter is required to determine the direction of the normal to the face. All edges comprising a loop are entered sequentially into the loop table.

3.4. Face table. Since a face may contain more than one (coplanar) loop, each entry in a face table includes the following:
- the index of the first entry of the first loop on the face;
- the loop table index of the last entry of the last loop on the face;
- a point defining the unit normal to the face;
- the distance from the world coordinate origin to the unbounded plane on which the current face is defined.

CSG trees further determine the relationships between various polyhedra that form objects; nodes in these trees contain pointers to sets of tables which define constituent polyhedra. Figure 5 provides an example of vertex, edge, loop, and face tables, which define a GDP polyhedron.

4. Conversion of Analytic Surfaces to Polyhedral Approximations.

4.1. Refinement algorithms. Several refinement algorithms were investigated. Symmetric patch splitting was used in the prototype implementation; an asymmetric splitting method, which can reduce model size at the expense of processing speed, was also examined.

Both splitting algorithms compare the distances between adjacent control points (i.e., chord length) with the arc lengths of the curve segments between the points. Until their ratio is within a user-defined tolerance for a pair of points, the midpoint of the curve segment is used as an additional control point.

Symmetric splitting preserves the rectangular nature of the original grid of points: if a curve segment is split, all segments in that row (or column) of the FCNet are split at the corresponding parameter value. Asymmetric splitting, on the other hand, subdivides only individual patches. Figure 6 shows recursive subdivision of a surface using symmetric splitting; Fig. 6a is before subdivision and Fig. 6b is after.

Although either refinement process insures that any natural boundary of a quadrilateral patch can be approximated by a linear segment, it does not insure coplanarity of the corner points of the patch. The refined surface is consequently triangulated as shown in Fig. 7. In the current implementation, newly introduced triangulating edges are arbitrarily oriented.

The refinement algorithms do not produce satisfactory results for patches that have significantly greater curvature at their center than at their edges. Application of the refinement criterion (i.e., chord length/arc length) to the diagonals of the patch could significantly improve the algorithm's performance by determining whether a patch required further refinement due to central curvature and the optimal orientation of the triangulating edge.

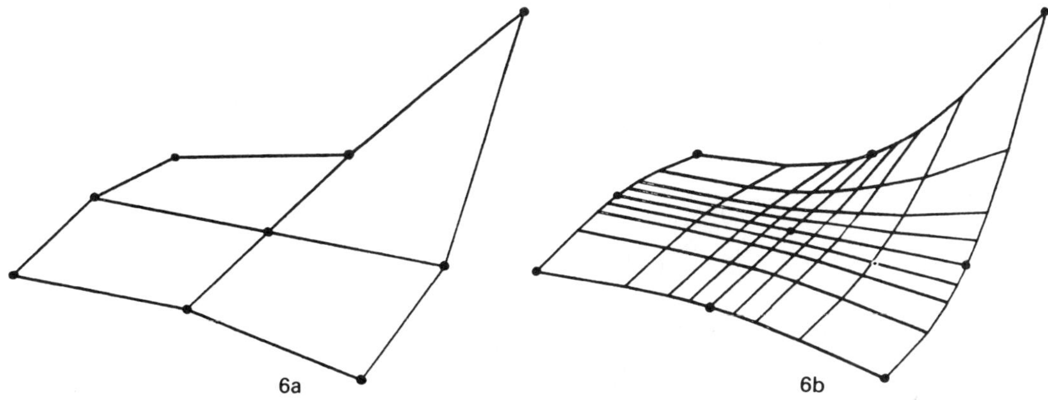

Figure 6.   Recursive subdivision of an approximated surface.

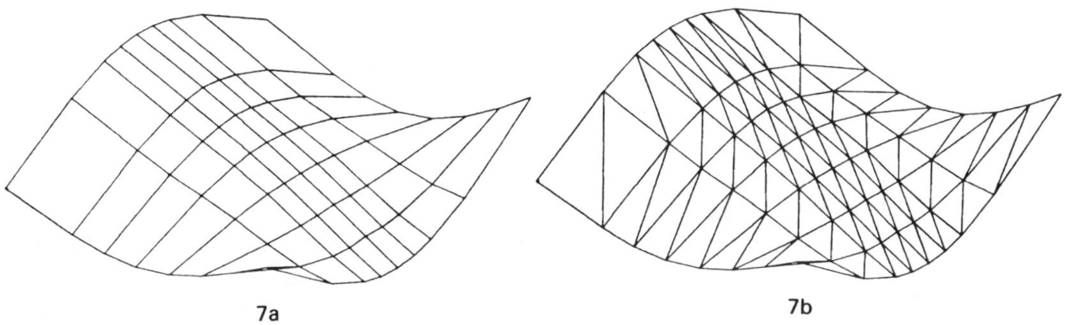

Figure 7.   Triangulation.

Another straightforward method of faceting surfaces was considered: recursively triangulating until the normals of adjacent facets were within a specified tolerance, thereby insuring smoothness of the faceted surface. It was not used since it is awkward to apply to surfaces with discontinuities or other nonsmooth features, and curves (rather than surfaces) could not be approximated using the same technique. Using two different algorithms—one for curves and one for surfaces—is undesirable because surfaces are often lofted through inviolable cross-sectional curves.

4.2.   Surface to volume conversion.   All surface interpolation and approximation algorithms discussed represent surfaces as volumeless entities in three-dimensional space. GDP, however, requires valid objects to be polyhedra enclosing finite volumes. A surface, therefore must first be given a volume in order to be a valid polyhedron.[3]

---

[3] A subsequent section describes a procedure to use wireframe "ghosts" to temporarily reduce model size and therefore processing time.

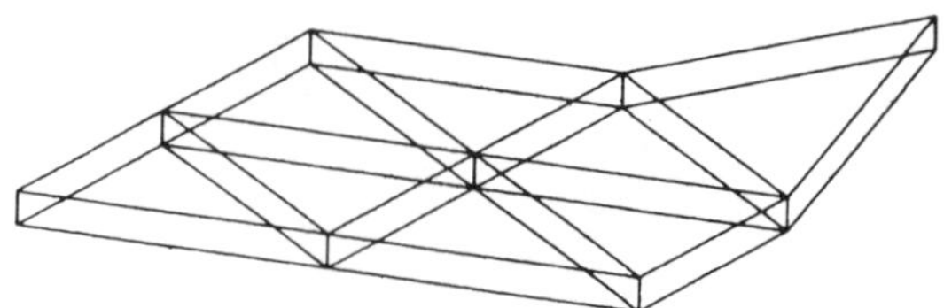

Figure 8. "Thin" polyhedron.

In order to create a "back side" of a faceted surface, the prototype implementation offsets the planar-faceted approximate surface by a constant distance in an arbitrary direction. Tangent information of the original surface is used to calculate new vertices on the offset surface: the average of the incoming and outgoing tangent vectors in the two parametric directions defines a plane at each vertex of the surface. Each vertex is offset by a (small) distance along the normal to the plane, and this value is used.

Finally, the two surfaces must be joined by introducing and triangulating faces along outer boundaries. This makes the object an enclosed volume and therefore a valid GDP polyhedron on which all GDP operations can be performed, as shown in Fig. 8.

If M and N are the number of rows and columns, respectively, of an FCNet after recursive subdivision, the resulting "thin" polyhedron has $4(M-1)(N-1) + 4(M-1) + 4(N-1)$ faces. The first term is the number of "top" and "bottom" faces created, and the other terms represent the number of faces around the "sides" joining the two surfaces. The unavoidable by-product is that the number of faces is more than doubled in order to create a polyhedron.

4.3. Volume to GDP polyhedron conversion. After a surface has been subdivided and converted into a "thin" volume, edges are entered into the loop table. As each rectangular face is traversed, the loop entries for the two triangular loops on the front side and the two triangular loops on the back side of each patch are entered into the table. The normal and distance from the origin of each face (front and back) are calculated and stored. Since each diagonal edge is shared by two triangular patches, one of the patches must be listed in reverse order from the other.

4.4. Coplanarity of loops and adjacent faces. GDP's algorithms operate on faces that contain one or more coplanar, nonadjacent loops. Therefore, two methods of data reduction are immediately possible: (1) combining nonadjacent coplanar faces into a single face, and (2) merging of coplanar adjacent loops into a single loop. Initially, each triangular loop is considered as a face, and appropriate information is entered into tables. After face and loop tables have been filled, they are scanned for coplanar loops. If any are found, all corresponding face table entries are removed and replaced by a single entry which references all coplanar loops. The face table is then reduced by the appropriate number of entries, and all loops which are coplanar are stored contiguously in the loop table and can be shared by a single face table entry.

Coplanarity is checked by comparing the normal and the distance from the origin of each face in a newly created face table. All faces must be compared to each other, and this process must be repeated each time a new polyhedron is generated.

Although the prototype implementation uses only triangular loops (to obtain planar facets), GDP supports loops with any number of sides. The second method for reducing model size—merging adjacent coplanar loops—will produce loops with three or more sides. Consequent trade-offs between model size, processing time and algorithmic complexity are currently being evaluated.

5. Surfaces as Wireframe Meshes. GDP's Boolean operators and hidden line algorithms recognize only polyhedra. For surface models, however, large model sizes significantly slow these procedures, and the model size cannot be reduced without decreasing the accuracy of the approximation. In an attempt to address this problem, we have investigated representing surfaces as wireframe meshes (GDP's "ghosts") which can be converted subsequently to polyhedra. This significantly reduces model size and processing time during surface design.

Although it is usually sufficient to represent surfaces by wireframe meshes, at some point, a wireframe model must be converted into a polyhedron. This process is straightforward since surfaces are originally defined by fitting procedures which return rectangular grids of points with tangent information. These points, which are then used as vertices for faceted approximations, are also used as vertices for wireframe meshes. Thus, to convert a wireframe surface into a polyhedron, only vertex information from the wireframe model is extracted and used to define the polyhedral "surface."

As discussed in § 3, a GDP polyhedron is represented by four tables: the vertex, edge, face, and loop tables. Only the vertex and edge tables are needed for a wireframe model. By temporarily eliminating loop and face tables, model size can be reduced by about half. Furthermore, wireframe representation does not require a "two-sided" surface, and since no face information is preserved in a wireframe, there is no need to triangulate quadrilateral faces.

Representing surfaces by wireframes, prior to converting them to polyhedra, offers several other advantages. In a wireframe, it is quite simple to move a single point in a model. This operation on a polyhedron, however, entails recalculating face information and checking for coplanarity between loops. A prototype implementation of wireframe surface definition, editing, and conversion to polyhedra is being completed.

6. Examples. Figure 9 is a 3-row by 4-column FCNet as originally defined in NGS and as approximated in GDP. Figures 10 and 11 show FCNet approximations in GDP in which surfaces are closed in one parametric direction.

7. Conclusions. The techniques used to implement surface definition in GDP have proven feasible, although they produce some undesirable side effects. The most significant of these are loss of accuracy and large model size.

A refinement procedure is repeated until the differences between chord lengths and arc lengths between control points are within a specific tolerance. In cases where patches have significantly greater curvature toward their center than near boundaries, this faceting algorithm does not produce satisfactory results.

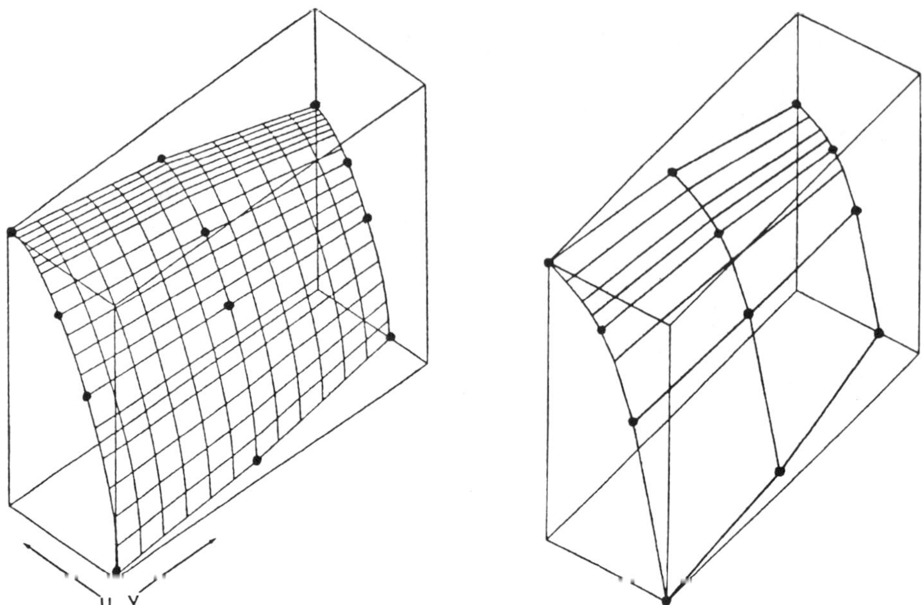

Figure 9. Surface modeled in NGS and approximated in GDP.

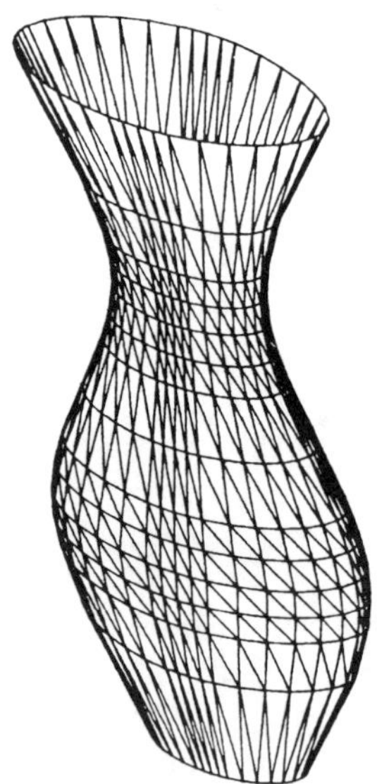

Figure 10. Closed GDP surface.

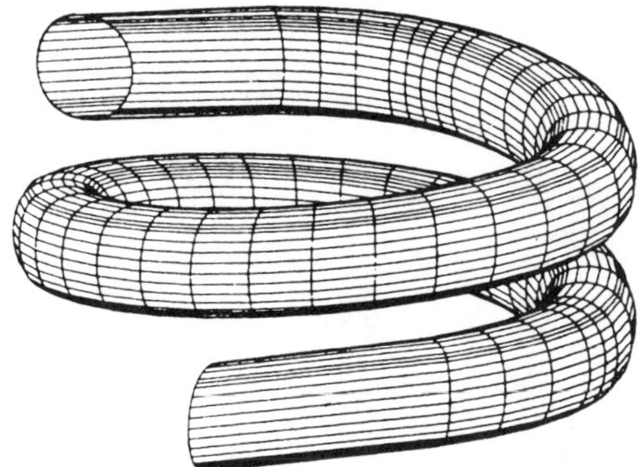

Figure 11. Pipe-like GDP surface.

A primary drawback of using faceted surfaces is that model sizes for even simple surfaces are quite large. For example, the surface shown in Fig. 9 was defined with only twelve control points but requires several hundred edge and loop table entries. For more complex surfaces, model size may be several megabytes.

In summary, we have shown that using faceted surfaces in a CSG system is possible. These surfaces are, in fact, "very thin" polyhedra on which all CSG operations can be performed. Work is currently under way to allow use of "cutting surfaces" to contour solids. Further research, however, is necessary to improve accuracy of faceting algorithms and to reduce model size.

Acknowledgment. The work described in this paper was performed in the Computer Aided Design Laboratory at the University of California, Los Angeles, California and was supported in part by the IBM Corporation.

REFERENCES

[1] S. A. COONS, *Surfaces for Computer Aided Design of Space Forms,* Project MAC, Massachusetts Institute of Technology, Cambridge, MA, TR-41, June 1964.

[2] G. DURAND, *An Algorithm for Converting Ferguson Surfaces into Polyhedral Representations,* Univ. Calif. Computer Science Dept., Los Angeles, CA, 1983.

[3] I. O. FAUX AND M. J. PRATT, *Computational Geometry for Design and Manufacture,* John Wiley, New York, 1979, pp. 126–137, 198–230.

[4] J. C. FERGUSON, *Multivariative curve interpolation,* J. Assoc. Comput. Mach., 11 (1964), pp. 221–228.

[5] W. J. FITZGERALD, F. GRACER AND R. N. WOLFE, *The* IBM SOLID *Modeling Users Guide,* IBM Manufacturing Research Department, Yorktown Heights, NY, 1982.

[6] R. FUHR, *A Technical Introduction to the Rational B-Spline Representation for Curves and Surfaces,* Boeing Aircraft Report, Seattle, WA, 1981.

[7] J. P. MAYFIELD AND R. M. BURKLEY, *Applications of a numerical geometry system in engineering,* Proc. IEEE Design Automation Conference, 13 (1976), pp. 25–33.

[8] M. MORTENSON, *Geometric Modeling,* John Wiley, New York, 1985.

[9] W. M. NEWMAN AND R. P. SPROULL, *Principles of Interactive Computer Graphics,* McGraw-Hill, New York, 1979, pp. 309–325.

[10] T. PAVLIDIS, *Algorithms for Graphics and Image Processing,* Computer Science Press, 1982, pp. 280–291.
[11] D. ROGERS AND J. A. ADAMS, *Mathematical Elements for Computer Graphics,* McGraw-Hill, New York, 1976, pp. 170–184.
[12] G. J. SILVERMAN AND L. LICHTEN, *Fitting Space Curves with Parametric Cubic Splines,* IBM Los Angeles Scientific Center, Report G320–2663, October 1974.

A New Look at Higher Dimensional Surfaces through Computer Graphics

ROBERT E. BARNHILL, GEORGE T. MAKATURA AND
SARAH E. STEAD

Abstract. Many physical and engineering data involve functions of three (or more) variables. Trivariate surfaces are difficult to construct and to display. However, recent advances in computer graphics and in the interpolation of surfaces now permit some progress in this subject. Specific problems from aerodynamics illustrate the possibilities.

1. Introduction. Many physical and engineering data involve functions of three independent variables. An example in the subject of combustion is temperature as a function of the three spatial variables. Such problems can be considered as the construction and display of a surface in four-space, colloquially called a "four-dimensional surface." "In the field" one has a variety of information so it may not be obvious that one actually has a "surface" problem. The need for multidimensional surface methods is discussed by Barnhill and Chang [4], who include a variety of examples.

Visualizing a four-dimensional surface is very difficult. However, four-dimensional surfaces can now be displayed by the judicious use of interactive color computer graphics. A convincing case for the use of computer graphics for a variety of three- and four-dimensional surfaces is made by Gregory and Carmichael [14]. A principal theme of this paper is to introduce a variety of possibilities for displaying four-dimensional surfaces. The term "new" in the title of this paper refers to the display of approximations to higher-dimensional data. A second theme is that the possibility of computer graphics illustrations has stimulated interest in developing new methods for the construction of four-dimensional surfaces, an example being the "multistage methods" discussed below.

The surface representation appropriate for a specific application depends upon that application. In general, the data are arbitrarily located, as in the examples that follow. The dichotomy of interpolation versus approximation (Davis [12]) is often useful. We use a synthesis of interpolation and approximation in that the methods are in the spirit of interpolation, but the final surfaces are really approximations.

A purpose of this paper is to point out that problems arising in applications can sometimes be interpreted as "surface" problems and hence are amenable to study by methods utilized by the computer aided geometric design community. We illustrate multidimensional surface approximations and their display by reporting on our continuing collaboration on some specific multidimensional problems with practical applications.

The organization of this paper is as follows: a brief history of multidimensional approximation relevant to surfaces is followed by solutions to some specific problems, which are illustrated by color computer graphics.

2. History of Multidimensional Surfaces.

2.1. Rectangular interpolants.   If the data for a multidimensional surface are in a rectangular grid, then a tensor product of univariate methods interpolates the data. Tensor products have the attractive features of being easily formed for arbitrary dimensions and of efficient calculation. However, tensor products have the geometric inflexibility of being applicable only to rectangularly gridded data. A more subtle inflexibility is that if derivative data are involved, then a tensor product, being the composition of operators, contains derivatives of higher order than the original data. An example of this inflexibility occurs in a method used later in this paper: the trivariate tensor product H of three univariate cubic Hermite interpolants, each involving first order derivatives, contains $(1, 1, 1)$ derivatives.

Tensor products have been known and studied for a long time. The tensor product of univariate cubic Hermite interpolants, mentioned above, has been applied to three-dimensional surface design since the mid-1960s. There is a hierarchy of methods for data with a rectangular structure, but not necessarily in a grid, an example being the rectangular "serendipity elements" of the finite element engineers. Finally, whole curves of data in a rectangular structure may be given, as in the automobile industry, for which Coons (see Barnhill [2]) invented his patches. Coons also considered higher-dimensional surfaces ("hypersurfaces"). The original Coons patches do not allow for possible twist incompatibilities. (Twists are the $(1, 1)$ derivatives that arise in composing two univariate projectors involving first derivatives.) Gregory's Square (Gregory [13]) is a "compatibly corrected" Coons patch solving this problem. A compatibly corrected Coons patch for n variables has recently been developed (Barnhill and Worsey [10]), for future use in a multistage method.

We now discuss interpolants for arbitrarily located data. Methods for arbitrarily located data do not, by definition, have the above-mentioned geometric inflexibility and some do not have the derivative inflexibility. There are two broad classes of interpolants for trivariate data:
* Distance-weighted interpolants,
* Tetrahedral interpolants.

2.2. Tetrahedral interpolants.   Finite element engineers have developed many tetrahedral elements. These correspond to interpolants defined over tetrahedra, the simplest example being piecewise linear interpolation. The construction of surfaces via tetrahedral interpolants involves the following steps:

(1) define the domain tetrahedra,
(2) estimate any necessary derivative data at the data sites,
(3) evaluate favorite tetrahedral interpolant.

Given a collection of points in 3-space, there is not a unique triangulation of them into tetrahedra. We suggest that an "optimal" triangulation be found, one example being to maximize over all possible triangulations the minimum of R_I/R_C, the ratio of the radii of the inscribed sphere to the circumscribed sphere (Little [15]). A second possibility of an optimal triangulation is the Delaunay triangulation corresponding to the Dirichlet tessellation (Bowyer [11]; Watson [21]). Tetrahedral interpolants require derivative information at the data sites, the exact requirements depending upon the interpolant's form. Triangular Shepard's method, multiquadrics, and Alfeld's minimization of functionals are three possible ways of determining these derivatives (Barnhill [3]; Stead [19]). Tetrahedral interpolants currently available include a discretization of a transfinite Barnhill, Birkhoff and Gordon scheme (Barnhill and Little [6]), C^1 Clough–Tocher schemes (Alfeld [1]; Worsey and Farin [22]) and Gregory's symmetric schemes (Gregory [13]).

2.3. Distance-weighted interpolants. "Distance-weighted" interpolants have the structure that the given data are weighted by a function of the distance from a point of evaluation to the data sites. We mention two distance-weighted interpolants:

- Shepard's formula,
- Hardy's multiquadrics.

Shepard's formula is given by

$$\sum_i w_i L_i F \quad \text{where } w_i = \frac{d_i^{-2}}{\sum_j d_j^{-2}},$$

d_i is the distance from an arbitrary point of evaluation to the ith data site and $L_i F$ is a truncated Taylor expansion at the ith data site. Shepard's formula is a "convex combination" of the data (the $L_i F$) (see Barnhill, Dube and Little [5]). Shepard's formula has flaws which subsequent research has corrected, but it is a good first distance-weighted interpolant to study.

The multiquadric used later in this paper is of the form

$$M_1 = \sum_i c_i w_i \quad \text{where } w_i = [d_i^2 + R]^{1/2},$$

R is a fixed number and the coefficients c_i are found by solving the Vandermonde system of interpolation conditions. We observe that the structure of the weight functions w_i appears to be very different for Shepard's method and for Hardy's multiquadrics, being inverse distance squared for the Shepard's formula above and a "radial" distance for the multiquadric shown. However, the general multiquadric

$$M_u = \sum_i c_i [d_i^2 + R]^{u/2}$$

is an inverse distance squared interpolant for the special case of $u = -2$. Moreover, the general Shepard's formula has weight function

$$w_i = w_{i,u} = \frac{d_i^{-u}}{\sum_j d_j^{-u}}$$

so that the inverse distance squared Shepard's formula corresponds to $u = 2$. The (similar) form of the weight functions for these two families of interpolants justifies calling both "distance-weighted."

3. Display of Trivariate Surfaces. The complete display of four-dimensional surfaces by means of two-dimensional images is impossible. However, some degree of visualization is possible by use of interactive color computer graphics. We have experimented with displaying various "sections" of four-dimensional surfaces, the simplest being "cross-sections" obtained by setting one independent variable equal to a constant, e.g., $w_0 = F(x, y, z_0)$ ("isoparametric surfaces"). Such a cross-section is a three-dimensional surface and has the advantage of being easy to calculate. Unfortunately, we have not found cross-sections to provide an adequate means of visualization of four-dimensional surfaces and we do not recommend their use.

Our current recommendation is to compute contours of four-dimensional surfaces in order to visualize the surfaces. We recall that a contour is the set of (x, y, z) such that $F(x, y, z) = $ some constant. Thus a contour of a four-dimensional surface is an implicitly defined three-dimensional surface which must be computed and then displayed. Examples are given in Barnhill [2], Barnhill and Stead [9], Barnhill [3], and Petersen, Piper and Worsey [17].

4. Combining the Construction and Display of Four-Dimensional Surfaces: Multistage Methods. Stead [18] had the idea of combining the construction and display of four-dimensional surfaces into the following stages:
   (1)   Interpolate to arbitrarily located data with an operator M.
   (2)   Evaluate M on a coarse grid, obtaining new data on the coarse grid.
   (3)   Interpolate to the data on the coarse grid with a tensor product interpolant H and evaluate on a fine grid.
The multistage method corresponds to the composition HM which has the following features:
   •   efficient evaluation,
   •   the smoothness of H,
   •   does not interpolate to the original data,
   •   function precision equal to the intersection of the function precisions of H and M.
Two choices of the first stage interpolant M are used below (Fig. 2):
   (1)   the multiquadric M_1 defined above, localized by using only data from the nearest n data sites, or
   (2)   a local least squares approximant L which uses only the nearest n data sites.

We observe the synthesis of interpolation and approximation methods in the preceding multistage method: an interpolation operator is applied first, followed by an approximation method that can be efficiently evaluated for displaying the surface.

5. Examples. The first example, Fig. 1 (see color insert), is a shaded, solid display of the tricubic example given in Barnhill [2]. That is, values of the function

$$F(x, y, z) = (x - 0.5)^3 + (y - 0.5)^3 + (z - 0.5)^3$$

are taken at 216 data sites arbitrarily located in the unit cube. The two contours -0.1 (yellow) and 0.1 (blue) are computed and displayed. The multistage method HM_1, with the multiquadric constant $R = 0.6$, using data from the nearest 20 data sites, and a coarse grid of $10 \times 10 \times 10$ points, was applied to the tricubic. To the resolution shown, HM_1 looks almost the same as the tricubic, so its picture is not included. This example is an analytic one for which there is an error surface. (We invite attention to the following question: make a meaningful display of such an error surface.) All the other examples in this paper involve approximations to data, not functions.

The next example (Stead and Webb [20]) illustrates differences between the two possibilities HM_1 and HL. The multiquadric M_1 and least squares scheme L are as above and H is the tensor product tricubic with all derivatives of order greater than one set equal to zero (Barnhill and Stead [9]). Pollen samples are taken from sediment cores taken from lakes in the midwestern United States. The goal is to determine the percentage of a given pollen as a function of the three variables: latitude, longitude, and time. The data are 684 samples of pollen. The 20% contour of prairie forb pollen has the physical significance of demarcating forest and prairie through time (10,000 years). This contour is computed by HM_1 in Fig. 2a (see color insert) and by HL in Fig. 2b (see color insert). The multiquadric produces a lumpy but accurate surface whereas the least squares approximant is more useful for showing trends.

A variety of rendering possibilities is displayed by the pressure contours on the NASA Space Shuttle in Figs. 3 (see color insert). Pressures are desired both on the surface of the shuttle and nearby and the corresponding "inner" and "outer" layers of the three-dimensional grid are shown in Fig. 3a. For the inner layer (the surface of the shuttle), wireframe contours are displayed in Fig. 3b. The corresponding shaded contours are shown in Fig. 3c. The term "angle of attack" refers to the angle that the craft makes with the horizontal, a positive number implying that the craft's nose is "up". A wireframe representation of the contours over both the inner and outer grid layers is shown in Fig. 3d. Contours are displayed on two cross-sections which are not grid layers in Fig. 3e. The (solid) contours shown in Figs. 3c and e are drawn using polygon fills for a dynamic display, a Silicon Graphics IRIS being used here. Transparent contours are scan-line rendered for a nondynamic display in Fig. 3f, for the same data as in Fig. 3d. Air density contours on a constant pressure contour are presented in Fig. 3g. The pressure contour is the same as in Fig. 3d. The air density "surface" is a function of four variables and hence may be considered as a five-dimensional surface.

The aerodynamic problem of pressure on the oblique wing is displayed in the three pictures in Figs. 4 (see color insert), using methods of Barnhill, Piper and Stead [8]. These three pictures are three "frames" of a visually continuous set which, rendered on a dynamic display, enables one to see the pressure change as the wing's angle of attack varies continuously. Two new classes of methods for the oblique wing problem are given in Barnhill, Piper and Rescorla [7].

The variety of three-dimensional contours of four-dimensional surfaces given above illustrates some of the current possibilities of displaying multidimensional surfaces by means of color computer graphics.

Acknowledgments.   This research was supported by the National Aeronautics and Space Administration Ames Research Center, Moffett Field, California under Interchange Number NCA2-1R821-401 with the University of Utah and by the Department of Energy with contract DE-C02-85ER12046 with the University of Utah. The wind tunnel tests in the examples were performed by the Advanced Aerodynamics Concepts Branch at NASA Ames. We thank Ralph Carmichael for providing the data for these examples.

REFERENCES

[1]   P. ALFELD, *A trivariate Clough–Tocher scheme for tetrahedral data,* Computer Aided Geometric Design, 1 (1984), pp. 169–181.

[2]   R. E. BARNHILL, *A survey of the representation and design of surfaces,* IEEE Computer Graphics Appl., 3 (1983) pp. 9–16.

[3]   ———, *Surfaces in computer aided geometric design: A survey with new results,* in Surfaces in Computer Aided Geometric Design '84, R. E. Barnhill and W. Boehm, eds., North-Holland, Amsterdam, 1985, and Computer Aided Geometric Design, 2 (1985), pp. 1–17.

[4]   R. E. BARNHILL AND R. CHANG, *Approximation,* in Program Directions for Computational Mathematics, R. E. Huddleston, ed., Dept. Energy, Washington, D.C., 1979.

[5]   R. E. BARNHILL, R. P. DUBE AND F. F. LITTLE, *Properties of Shepard's surfaces,* Rocky Mountain J. Math., 13 (1983), pp. 365–382.

[6]   R. E. BARNHILL AND F. F. LITTLE, *Three- and four-dimensional surfaces,* Rocky Mountain J. Math., 14 (1984), pp. 77–102.

[7]   R. E. BARNHILL, B. R. PIPER AND K. L. RESCORLA, *Interpolation to arbitrary data on a surface,* this Volume, 1987. An extended form of this paper appears as: *Representation for the graphical display of structured data,* Final Report on Joint Research Interchange, NASA-Ames, Moffett Field, CA.

[8]   R. E. BARNHILL, B. R. PIPER AND S. E. STEAD, *Surface representation for the graphical display of structured data,* Visual Computer, 1 (1985), pp. 108–111.

[9]   R. E. BARNHILL AND S. E. STEAD, *Multistage trivariate surfaces,* Rocky Mountain J. Math., 14 (1984), pp. 103–118.

[10]   R. E. BARNHILL AND A. J. WORSEY, *Smooth interpolation over hypercubes,* Computer Aided Geometric Design, 1 (1985), pp. 101–113.

[11]   A. BOWYER, *Computing Dirichlet tessellations,* Computer J., 24 (1981), pp. 162–166.

[12]   P. J. DAVIS, *Interpolation and Approximation,* Dover, New York, 1975.

[13]   J. A. GREGORY, *Interpolation to boundary data on the simplex,* in Surfaces in Computed Aided Geometric Design '84, R. E. Barnhill and W. Boehm, eds., North-Holland, Amsterdam, 1985, and Computer Aided Geometric Design, 2 (1985), pp. 1–17.

[14]   T. J. GREGORY AND R. L. CARMICHAEL, *Getting the picture through computer graphics,* AIAA Astronautics and Aeronautics, 21 (1983), pp. 16–29.

[15]   F. F. LITTLE, *Tessellation of tetrahedra,* Math CAGD Seminars, Univ. Utah, Salt Lake City, UT, 1981.

[16] C. S. PETERSEN, *Contours of three and four dimensional surfaces*, Masters thesis, Dept. Mathematics, Univ. Utah, Salt Lake City, UT, 1983.

[17] C. S. PETERSEN, B. PIPER AND A. J. WORSEY, *Adaptive contouring of a trivariate interpolant*, this Volume, 1987.

[18] S. E. STEAD, *Smooth multistage multivariate approximation*, Ph.D. thesis, Div. Applied Mathematics, Brown Univ., Providence, RI, 1983.

[19] ———, (1984), *Estimation of gradients from scattered data*, Rocky Mountain J. Math., 14 (1984), pp. 265–280.

[20] S. E. STEAD AND T. WEBB, III, *Representation and display of trivariate arbitrarily-spaced discrete pollen data*, submitted for publication.

[21] D. F. WATSON, *Computing the n-dimensional Delaunay tessellation with application to Voronoi polytopes*, Computer J., 24 (1981), pp. 167–172.

[22] A. J. WORSEY AND G. FARIN, *An n-dimensional Clough–Tocher interpolant*, Constructive Approximation, to appear.

B-Form Basics

CARL de BOOR

Abstract. The basic facts about the B(arycentric, -ernstein, -ézier)-form of a multivariate polynomial are recorded and, in part, proved. These include: evaluation (de Casteljau's algorithm), differentiation and integration, product, degree raising, change of the underlying simplex, and the behavior on the boundary of the underlying simplex, with application to the construction of smooth pp functions on a given "triangulation."

Some effort has gone into making the notation fully reflect the symmetries and structure of this form. In particular, the description of this form in terms of difference operators is stressed.

Introduction; Notation. This paper lists the essential facts about the representation of polynomials in m variables as **Bernstein polynomials.** An expanded version may appear elsewhere.

While univariate Bernstein polynomials are well studied (see, e.g., Lorentz' classical book, Lorentz [11]), the multivariate version has only attracted attention sporadically. Lorentz' book devotes just one page to the two most direct generalizations: the tensor product or coordinate degree generalization, and the total degree generalization which is the topic of the present paper.

Motivation for the paper comes from computer aided geometric design where, through the initiative of de Casteljau and Bézier, the Bernstein polynomials of mostly one variable have become the main tool for the representation and computational use of **pp** (:= piecewise polynomial) functions. Farin's work (Farin [9]; Farin [10]) brought popularity and understanding to the use of bivariate Bernstein polynomials, and my own understanding starts from that work. My own interest has been started and repeatedly reinforced by work with smooth pp functions in two or more variables (de Boor and Höllig [5], [6]), in which their representation in terms of Bernstein polynomials, i.e., their **B-net**, for short, plays an essential role, since it reflects so nicely, and far better than other standard representations, the interplay between the geometry of the underlying triangular partition and the smoothness requirements.

For the sake of brevity, and since there are several people and ideas responsible, I am proposing here the term **B-form** (and correspondingly, B-net) for what would, more properly, be called the **barycentric-Bernstein-de Casteljau-Bézier-Farin-...-form.** I apologize to de Casteljau and Farin and . . . for the slight they might feel.

While the bivariate and trivariate situation is of most practical interest, I have chosen here to record the facts in the general m-dimensional context. This forces careful consideration of notation and brings out the essential mathematical aspects and surprising beauty of the B-form.

I will adhere to the following notational conventions: I will not bother with boldface, arrows, or underlines to distinguish points in $\mathbb{R}^m$ from other objects. The jth component of a point $x \in \mathbb{R}^m$ I will denote by $x(j)$ (rather than x_j). I will use standard multi-index notation throughout. Thus

$$x^\alpha := x(1)^{\alpha(1)}x(2)^{\alpha(2)} \cdots x(m)^{\alpha(m)}$$

with $\alpha \in \mathbb{Z}^m$, i.e., α an m-vector with integer entries. The **normalized monomial** is so handy a function that it deserves a special symbol:

$$(0.1) \qquad [\![x]\!]^\alpha := x^\alpha/\alpha! = \prod_{j=1}^{m} x(j)^{\alpha(j)}/\alpha(j)!,$$

hence $[\![x]\!]^\alpha = [\![x(1)]\!]^{\alpha(1)}[\![x(2)]\!]^{\alpha(2)} \cdots [\![x(m)]\!]^{\alpha(m)}$, with the conventions

$$[\![z]\!]^n = \begin{cases} 1 & \text{if } n = 0, \\ 0 & \text{if } n < 0. \end{cases}$$

In these terms, the **multinomial theorem** (for the power of a vector sum) takes the very simple form

$$[\![x + y + \cdots + z]\!]^\alpha = \sum_{\xi + v + \cdots + \zeta = \alpha} [\![x]\!]^\xi [\![y]\!]^v \cdots [\![z]\!]^\zeta,$$

whose proof by induction on the **length**

$$|\alpha| := \sum_{j=1}^{m} \alpha(j)$$

of α is immediate. It is possible (and a useful exercise) to build the entire discussion of the B-form on this identity.

The normalization of the monomials used here also makes differentiation neat. With

$$D^\alpha := D_1^{\alpha(1)}D_2^{\alpha(2)} \cdots D_m^{\alpha(m)}$$

and $D_j f$ the partial derivative of $f : \mathbb{R}^m \to \mathbb{R}$ with respect to its jth argument, we have

$$D^\alpha [\![\ ]\!]^\beta = [\![\ ]\!]^{\beta - \alpha}.$$

Hence

$$(0.2) \qquad D^\alpha [\![0]\!]^\beta = \delta_{\alpha\beta},$$

showing that $\{[\![\ ]\!]^\alpha : |\alpha| \le k,\ \alpha \in \mathbb{Z}_+^m\}$ is linearly independent. Their span

$$\pi_k := \operatorname{span} \{[\![\ ]\!]^\alpha : |\alpha| \le k,\ \alpha \in \mathbb{Z}_+^m\}$$

is, by definition, the collection of all polynomials of degree $\leq k$. We conclude that

$$(0.3) \qquad \dim \pi_k = \#\{[\![\ ]\!]^\alpha : |\alpha| \leq k,\ \alpha \in \mathbb{Z}_+^m\} = \binom{m+k}{k}.$$

While the **power form**

$$p = \sum_{|\alpha| \leq k} [\![\ ]\!]^\alpha c(\alpha)$$

(with $c(\alpha) = (D^\alpha p)(0)$, by (0.2)) is the standard mathematical representation for $p \in \pi_k$, it is not suited for work with pp functions since it provides explicit information only about the behavior of p near 0. By contrast, the B-form (with respect to some $(m+1)$-set $V \subset \mathbf{R}^m$) provides explicit information about the behavior of p at all the faces of the **convex hull**

$$[V]$$

of V. This makes it appropriate for the representation of smooth multivariate pp functions over a triangular partition.

1.   Linear Interpolation.   A set V of $m+1$ points in $\mathbb{R}^m$ is said to be **in general position** in case every linear polynomial on $\mathbb{R}^m$ can be written in terms of its values on V, i.e.,

$$(1.1) \qquad \forall p \in \pi_1 \qquad p = \sum_{v \in V} \xi_v p(v).$$

There are equivalent definitions of this term. For example, V is in general position in case its affine hull is all of $\mathbb{R}^m$, or, in case the simplex $[V]$ is proper, or if the $m+1$ vectors $(v \mid 1)$, $v \in V$, in $\mathbb{R}^{m+1}$ are linearly independent, etc. But I stick with the above definition since it uses the property of immediate interest here. In this way we associate with each such V $m+1$ linear polynomials ξ_v, $v \in V$, characterized by the fact that (1.1) holds. In particular, with p the constant polynomial, we find that

$$(1.2) \qquad 1 = \sum_{v \in V} \xi_v,$$

while, with $p : x \mapsto x(j)$, we get

$$x(j) = \sum_{v \in V} \xi_v(x) v(j)$$

for $j = 1, \cdots, m$, hence

$$(1.3) \qquad \forall x \quad x = \sum_{v \in V} \xi_v(x) v.$$

We conclude that the vector

$$(1.4) \qquad \xi(x) := (\xi_v(x))_{v \in V}$$

provides the **barycentric coordinates** for x with respect to V. Note that I have chosen here to use the points in V (rather than the integers from 0 to m, or from 1 to $m+1$) to index the components of the vector $\xi(x)$. This unorthodox notation is more to the point since it does not impose some artificial order on the **vertices** v; it also simplifies notation.

Since $\dim \pi_1 = m + 1$, we conclude from (1.1) that $(\xi_v)_{v \in V}$ is a basis for π_1. In particular, the representation (1.1) is unique. Therefore

$$(1.5) \qquad p = \sum_{v \in V} \xi_v a(v) \quad \text{implies that } a(v) = p(v), \quad v \in V.$$

We conclude that

$$(1.6) \qquad \xi_w(v) = \delta_{wv},$$

hence

$$(1.7) \qquad \text{affine } (V \backslash w) = \ker \xi_w := \{x \in \mathbb{R}^m : \xi_w(x) = 0\}.$$

2.   Definition of the B-Form.   The B-form for $p \in \pi_k$ is a somewhat unexpected generalization of the linear interpolation formula (1.1), viz.

$$(2.1) \qquad p = (\xi E)^k c(0).$$

Here, c is a **mesh function** and ξE is a **difference operator,** and the formula is to be read as an instruction: "Apply the difference operator ξE k times, starting with the mesh function c, then evaluate the resulting mesh function at the mesh point 0."

This definition of the B-form is unorthodox. In effect, I propose here to use de Casteljau's algorithm (de Casteljau [7]) as the definition, and to derive the other properties of the B-form from this algorithm. Implicit in this is the claim that this provides a more efficient path to these properties than standard approaches.

Consider now this definition more explicitly. The c appearing in (2.1) is a **mesh function,** i.e., defined on the mesh of nonnegative integer points

$$\alpha := (\alpha(v))_{v \in V} \in \mathbb{Z}_+^V,$$

and the difference operator ξE acts on mesh functions by the rule

$$(2.2) \qquad (\xi E)c(\alpha) := \sum_{v \in V} \xi_v c(\alpha + e_v),$$

with e_v the unit vector given by $e_v(w) := \delta_{vw}$.

If $k = 0$, then we are to apply the difference operator no times, i.e., then

$$p = c(0).$$

If $k = 1$, then we are to apply the difference operator one time, i.e., then

$$p = \sum_{v \in V} \xi_v c(e_v).$$

This is just (1.1) again, in slightly changed notation, i.e., $c(e_v) = p(v)$, $v \in V$.

If $k = 2$, then we are to apply the difference operator two times, i.e., then

$$p = \sum_{v \in V} \sum_{w \in V} \xi_v \xi_w c(e_v + e_w).$$

For general k,

$$(2.3) \qquad p = \sum_{u \in V} \sum_{v \in V} \cdots \sum_{w \in V} \xi_u \xi_v \cdots \xi_w c(e_u + e_v + \cdots + e_w),$$

and this shows that **the function p given by (2.1) is a polynomial of degree** $\leq k$ since it shows that p is a linear combination of products of k linear polynomials. This also shows that, for the purpose of the definition (2.1), we only need to know c at meshpoints of the form

$$e_u + e_v + \cdots + e_w$$

involving exactly k summands, i.e., at all meshpoints $\alpha \in \mathbb{Z}_+^V$ with

$$|\alpha| := \sum_{v \in V} \alpha(v) = k.$$

Writing (2.3) in terms of distinct meshpoints $\alpha \in \mathbb{Z}_+^V$, we get

$$(2.4) \qquad p = \sum_{|\alpha|=k} B_\alpha c(\alpha),$$

with

$$B_\alpha := |\alpha|! \, [\![\xi^\alpha]\!] = \binom{|\alpha|}{\alpha} \xi^\alpha$$

and

$$[\![\xi]\!]^\alpha := \prod_{v \in V} [\![\xi_v]\!]^{\alpha(v)} = \prod_{v \in V} \xi_v^{\alpha(v)} / \alpha(v)!.$$

The **multinomial coefficient** $\binom{|\alpha|}{\alpha} = |\alpha|!/\Pi_{v \in V} \alpha(v)!$ appears here in accordance with the multinomial theorem, but its precise value is not important here. The only thing that matters is that there are exactly

$$\#\{\alpha \in \mathbb{Z}_+^V : |\alpha| = k\} = \#\{\beta \in \mathbb{Z}_+^m : |\beta| \leq k\} = \dim \pi_k$$

summands in (2.4). Hence, necessarily

$$(B_\alpha)_{|\alpha|=k} \text{ and } (\xi^\alpha)_{|\alpha|=k} \text{ are both bases for } \pi_k,$$

provided we can convince ourselves that **every $p \in \pi_k$ can be written in the form (2.3) or (2.4).** But that is not hard to do. Observe that every $p \in \pi_k$ can be written as a sum of products of k linear polynomials—e.g.,

$$\forall |\beta| \leq k \qquad x^\beta = \left(\prod_{\beta(j)>0} (x(j))^{\beta(j)} \right) (1)^{k-|\beta|}$$

—and each linear polynomial can be written as a linear combination of the ξ_v as in (1.1). Thus, for every $p \in \pi_k$, there is a collection of k-tuples $(q, r, \cdots, s)$ of linear polynomials so that

$$p = \sum_{(q,r,\cdots,s)} qr \cdots s$$

$$= \sum_{(q,r,\cdots,s)} \left(\sum_{u \in V} \xi_u q(u) \right) \left(\sum_{v \in V} \xi_v r(v) \right) \cdots \left(\sum_{w \in V} \xi_w s(w) \right)$$

$$= \sum_{u,v,\cdots,w} \xi_u \xi_v \cdots \xi_w \sum_{(q,r,\cdots,s)} q(u) r(v) \cdots s(w)$$

which shows our claim.

In particular, the representation (2.4) is unique, i.e., for each $p \in \pi_k$, there is exactly one choice for $c(\alpha)$, $|\alpha| = k$, so that (2.1) (or, equivalently (2.4)) holds.

Since $\sum_{|\alpha|=k} B_\alpha = (\sum_{v \in V} \xi_v)^k = 1$, the basis $(B_\alpha)_{|\alpha|=k}$ forms a **partition of unity**, and this partition of unity is nonnegative on $[V]$ since the barycentric coordinates are nonnegative there.

Further, B_α **is unimodular on** $[V]$, **and the coefficient** $c(\alpha)$ **has its biggest influence on** p **in** $[V]$ **at the point**

$$(2.5) \qquad v_\alpha := \sum_{v \in V} v\alpha(v)/|\alpha|$$

since B_α, and equivalently ξ^α, **takes its maximum over** $[V]$ **at** v_α. For this and many other reasons, the pointset

$$(2.6) \qquad \{(v_\alpha, c(\alpha)) \in \mathbb{R}^{m+1} : |\alpha| = k, \ \alpha \in \mathbb{Z}_+^V\}$$

in $\mathbb{R}^{m+1}$ is given special attention. It is called the **B-net** for p. The points which make up the B-net are often called the **Bézier control points** for p.

The point of the formulation (2.1) is to avoid having to deal with expressions like (2.3) and to operate, calculate and reason directly with the simple expression (2.1) according to the operations it prescribes. The next two sections may make this clearer.

3. Evaluation of the B-Form. Evaluation of the B-form (2.1) at some point x requires k-fold application of the difference operator $\xi(x)E$. Since we are only interested in $(\xi(x)E)^k c$ at the meshpoint 0, we only need to apply the difference operator "at" certain meshpoints. Precisely, we calculate

$$c_1(\alpha) := (\xi(x)E)c(\alpha) = \sum_{v \in V} \xi_v(x)c(\alpha + e_v) \quad \text{for } |\alpha| = k - 1$$

and this only requires knowledge of $c(\alpha)$ for $|\alpha| = k$. Then we calculate

$$c_2(\alpha) := (\xi(x)E)c_1(\alpha) = \sum_{v \in V} \xi_v(x)c_1(\alpha + e_v) \quad \text{for } |\alpha| = k - 2$$

and this only requires knowledge of $c_1(\alpha)$ for $|\alpha| = k - 1$. In this way, we calculate

$$c_j(\alpha) := (\xi(x)E)c_{j-1}(\alpha) = \sum_{v \in V} \xi_v(x)c_{j-1}(\alpha + e_v) \quad \text{for} \quad |\alpha| = k - j$$

for $j = 1, \cdots, k$ (and with $c_0 = c$), and the final calculation gives

$$p(x) = c_k(0) = (\xi(x)E)c_{k-1}(0) = \sum_{v \in V} \xi_v(x)c_{k-1}(e_v).$$

In this description, I used different meshfunctions c_j to avoid confusion. But, since c_j is only considered and generated at meshpoints α with $|\alpha| = k - j$, we might as well use the same letter c for all of them. Thus, the evaluation amounts to generating the whole $(m + 1)$-simplex

$$c(\alpha), \ |\alpha| \le k,$$

of numbers from its base

$$c(\alpha), \ |\alpha| = k,$$

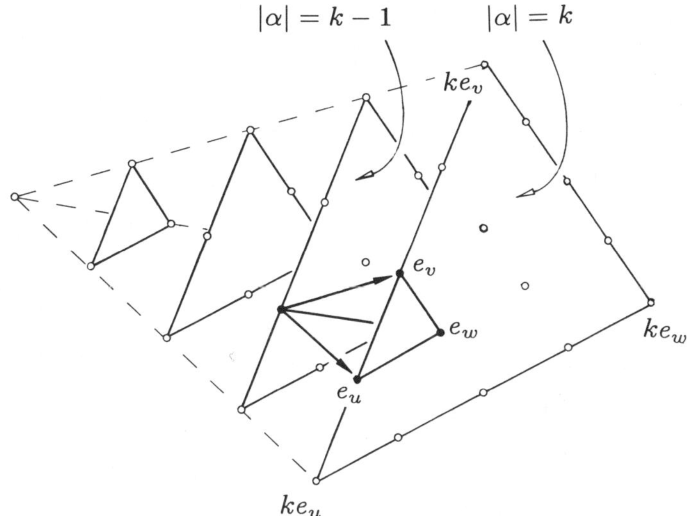

Figure 1.   The meshpoint simplex for evaluation.

by the calculation

(3.1)   for $j = 1, \cdots, k$, do:

$$c(\alpha) = (\xi(x)E)c(\alpha) = \sum_{v \in V} \xi_v(x)c(\alpha + e_v), \qquad |\alpha| = k - j.$$

While the base of this discrete simplex never changes, the layers built upon it do depend on x. Its apex, $c(0)$, provides the desired number $p(x)$. We will see later that the numbers $c(\alpha)$, $|\alpha| < k$, generated here also provide useful information about p.

Figure 1 shows the meshpoints of interest for the case $m = 2$, i.e., the **bivariate** case. In this case, the meshpoints have **three** components, corresponding to the fact that a two-dimensional simplex has three vertices. Correspondingly, each of the mesh-point layers $|\alpha| = k - j$ of interest forms a triangle in this case, and the total set forms a tetrahedron. For $m = 1$, the meshpoints of interest would form a triangle. This is quite familiar from the evaluation of univariate polynomials, e.g., from their Newton form in which one also generates a triangular array of numbers.

Figure 1 also shows the typical stencil of the difference operator which, for the calculation of

$$(\xi(x)E)c(\beta) = \sum_{v \in V} \xi_v(x)c(\beta + e_v),$$

requires one to go from meshpoint β into each of the $m + 1$ coordinate directions, picking up the value of c at the $m + 1$ meshpoints $\beta + e_v$, $v \in V$, reached, multiplying that value with the number $\xi_v(x)$ and then summing these products over v to obtain the value of $(\xi(x)E)c$ at β. Recall that $\sum_{v \in V} \xi_v(x) = 1$, hence application of the difference operator $\xi(x)E$ always amounts to **averaging.** If $x \in [V]$, then this average is proper since then $\xi(x) \geq 0$. Thus,

(3.2) $p(x)$ *is a convex combination of* $\{c(\alpha), |\alpha| = k\}$ *in case* $x \in [V]$.

As an example, consider the calculation of $p(w)$ for some $w \in V$. Since $\xi(w) = e_w$, the difference operator simplifies in this case to

$$(\xi(w)E)c(\beta) = c(\beta + e_w),$$

i.e., we merely pick up the value at the next meshpoint in the w-direction. We conclude that therefore

(3.3) $p(w) = c(e_w)$ for $w \in V$.

As another example, consider the calculation of $p(x)$ for some $x \in [V \backslash u]$. Now $\xi_u(x) = 0$, hence

$$(\xi(x)E)c(\beta) = \sum_{v \in (V \backslash u)} \xi_v(x)c(\beta + e_v).$$

We conclude that, in this case, $p(x)$ is a convex combination of just the coefficients $c(\alpha)$ with $|\alpha| = k$ and $\alpha(u) = 0$. More generally,

(3.4) $x \in [V \backslash U]$ $\Rightarrow$ $p(x) \in [c(\alpha) : |\alpha| = k,\ \text{supp } \alpha \subset (V \backslash U)]$.

To put it differently: **For** $W \subset V$, **the coefficients for the B-form** (with respect to W) **of the restriction of** p **to the affine hull of** W **are provided by the restriction of** c **to the corresponding mesh "simplex"** $\{\alpha \in \mathbb{Z}_+^V : |\alpha| = k,\ \text{supp } \alpha \subset W\}$. In these terms, (3.3) provides the extreme case $W = \{w\}$.

4.   Differentiation of the B-Form.   The **directional derivative** $D_y f$ of the function f on $\mathbb{R}^m$ in the direction y is, by definition, given by the rule

$$(D_y f)(x) := \lim_{t \to 0} \frac{f(x + ty) - f(x)}{t}.$$

Hence, in terms of the partial derivatives,

$$D_y = \sum_{j=1}^{m} y(j)D_j.$$

(This would suggest the alternative notation yD for the operator D_y or of E_ξ for ξE.)
It requires nothing more than the chain rule to differentiate the B-form:

(4.1) $D_y p = D_y(\xi E)^k c(0) = (\xi E)^{k-1} k (D_y \xi E) c(0).$

Since $\xi(x)$ is the unique solution of the linear system

$$\sum_v \xi_v(x) = 1, \qquad \sum_v \xi_v(x)v = x,$$

we have $\xi(x + ty) - \xi(x) = t\eta(y)$, with $\eta(y)$ the unique solution of

$$\sum_v \eta_v(y) = 0, \qquad \sum_v \eta_v(y)v = y.$$

For example,

$$(4.2) \qquad \eta(w - v) = e_w - e_v, \qquad w \in V \backslash v.$$

Thus, **with**

$$(4.3a) \qquad c_y := (\eta(y)E)c,$$

we obtain explicitly the B-form

$$(4.3b) \qquad D_y p = (\xi E)^{k-1} k c_y(0)$$

for the polynomial $D_y p$. Therefore, at a vertex,

$$D_y p(v) = k c_y((k-1)e_v) = k(\eta(y)E)c((k-1)e_v), \qquad v \in V.$$

In particular, from (4.2),

$$D_{w-v} p(v) = \frac{c((k-1)e_v + e_w) - c(ke_v)}{1/k}, \qquad w \in V \backslash v,$$

hence **the** $m + 1$ **distinct points**

$$\{(v_\alpha, c(\alpha)) \in \mathbb{R}^{m+1} : \alpha = (k-1)e_v + e_w, w \in V\}$$

with

$$(4.4) \qquad v_\alpha := \sum_{v \in V} v\alpha(v)/|\alpha|$$

all lie on the tangent plane to p **at** v and therefore determine that plane. This is a further indication of the usefulness of the **B-net** for p, which, to recall from (2.6), is the collection of points

$$\{(v_\alpha, c(\alpha)) \in \mathbb{R}^{m+1} : |\alpha| = k, \alpha \in \mathbb{Z}_+^V\}.$$

Higher derivatives are obtained by iteration of this process. If $Y \subset \mathbb{R}^m \backslash 0$ contains r points, then

$$(4.5a) \qquad D_Y p := \left(\prod_{y \in Y} D_y\right) p = \frac{k!}{(k-r)!} (\xi E)^{k-r} c_Y(0)$$

with

$$(4.5b) \qquad c_Y(\alpha) := \prod_{y \in Y} (\eta(y)E)c(\alpha), \qquad |\alpha| = k - r.$$

The evaluation of such a derivative at some point proceeds just as the evaluation of p itself, except that different difference operators are to be used at different stages. The order in which these are applied is immaterial since all (constant coefficient) difference operators commute.

5. A Taylor Formula. By the binomial theorem and the fact that $\xi(x + y) = \xi(x) + \eta(y)$,

$$(5.1) \qquad p(x + y) = (\xi(x+y)E)^k c(0) = \sum_{r=0}^{k} \binom{k}{r} (\xi(x)E)^{k-r} (\eta(y)E)^r c(0).$$

Since

$$\binom{k}{r}(\xi(x)E)^{k-r}(\eta(y)E)^{r}c(0) = \frac{1}{r!}(D_{y}^{r}p)(x)$$

by (4.5), we recognize in (5.1) the standard Taylor formula

$$p(x+y) = p(x) + (D_{y}p)p(x) + \tfrac{1}{2}(D_{y}^{2})(x) + \cdots.$$

6.   Invariance Under an Affine Change of Variables.   Any affine change of variables leaves the B-form unchanged. More precisely, if also W is an $(m+1)$-point set in $\mathbb{R}^{m}$ in general position, and f is an affine map so that $f(W) = V$, then the B-form for the polynomial $q := p \circ f$ with respect to W has again c as its coefficient sequence, in the following sense. With $\xi'(x)$ the barycentric coordinates of x with respect to W,

$$q(x) = p(f(x)) = (\xi(x)E)^{k}c(0) = (\xi'(x)E)^{k}c^{f}(0)$$

where

$$c^{f}(\alpha) := c(\alpha \circ f^{-1}).$$

7.   Integration of the B-Form.   It is possible to express the integral of p over $[V]$ quite simply in terms of the B-form:

$$(7.1) \qquad \int_{[V]} p = \frac{\mathrm{vol}_{m}[V]}{\binom{k+m}{k}} \sum_{|\alpha|=k} c(\alpha).$$

In effect, this identity claims that $\int_{[V]} B_{\alpha}$ only depends on $|\alpha|$. The quickest way to see this is to realize that

$$\int_{[V]} B_{\alpha} = \int_{[W]} 1,$$

with W the $(m+1+k)$-set in $\mathbb{R}^{m+k}$ generated from V and α by thinking of V as a subset of $\mathbb{R}^{m+k}$ and adjoining to it, for each $v \in V$, $\alpha(v)$ points of the form (v, e_{j}), with the unit vectors $e_{j} \in \mathbb{R}^{k}$ different from point to point.

8.   Product of Two B-Forms.   The B-form of the product of two polynomials is obtainable from their B-forms with the aid of a few factorials:

$$(\xi E)^{k}c(0)(\xi E)^{h}d(0) = (\xi E)^{k+h}c * d(0)$$

with

$$(8.1) \qquad c * d(\gamma) := \sum_{\alpha+\beta=\gamma} c(\alpha)d(\beta)C_{\alpha\beta}$$

and

$$C_{\alpha\beta} := \frac{\binom{|\alpha|}{\alpha}\binom{|\beta|}{\beta}}{\binom{|\alpha+\beta|}{\alpha+\beta}},$$

since
$$B_\alpha B_\beta = \binom{|\alpha|}{\alpha}\xi^\alpha \binom{|\beta|}{\beta}\xi^\beta = C_{\alpha\beta}B_{\alpha+\beta}.$$

The special choice $d(\beta) = 1$, $|\beta| = h$ gives the formula
$$(\xi E)^k c(0) = (\xi E)^{k+h} c'(0)$$

with
$$c'(\gamma) := \sum_{\alpha+\beta=\gamma} c(\alpha)C_{\alpha\beta}, \quad |\gamma| = k + h.$$

The particular case $h = 1$ of such **degree raising** has received special attention.

9. Degree Raising. Observe that
$$[\![\xi]\!]^\alpha = \left(\sum_v \xi_v\right)[\![\xi]\!]^\alpha = \sum_v (\alpha(v) + 1)[\![\xi]\!]^{\alpha+e_v}.$$

Hence
$$\sum_{|\alpha|=k} B_\alpha c(\alpha) = k! \sum_{|\alpha|=k} \sum_v (\alpha(v) + 1)[\![\xi]\!]^{\alpha+e_v} c(\alpha)$$
$$= k! \sum_{|\alpha|=k+1} [\![\xi]\!]^\alpha \sum_v \alpha(v)c(\alpha - e_v).$$

Conclude that

(9.1)
$$\sum_{|\alpha|=k} B_\alpha c(\alpha) = \sum_{|\alpha|=k+1} B_\alpha (Rc)(\alpha)$$

with

(9.2)
$$(Rc)(\alpha) := \sum_{v \in V} c(\alpha - e_v)\alpha(v)/|\alpha|.$$

Note that (9.1)–(9.2) requires knowledge of $c(\beta)$ for certain β with a negative entry. This presents no difficulty, though, since all such values are multiplied by zero, hence are not really needed.

Formula (9.2) can be interpreted as **linear interpolation** at the point v_α of the plane or linear polynomial through the points
$$(v_\beta, c(\beta)) \text{ with } \beta = \alpha - e_v, v \in V.$$

This draws further attention to the **control polytopes** for p, i.e., the piecewise linear functions obtained from the B-net
$$\{(v_\alpha, c(\alpha)) : |\alpha| = k\}$$

by local linear interpolation.

10. The Bernstein Polynomial. The **Bernstein polynomial for f of order k with respect to V** is, by definition, the particular polynomial

(10.1)
$$B_k f := \sum_{|\alpha|=k} B_\alpha f(v_\alpha).$$

The Bernstein polynomial provides an approximation to f which, on $[V]$, converges uniformly to f as $k \to \infty$ in case f is continuous (cf. Lorentz [11, p. 51]). The convergence is monotone in case f is **V-convex** in the sense that each of the univariate functions $t \mapsto f(x + t(v - w))$ is convex (Berens [1]). Moreover, in this case, $B_k f$ is also V-convex. But $B_k f$ need not be convex even if f is (Stancu [12]; Berens [1]; Chang and Davis [8]).

The B-form $(\xi E)^k c(0)$ for $p \in \pi_k$ (with respect to V) provides the essential information about any f for which $p = B_k f$. We have

$$(10.2) \qquad p = B_k f \text{ if and only if } f(v_\alpha) = c(\alpha) \text{ for all } |\alpha| = k.$$

The simplest such functions f are the **control polytopes** for p, i.e., the linear interpolants to the data

$$(10.3) \qquad (v_\alpha, c(\alpha)), \qquad |\alpha| = k.$$

For this reason, we denote any such control polytope by

$$(10.4) \qquad B_k^{-1} p.$$

I have used the plural here advisedly since, for $m > 2$, there are several equally reasonable piecewise linear interpolants, as has been rightfully stressed and detailed by Dahmen and Micchelli in [13]. Different interpolants differ in how the points $\{v_\alpha : |\alpha| = k\}$ are connected to produce a **triangulation**, i.e., a partition into simplices, for the simplex $[V]$. The typical triangulation is obtained by choosing an ordering $v_0, v_1, \cdots, v_m$ of the vertex set V, thus obtaining the directions $d_i := v_i - v_{i-1}$, $i = 1, \cdots, m$. The corresponding triangulation with meshpoints $\{v_\alpha : |\alpha| = k\}$ for V consists of all simplices in $[V]$ of the form

$$\sigma_{\alpha,q} := v_\alpha + [0, d_{q(1)}, d_{q(1)} + d_{q(2)}, \cdots, d_{q(1)} + \cdots + d_{q(m)}]/k$$

with $|\alpha| = k$ and q a permutation of the first m integers. Thus,

$$\sigma_{\alpha,q} = [v_{\alpha_0}, \cdots, v_{\alpha_m}],$$

with

$$\alpha_0 := \alpha,$$
$$\alpha_j := \alpha_{j-1} + e_{v_{q(j)}} - e_{v_{q(j)-1}}, \qquad j = 1, \cdots, m.$$

A simplex may appear in the triangulation only for certain orderings of V and not for others. To see this, observe that two points v_α and v_β will be vertices for the same simplex if and only if their difference can be written as a sum of some of the vectors d_j/k. If we order the entries of α and β to correspond to the ordering of the vertex set used, writing, e.g., $\alpha(j)$ instead of $\alpha(v_j)$, this means that either $\beta - \alpha$ or else $\alpha - \beta$ must be writable as a sum of distinct vectors of the form $e_j - e_{j-1}$. For example, for $m = 3$ and $k = 2$, the two vertices $v_{(1,0,0,1)}$ and $v_{(0,1,1,0)}$ are not connected by a meshline (in the triangulation corresponding to the ordering used), while the two vertices $v_{(0,1,0,1)}$ and $v_{(1,0,1,0)}$ are. This shows that the reordering v_1, v_0, v_2, v_3 would connect the former and disconnect the latter.

On the other hand, the simplices

$$[v_{\beta-e_v}:v\in V]\quad\text{with }|\beta|=k+1$$

involved in degree raising are part of any such triangulation since, in terms of the particular ordering used,

$$[v_{\beta-e_v}:v\in V]=\sigma_{\alpha,q}$$

with $\alpha=\beta-e_{v_m}$ and $q(j)=m+1-j$, $j=1,\cdots,m$. Thus, regardless of the particular ordering of the vertex set V used, the resulting piecewise linear interpolant $B_k^{-1}p$ to the data (10.3) will agree with $B_{k+1}^{-1}p$ at the basepoints v_β, $|\beta|=k+1$, as we saw in §9. This implies, by induction, that the Lipschitz constant (over $[V]$) for any $B_{k+n}^{-1}p$, $n>0$, is no bigger than that for $B_k^{-1}p$, hence the sequence $(B_{k+n}^{-1}p)$ has uniform limit points. It implies further that $B_{k+n}^{-1}p$ converges pointwise to some function f as $n\to\infty$, hence f is the uniform limit of $B_{k+n}^{-1}p$. But this limit function is necessarily p since

$$p=B_{k+n}B_{k+n}^{-1}p=B_{k+n}f+B_{k+n}(B_{k+n}^{-1}p-f)\to f\quad\text{as }n\to\infty,$$

using the facts that $B_{k+n}f$ converges to f and $\|B_{k+n}(B_{k+n}^{-1}p-f)\|\le\|B_{k+n}^{-1}p-f\|\to0$.

Since local linear interpolation preserves convexity, we conclude that p is convex in case its control polytope $B_k^{-1}p$ is.

11. Boundary Behavior. We now come to the heart of the B-form. We consider how to extract from the B-form of p information about its behavior on the boundary of the simplex $[V]$.

The boundary of $[V]$ is made up of faces, i.e., of convex hulls of subsets of V. For any $W\subset V$, we call $[W]$ the W-**face** of $[V]$. The (proper) faces of highest dimension are the **facets** of $[V]$. We find it convenient to call the $(V\setminus w)$-face of $[V]$ the w-**facet** of $[V]$. In other words, we identify the faces of the simplex by the set of vertices contained in them, but identify a facet by the sole vertex not contained in it.

Recall from §3 that, on the W-face, p is entirely determined by $c(\alpha)$ with supp $\alpha\subset W$. Recall from §4 that the tangent plane for p at the vertex w is entirely determined by $c(\alpha)$ with $\alpha=(k-1)e_w+e_v$, $v\in V$. We can describe this last set also as

$$\{\alpha:|\alpha|_{V\setminus w}\le1\},$$

using the abbreviation

$$|\alpha|_W:=\sum_{v\in W}\alpha(v).$$

This makes it easy to recognize both of these facts as special cases of the following theorem.

THEOREM 11.   *All derivatives of p of order $\le\rho$ on the W-face are determined by*

$$(11.1)\qquad\qquad\qquad c(\alpha),\ |\alpha|_{\setminus W}\le\rho.$$

If the W-face in question is a facet, say the w-facet, then the coefficients involved are those $c(\alpha)$ with $\alpha(w)\le\rho$. In the whole coefficient-"simplex," these occupy layers $0,1,\cdots,\rho$ "parallel" to the w-facet, i.e., the layers $c(\alpha)$, $\alpha(w)=j$, with $j=0,1,\cdots,\rho$.

For the general W-face, the relevant coefficients are those no more than ρ steps away from the corresponding coefficient "facet" $c(\alpha)$, supp $\alpha \subset W$.

For the **proof,** observe that the theorem's claim is equivalent to the statement that p vanishes $\rho + 1$-fold on $[W]$ iff $c(\alpha) = 0$ for $|\alpha|_{\backslash W} \le \rho$. But this follows from the fact that $|\alpha|_{\backslash W} + 1$ gives the order to which ξ^α vanishes on $[W]$. $\square$

In terms of the B-net

$$b_p := \{(v_\alpha, c(\alpha)) : \alpha \in \mathbb{Z}_+^V\}$$

for p introduced in § 4, the theorem states that knowing all derivatives of order $\le \rho$ on the W-face is the same as knowing b_p on all v_α within ρ steps from that face. It is part of the attraction of the B-net that it makes such neat geometric statements possible.

For the application of this theorem to the problem of smoothly fitting together polynomial pieces, we must be prepared to express two such polynomial pieces in B-form with respect to the same simplex. In approaching this problem, we give another proof of the theorem. The approach makes use of the polynomials whose B-form with respect to V is part of the B-form for p.

12. The Subpolynomials. The evaluation of $p \in \pi_k$ and its derivatives from the B-form proceeds by repeated **differencing**. It is a remarkable fact that this differencing is **uniform**. Regardless of the meshpoint at which it is applied, the difference operator is the same. This implies that, **during the calculation of some information about p, we are simultaneously computing the same information for** a whole host of polynomials, viz. **all polynomials whose B-form coefficients** (with respect to V) **form a subsimplex of those for p.** These are the polynomials

$$(12.1) \qquad p_\alpha := (\xi E)^{k-|\alpha|} c(\alpha), \qquad |\alpha| \le k.$$

For $|\alpha| = k$, p_α is the constant $c(\alpha)$, while, at the other extreme, $p_0 = p$.

Since the B-form coefficients for such a p_α form a subsimplex of the coefficient simplex for p, the B-form coefficients of its derivative $D_Y p_\alpha$ form a corresponding subsimplex of those for $D_Y p$, up to a scalar factor. Precisely, from (4.5), with $Y \subset \mathbb{R}^m \backslash 0$ and $r := \#Y$, and $|\alpha| \le k - r$,

$$(12.2) \qquad D_Y p_\alpha = \frac{(k - |\alpha|)!}{(k - |\alpha| - r)!} (\xi E)^{k-|\alpha|-r} c_Y(\alpha)$$

with

$$(12.3) \qquad c_Y := \left(\prod_{y \in Y} \eta(y) E \right) c,$$

while

$$D_Y p = D_Y p_0 = \frac{k!}{(k-r)!} (\xi E)^{k-r} c_Y(0).$$

This shows that, on the W-face, $D_Y p_\alpha$ is determined by the numbers

$$(12.4) \qquad c_Y(\alpha + \beta), \quad \text{supp } \beta \subset W, \quad |\beta| = k - |\alpha| - r,$$

while $D_Y p$ is determined there by the numbers

$$(12.5) \qquad c_Y(\gamma), \quad \text{supp } \gamma \subset W, \quad |\gamma| = k - r.$$

Now note that (12.4) is a subset of (12.5) exactly when supp $\alpha \subset W$. Since p_α is of degree $\leq k - |\alpha|$, this implies that, for any α with supp $\alpha \subset W$, we know **all** derivatives of p_α on the W-face as soon as we know there all derivatives of p of order $\leq k - |\alpha|$. But, knowing all the derivatives of a polynomial even at just one point determines that polynomial entirely. This proves the following.

PROPOSITION 12. *Each p_α depends linearly on p and its derivatives of order $\leq k - |\alpha|$ on* [supp α].

Conversely, (12.5) is the union of all the sets (12.4) with supp $\alpha \subset W$. This proves the following restatement of Theorem 11.

THEOREM 12. *Let p, $q \in \pi_k$, $W \subset V$. Then*

$$\forall(Y \subset \mathbb{R}^m \backslash 0,\ \#Y \leq r)\quad D_Y p = D_Y q \text{ on the } W\text{-face}$$

$$\Leftrightarrow\quad \forall(|\alpha| = k - r, \text{ supp } \alpha \subset W)\quad p_\alpha = q_\alpha.$$

13. Change of V. The subpolynomials p_α introduced in the preceding section depend on V. This is reflected in the notation since, after all, α is defined on V. But, by Proposition 12, p_α depends, more precisely, only on the points in supp α. To say it differently:

PROPOSITION 13. *If also V' is an $(m + 1)$-set in $\mathbb{R}^m$ in general position, and $\alpha \in \mathbb{Z}_+^V$ has its support in $V \cap V'$, then*

$$p_\alpha = p_{\alpha'},$$

with

$$\alpha' : V' \to \mathbb{Z}_+ : v \mapsto \begin{cases} \alpha(v) & \text{if } v \in V \cap V', \\ 0 & \text{otherwise.} \end{cases}$$

This is so because, by the proposition, p_α only depends on p and its derivatives on [supp α]. This suggests the identification of any two α, α' which agree on their support, and we will follow this suggestion from now on. In effect, we think of α as defined at all the vertices that might enter the discussion, but to be zero on all but at most $m + 1$ of them.

This makes it easy to describe the change of V, i.e., the derivation of the B-form $p := (\xi' E)^k c'(0)$ for p with respect to V' from the B-form with respect to V. It is sufficient to consider the case

$$V' = (V \backslash w) \cup w',$$

since an arbitrary V' can be reached from V as the $(m + 1)$st in a chain of $(m + 1)$-sets whose neighbors only differ by one point.

The crucial observation is the following. The coefficient $c'(\alpha)$ is the extreme coefficient (associated with the vertex w') for the subpolynomial p_β with $\beta = \alpha - \alpha(w')e_{w'}$. In other words,

$$(13.1) \qquad\qquad c'(\alpha) = p_\beta(w') \quad \text{with } \beta := \alpha - \alpha(w')e_{w'}.$$

On the other hand, supp $\beta \subset (V \backslash w)$, hence

$$(13.2) \qquad\qquad p_\beta = (\xi E)^{\alpha(w')} c(\beta).$$

This implies that p_β is evaluated during the course of evaluation of p from its B-form with respect to V. Specifically, we find $c'(\alpha) = p_\beta(w')$ at position β in the $(m+1)$-simplex $c(\beta)$, $|\beta| \le k$, generated during the evaluation of p at w', i.e.,

$$(13.3) \qquad c'(\alpha) = c(\alpha - \alpha(w')e_{w'}) \quad \text{for } |\alpha| = k.$$

In fact, since the evaluation of p at w' from the B-form with respect to V proceeds without any special attention paid to the vertex w, it follows that we are generating simultaneously the B-form coefficients for p with respect to every one of the $(m+1)$ sets V' obtainable from V by an exchange of some $w \in V$ for w'. This provides a **subdivision algorithm**. Choosing w' somewhere in $[V]$, we obtain a triangulation of $[V]$ into at most $m+1$ nontrivial simplices $[V_w]$, with $V_w := (V \backslash w) \cup w'$, and the coefficient simplex for the B-form of p with respect to V_w is to be found in the w-facet of the $(m+1)$-simplex $c(\beta)$, $|\beta| \le k$, generated during the evaluation of p at w'.

14. Smoothness Across an Interface. The matching of derivatives of polynomial pieces across an interface between two simplices is easily described in terms of the subpolynomials associated with that interface, since these describe completely the behavior of a polynomial and its derivatives on that interface, by Theorem 12. The precise statement of the smoothness conditions is made quite simple by our agreement to think of meshpoints α as defined on all vertices that might appear in the discussion, with its value usually zero, with at most $m+1$ exceptions.

THEOREM 14. *Let p, $q \in \pi_k$; let $\rho \le k$; and let V, V' be the vertex sets of two simplices in a triangulation. Then the pp function*

$$f : [V] \cup [V'] \to \mathbb{R} : x \mapsto \begin{cases} p(x) & \text{if } x \in [V], \\ q(x) & \text{if } x \in [V'], \end{cases}$$

is in C^ρ if and only if

$$(14.1) \qquad \forall(\text{supp } \alpha \subset V \cap V', \ |\alpha| = k - \rho), \qquad p_\alpha = q_\alpha.$$

If V and V' differ by just one point,

$$V' = (V \backslash w) \cup w',$$

say, and $q = (\xi' E)^k c'(0)$ is the B-form for q with respect to V', then the condition (14.1) reads more explicitly

$$(14.2) \quad \forall(\text{supp } \alpha \subset (V \backslash w), \ k - \rho \le |\alpha| \le k) \quad (\xi(w')E)^{k-|\alpha|}c(\alpha) = c'(\alpha + (k - |\alpha|)e_{w'}).$$

Note that **these conditions are independent of k** and depend on α only in the sense that the weights in the linear relations between c and c' in (14.2) depend on ρ or $k - |\alpha|$, i.e., on the order of the derivatives being constrained to be continuous. This means that, in studying a linear system of such conditions across one or more neighboring facets, we can choose k at will, e.g., $k = \rho$.

In general, C^ρ-continuity across the w-facet of $[V]$ imposes conditions which connect $c(\alpha)$ for $\alpha(w) \leq \rho$ with $c'(\alpha)$ for $\alpha(w') \leq \rho$. The form (14.2) makes explicit that this involves exactly

$$\#\{\alpha \in \mathbb{Z}_+^V : \alpha(w) = 0, \, k - \rho \leq |\alpha| \leq k\}$$

linearly independent conditions, i.e., exactly as many conditions as there are degrees of freedom in p and its directional derivatives of order $\leq \rho$ on the w-facet in some fixed direction transversal to that facet.

15.   The B-Net.   Let Δ be a triangulation of some domain in $\mathbb{R}^m$. This means that Δ consists of simplices δ, with the intersection $\delta \cap \delta'$ of any two always a face (possibly the empty face) of both of them.

I denote by V_δ the vertex set of the simplex $\delta \in \Delta$, and by

$$V := \bigcup_{\delta \in \Delta} V_\delta$$

the totality of the vertices of simplices of Δ. Denote by

$$A = A_{k,\Delta} := \{\alpha \in \mathbb{Z}_+^V : |\alpha| = k, \, \exists \delta \in \Delta \quad \operatorname{supp} \alpha \subset V_\delta\}$$

the corresponding set of index meshpoints of interest. In words, these are elements of $\mathbb{Z}_+^V$, i.e., defined on V and with nonnegative integer entries. In addition, each $\alpha \in A$ has support only on some V_δ and has length $|\alpha| = k$.

Consider now the space

$$S := \pi_{k,\Delta}^\rho$$

of pp functions of degree $\leq k$ on the triangulation Δ and in C^ρ. This means that each $f \in S$ agrees on each simplex in Δ with some polynomial of degree $\leq k$, and these polynomial pieces fit together to form a function with ρ continuous derivatives.

Consider specifically

$$S_0 := \pi_{k,\Delta}^0,$$

the space of continuous pp functions (of degree $\leq k$) on Δ. Since two of its polynomial pieces on neighboring simplices fit together continuously exactly when their B-form coefficients associated with the common face coincide, it is possible to describe an element f of S_0 by the mesh function c defined on the mesh A and providing in

$$c(\alpha), \, \operatorname{supp} \alpha \subset V_\delta,$$

the B-form coefficients for the polynomial piece $f_{|\delta}$ with respect to the vertex set V_δ of δ.

The **B-net** for such f is, by definition, the collection of points

$$(v_\alpha, c(\alpha)), \quad \alpha \in A,$$

with

$$v_\alpha := \sum_{v \in V} v\alpha(v)/|\alpha|, \quad \alpha \in A.$$

While it is satisfactory to deal with the mesh function c, the B-net reflects more explicitly the geometry of the situation. We think of the B-net as the function

$$b_f : V_A \to \mathbb{R} : v_\alpha \mapsto c(\alpha)$$

on the discrete set

$$V_A := \{v_\alpha : \alpha \in A\},$$

which is a subset of the domain of $f \in S_0$. The values of this discrete function at all the points in some face of some δ determine f on that face. In particular,

$$f(v) = c(ke_v) = b_f(v), \quad v \in V.$$

Further, C^ρ-continuity of f is equivalent to certain linear relations involving b_f on points at most ρ layers away from the facets of the δ. For example, C^1-continuity is equivalent to having each $(m+2)$-tuple

$$(v_\beta, b_f(v_\beta)), \quad \beta = \alpha + e_v, \quad v \in W$$

lie on a plane, with W the vertices of any two simplices having a facet in common, and $|\beta| = k - 1$ with support only on the vertices common to both simplices. This localizes the effect of such continuity conditions as much as possible.

Acknowledgment.   This work was supported by the U.S. Army under contract DAAG29-80-C-0041.

REFERENCES

[1] H. BERENS, *Über Bernsteinpolynome in mehreren Veränderlichen*, talk given at the April 1976 Oberwolfach meeting organized by Zeller and Schempp, 1976.

[2] S. BERNSTEIN, *Démonstration du théorème de Weierstrass, fondée sur le calcul des probabilités*, Comm. Soc. Math. Kharkov (2), 13 (1912–13), pp. 1–2.

[3] P. BÉZIER *Numerical Control—Mathematics and Applications*, John Wiley, London, 1972.

[4] W. BOEHM, G. FARIN AND J. KAHMANN, *A survey of curve and surface methods in CAGD*, Computer Aided Geometric Design 1 (1984), pp. 1–60.

[5] C. DE BOOR AND K. HÖLLIG, *Approximation order from bivariate C^1-cubics: A counterexample*, Proc. Amer. Math. Soc., 87 (1983), pp. 649–655.

[6] ———, *Approximation order from smooth bivariate pp functions*, manuscript in preparation.

[7] P. DE CASTELJAU, *Courbes et surfâces a pôles*, André Citroën Automobiles SA, Paris, 1959.

[8] G.-Z. CHANG AND P. J. DAVIS, *A new proof for the convexity of Bernstein–Bézier surfaces over triangles*, J. Approx. Theory, 40 (1984), pp. 11–28.

[9] G. FARIN, *Subsplines über Dreiecken*, Dissertation, Braunschweig, West Germany, 1979.

[10] ———, *Bézier polynomials over triangles and the construction of piecewise C^r-polynomials*, TR/91, Dept. Mathematics, Brunel Univ., Uxbridge, Middlesex, UK, 1980.

[11] G. G. LORENTZ, *Bernstein Polynomials*, University of Toronto Press, Toronto, Canada, 1953.

[12] D. D. STANCU, *De l'approximation, par des polynômes du type Bernstein, des fonctions de deux variables*, Comm. Akad. R. P. Romine, 9 (1959), pp. 773–777.

[13] W. DAHMEN AND C. A. MICCHELLI, *Convexity of multivariate Bernstein polynomials and box spline surfaces*, Research Report RC 11176 (#50344) 5/31/85, IBM T. J. Watson Research Center, Yorktown Heights, NY, May 1985.

A Case Study of Multivariate Piecewise Polynomials

PETER ALFELD

Abstract. In this article once or twice differentiable piecewise polynomial functions defined on a triangulation or a tetrahedralization are considered and several examples where the dimension of the corresponding function space changes with the geometry of the tessellation are given.

1. Introduction. Fundamental in the solution of numerical problems involving functions of a *single* independent variable are piecewise polynomial functions of a certain degree of global smoothness. It is natural to contemplate the use of similar functions in problems involving *several* independent variables.

In this paper, we consider piecewise polynomial functions of two or three variables that are defined on a triangulation or a tetrahedralization, respectively, and that are globally once or twice differentiable. Necessary for any utilization of such functions is a knowledge of the dimension of the relevant function space.

The papers by Schumaker [10], [11] or the one by Alfeld [1] should be consulted for background information and for a brief outline of the history.

The fundamental difficulty with piecewise polynomial functions is that their structure depends not just on the *topology* of the tessellation, but also on its precise *geometry*! That dependence is very complicated and a comprehensive theory seems unlikely to emerge in the foreseeable future. At this stage, only upper and lower bounds are available (see Schumaker [10], [11]) that disagree in many instances. As a further step towards a general theory it appears necessary to build a body of examples that illustrates as many types of dependence as possible. In this paper, several illustrative examples are given.

We consider a tesselation of a domain D with (boundary and interior) vertices $V_1, V_2, \cdots, V_N \in \mathcal{R}^d$ (where $d \in \{2, 3\}$) by simplices Δ_i, i.e., triangles (if $d = 2$) or tetrahedra (if $d = 3$). The spaces of interest are

$$S_d^{k,m} := \{p \in C^k(D), \ p \text{ is a polynomial of degree } m \text{ on each } \Delta_i\}.$$

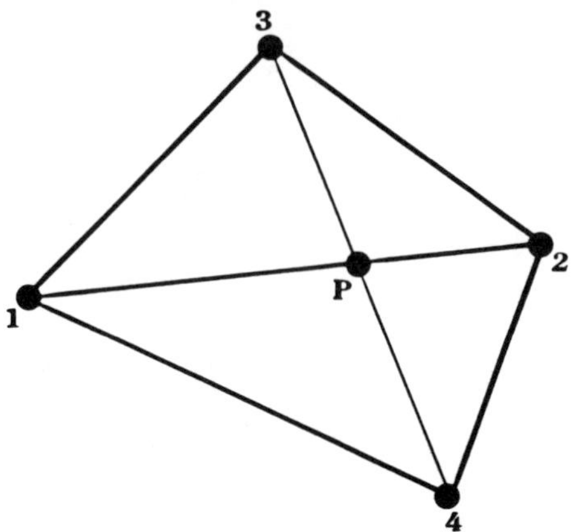

Figure 1.   Bivariate smoothness conditions.

The functions in these spaces are characterized by systems of linear equations that are derived in §§ 2.1 and 3.1. The linear equations, and hence the dimensions of $S_d^{k,m}$, are affinely invariant. Thus we express the location of all vertices in terms of the vertices $V_1, V_2, \cdots, V_{d+1}$ which must be in general position, but are otherwise arbitrary. (The vertices $V_1, V_2, \cdots, V_{d+1}$ are *in general position* if their affine hull is all of $\mathcal{R}^d$. For several alternative definitions see de Boor [4].)

The linear equations were analyzed using the symbol manipulation language REDUCE, which employs exact arithmetic. Thus there are no ambiguities due to round-off errors.

For reproducibility the exact location of all points is given for all examples.

We will denote edges by $e_{ij} := V_j - V_i$.

2. The Bivariate Case.   The best known example of a geometric degeneracy is a *singular vertex* (see any of the papers by Morgan and Scott [8], [9]; Schumaker [10], [11]; or Alfeld, [1]). Until recently, the only other known example was the Morgan/Scott split (see below) which causes degeneracies for C^1 *quadratics*. In Meyling and Pfluger [7], and in Chou et al. [5], some additional degeneracies for C^1 *quadratics* are described. Below we give more examples for C^1 *quadratics* and also for C^2 functions. No degeneracies other than singular vertices are known for $S_2^{1,m}$ where $m > 2$, and Morgan and Scott [8] showed that no others exist if $m > 4$.

2.1. The smoothness conditions.   In this subsection, we derive algebraic smoothness conditions that are expressed in terms of easily computed constants characterizing the geometry of a triangulation. The derivation is based on the generalized Bézier–Bernstein form of a bivariate polynomial; see Alfeld [2], Boehm, Farin and Kahmann [3], or de Boor [4], for an introduction.

Consider two triangles as given in Fig. 1. The location of a point x in either of the triangles is expressed as

$$x = \sum_{i=1}^{4} b_i V_i$$

where the b_i are the piecewise linear cardinal functions associated with V_i. A *continuous* piecewise degree m polynomial function p is then expressed as

$$p(x) = \sum_{i+j+k+l=m} \frac{m!}{i!\,j!\,k!\,l!} c_{ijkl} b_1^i b_2^j b_3^k b_4^l.$$

Since p is already continuous it is sufficient for differentiability to force continuity of any derivative across the common edge e_{12}. It is natural to differentiate in the direction of the line segment $V_4 - V_3$. Let P denote the intersection of e_{12} and $V_4 - V_3$ and write

$$P = \alpha_1 V_1 + \alpha_2 V_2 = \sigma_1 V_3 + \sigma_2 V_4 \quad \text{where } \alpha_1 + \alpha_2 = \sigma_1 + \sigma_2 = 1.$$

Also let

$$\gamma := \frac{\sigma_1}{\sigma_2} = \frac{\|V_4 - P\|}{\|V_3 - P\|} \neq 0.$$

Then differentiating on each triangle and equating the derivatives along edge e_{12} yields the conditions

$$(1) \qquad 0 = \left(2 + \gamma + \frac{1}{\gamma}\right)(\alpha_1 c_{i+1,j00} + \alpha_2 c_{i,j+1,00}) - (1 + \gamma)c_{ij10} - \left(1 + \frac{1}{\gamma}\right)c_{ij01}$$

for all i and j such that $i + j = m - 1$ and $i, j \geq 0$. (Subscripts are separated by commas only if they comprise more than one character.) Note that the conditions in (1) are well defined since the ratio γ never vanishes.

For second order differentiability it is sufficient to consider the second derivative in the direction of $V_4 - V_3$. Proceeding similarly as before we obtain the conditions:

$$
\begin{aligned}
(2) \quad 0 = {}& \left[(1 + \gamma)^2 - \left(1 + \frac{1}{\gamma}\right)^2\right](\alpha_1^2 c_{i+2,j00} + \alpha_2^2 c_{i,j+2,00} + 2\alpha_1\alpha_2 c_{i+1,j+1,00}) \\
& + (1 + \gamma)^2(c_{ij20} - 2\alpha_1 c_{i+1,j10} - 2\alpha_2 c_{i,j+1,10}) \\
& - \left(1 + \frac{1}{\gamma}\right)^2 (c_{ij02} - 2\alpha_1 c_{i+1,j01} - 2\alpha_2 c_{i,j+1,01})
\end{aligned}
$$

for all i and j such that $i + j = m - 2$ and $i, j \geq 0$.

Example. (Clough–Tocher Split). Consider a triangle with vertices V_1, V_2, and V_3 that has been split about its centroid $V_4 = (V_1 + V_2 + V_3)/3$. Then, labeling the vertices appropriately, one obtains for all interior edges $\alpha_1 = \frac{3}{2}$, $\alpha_2 = -\frac{1}{2}$, $\sigma_1 = \sigma_2 = \frac{1}{2}$, and $\gamma = 1$. The C^1 conditions (1), for example, across e_{14} turn into the simple averaging conditions

$$0 = 6c_{i,j+1,00} - 2c_{i,j+1,00} - 2c_{ij10} - 2c_{ij01}$$

and the C^2 conditions become

$$c_{ij20} - 3c_{i+1,j10} + c_{i,j+1,10} = c_{ij02} - 3c_{i+1,j01} + c_{i,j+1,01}.$$

THEOREM 1. *The smoothness conditions (1) and (2) are invariant under affine transformations.*

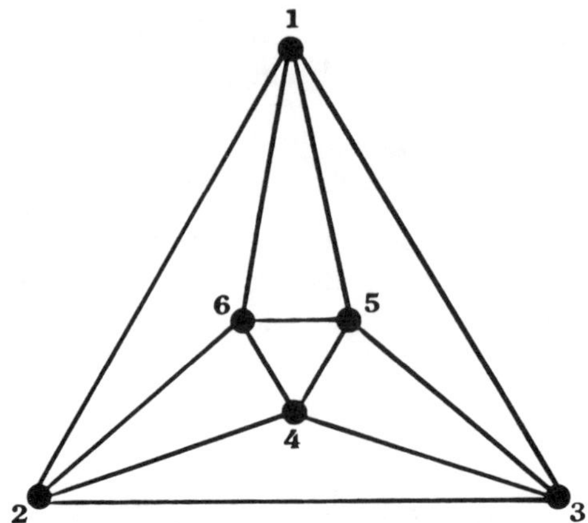

Figure 2.   The Morgan/Scott example.

Proof.   The conditions (1) and (2) depend only on the parameters α_1 and σ_1. These are determined by the linear system of equations

$$(3) \qquad [(V_1 - V_2), (V_4 - V_3)]\begin{bmatrix} \alpha_1 \\ \sigma_1 \end{bmatrix} = V_4 - V_2.$$

Now consider a transformation $V_i \mapsto AV_i + s$ where A is a nonsingular matrix and s is a constant vector. The corresponding system can be obtained from (3) by multiplying both sides with A which does not alter the solution of the linear system. $\square$

COROLLARY.   *The dimension of $S_2^{k,m}$ is affinely invariant.*

2.2.   The Morgan/Scott example.   One of the most fruitful sources of instructive examples is a split of a triangle first given by Morgan and Scott [9], and also described in Schumaker [11] (see Fig. 2).

For illustration we consider the symmetric version of the Morgan/Scott split in some detail. The vertices of the inner triangle are given by

$$(4) \qquad V_{3+i} = \frac{V_i + 2V_{i+1} + 2V_{i+2}}{5}, \qquad i = 1, 2, 3$$

where the index arithmetic on the right-hand side is modulo 3. Table 1 lists the relevant parameters for setting up the smoothness conditions (1) and (2). The first four columns of the table give the relabeling of the vertices that has to be applied in order for the formulas (1) and (2) to be valid.

Existing lower bounds on $\dim(S_2^{k,m})$ depend on the number of edges of *different slope* emanating from interior vertices (see Schumaker [11] or Alfeld [1]). Therefore, in the following examples we emphasize configurations where different edges emanating from the same vertex are parallel, i.e., where appropriate triples of vertices are collinear.

Table 2 lists the dimensions of some spaces for the following geometric configurations of the Morgan/Scott split:

Table 1. The symmetric Morgan/Scott split.

V_1	V_2	V_3	V_4	α_1	γ
1	5	6	3	$-1/4$	$5/3$
1	6	2	5	$-1/4$	$3/5$
2	6	4	1	$-1/4$	$5/3$
2	4	3	6	$-1/4$	$3/5$
3	4	5	2	$-1/4$	$5/3$
3	5	1	4	$-1/4$	$3/5$
5	6	1	4	$1/2$	$1/3$
6	4	2	5	$1/2$	$1/3$
4	5	3	6	$1/2$	$1/3$

(1)   (Generic configuration, i.e., there are no recognizable artifacts of the geometry.) $V_4 = (V_1 + 5V_2 + 4V_3)/10$, $V_5 = (45V_1 + 10V_2 + 45V_3)/100$ and $V_6 = (35V_1 + 45V_2 + 20V_3)/100$.

(2)   (The triple of vertices $\{V_1, V_5, V_4\}$ is collinear.) $V_4 = (V_1 + 2V_2 + 2V_3)/5$, $V_5 = V_4 + (V_1 - V_4)/5$ and $V_6 = (V_1 + V_2 + V_4)/3$.

(3)   (The two triples of vertices $\{V_1, V_5, V_4\}$ and $\{V_3, V_4, V_6\}$ are each collinear.) $V_5 = (3V_1 + V_2 + V_3)/5$, $V_4 = V_5 + (V_5 - V_1)/5$ and $V_6 = V_4 + (V_4 - V_3)/5$.

(4)   (The two triples of vertices $\{V_2, V_4, V_5\}$ and $\{V_3, V_4, V_6\}$ are each collinear, i.e., V_4 is a singular vertex.) $V_4 = (V_1 + 2V_2 + 2V_3)/5$, $V_5 = V_2 + 6(V_4 - V_2)/5$ and $V_6 = V_3 + 6(V_4 - V_3)/5$.

(5)   (The three triples of vertices $\{V_1, V_6, V_4\}$, $\{V_2, V_4, V_5\}$, and $\{V_3, V_5, V_6\}$ are each collinear.) $V_4 = (3V_3 + 9V_2 + V_1)/13$, $V_5 = (9V_3 + V_2 + 3V_1)/13$ and $V_6 = (V_3 + 3V_2 + 9V_1)/13$.

(6)   (Fully symmetric, i.e., as described in (4) and shown in Fig. 2.)

(7)   (Symmetry about the line e_{14}.) $V_4 = (11V_1 + 20V_2 + 20V_3)/51$, $V_5 = (2V_1 + V_2 + 2V_3)/5$ and $V_6 = (2V_1 + 2V_2 + V_3)/5$.

The columns headed "lb" and "ub" give Schumaker's lower and upper bounds, respectively, on the dimensions. For the lower bound, a generic configuration is assumed. For the upper bound, V_6 is assumed to be a singular vertex, and the vertices are ordered V_4, V_5, V_6, V_1, V_2, V_3 (see Schumaker [11]). Note, however, that Schumaker's bounds do take into account the geometry of the triangulation. In

Table 2. Dimensions for the bivariate Morgan/Scott example.

Configuration:	1	2	3	4	5	6	7	lb	ub
$S_2^{1,2}$	6	6	6	7	6	7	7	6	7
$S_2^{1,3}$	16	16	16	17	16	16	16	16	17
$S_2^{2,4}$	15	16	17	18	18	16	16	15	19
$S_2^{2,5}$	30	31	32	33	33	30	30	30	34

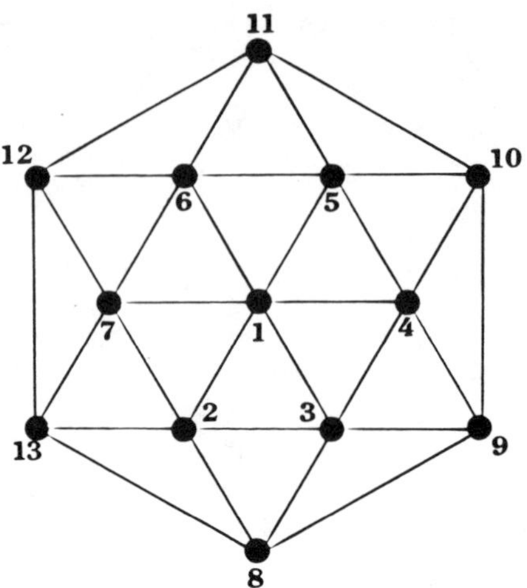

Figure 3.   A hexagonal triangulation.

particular, they fully incorporate singular vertices. Taking the geometry into account, the actual dimension equals the lower bound in all cases except configurations 6 and 7 in $S_2^{1,2}$ and in $S_2^{2,4}$.

Remarks.

(1)   The dimensions of $S_2^{1,3}$ and $S_2^{2,5}$ equal the lower bounds given in Alfeld [1], and Schumaker [11].

(2)   The increase in the dimensions of $S_2^{1,2}$ and $S_2^{2,4}$ for configurations 6 and 7 cannot be explained in terms of a configuration with a single interior vertex.

(3)   For configuration 2, all C^2 *quartics* are in fact quartic polynomials. The same is true for all triangulations obtained by applying this split recursively to any number of subtriangles.

(4)   Dimensions of $S_2^{2,3}$ have not been listed since they are always 10. Thus C^2 *cubics* are cubic polynomials on all of the above configurations.

2.3.   Other examples.   We consider in detail a configuration with 13 vertices as in Fig. 3 with the following geometric configurations:

(1)   (Fully symmetric as shown in Fig. 3.) $V_4 = V_1 + V_3 - V_2$, $V_5 = V_1 + V_4 - V_3$, $V_6 = V_1 + V_5 - V_4$, $V_7 = V_1 + V_6 - V_5$, $V_8 = V_2 + V_3 - V_1$, $V_9 = V_3 + V_4 - V_1$, $V_{10} = V_4 + V_5 - V_1$, $V_{11} = V_5 + V_6 - V_1$, $V_{12} = V_6 + V_7 - V_1$ and $V_{13} = V_7 + V_2 - V_1$. Note that, for example, the triples $\{V_2, V_1, V_5\}$ and $\{V_7, V_2, V_8\}$ are collinear. Thus the vertices $V_1, V_2 \cdots V_7$ are all *Clough–Tocher vertices*, i.e., precisely three edges of *different* slope emanate from each of them. Each Clough–Tocher vertex adds 1 to the lower bound on the dimension of C^2 piecewise polynomial functions on a generic triangulation (see Alfeld [1] or Schumaker [11]).

(2)   ($V_2, V_3 \cdots V_7$ are Clough–Tocher vertices.) $V_4 = (21V_3 - 21V_2 + 20V_1)/20$, $V_5 = (V_3 - 11V_2 + 20V_1)/10$, $V_6 = (-9V_3 - V_2 + 20V_1)/10$, $V_7 = (-19V_3 + 19V_2 + 20V_1)/20$, $V_8 = (19V_3 + 21V_2 - 20V_1)/20$, $V_9 = 2V_3 - V_2$, $V_{10} = (11V_3 - 21V_2 + 20V_1)/10$, $V_{11} = (-17V_3 - 23V_2 + 60V_1)/20$, $V_{12} = (-19V_3 + 9V_2 + 20V_1)/10$ and $V_{13} = -V_3 + 2V_2$.

(3) $(V_1, V_4, V_5, V_6, V_7$ are Clough–Tocher vertices.) $V_4 = V_3 - V_2 + V_1$, $V_5 = -V_2 + 2V_1$, $V_6 = -V_3 + 2V_1$, $V_7 = -V_3 + V_2 + V_1$, $V_8 = (11V_3 + 9V_2 - 10V_1)/10$, $V_9 = 2V_3 - V_2$, $V_{10} = V_3 - 2V_2 + 2V_1$, $V_{11} = -V_3 - V_2 + 3V_1$, $V_{12} = -2V_3 + V_2 + 2V_1$ and $V_{13} = -V_3 + 2V_2$.

(4) $(V_1, V_5, V_6, V_7$ are Clough–Tocher vertices.) $V_4 = V_3 - V_2 + V_1$, $V_5 = -V_2 + 2V_1$, $V_6 = -V_3 + 2V_1$, $V_7 = -V_3 + V_2 + V_1$, $V_8 = (10V_3 + 9V_2 - 9V_1)/10$, $V_9 = (19V_3 - 10V_2 + V_1)/10$, $V_{10} = V_3 - 2V_2 + 2V_1$, $V_{11} = -V_3 - V_2 + 3V_1$, $V_{12} = -2V_3 + V_2 + 2V_1$ and $V_{13} = -V_3 + 2V_2$.

(5) $(V_1, V_6, V_7$ are Clough–Tocher vertices.) $V_4 = V_3 - V_2 + V_1$, $V_5 = -V_2 + 2V_1$, $V_6 = -V_3 + 2V_1$, $V_7 = -V_3 + V_2 + V_1$, $V_8 = (10V_3 + 9V_2 - 9V_1)/10$, $V_9 = (19V_3 - 10V_2 + V_1)/10$, $V_{10} = (9V_3 - 19V_2 + 20V_1)/10$, $V_{11} = -V_3 - V_2 + 3V_1$, $V_{12} = -2V_3 + V_2 + 2V_1$ and $V_{13} = -V_3 + 2V_2$.

(6) $(V_1$ and V_7 are Clough–Tocher vertices.) $V_4 = (20V_3 - 19V_2 + 19V_1)/20$, $V_5 = -V_2 + 2V_1$, $V_6 = -V_3 + 2V_1$, $V_7 = (-20V_3 + 19V_2 + 21V_1)/20$, $V_8 = (20V_3 + 19V_2 - 19V_1)/20$, $V_9 = (19V_3 - 10V_2 + V_1)/10$, $V_{10} = (180V_3 - 361V_2 + 381V_1)/200$, $V_{11} = (-20V_3 - 19V_2 + 59V_1)/20$, $V_{12} = (-20V_3 + 9V_2 + 21V_1)/10$ and $V_{13} = (-10V_3 + 19V_2 + V_1)/10$.

(7) $(V_1$ is a Clough–Tocher vertex.) $V_4 = (20V_3 - 19V_2 + 19V_1)/20$, $V_5 = -V_2 + 2V_1$, $V_6 = -V_3 + 2V_1$, $V_7 = (-20V_3 + 19V_2 + 21V_1)/20$, $V_8 = (20V_3 + 19V_2 - 19V_1)/20$, $V_9 = (19V_3 - 10V_2 + V_1)/10$, $V_{10} = (180V_3 - 361V_2 + 381V_1)/200$, $V_{11} = (-20V_3 - 19V_2 + 59V_1)/20$, $V_{12} = (-20V_3 + 9V_2 + 21V_1)/10$ and $V_{13} = (-180V_3 + 361V_2 + 19V_1)/200$.

(8) (There are no Clough–Tocher vertices.) $V_4 = (18V_3 - 17V_2 + 19V_1)/20$, $V_5 = -V_2 + 2V_1$, $V_6 = (-9V_3 - V_2 + 20V_1)/10$, $V_7 = (-18V_3 + 17V_2 + 21V_1)/20$, $V_8 = (18V_3 + 21V_2 - 19V_1)/20$, $V_9 = (19V_3 - 10V_2 + V_1)/10$, $V_{10} = (142V_3 - 323V_2 + 381V_1)/200$, $V_{11} = (-18V_3 - 21V_2 + 59V_1)/20$, $V_{12} = (-18V_3 + 7V_2 + 21V_1)/10$ and $V_{13} = (-162V_3 + 343V_2 + 19V_1)/200$.

Table 3 shows some dimensions obtained. As before, lb and ub give lower and upper bounds. For the upper bound, the vertices are in their natural order, as given in Fig. 3, and all interior vertices are assumed to be Clough–Tocher vertices. For each configuration, η denotes the number of Clough–Tocher vertices. Schumaker's bounds for C^2 functions do take into account Clough–Tocher vertices. Thus the actual dimensions always exceed the lower bounds except for configurations 5, 6, 7, 8 in $S_2^{1,2}$, configurations 6, 7, 8 in $S_2^{2,4}$, and configurations 1, 2, 3 in $S_2^{2,5}$.

An asterisk indicates a configuration where the size of the linear system exceeded the core space available.

Table 3.   Dimensions for a hexagonal configuration.

Configuration:	1	2	3	4	5	6	7	8	lb	ub
η:	7	6	5	4	3	2	1	0		7
$S_2^{1,2}$	12	12	10	10	9	9	9	9	9	17
$S_2^{2,3}$	16	16	14	14	13	11	10	10	$6 + \eta$	21
$S_2^{2,4}$	34	33	30	30	28	26	25	24	$24 + \eta$	51
$S_2^{2,5}$	67	66	65	65	64	*	*	*	$60 + \eta$	93

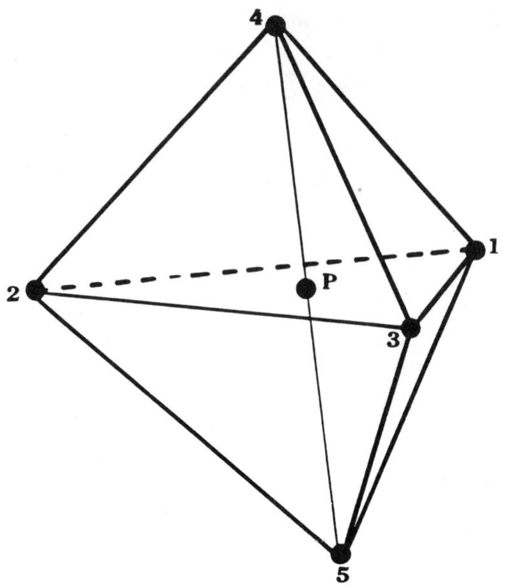

Figure 4. Trivariate smoothness conditions.

Remarks.

(1) For all of the above configurations, the dimension of $S_2^{1,3}$ is 33, which equals the lower bound given by Schumaker [11] and Alfeld [1].

(2) Geometric degeneracies occur for C^1 *quadratics* as well as for C^2 functions of all investigated polynomial degrees through $m = 5$.

(3) Although there is a tendency for the dimension to increase with the number of Clough–Tocher vertices, no simple and unambiguous relation emerges.

(4) Some configurations could not be analyzed for $m = 5$ with current computing resources. This is because the complexity of the linear system increases not just with the number of equations and unknowns, but also with the complexity of the coefficients, which in all cases considered were rational numbers. Since the calculations were done in exact arithmetic the sizes of the numerators and denominators increased in some cases beyond the available computer resources.

3. The Trivariate Case. Lower bounds on the dimensions of $\dim(S_3^{k,m})$, $k \in \{1, 2\}$, $m \geq 2k + 1$, are given in Alfeld [1]. As a natural generalization of the bivariate case, we emphasize configurations where appropriate quadruples of points are coplanar.

3.1. The smoothness conditions. We proceed similarly as in the bivariate case. Thus we consider two tetrahedra sharing a face with vertices V_1, V_2 and V_3, and consider the cross derivative in the direction of $V_5 - V_4$ (cf. Fig. 4). The location of a point x is expressed as

$$x = \sum_{i=1}^{5} b_i V_i$$

and the continuous piecewise polynomial function of degree m is given by

$$p(x) = \sum_{h+i+j+k+l=m} \frac{m!}{h!\,i!\,j!\,k!\,l!}\, c_{hijkl}\, b_1^h b_2^i b_3^j b_4^k b_5^l.$$

Let P denote the intersection of the line segment $V_5 - V_4$ and the common face and write

$$P = \sum_{i=1}^{3} \alpha_i V_i = \sigma_1 V_4 + \sigma_2 V_5 \quad \text{where} \quad \sum_{i=1}^{3} \alpha_i = \sigma_1 + \sigma_2 = 1.$$

As before we define

$$\gamma := \frac{\sigma_1}{\sigma_2} = \frac{\|V_5 - P\|}{\|V_4 - P\|} \neq 0.$$

Then we obtain the first order conditions

$$(5) \quad 0 = \left(2 + \gamma + \frac{1}{\gamma}\right)\left(\alpha_1 c_{i+1,jk00} + \alpha_2 c_{i,j+1,k00} + \alpha_3 c_{ij,k+1,00}\right) - (1+\gamma)c_{ijk10} - \left(1 + \frac{1}{\gamma}\right)c_{ijk01}$$

(for all i, j, k such that $i+j+k = m-1$ and $i, j, k \geq 0$) and the second order conditions

$$0 = \left[(1+\gamma)^2 - \left(1 + \frac{1}{\gamma}\right)^2\right]\left(\alpha_1^2 c_{i+2,jk00} + 2\alpha_1\alpha_2 c_{i+1,j+1,k00} + 2\alpha_1\alpha_3 c_{i+1,j,k+1,00}\right.$$

$$+ \left.\alpha_2^2 c_{i,j+2,k00} + 2\alpha_2\alpha_3 c_{i,j+1,k+1,00} + \alpha_3^2 c_{ij,k+2,00}\right)$$

$$(6) \qquad + (\gamma+1)^2\left(c_{ijk20} - 2\alpha_1 c_{i+1,jk10} - 2\alpha_2 c_{i,j+1,k10} - 2\alpha_3 c_{ij,k+1,10}\right)$$

$$- \left(1 + \frac{1}{\gamma}\right)^2\left(c_{ijk02} - 2\alpha_1 c_{i+1,jk01} - 2\alpha_2 c_{i,j+1,k01} - 2\alpha_3 c_{ij,k+1,01}\right)$$

(for all i, j, k such that $i+j+k = m-2$ and $i, j, k \geq 0$).

Similarly as in the previous section we obtain:

THEOREM 2. *The smoothness conditions (5) and (6) are invariant under affine transformations.*

COROLLARY. *The dimension of $S_3^{k,m}$ is affinely invariant.*

3.2. The Morgan/Scott example. The tessellation given in Fig. 5 is a natural generalization of the Morgan/Scott split to tetrahedralizations. Thus we consider the tessellation of a tetrahedron with vertices V_1, V_2, V_3, V_4 by the 15 tetrahedra $\{5,6,7,8\}$, $\{5,6,7,4\}$, $\{5,6,3,8\}$, $\{5,2,7,8\}$, $\{1,6,7,8\}$, $\{5,6,3,4\}$, $\{5,7,2,4\}$, $\{5,8,2,3\}$, $\{6,7,1,4\}$, $\{6,8,1,3\}$, $\{7,8,1,2\}$, $\{1,2,3,8\}$, $\{1,2,7,4\}$, $\{1,6,3,4\}$, and $\{5,2,3,4\}$. As in the bivariate case, there is one inner simplex—namely $\{5,6,7,8\}$—connected to the boundary of the outer simplex by other simplices. The sequence of tetrahedra given here might be used to build the tessellation, starting at the inner tetrahedron and working outwards toward the boundary.

We consider the following configurations:

(1) (Fully symmetric.) $V_5 = (V_1 + 2V_2 + 2V_3 + 2V_4)/7$, $V_6 = (2V_1 + V_2 + 2V_3 + 2V_4)/7$, $V_7 = (2V_1 + 2V_2 + V_3 + 2V_4)/7$ and $V_8 = (2V_1 + 2V_2 + 2V_3 + V_4)/7$.

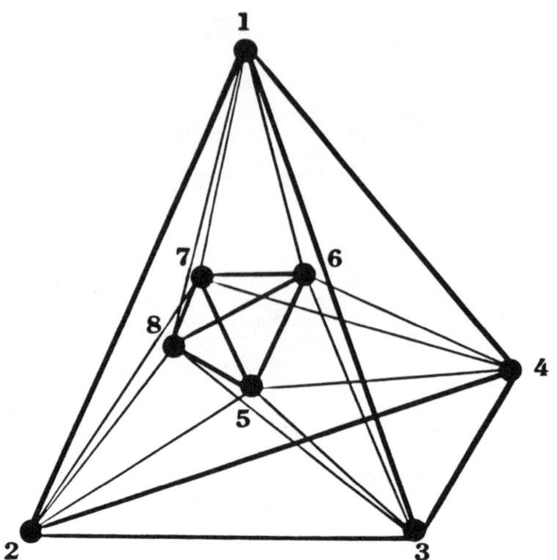

Figure 5.   The trivariate Morgan/Scott example.

(2)   (Generic, i.e., no recognizable artifacts.) $V_5 = (11V_1 + 20V_2 + 20V_3 + 20V_4)/71$, $V_6 = (21V_1 + 10V_2 + 20V_3 + 20V_4)/71$, $V_7 = (19V_1 + 22V_2 + 10V_3 + 20V_4)/71$ and $V_8 = (20V_1 + 18V_2 + 23V_3 + 10V_4)/71$.

(3)   (The vertex V_8 has been pulled towards V_4.) $V_5 = (V_1 + 2V_2 + 2V_3 + 2V_4)/7$, $V_6 = (2V_1 + V_2 + 2V_3 + 2V_4)/7$, $V_7 = (2V_1 + 2V_2 + V_3 + 2V_4)/7$ and $V_8 = (20V_1 + 20V_2 + 20V_3 + 11V_4)/71$.

(4)   (Similar to 3; additionally V_7 has been pulled towards V_3.) $V_5 = (V_1 + 2V_2 + 2V_3 + 2V_4)/7$, $V_6 = (2V_1 + V_2 + 2V_3 + 2V_4)/7$, $V_7 = (20V_1 + 20V_2 + 11V_3 + 20V_4)/71$ and $V_8 = (20V_1 + 20V_2 + 20V_3 + 11V_4)/71$.

(5)   (The vertex V_8 is completely asymmetric.) $V_5 = (V_1 + 2V_2 + 2V_3 + 2V_4)/7$, $V_6 = (2V_1 + V_2 + 2V_3 + 2V_4)/7$, $V_7 = (2V_1 + 2V_2 + V_3 + 2V_4)/7$ and $V_8 = (21V_1 + 20V_2 + 20V_3 + 11V_4)/72$.

(6)   (The quadruple $\{4, 6, 7, 8\}$ is coplanar.) $V_5 = (V_1 + 2V_2 + 2V_3 + 2V_4)/7$, $V_6 = (2V_1 + V_2 + 2V_3 + 2V_4)/7$, $V_7 = (2V_1 + 2V_2 + V_3 + 2V_4)/7$ and $V_8 = V + (V - V_4)/4$ where $V = (V_6 + V_7)/2$.

(7)   (The edge e_{56} is singular, cf. Alfeld [1].) $V_5 = (V_1 + 2V_2 + 2V_3 + 2V_4)/7$, $V_6 = (2V_1 + V_2 + 2V_3 + 2V_4)/7$, $V_7 = V + (V - V_3)/4$ and $V_8 = V + (V - V_4)/4$ where $V = (V_5 + V_6)/2$.

(8)   (The two quadruples $\{3, 5, 6, 7\}$ and $\{4, 6, 7, 8\}$ are each coplanar.) $V_5 = (V_1 + 2V_2 + 2V_3 + 2V_4)/7$, $V_6 = (2V_1 + V_2 + 2V_3 + 2V_4)/7$, $V_7 = V + (V - V_3)/4$ and $V_8 = V + (V - V_4)/4$ where $V = (V_6 + V_7)/2$.

(9)   (The three quadruples $\{1, 5, 6, 8\}$, $\{2, 5, 6, 7\}$ and $\{3, 6, 7, 8\}$ are each coplanar.) $V_5 = (V_1 + 2V_2 + 2V_3 + 2V_4)/7$, $V_6 = (75V_1 + 24V_2 + 45V_3 + 80V_4)/224$, $V_7 = (11V_1 + 8V_2 + V_3 + 8V_4)/28$ and $V_8 = (47V_1 + 56V_2 + 41V_3 + 16V_4)/160$.

(10)   (The four quadruples $\{1, 5, 6, 8\}$, $\{2, 5, 6, 7\}$, $\{3, 6, 7, 8\}$ and $\{4, 5, 7, 8\}$ are each coplanar.) $V_5 = (7V_1 + 20V_2 + 25V_3 + 20V_4)/72$, $V_6 = (20V_1 + 7V_2 + 20V_3 + 25V_4)/72$, $V_7 = (25V_1 + 20V_2 + 7V_3 + 20V_4)/72$ and $V_8 = (20V_1 + 25V_2 + 20V_3 + 7V_4)/72$.

Table 4. Dimensions for the trivariate Morgan/Scott example.

Configuration:	1	2	3	4	5	6	7	8	9	10	lb
$S_2^{1,2}$	11	10	11	10	10	10	11	10	10	11	
$S_2^{1,3}$	25	20	25	25	21	21	24	20	20	24	20
$S_2^{1,4}$	54	43	52	47	47	47	53	44	46	54	43
$S_2^{1,5}$	113	*	*	*	*	*	*	*	*	*	104

Table 4 lists some dimensions obtained. The column headed "lb" gives the lower bound for a generic tetrahedralization, as derived in Alfeld [1].

Remarks.

(1)   The possible range of dimensions is much larger than in the corresponding bivariate example (cf. Table 2).

(2)   Geometric degeneracies are apparent for all investigated polynomial degrees through $m = 5$. This contrasts with the bivariate case (see the beginning of § 2).

(3)   Similarly as in the bivariate case, no clear correspondence between the dimension and the number of coplanar quadruples arises.

(4)   The C^2 case could not be analyzed with current computer resources.

Acknowledgments.   This research was supported by the Department of Energy under contract DEAC02-82-ER-12046 to the University of Utah, Salt Lake City, Utah. The author has benefited greatly from the stimulating environments provided by the Mathematics Computer Aided Geometric Design Group at the University of Utah. The computation of the dimensions for so many configurations would not have been possible without the excellent computing facilities at the University of Utah.

REFERENCES

[1]   P. ALFELD, *On the dimension of piecewise polynomial functions,* Proc. of the Biennial Dundee Conference on Numerical Analysis, June 25–28, 1985, Pitman Publishers, Boston, 1985.

[2]   ——, *A trivariate Clough–Tocher scheme for tetrahedral data,* Computer Aided Geometric Design J., 1 (1984), pp. 169–181.

[3]   W. BOEHM, G. FARIN AND J. KAHMANN, *A survey of curve and surface methods in* CAGD, Computer Aided Geometric Design J., 1 (1984), pp. 1–60.

[4]   C. DE BOOR, *B-form basics,* this Volume, 1987.

[5]   Y. S. CHOU, LO-YUNG SU AND R. H. WANG, *The dimensions of bivariate spline spaces over triangulations,* in Multivariate Approximation III, W. Schempp and K. Zeller, eds., Birkhäuser Verlag, Basel, 1985, pp. 71–83.

[6]   A. C. HEARN, REDUCE *User's Manual, Version* 3.0; The Rand Corporation, Santa Monica, CA, 1983.

[7]   R. H. J. GMELLIG MEYLING AND P. R. PFLUGER, *On the dimension of the spline space* $S_2^1(\Delta)$ *in special cases,* manuscript.

[8]   J. MORGAN AND R. SCOTT, *A nodal basis for* C^1 *piecewise polynomials of degree* $n \geq 5$, Math. Comput., 29 (1975), pp. 736–740.

[9]   ——, *The dimension of the space of* C^1 *piecewise polynomials,* manuscript, 1975.

[10]   L. L. SCHUMAKER, *On the dimension of piecewise polynomials in two variables,* in Multivariate Approximation, W. Schempp and K. Zeller, eds., Birkhäuser Verlag, Basel, 1979, pp. 251–264.

[11]   ——, *Bounds on the dimension of spaces of multivariate piecewise polynomials,* Rocky Mountain J. Math., 14 (1984), pp. 251–264.

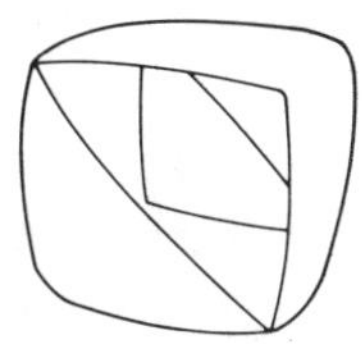

Part III: Shape

Many curve and surface algorithms generate surfaces whose shape is an artifact of the algorithm rather than a design parameter. Consequently, a designer might feel that the generated result is not what he or she had in mind. In order to overcome this problem, it is necessary to quantify the idea of shape enough for it to become programmable.

Some of the most important efforts in that direction deal with the concept of visual continuity, *also called* geometric continuity. *This has become a field of very active research in CAGD, as the number of papers concerned with that topic indicates. The concept of visual continuity is a generalization of strictly parametric continuity. While the latter concept is meaningful to a mathematician, a designer would be more interested in the smoothness of a curve or surface that can actually be seen (hence the term "visual"). Thus, visual continuity employs geometric terms like curvature continuity, osculating plane continuity, etc. As an important example, V^1 continuity (short for first-order visual continuity) can be achieved for curves and surfaces that are not even differentiable. The article by G. Herron contains many results about visual continuity for curves and surfaces.*

Historically, both CAGD pioneers S. Coons and P. Bézier have considered V^1 surfaces, yet the given formulas were more of a descriptive than a constructive nature. In 1974 G. Nielson presented the concept of v-splines—C^1 cubic splines whose shape could be altered by means of so-called tension parameters. The scheme had an interesting (actually unintended) property: although the curves it generated were not C^2,

Nielson showed that they were curvature-continuous. In 1981 B. Barsky presented a class of V^2 cubic B-spline-like curves, so-called β-splines, but it has taken several years to remove some inherent shortcomings of this curve scheme. W. Boehm used a completely different, geometric approach to the same problem, which led to an elegant curve scheme; it is described by Boehm in this section.

Another class of curves that have a long history are so-called Wilson–Fowler splines. There is a renewed interest in these splines because they are V^2 spline curves that might help us to gain more insight into v-spline interpolation, namely into the problem of how to determine the tension parameters automatically.

As an indication of what areas are currently of importance in CAGD, three papers address the same problem: given point data in $\mathbf{R}^3$ that have been triangulated so as to form a piecewise linear surface, find a smooth (V^1) surface that interpolates to all data points. It is of importance here that the triangulation may be closed in order to yield closed surfaces. Because using rectangular patches will result in degenerate parameterizations, all three methods employ triangular patches. The first two, by B. Piper and T. Jensen, utilize triangular Bernstein–Bézier patches, and, although both methods look very similar, they differ considerably in their theoretical approaches. The third, by G. Nielson, uses a rational scheme that has the flavor of a discretized Coons patch.

A problem that is more complicated than "simple" interpolation is that of constrained interpolation: here one requires the interpolant to reproduce certain shape properties of the given data, convexity being the most commonly encountered. Such methods are usually called "shape-preserving." The papers by K. Bosworth and A. Jones address the problem of shape preservation; while Bosworth outlines a very general algorithm, Jones gives a more detailed description of an algorithm that optimizes the shape of B-spline curves and surfaces.

As an example of how shape considerations can influence the design of an algorithm, R. Barnhill, B. Piper and K. Rescorla describe two methods to interpolate to data given on a surface.

While the above methods generate surfaces, the article by F. Munchmeyer addresses the problem of "surface interrogation": Once a surface is constructed, how can we decide if it has acceptable shape characteristics? Munchmeyer considers the examination of various "curvature plots" of a surface as an aid in the interrogation problem.

The study of interrogation methods should eventually lead to curve and surface schemes that will guarantee that a designer's idea of "nice shape" is automatically satisfied. Presently, we are only trying to quantify those ideas.

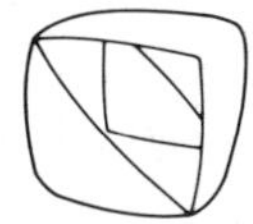

Techniques for Visual Continuity

GARY HERRON

Abstract. In CAGD, surfaces are usually formed from collections of patches, making joining conditions between adjacent patches an important consideration. Because the usual insistence of C^1 or C^2 continuity is too strict in many cases, we often relax continuity constraints to "Visual C^1" and "Visual C^2" continuity, called VC^1 and VC^2, respectively. This paper defines VC^1 and VC^2 continuity as the preservation of certain geometric quantities across the boundary. It is shown that these definitions imply that either patch can be reparameterized to give a true C^1 or C^2 join, and this suggests a viable generalization to VC^n continuity. Following this is a survey of methods investigators have used to achieve VC^n continuity for curves and surfaces.

1. Introduction. When a complex surface is being designed using methods from CAGD, it is usual to build the surface from separate patches, and to require that the patches meet smoothly. The usual definitions of smoothness involve C^1 or C^2 continuity, but these force two joining surfaces not only to be smooth across the join, but also force the parameterization across the join to be C^1 or C^2. This normally serves no useful purpose, since in a final manufactured surface, the parameterization is no longer evident.

In order to alleviate the unnecessary restrictions of strict C^1 or C^2 continuity, investigators have defined VC^1 and VC^2 continuity which attempts to quantify the smoothness of the join between two surfaces by looking at the *shape* of the surfaces without regard to their parameterization. This is done by defining geometric quantities associated with a surface which are independent of parameterization and which quantify the shape of a surface. Two surfaces are then considered visually continuous if they share these shape indicators.

It would appear that visual continuity is a relaxation of strict continuity, in the sense that a visually continuous surface will not, in general, be truly continuous. However, a major point of this paper is that with a visually continuous join between two surfaces, a reparameterization of one of the surfaces will result in a truly continuous join. Thus we will see that visual continuity allows a relaxation in the parameterization of joining surfaces, and not in their shape.

In this paper, we introduce, in four separate sections, VC^1 continuity for curves, VC^2 continuity for curves, VC^1 continuity for surfaces, and VC^2 continuity for surfaces. Each of these four sections will have the same organization: we will (1) introduce the needed geometric quantities, (2) prove that the quantities are independent of the parameterization, (3) define the type of VC^n continuity under consideration and (4) prove that if two curves/surfaces are VC^n, then one can be reparameterized to yield a true C^n join. Steps (1) and (2) are standard results from differential geometry. For the most part, the derivations given here are taken from Faux and Pratt [7]. Similar derivations can also be found in Bézier [3], and more thorough treatments can be found in standard differential geometry texts such as Guggenheimer [9].

Following this we will introduce a possible generalization to higher order VC^n continuity, followed by a review of the methods of achieving visual continuity to be found in the literature.

The following notation will be used. P will be a function of t for a curve, or u and v for a surface. The derivatives of a curve will be represented by P' and P'', and the partial derivatives of a surface will be indicated by P_u, P_v, P_{uu}, P_{uv}, and P_{vv}. Various vector functions T and N, and real functions s and κ, will be derived from P. In most cases, the variables over which these functions are defined will not be written. $\bar{P}$ will be an adjacent curve or surface at some point, and the join between the two will be in question. When we wish to reparameterize P we will write $\tilde{P}(t) = P(r(t))$, or $\tilde{P}(u, v) = P(\bar{u}(u, v), \bar{v}(u, v))$. In each case $\bar{P}$ and $\tilde{P}$ will have their own set of derived functions $\bar{s}$, $\tilde{s}$, and so on.

In order to avoid pathological cases, we will assume that all curves and surfaces dealt with in this paper are piecewise smooth. We will also assume that all derivatives do not degenerate to zero, and that P_u and P_v are not collinear, and that $P_{uv} = P_{vu}$.

2. Visual Continuity for Curves. Let $P(t)$ be a curve from $\mathbb{R}^1$ into $\mathbb{R}^3$. The quantity used to embody VC^1 continuity will be the unit tangent vector at a point on the curve, and the quantities used for VC^2 continuity will be the curve normal and its curvature at a point. In geometric terms, the center of the osculating circle will be in the direction of the curve normal, and its radius will be the inverse of the curvature.

2.1. VC^1 for curves. Let $s = |P'|$, be a measure of the arc length (to be precise, s is the derivative of arc length), and define the unit tangent vector T to be $T = P'/s$. Rewriting this yields

$$(1) \qquad\qquad P' = sT,$$

an identity of fundamental importance.

THEOREM 2.1.   *The unit tangent vector T of P is independent of parameterization up to orientation.*

Proof.   If $\tilde{P} = P(r(t))$ is a reparameterization, then $\tilde{s} = |r'| |P'|$, and $\tilde{P}' = r'P'$, so $\tilde{T} = T$.   □

DEFINITION 2.1.   Two curves P and $\bar{P}$ are said to be VC^1 if their unit tangent vectors T and $\bar{T}$ are equal.

THEOREM 2.2. *If P and $\bar{P}$ are two curves that join in a VC^1 manner at $P(t_0) = \bar{P}(\bar{t}_0)$, then P can be reparameterized by $\tilde{P}(t) = P(r(t))$ so that $\tilde{P}$ and $\bar{P}$ form a true C^1 join at $\bar{P}(t_0) = \bar{P}(\bar{t}_0)$.*

Proof. Choose $r(t)$ to be a linear polynomial which satisfies $r(t_0) = t_0$, and $r'(t_0) = \bar{s}(\bar{t}_0)/s(t_0)$. Then using (1) twice gives $\tilde{P}'(t_0) = P'(r(t_0))r'(t_0) = r'(t_0)s(t_0)T(t_0) = \bar{s}(\bar{t}_0)\bar{T}(\bar{t}_0) = \bar{P}'(\bar{t}_0)$. $\quad\square$

2.2. VC^2 for curves. Continuing with P, and its s and T functions from the last section, consider T'. Choose N, the principal *normal vector*, and κ the *curvature*, so that N has unit length, $\kappa \geq 0$, and $T' = s\kappa N$. Since T is a vector function with unit length, T', and therefore N, are perpendicular to T. Differentiating (1) yields this second fundamental equation

$$(2) \qquad\qquad P'' = sT' + s'T = s^2\kappa N + s'T.$$

It can be shown that $1/\kappa$ is the radius of the osculating circle, and its center is in the direction of N.

THEOREM 2.3. *The normal direction N and the curvature κ of P are independent of parameterization.*

Proof. This is clear since T is independent of parameterization, and both N and κ are dependent solely on T. $\quad\square$

DEFINITION 2.2. Two curves P and $\bar{P}$ are said to be VC^2 if they share tangent vectors, normal vectors and curvatures.

THEOREM 2.4. *If P and $\bar{P}$ are two curves that join in a VC^2 manner at $P(t_0) = \bar{P}(\bar{t}_0)$, then P can be reparameterized by $\tilde{P}(t) = P(r(t))$ so that $\tilde{P}$ and $\bar{P}$ form a true C^2 join at their common point.*

Proof. Choose $r(t)$ to be a quadratic polynomial which satisfies $r(t_0) = t_0$, $r'(t_0) = \bar{s}(\bar{t}_0)/s(t_0)$, and $r''(t_0) = (\bar{s}'(\bar{t}_0)s(t_0) - \bar{s}(\bar{t}_0)s'(t_0))/s^2(t_0)$. Then $\tilde{P}'(t_0) = \bar{P}'(\bar{t}_0)$ according to Theorem 2.2, and

$$\begin{aligned}
\tilde{P}''(t_0) &= P''(r(t_0))r'^2(t_0) + P'(r(t_0))r''(t_0) \\
&= (s^2(t_0)r'^2(t_0)\kappa)N + (s'(t_0)r'^2(t_0) + s(t_0)r''(t_0))T(t_0) \\
&= (\bar{s}^2(\bar{t}_0)\kappa)N + \bar{s}'(\bar{t}_0)T(t_0) \\
&= \bar{P}''(\bar{t}_0). \qquad\qquad\qquad\qquad\qquad\qquad\square
\end{aligned}$$

3. Visual Continuity for Surfaces. Let $P(u, v)$ be a surface from $\mathbb{R}^2$ into $\mathbb{R}^3$. The quantity to be defined for VC^1 continuity is the unit surface normal N, and the quantities for VC^2 continuity are the minimum and maximum curvature κ_1 and κ_2, and the directions on the surface in which these curvatures are attained. The curvature of a surface at some point in some direction is calculated to be the curvature of a curve going through that point in the specified direction in such a way that the curve's normal equals the surface normal N.

3.1. VC^1 for surfaces. The surface normal is

$$N = \pm \frac{P_u \times P_v}{|P_u \times P_v|}.$$

THEOREM 3.1.   *The surface normal is independent of parameterization.*
Proof.   Let $\tilde{P}(u, v) = P(\tilde{u}(u, v), \tilde{v}(u, v))$. Then

$$\tilde{P}_u = \tilde{u}_u P_u(\tilde{u}(u, v), \tilde{v}(u, v)) + \tilde{v}_u P_v(\tilde{u}(u, v), \tilde{v}(u, v)),$$

$$\tilde{P}_v = \tilde{u}_v P_u(\tilde{u}(u, v), \tilde{v}(u, v)) + \tilde{v}_v P_v(\tilde{u}(u, v), \tilde{v}(u, v)).$$

Since $\tilde{P}_u$ and $\tilde{P}_v$ are linear combinations of P_u and P_v, the corresponding cross-products will be collinear.   □

DEFINITION 3.1.   Two surfaces P and $\tilde{P}$ are said to be VC^1 if their surface normals N and $\tilde{N}$ are equal.

THEOREM 3.2.   *If P and $\tilde{P}$ are VC^1 along a line, then P can be reparameterized to $\tilde{P}$ so that $\tilde{P}$ and $\tilde{P}$ form a true C^1 join along that line.*

Proof.   Firstly, assume that the shared line between the two surfaces is $(0, v)$, $0 \le v \le 1$. We can do this without loss of generality with simple linear reparameterizations of each surface, which by a previous theorem does not affect the VC^1 property of their join. Thus we have

$$(3) \qquad \tilde{P}(0, v) = P(0, v) \quad \text{and} \quad \tilde{P}_v(0, v) = P_v(0, v).$$

Choose functions of v, $r_1(v)$ and $r_2(v)$, so that

$$(4) \qquad \tilde{P}_u(0, v) = r_1(v)P_u(0, v) + r_2(v)P_v(0, v).$$

This can be done since both surfaces have identical tangent planes for all v. With this, we form the reparameterization

$$\tilde{P}(u, v) = P(\tilde{u}(u, v), \tilde{v}(u, v))$$

where

$$\tilde{u}(u, v) = ur_1(v) \quad \text{and} \quad \tilde{v}(u, v) = v + ur_2(v).$$

To show C^1 continuity we compute, for the u-partial,

$$\tilde{P}_u(0, v) = r_1(v)P_u(0, v) + r_2(v)P_v(0, v) = \tilde{P}_u(0, v)$$

by first applying the chain rule to $\tilde{P}$, and then (4). For the v-partial, we get from (3)

$$\tilde{P}_v(0, v) = P_v(0, v) = \tilde{P}_v(0, v). \qquad \qquad □$$

3.2. VC^2 for surfaces.   Let N be the surface normal, and

$$A = \begin{pmatrix} P_u \\ P_v \end{pmatrix}.$$

The *first fundamental matrix* of the surface P is

$$G = AA^T = \begin{pmatrix} P_u \cdot P_u & P_u \cdot P_v \\ P_v \cdot P_u & P_v \cdot P_v \end{pmatrix}.$$

The *second fundamental matrix* of the surface P is

$$D = \begin{pmatrix} N \cdot P_{uu} & N \cdot P_{uv} \\ N \cdot P_{vu} & N \cdot P_{vv} \end{pmatrix}.$$

Both G and D are symmetric matrices. The normal curvature of the surface P with respect to direction $c = (u, v)$ is

$$\kappa(c) = \frac{cDc^T}{cGc^T}.$$

The directions c_1 and c_2 which produce the minimum and maximum normal curvatures $\kappa(c_1)$, and $\kappa(c_2)$ can be found as follows: because of the symmetry of D, we compute $\nabla(cDc^T) = 2cD$. The minimum and maximum occur when

$$\frac{\partial \kappa}{\partial c} = \frac{2(cD)(cGc^T) - 2(cDc^T)(cG)}{(cGc^T)^2} = (0, 0).$$

This reduces to

(5)
$$c(D - \kappa G) = (0, 0),$$

which is satisfied whenever $|D - \kappa G| = 0$. The two solutions κ_1 and κ_2 to this quadratic function of κ are called the *principal curvatures,* and if c_1 and c_2 are their corresponding solutions to (5), then the directions of the vectors $c_1 A$ and $c_2 A$ are called the *principal directions*. These principal directions are the directions on the surface in which the minimum and maximum curvature occur. It can be shown that the principal directions are perpendicular when $\kappa_1 \neq \kappa_2$.

THEOREM 3.3.   *The principal normal curvatures and their principal directions are independent of parameterization.*

Proof.   Let $\tilde{P}(u, v) = P(\bar{u}(u, v), \bar{v}(u, v))$ be   any   reparameterization   of P. Applications of the chain rule of differentiation yield

$$\tilde{P}_u = \bar{u}_u P_u + \bar{v}_u P_v,$$
$$\tilde{P}_v = \bar{u}_v P_u + \bar{v}_v P_v,$$
$$\tilde{P}_{uu} = \bar{u}_u^2 P_{uu} + 2\bar{u}_u \bar{v}_u P_{uv} + \bar{v}_u^2 P_{vv} + 2\bar{u}_{uu} P_u + 2\bar{v}_{uu} P_v,$$
$$\tilde{P}_{uv} = \bar{u}_u \bar{u}_v P_{uu} + (\bar{u}_u \bar{v}_v + \bar{u}_v \bar{v}_u) P_{uv} + \bar{v}_u \bar{v}_v P_{vv} + \bar{u}_{uv} P_u + \bar{v}_{uv} P_v,$$
$$\tilde{P}_{vv} = \bar{u}_v^2 P_{uu} + 2\bar{u}_v \bar{v}_v P_{uv} + \bar{v}_v^2 P_{vv} + 2\bar{u}_{vv} P_u + 2\bar{v}_{vv} P_v.$$

If we define

$$R = \begin{pmatrix} \bar{u}_u & \bar{v}_u \\ \bar{u}_v & \bar{v}_v \end{pmatrix},$$

then we can calculate $\tilde{A} = RA$, $\tilde{G} = RGR^T$, and $\tilde{D} = RDR^T$, where $\tilde{A}$ and A are the partial derivative matrices, $\tilde{G}$ and G are the first fundamental matrices, and $\tilde{D}$ and D are the second fundamental matrices of $\tilde{P}$ and P, respectively. Thus (5) becomes

$$\tilde{c}(\tilde{D} - \kappa \tilde{G}) = \tilde{c} R(D - \kappa G) = (0, 0)$$

which has the same solutions for κ as $c(D - \kappa G) = (0, 0)$, and if $\tilde{c}_i$ is a solution, its principal direction $\tilde{c}_i \tilde{A} = \tilde{c}_i RA$ will yield $\tilde{c}_i R$ as a solution to the other. $\square$

DEFINITION 3.2.   Two surfaces P and $\tilde{P}$ are said to be VC^2 continuous at a common point if they share the same surface normal, share the same principal curvatures κ_1, and κ_2, and in the case that $\kappa_1 \neq \kappa_2$, share the same principal directions.

We produce several necessary results before proving that VC^2 surfaces can be reparameterized to C^2 continuity. The first result shows how the first fundamental matrix and second fundamental matrix are related in VC^2 surfaces.

THEOREM 3.4.   *Let two surfaces P and $\bar{P}$ be VC^2. If G and $\bar{G}$ and D and $\bar{D}$ are the first and second fundamental matrices, respectively, of P and $\bar{P}$, then there exists a 2×2 matrix R such that $\bar{G} = RGR^T$ and $\bar{D} = RDR^T$.*

Proof.   Let

$$A = \begin{pmatrix} P_u \\ P_v \end{pmatrix} \quad \text{and} \quad \bar{A} = \begin{pmatrix} \bar{P}_u \\ \bar{P}_v \end{pmatrix}.$$

Then

$$G = AA^T \quad \text{and} \quad \bar{G} = \bar{A}\bar{A}^T.$$

Since both surfaces are nondegenerate, and share the same normal, we can find a 2×2 matrix R, such that $\bar{A} = RA$. Then

$$\bar{G} = \bar{A}\bar{A}^T = RAA^TR^T = RGR^T. \qquad \square$$

To show that $\bar{D} = RDR^T$, we divide our reasoning into two cases.

Case 1 $(\kappa_1 \neq \kappa_2)$.   Let the shared principal directions be $\bar{c}_1\bar{A} = c_1A$, and $\bar{c}_2\bar{A} = c_2A$. Since $c_1A = \bar{c}_1\bar{A} = \bar{c}_1RA$ and the rows of A are linearly independent, we conclude $c_1 = \bar{c}_1R$. From this and several applications of (5), we can compute

$$c_1(D - \kappa_1G) = 0 \;\Rightarrow\; \bar{c}_1RDR^T = \kappa_1\bar{c}_1\bar{G},$$

and

$$\bar{c}_1(\bar{D} - \kappa_1\bar{G}) = 0 \;\Rightarrow\; \bar{c}_1\bar{D} = \kappa_1\bar{c}_1\bar{G}$$

and conclude

$$\bar{c}_1RDR^T = \bar{c}_1\bar{D}.$$

Similarly, we conclude

$$\bar{c}_2RDR^T = \bar{c}_2\bar{D}.$$

Finally, since $\bar{c}_1$ and $\bar{c}_2$ are perpendicular for this case, we conclude from these last two identities that

$$RDR^T = \bar{D}.$$

Case 2 $(\kappa_1 = \kappa_2)$.   Since κ_1 and κ_2 are the min and max of $\kappa(c)$, we compute

$$c(D - \kappa_1G) = c(D - \kappa(c)G) = 0$$

for all c. Thus $D - \kappa_1G = 0$, and similarly $\bar{D} - \kappa_1\bar{G} = 0$. Multiplying the first of these by R and R^T on the left and right, respectively, and combining with the second gives $RDR^T = \bar{D}$. $\square$

LEMMA 3.1.   *If P_u and P_v are the u-partial and v-partial of a surface P at a point, and N is the surface normal at that point, then N, P_u and P_v form a basis for $\mathbb{R}^3$, and furthermore, if V is written as a linear combination of N, P_u and P_v, then the coefficient of N is $N \cdot V$.*

Proof.   Let $V = \alpha N + \beta P_u + \gamma P_v$. Then by the definition of N, $N \cdot P_u = N \cdot P_v = 0$, and $N \cdot N = 1$, so we have $N \cdot V = \alpha$. $\square$

THEOREM 3.5.   *If P and $\bar{P}$ are VC^2 along a line, then P can be reparameterized to $\bar{P}$ so that $\tilde{P}$ and $\bar{P}$ form a true C^2 join along that line.*

Proof.   Firstly, assume that the shared line between the two surfaces is $(0, v)$, $0 \le v \le 1$. We can do this without loss of generality with simple linear reparameterizations of each surface, which by a previous theorem does not affect the VC^2 property of their join. Thus we have

$$(6) \qquad \bar{P}(0, v) = P(0, v), \quad \bar{P}_v(0, v) = P_v(0, v) \quad \text{and} \quad \bar{P}_{vv}(0, v) = P_{vv}(0, v).$$

Choose functions of v, $r_1(v)$, $r_2(v)$, $r_3(v)$, and $r_4(v)$, so that

$$(7) \qquad \begin{aligned} \bar{P}_u(0, v) &= r_1(v)P_u(0, v) + r_2(v)P_v(0, v), \\ \bar{P}_{uu}(0, v) &= (N \cdot \bar{P}_{uu}(0, v))N + 2r_3(v)P_u(0, v) + 2r_4(v)P_v(0, v). \end{aligned}$$

The first can be done since both surfaces have identical tangent planes for all v, and the second can be done since N, P_u, and P_v form a basis, and the previous lemma gives us the coefficient of N.

With this, we form the reparameterization

$$\tilde{P}(u, v) = P(\bar{u}(u, v), \tilde{v}(u, v))$$

where

$$\bar{u}(u, v) = ur_1(v) + \tfrac{1}{2}u^2 r_2(v) \quad \text{and} \quad \tilde{v}(u, v) = v + ur_3(v) + \tfrac{1}{2}u^2 r_4(v).$$

To show C^1 continuity we compute, for the u-partial, from the chain rule on $\tilde{P}$ and from (7)

$$\tilde{P}_u(0, v) = r_1(v)P_u(0, v) + r_2(v)P_v(0, v) = \bar{P}_u(0, v),$$

and for the v-partial, we compute from (6)

$$\tilde{P}_v(0, v) = P_v(0, v) = \bar{P}_v(0, v).$$

To show C^2 continuity we compute

$$\begin{aligned} \tilde{P}_{uu}(0, v) &= r_1^2(v)P_{uu}(0, v) + 2r_1(v)r_2(v)P_{uv}(0, v) \\ &\quad + r_2^2(v)P_{vv}(0, v) + 2r_3(v)P_u(0, v) + 2r_4(v)P_v(0, v) \\ &= (N \cdot \bar{P}_{uu})N + 2r_3(v)P_u(0, v) + 2r_4(v)P_v(0, v) \\ &= \bar{P}_{uu} \end{aligned}$$

by first applying the chain rule, then the lemma, and then one of the above identities. Also from $\tilde{P}_u(0, v) = \bar{P}_u(0, v)$ we compute $\tilde{P}_{uv}(0, v) = \bar{P}_{uv}(0, v)$ and from $\tilde{P}_v(0, v) = \bar{P}_v(0, v)$ we compute $\tilde{P}_{vv}(0, v) = \bar{P}_{vv}(0, v)$. $\square$

4. Higher Order Visual Continuity. We would like to define VC^n continuity in a similar manner by finding geometric quantities which are independent of parameterization and embody the information of the nth order derivatives. Such quantities are not known to exist in forms simple enough to work with. However, the equivalence of VC^1 and VC^2 continuity to an appropriate reparameterization makes the following definition of VC^n continuity reasonable and expected.

DEFINITION 4.1.   Two surfaces are said to be VC^n continuous if one can be reparameterized to yield a true C^n join between the two.

By the results of the previous sections, this definition incorporates the definitions of VC^1 and VC^2 continuity as special cases. See Barsky and DeRose [1], [2], and DeRose [5] for further discussion of this definition, and Veron et al. [19] for a similar definition. It is admitted that this definition, while being quite reasonable, is not very practical in terms of building interpolants with VC^n behavior.

5.   A Survey of Methods of Achieving Visual Continuity.   In this section we present a survey of the many methods of achieving visual continuity of various orders for curves in the first subsection and surfaces in the second.

5.1.   Visual continuity for curves.   Many investigators (Faux and Pratt [7], Bézier [3], Fowler and Wilson [8], Manning [14], Nielson [15] and Sabin [18]) have defined VC^1 continuity by requiring unit tangents to be equal. This is usually achieved by requiring that the curve defining data satisfy a constraint which results in collinear tangent vectors.

In the case of Nielson [15], his v-splines automatically preserve unit tangent and curvature across knots thus achieving both VC^1 and VC^2 continuity.

Barsky and DeRose [1], [2], and DeRose [5] have achieved VC^n continuity for curves through a direct manipulation of the reparameterization discussed throughout this paper. For instance, if $r(t)$ is the reparameterization function which makes two curves $P(t)$ and $\bar{P}(t)$ meet in a true C^n manner, then successive applications of the chain rule of differentiation yield

$$
\begin{aligned}
P(t) &= \bar{P}(r(t)), \\
P'(t) &= r'(t)\bar{P}'(r(t)), \\
P''(t) &= r''(t)\bar{P}'(r(t)) + r'^2(t)\bar{P}''(r(t)), \\
P'''(t) &= r'''(t)\bar{P}'(r(t)) + 3r'(t)r''(t)\bar{P}''(r(t)) + r'^3(t)\bar{P}'''(r(t)), \\
P^{(4)}(t) &= \cdots.
\end{aligned}
$$

If $\bar{P}$ and its derivatives are known, and if the values of r' and r'' are chosen as design parameters by a designer, the value of P and its derivatives can be computed from the above. In such a case, the two curves will meet in a VC^n manner because the function r provides the needed reparameterization to true C^n continuity.

In Boehm [4] it is shown that torsion continuity is a valid generalization to VC^3 continuity. He shows that torsion continuity is achieved when

$$
P'''(t) = \bar{P}'''(t) + \lambda \bar{P}''(t) + \mu \bar{P}'(t),
$$

and that this is a more general relation on the derivatives of P and $\bar{P}$ than the relation induced by reparameterization (see $P'''(t)$ in the preceding paragraph).

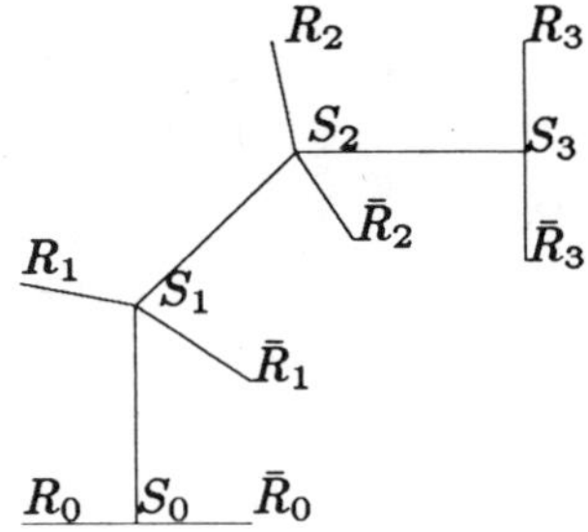

Figure 1.   Boundary and cross boundary data for two Bézier surfaces.

5.2.   Visual continuity for surfaces.   Farin [6] developed a set of constraints between two Bézier patches which must be satisfied in order to ensure VC^1 continuity. The patches may be either square, triangular, or one of each, because the cross-boundary derivative can be written in an identical form for either type of patch. Farin achieves VC^1 continuity by insisting that the partial derivatives of the two patches be related by three polynomials of v,

$$(8) \qquad \mu(v)P_u + \alpha(v)P_u + \lambda(v)P_v = 0.$$

To make calculations manageable, he restricts λ to be linear, which implies μ and α must be constant, and α may be chosen to be one. With these conditions, we can arrive at the following condition: if S_i are the Bézier defining points on the shared boundary, and R_i, and $\bar{R}_i$ are the interior quantities defining the cross boundary derivative, respectively, of P and $\bar{P}$ (see Fig. 1) then we can perform the following calculations. The compatibility condition that R_0, $\bar{R}_0$, S_0, and S_1 be coplanar allows us to write $\bar{R}_0$ as a linear combination of the other three as

$$\bar{R}_0 = \alpha_0 R_0 + \alpha_1 S_0 + \alpha_2 S_1.$$

Similarly, on the other end of the shared boundary, we can write

$$\bar{R}_n = \alpha_0 R_n + \alpha_3 S_{n-1} + \alpha_4 S_n,$$

and finally, the VC^1 condition can be written

$$\bar{R}_i = \frac{n-i}{n}(\alpha_0 R_i + \alpha_1 S_i + \alpha_2 S_{i+1}) + \frac{i}{n}(\alpha_0 R_i + \alpha_3 S_{i-1} + \alpha_4 S_i).$$

Sabin [18] derives an analogous but different set of conditions from the same assumptions on λ, μ, and α.

Piper [17] is able to derive constraints in a similar manner, but with fewer restrictions than found in Farin's scheme.

In another method, developed by Veron et al. [19], the surface designer specifies, as two design parameters, polynomials of v, $\mu(v)$, and $\lambda(v)$ which specify how the cross-boundary derivative of $\bar{P}$ is related to the two first derivatives of P. That is VC^1 continuity is achieved by specifying that

$$\bar{P}_u = \mu(v)P_u + \lambda(v)P_v.$$

If $\bar{P}(u, v)$ is a general polynomial of the form

$$\bar{P}(u, v) = (1 \quad u \quad \cdots \quad u^p)B(1 \quad v \quad \cdots \quad v^q)^T$$

then the C^0 condition and this VC^1 condition will determine the first two rows of the coefficient matrix B. For VC^2, (Veron et al. [19]) continuity is achieved by requiring that normal curvature in all directions be shared. This is achieved by requiring that the second derivatives satisfy

$$\bar{P}_{uv} = \mu P_{uv} + \lambda P_{vv},$$
$$\bar{P}_{uu} = \mu^2 P_{uv} + 2\lambda\mu P_{uv} + \lambda^2 P_{vv}.$$

These conditions determine the third row of B, and are shown to be equivalent to the VC^2 conditions of this paper.

Both Kahmann [13] and Veron et al. [19] show that sharing normal curvature in all directions is equivalent to requiring that the *Dupin indicatrix* of each patch agree along the boundary, and that this in turn is equivalent to the definition of VC^2 continuity of this paper. The Dupin indicatrix is a conic which characterizes the curvature of a surface at a point by incorporating the principal directions and curvatures. Kahmann [13] goes on to derive conditions, in a manner similar to Farin, on the Bézier points between two rectangular patches which achieves VC^2 continuity between two rectangular Bézier patches.

Herron [10] develops a cross boundary derivative function $R(t)$ which satisfies three properties: (1) R can be interpolated by P, (2) R depends only on the defining data of patch P, and (3) the direction of R is well determined enough to ensure that a similar process on an adjacent patch will form a VC^1 join. The original development was for triangular patches, but in Herron [10], this has been extended to patches over n-sided domains. If R_0 and R_1 are chosen perpendicular to a patch's tangential boundary derivative, and in the tangent plane of the patch at the two endpoints of the edge, the function

$$R(t) = [(1 - t)\,\|R_0\| + t\,\|R_1\|]\left[(1 - t)\frac{R_0}{\|R_0\|} + t\frac{R_1}{\|R_1\|}\right]$$

satisfies properties (2) and (3). A careful choice of the lengths of R_0 and R_1 allows a patch to interpolate $R(t)$ along the full boundary.

Jensen [12] achieves a similar effect by modifying the above cross-boundary derivative to the cubic,

$$R(t) = \left[(2t^2 - 3t + 1)\,\|R_0\| + \frac{1}{2}C(4t - t^2)(\|R_0\| + \|R_1\|) + (2t^2 - t)\,\|R_1\|\right]$$
$$\cdot\left[(1 - t)\frac{R_0}{\|R_0\|} + t\frac{R_1}{\|R_1\|}\right]$$

and interpolating it with a piecewise quartic polynomial patch. C is chosen by experiment to be about 1.2.

In Nielson [16] we see a similar method used to construct a surface normal along an edge of a triangular interpolant. If $T(u)$ is the tangential derivative along an edge and N_0 and N_1 are unit surface normals at the two endpoints, then

$$N(t) = \frac{T(u) \times [(1-t)N_0 \times T(0) + tN_1 \times T(1)]}{|T(u) \times [(1-t)N_0 \times T(0) + tN_1 \times T(1)]|}$$

can be used as a surface normal along the edge. This surface normal will agree with similarly defined normals along adjacent patches. Nielson goes on to define an interpolant to this unit surface normal.

Barsky and DeRose [1], [2] and DeRose [5] have achieved VC^n continuity by the following method. If two patches P and $\bar{P}$ are to join in a VC^n manner, apply the chain rule of differentiation to $P(u, v) = \bar{P}(\bar{u}(u, v), \bar{v}(u, v))$ n times. Let the derivatives of $\bar{u}$ and $\bar{v}$ be chosen by a designer in an attempt to relate the derivatives of the patches P and $\bar{P}$. Then VC^n continuity is assured because the reparameterization functions could be back calculated from their derivatives.

6. Conclusion. We have defined VC^1 and VC^2 continuity for both curves and surfaces, using standard quantities from differential geometry. These quantities (tangent, curvature, normal, principal normal curvature and principal directions) are shown to be independent of curve/surface parameterization. In each case, visual continuity is defined to be the preservation of these quantities across a boundary. The major thesis of this paper is that this preservation of geometric quantities implies that one curve/surface can be reparameterized to form a true C^1 or C^2 join with the other. Thus visual continuity is a relaxation of parameterization, and not a relaxation of "smoothness." This fact was used to extend the definition of VC^n continuity to $n > 2$, where geometric quantities are of little or no use.

Finally, we have demonstrated a number of methods from the literature to achieve visual continuity of various orders for both curves and surfaces.

REFERENCES

[1] B. A. BARSKY AND T. D. DEROSE, *Geometric continuity of parametric curves*, Report No. UCB/CSD84/205, 1984.

[2] ———, *An intuitive approach to geometric continuity for parametric curves and surfaces*, Graphics Interface '85, 1985.

[3] P. BÉZIER, *Numerical Control*, John Wiley, New York, 1970.

[4] W. BOEHM, *Smooth curves and surfaces*, this Volume, 1987.

[5] T. D. DEROSE, *Geometric continuity: A parameterization independent measure of continuity for computer aided geometric design*. Ph.D. thesis, Univ. California, Berkeley, CA.

[6] G. FARIN, *A construction for visual continuity of polynomial surface patches*, Computer Graphics and Image Processing, 1982.

[7] I. D. FAUX AND M. J. PRATT, *Computational Geometry for Design and Manufacture*, Ellis Horwood, 1979.

[8] A. H. FOWLER AND C. W. WILSON, *Cubic spline, a curve fitting routine*, Union Carbide Corporation Report Y-1400 (Rev. I), 1966.

[9] H. W. GUGGENHEIMER, *Differential Geometry*, McGraw-Hill, New York, 1963.

[10] G. J. HERRON, *Smooth closed surfaces with discrete triangular interpolants*, Computer Aided Geometric Design, 2 (1985), pp. 297–306.

[11] ———, *Visual continuity for multi-sided interpolants*, in preparation.

[12] T. JENSEN, *Assembling triangular and rectangular patches and multivariate splines*, this Volume, 1987.

[13] J. KAHMANN, *Continuity of curvature between adjacent Bézier patches*, in Surfaces in Computer Aided Geometric Design, R. E. Barnhill and W. Boehm, eds., North-Holland, Amsterdam, 1983.

[14] J. R. MANNING, *Continuity conditions for spline curves*, Computer J., 17 (1974).

[15] G. M. NIELSON, *Some piecewise polynomial alternatives to splines under tension*, Computer Aided Geometric Design, 1974.

[16] ———, *A transfinite, visually continuous, triangular interpolant*, this Volume, 1987.

[17] B. PIPER, *Visually smooth interpolation with triangular Bézier patches*, this Volume, 1987.

[18] M. A. SABIN, *Parametric splines in tension*, BAC Tech Report No. VTO/MS/160, 1970.

[19] M. VERON, G. RIS AND J.-P. MUSSE, *Continuity of biparametric surface patches*, Computer Aided Design, 8 (1976), pp. 267–273.

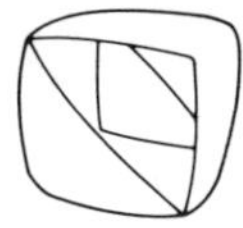

Smooth Curves and Surfaces

WOLFGANG BOEHM

Abstract. Geometric and algebraic continuity of curves and surfaces are discussed. Simple constructions of curvature and torsion continuous curves and corresponding surfaces from control points are given. Weight factors are used to "fine tune" the local shape. The Bernstein–Bézier representation is preferred.

1. Introduction. In CAD the smoothness of curves and surfaces plays an important role. Smoothness is a global as well as a local property. However, an exact definition has not been agreed upon yet. The simplest way to define local smoothness is to demand C^r continuity; however, this depends on the parameterization. More generally, one can claim C^r continuity for any parameterization (see, e.g., Barsky and de Rose [2] and Herron [8]), but this is more algebraic than geometric, if $r > 2$. More geometric are tangential, curvature and torsion continuity.

Let $\mathbf{x} = \mathbf{x}(u) \subset \mathbb{R}^3$ for all $u \in [a, b]$ be a continuous curve. Let $\dot{\mathbf{x}}(u)$ be nonvanishing, where "·" denotes differentiation with respect to u. Curvature κ and torsion τ are measures of the deviations of the curve from a straight line or plane, respectively, and are defined by

$$\kappa = \frac{|\dot{\mathbf{x}} \wedge \ddot{\mathbf{x}}|}{|\dot{\mathbf{x}}|^3}, \qquad \tau = \frac{\det(\dot{\mathbf{x}}, \ddot{\mathbf{x}}, \dddot{\mathbf{x}})}{|\dot{\mathbf{x}} \wedge \ddot{\mathbf{x}}|},$$

where "| |" denotes the Euclidean norm and " $\wedge$ " denotes the vector product (see, e.g., do Carmo [6]).

Let the subscripts "−" and "+" denote left and right limits at a point $u = u_0$, respectively. By definition the curve is

(1) tangent continuous at u_0, if $\dot{\mathbf{x}}_+ = \alpha \dot{\mathbf{x}}_-$, $\alpha > 0$,

(2) curvature continuous at u_0, if in addition $\kappa_+ = \kappa_-$ there, more exactly if $(\dot{\mathbf{x}} \wedge \ddot{\mathbf{x}})_+ = \alpha^3 (\dot{\mathbf{x}} \wedge \ddot{\mathbf{x}})_-$,

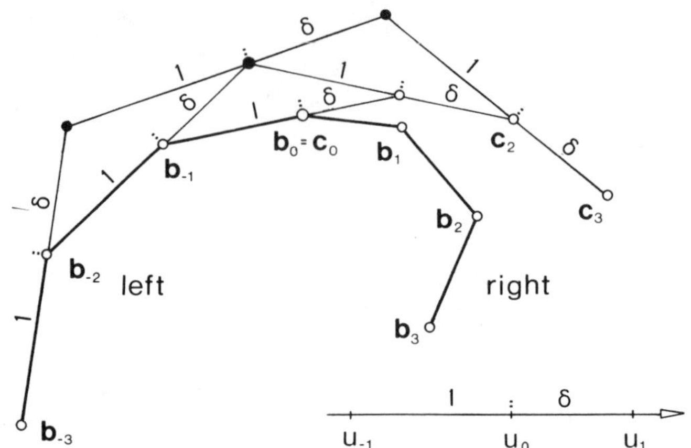

Figure 1.   Extension of the left segment together with the right segment.

(3) torsion continuous at u_0, if in addition $\tau_+ = \tau_-$ there, more exactly if $\tau_+ = \tau_-$ and $\kappa \neq 0$.

If $\kappa = 0$, τ is undefined according to the standard definition, but torsion continuity may be redefined by $(\dot{\mathbf{x}} \wedge \ddot{\mathbf{x}})_+ = \alpha^4 (\dot{\mathbf{x}} \wedge \ddot{\mathbf{x}})_-$. Note that (2) implies continuity of the osculating plane.

We will now concentrate on piecewise polynomial curves. It is sufficient to consider C^1 curves only, since any tangent continuous piecewise polynomial curve can be reparameterized such that $\alpha = 1$ without changing the polynomial degree. Let $\mathbf{b}_{-n}, \cdots, \mathbf{b}_{-1}, \mathbf{b}_0$ and $\mathbf{b}_0, \mathbf{b}_1, \cdots, \mathbf{b}_n$ be the Bézier points of two adjacent segments of such a (reparameterized) curve, corresponding to intervals $[u_{-1}, u_0]$, $[u_0, u_1]$ of the global parameter u. Further, let $\mathbf{c}_0, \mathbf{c}_1, \cdots, \mathbf{c}_n$ be the Bézier points of the C^n extension of the first (left) segment onto the interval $[u_0, u_1]$, constructed from $\mathbf{b}_{-n}, \cdots, \mathbf{b}_0$ by the algorithm of de Casteljau for $u = u_1$, as illustrated in Fig. 1 (see also Boehm, Farin and Kahmann [4]). Let $\delta = \Delta_1/\Delta_0$, where $\Delta_i = u_{i+1} - u_i$.

Note that the tangent T_- at u_0 is already spanned by $\mathbf{b}_{-1}, \mathbf{b}_0$, if $\dot{\mathbf{x}}_- \neq \mathbf{o}$. The osculating plane ω_- at u_0 is spanned by $\mathbf{b}_{-2}, \mathbf{b}_{-1}, \mathbf{b}_0$, if $\kappa_- \neq 0$. Note that ω_- degenerates to the line T_-, if $\kappa_- = 0$.

2. Curvature Continuous Curves.   The derivatives of $\mathbf{x}(u)$ at $u = u_0$ can easily be expressed in terms of the Bézier points. For example, one has

$$\dot{\mathbf{x}}_- = n(\mathbf{c}_1 - \mathbf{c}_0)/\Delta_1,$$
$$\ddot{\mathbf{x}}_- = n(n-1)(\mathbf{c}_2 - 2\mathbf{c}_1 + \mathbf{c}_0)/\Delta_1^2$$

(see Boehm, Farin and Kahmann [4]). From this it is easy to check that for any C^1 continuous curve, as above, the following conditions are equivalent (Nielson [9], Boehm [5], Farin [7]), see Fig. 2.

(2a) The curve is curvature continuous at u_0.

(2b) If $\mathbf{b}_{-2}$ has the distance 1 from the tangent T, $\mathbf{b}_2$ lies in the osculating plane ω on the same side of T in the distance δ^2.

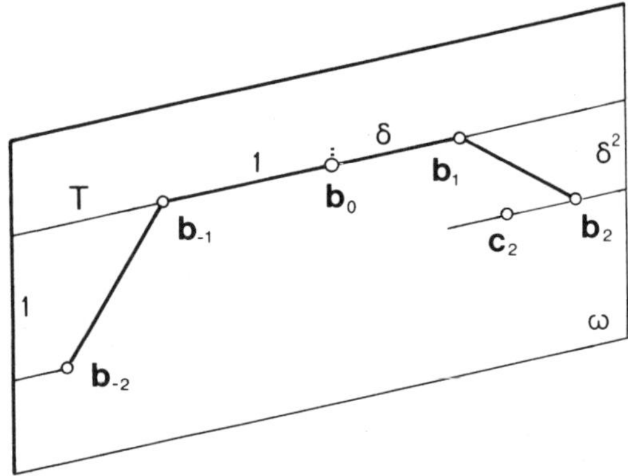

Figure 2. The curvature conditions (2b) and (2c).

(2c) $\mathbf{b}_2 = \mathbf{c}_2 + \beta_0\mathbf{b}_0 + \beta_1\mathbf{b}_1$ where $\beta_0 + \beta_1 = 0$.

(2d) $\ddot{\mathbf{x}}_+ = \ddot{\mathbf{x}}_- + \nu\dot{\mathbf{x}}_-$.

In particular, if the two straight lines spanned by $\mathbf{b}_{-2}$, $\mathbf{b}_{-1}$ and by $\mathbf{b}_1$, $\mathbf{b}_2$, respectively, meet in a finite point $\mathbf{d}$, condition (2b) may affinely be written as

(2e) $\delta\mathbf{d} = \delta((1 + \gamma\delta)\mathbf{b}_{-1} - \gamma\delta\mathbf{b}_{-2}) = (\gamma + \delta)\mathbf{b}_1 - \gamma\mathbf{b}_2$

where γ is a weight factor. The used ratios are shown in Fig. 3.

It suggests itself to use such points $\mathbf{d}_i$ as control points of a curve. Thus one gets the following simple construction of the Bézier points of a curvature continuous cubic curve (Boehm [5]) from its given control points $\mathbf{d}_1, \cdots, \mathbf{d}_m$, given knots $u_0, \cdots, u_{m+1}$, and given weight factors $\gamma_1, \cdots, \gamma_m$, attached to the $\mathbf{d}_i$, as illustrated in Fig. 4:

(i)   For $j = 1, \cdots, m - 1$ determine all inner Bézier points $\mathbf{b}_{3j+1}$ and $\mathbf{b}_{3j+2}$ from

$$(\gamma_j\Delta_{j-1} + \Delta_j + \gamma_{j+1}\Delta_{j+1})\mathbf{b}_{3j+1} = \gamma_j\Delta_{j-1}\mathbf{d}_{j+1} + (\Delta_j + \gamma_{j+1}\Delta_{j+1})\mathbf{d}_j,$$

$$(\gamma_j\Delta_{j-1} + \Delta_j + \gamma_{j+1}\Delta_{j+1})\mathbf{b}_{3j+2} = (\gamma_j\Delta_{j-1} + \Delta_j)\mathbf{d}_{j+1} + \gamma_{j+1}\Delta_{j+1}\mathbf{d}_j.$$

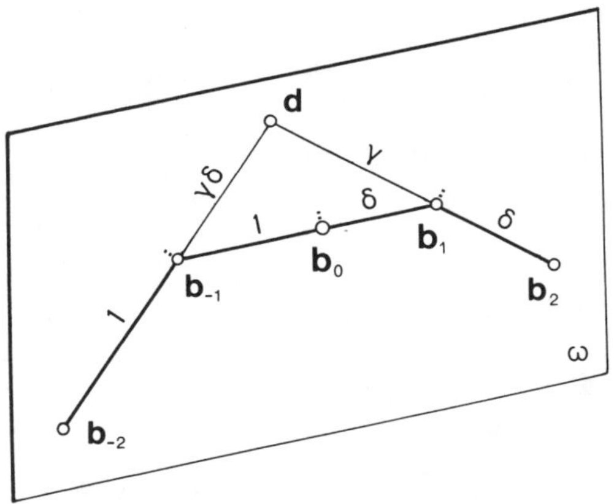

Figure 3. The curvature condition (2e).

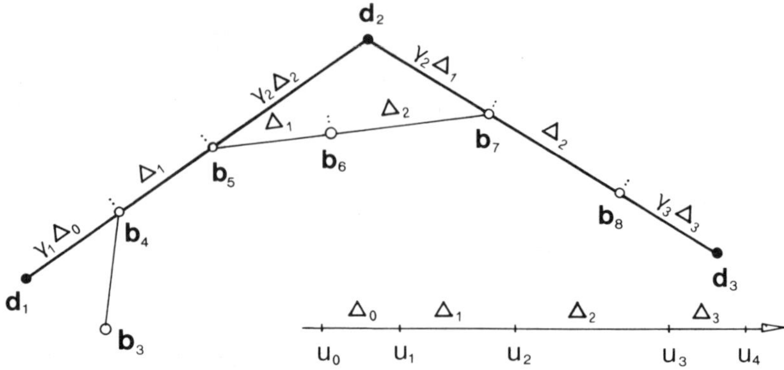

Figure 4. Construction of a curvature continuous cubic.

(ii)   For $j = 2, \cdots, m - 1$ determine all junction points $\mathbf{b}_{3j}$ from

$$(\Delta_{j-1} + \Delta_j)\mathbf{b}_{3j} - \Delta_{j-1}\mathbf{b}_{3j+1} | \Delta_j\mathbf{b}_{3j-1}.$$

(iii)   Choose endpoints $\mathbf{b}_3$ and $\mathbf{b}_{3m}$.
Note, if $\gamma_1\Delta_0 = 0$, then $\mathbf{b}_4 = \mathbf{d}_1$, and if $\gamma_m\Delta_m = 0$, then $\mathbf{b}_{3m-1} = \mathbf{d}_m$.

Now the curve can be constructed by the use of the well-known Bernstein–Bézier technique (see, e.g., Boehm, Farin and Kahmann [4]).

The shape of the control polygon roughly determines the shape of the curve, while the γ_i can be used to "fine tune" the curve shape. In particular, if all $\gamma_i = 1$, the curve is a usual C^2 cubic spline. If all $\gamma_i = 0$ and $\mathbf{b}_3 = \mathbf{d}_1$, $\mathbf{b}_{3m} = \mathbf{d}_m$, the curve degenerates to the polygon itself, although with an unusual (cubic) parameterization (Boehm [5]).

3. Torsion Continuous Curves. In the same way as above it is easy to check that the following conditions for a curvature continuous curve are equivalent (see Fig. 5):

(3a)   The curve is torsion continuous at u_0.

(3b)   If $\mathbf{b}_{-3}$ has the distance 1 from the osculating plane ω, $\mathbf{b}_3$ lies on the other side of ω in the distance δ^3.

(3c) $\mathbf{b}_3 = \mathbf{c}_3 + \beta_0\mathbf{b}_0 + \beta_1\mathbf{b}_1 + \beta_2\mathbf{b}_2$ where $\beta_0 + \beta_1 + \beta_2 = 0$.

(3d) $\dddot{\mathbf{x}}_+ = \dddot{\mathbf{x}}_- + \lambda\ddot{\mathbf{x}}_- + \mu\dot{\mathbf{x}}_-.$

In particular, if the involved points $\mathbf{l}$, $\mathbf{m}$, $\mathbf{r}$ (see Fig. 6) are finite, condition (3b) may affinely be written

(3e) $(1 + \alpha\delta)\mathbf{b}_{-2} = \mathbf{l} + \alpha\delta\mathbf{b}_{-3},$

$(\beta + \delta)\mathbf{b}_2 = \beta\mathbf{b}_3 + \delta\mathbf{r},$

$(\alpha + \beta\delta)\mathbf{m} = \alpha\mathbf{r} + \delta\beta\mathbf{l},$

while (2b) now will be written as

$$\delta\mathbf{m} = \delta((1 + \gamma\delta)\mathbf{b}_{-1} - \gamma\delta\mathbf{b}_{-2}) = (\gamma + \delta)\mathbf{b}_1 - \gamma\mathbf{b}_2.$$

The points and used ratios are shown in Fig. 6.

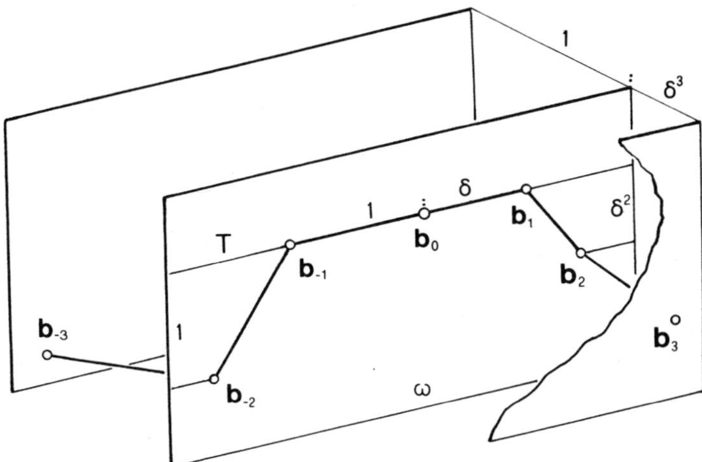

Figure 5.   The torsion conditions (3b) and (3c).

We shall now focus our attention on torsion continuous composite quartics. Let α_i, β_i, γ_i be weight factors corresponding to the junction point $\mathbf{b}_{4i}$, and let $\mathbf{l}_{i-1}$, $\mathbf{m}_i$, $\mathbf{r}_i$, defined as $\mathbf{l}$, $\mathbf{m}$, $\mathbf{r}$ above. Let L_i be the straight line spanned by $\mathbf{l}_{i-1}$, $\mathbf{r}_i$. By construction the lines L_i and L_{i+1} are coplanar and meet in a point $\mathbf{d}_i$ or are parallel. If they meet in a point, some calculations with barycentric coordinates with respect to $\mathbf{m}_{i+1}$, $\mathbf{d}_i$, $\mathbf{m}_i$, gives, e.g.,

$$(\beta_i\Delta_i + \rho_{i+1}\Delta_{i+1})\mathbf{r}_i = \beta_i\Delta_i\mathbf{d}_i + \rho_{i+1}\Delta_{i+1}\mathbf{m}_i,$$
$$(\lambda_i\Delta_{i-1} + \alpha_{i+1}\Delta_i)\mathbf{l}_i = \lambda_i\Delta_{i-1}\mathbf{m}_{i+1} + \alpha_{i+1}\Delta_i\mathbf{d}_i,$$

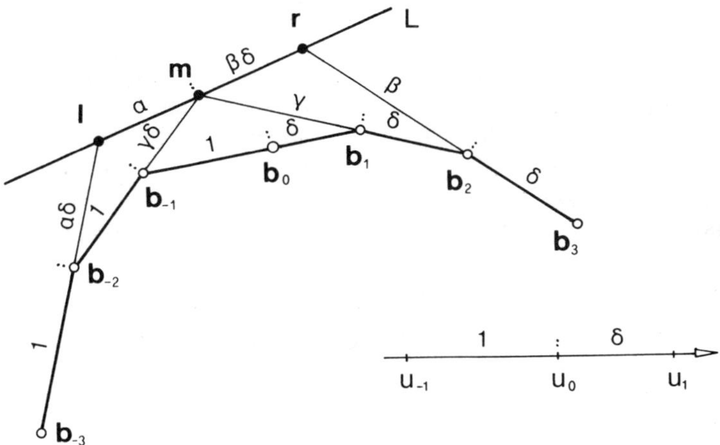

Figure 6.   The torsion condition (3e).

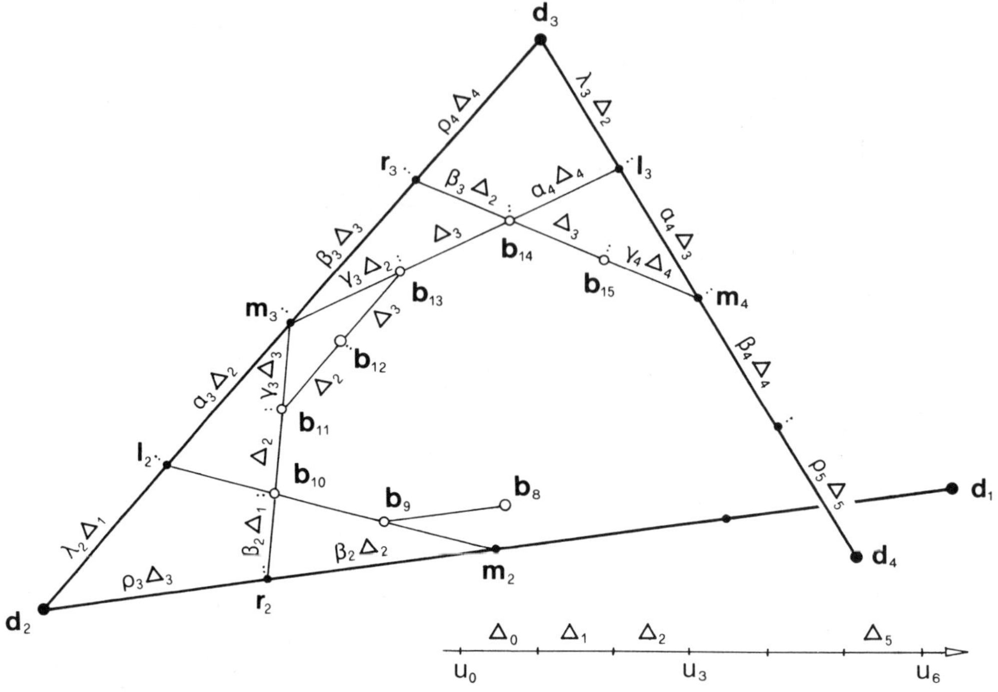

Figure 7. Construction of a torsion continuous quartic.

where

$$\rho_{i+1} = \frac{(\beta_i \Delta_{i-1} + \Delta_i + \gamma_{i+1}\Delta_{i+1})\alpha_{i+1}\beta_i\Delta_i}{(\gamma_i\Delta_{i-1} + \Delta_i)(\Delta_i + \gamma_{i+1}\Delta_{i+1}) - \beta_i\Delta_{i-1}\alpha_{i+1}\Delta_{i+1}},$$

$$\lambda_i = \frac{(\gamma_i\Delta_{i-1} + \Delta_i + \alpha_{i+1}\Delta_{i+1})\alpha_{i+1}\beta_i\Delta_i}{(\gamma_i\Delta_{i-1} + \Delta_i)(\Delta_i + \gamma_{i+1}\Delta_{i+1}) - \beta_i\Delta_{i-1}\alpha_{i+1}\Delta_{i+1}}.$$

The points and used ratios are shown in Fig. 7.

The well-known construction of the Bézier points of a spline from control points (see, e.g., Boehm [3], Boehm, Farin and Kahmann [4]) suggests to use the intersections $\mathbf{d}_i$ as control points again. This gives the following simple construction of the Bézier points of a torsion continuous composite quartic from given control points $\mathbf{d}_1, \cdots, \mathbf{d}_m$, given knots $u_0, \cdots, u_{m+2}$, and any given weight factors α_i, β_i, γ_i, attached now to the polygon legs $L_2, \cdots, L_m$ (i.e., corresponding to the knots $u_2, \cdots, u_m$):

(i)   For $j = 2, \cdots, m$ determine points $\mathbf{m}_j$ from

$$(\lambda_{j-1}\Delta_{j-2} + \alpha_j\Delta_{j-1} + \beta_j\Delta_j + \rho_{j+1}\Delta_{j+1})\mathbf{m}_j$$
$$= (\lambda_{j-1}\Delta_{j-2} + \alpha_j\Delta_{j-1})\mathbf{d}_j + (\beta_j\Delta_j + \rho_{j+1}\Delta_{j+1})\mathbf{d}_{j-1}.$$

(ii) For $j = 2, \cdots, m - 1$ determine all shoulder Bézier points

$$\mathbf{b}_{4j+2} = \xi_j\mathbf{m}_{j+1} + \eta_j\mathbf{d}_j + \zeta_j\mathbf{m}_j$$

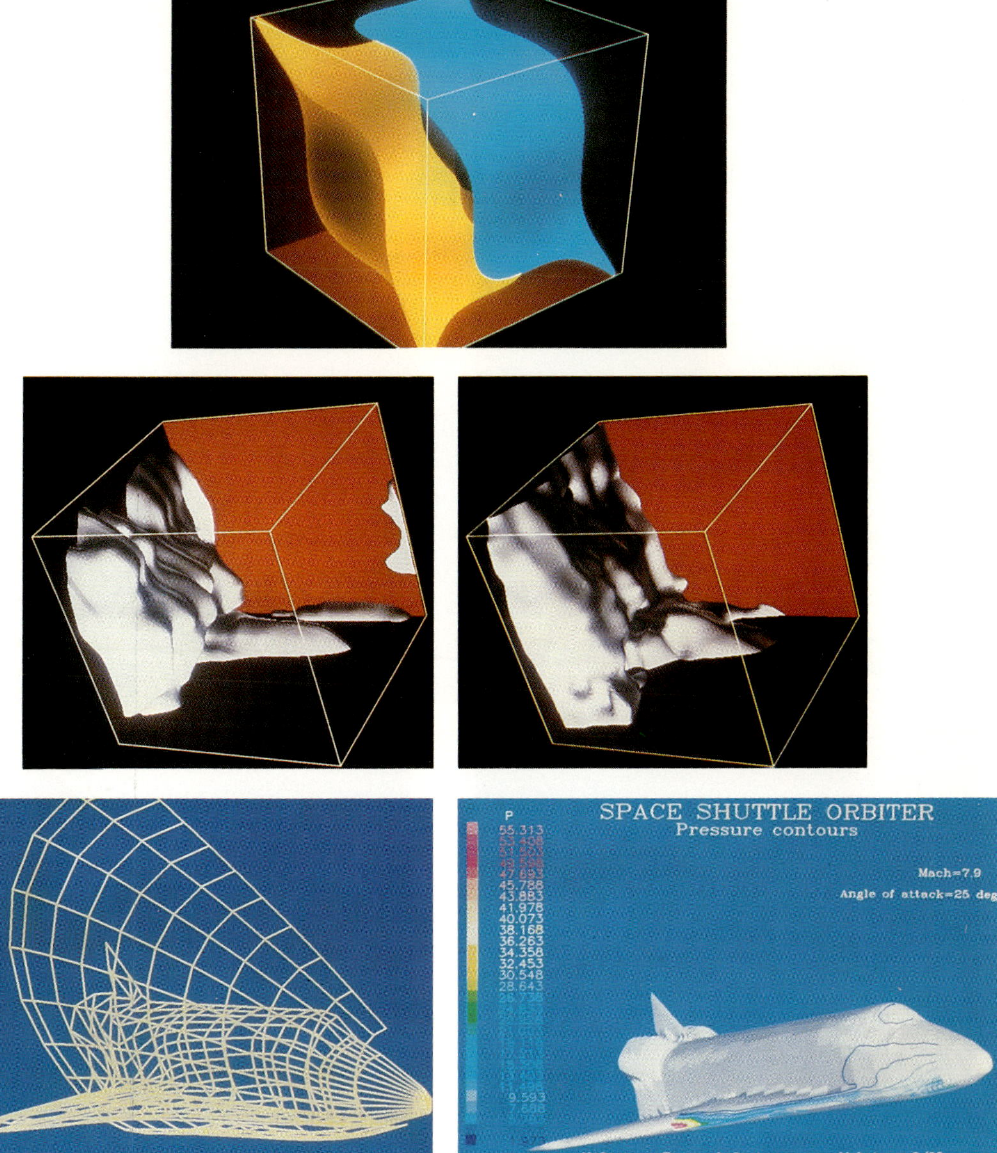

Figure 1. (top) Two contours of a trivariate tricubic function. (See page 127.)

Figure 2a. (middle left) 20% pollen contour of the multiquadric approximation HM_1. (See page 127.)

Figure 2b. (middle right) 20% pollen contour of the least squares approximation HL. (See page 127.)

Figure 3a. (bottom left) Three-dimensional grid for pressures on and near the NASA Space Shuttle. (See page 127.)

Figure 3b. (bottom right) Wireframe representation of pressure contours on the shuttle. (See page 127.)

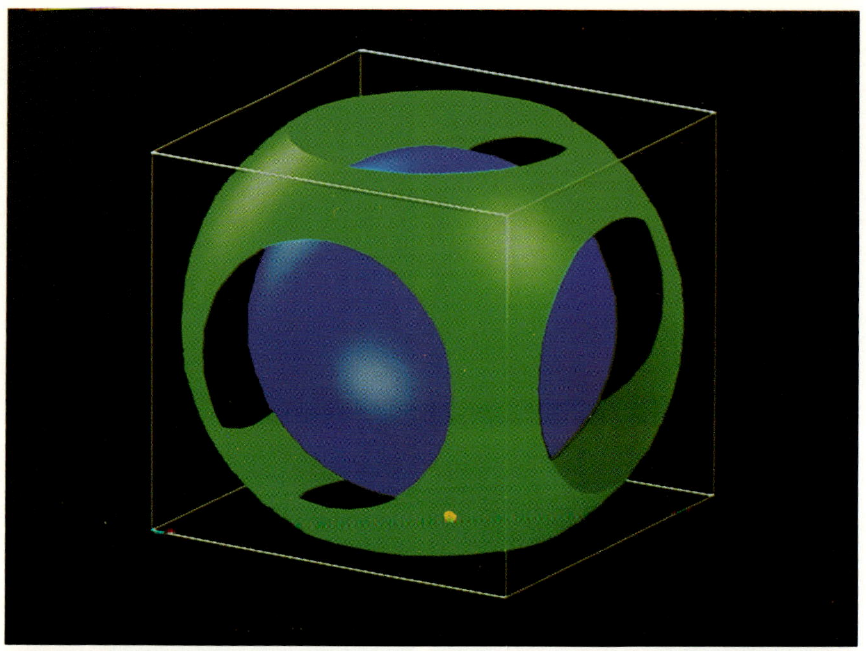

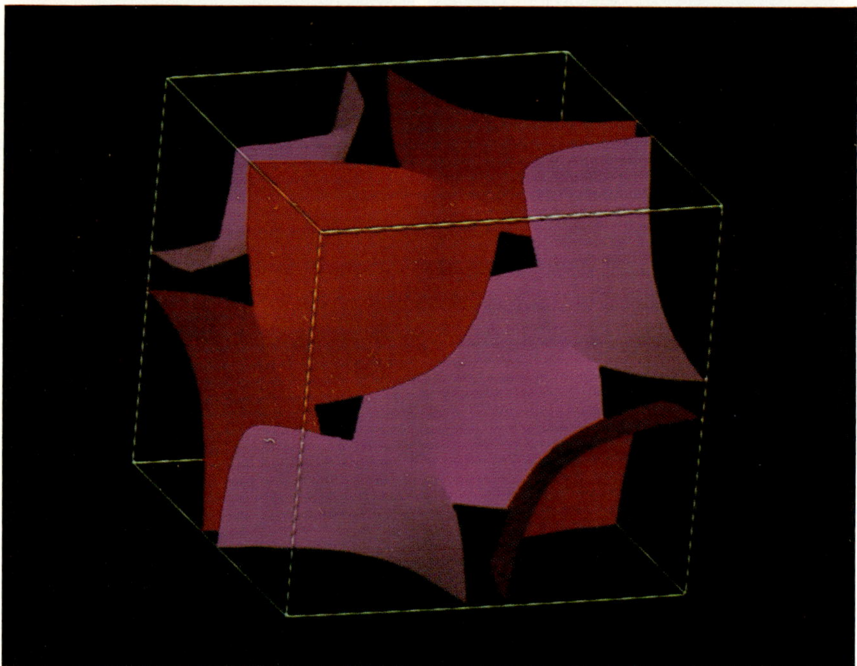

Figure 5.1. (top) Computed contours for Example 1. (See page 392.)
Figure 5.2. (bottom) Computed contours for Example 2. (See page 392.)

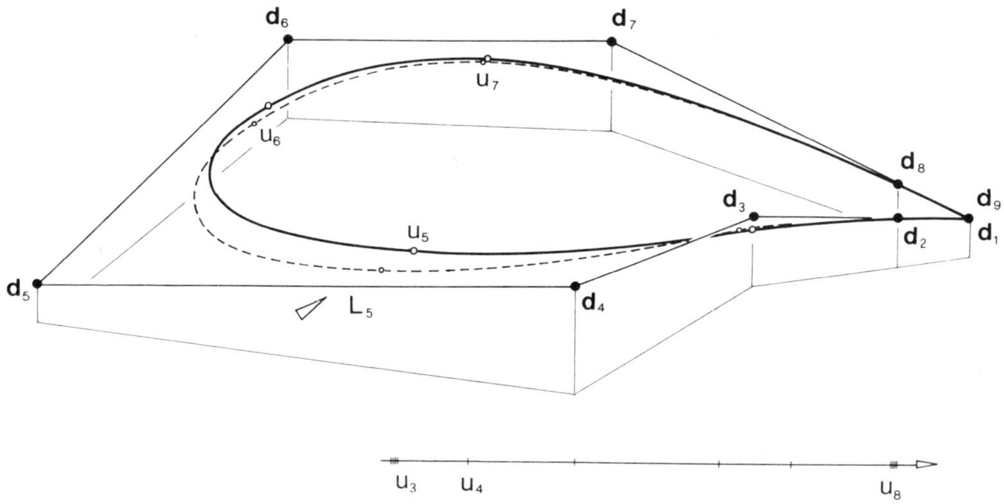

Figure 8. C^3-spline and torsion continuous quartic, example.

where $\xi_j + \eta_j + \zeta_j = 1$ and

$$\xi_j = \frac{\beta_j \Delta_{j-1}}{\beta_j \Delta_{j-1} + \Delta_j + \gamma_{j+1} \Delta_{j+1}}, \qquad \zeta_j = \frac{\alpha_{j+1} \Delta_{j+1}}{\gamma_j \Delta_{j-1} + \Delta_j + \alpha_{j+1} \Delta_{j+1}}.$$

(iii)   For $j = 2, \cdots, m - 1$ determine the Bézier points $\mathbf{b}_{4j+1}$, $\mathbf{b}_{4j+3}$ from

$$(\gamma_j \Delta_{j-1} + \Delta_j)\mathbf{b}_{4j+1} = \gamma_j \Delta_{j-1}\mathbf{b}_{4j+2} + \Delta_j \mathbf{m}_j,$$
$$(\Delta_j + \gamma_{j+1} \Delta_{j+1})\mathbf{b}_{4j+3} = \Delta_j \mathbf{m}_{j+1} + \gamma_{j+1} \Delta_{j+1}\mathbf{b}_{4j+2}.$$

(iv)   For $j = 3, \cdots, m - 1$ determine the junction points $\mathbf{b}_{4j}$ from

$$(\Delta_{j-1} + \Delta_j)\mathbf{b}_{4j} = \Delta_{j-1}\mathbf{b}_{4j+1} + \Delta_j \mathbf{b}_{4j-1}.$$

(v)   Choose end points $\mathbf{b}_8$ and $\mathbf{b}_{4m}$.

The given parameters α_i, β_i, γ_i control the local shape of the curve. In particular, if all $\alpha_i = \beta_i = \gamma_i = 1$, one gets the Bézier points of a usual quartic C^3 spline. In the general case, however, these are too many parameters to be considered. To simplify the variety of shapes one may set $\alpha_i = \beta_i = \gamma_i$ for all i. This simplifies the construction of the used ratios and points considerably. One gets, e.g.,

$$\rho_{i+1} = \gamma_i \gamma_{i+1} = \lambda_i \quad \text{and} \quad \xi_i : \eta_i : \zeta_i = \gamma_i \Delta_{i-1} : \Delta_i : \gamma_{i+1} \Delta_{i+1}.$$

An example is shown in Fig. 8. The corresponding C^3 spline is dashed, and the leg L_5 was imposed with a weight factor $\gamma_5 = 2$. This pushes the curve away from L_5. As an alternative to the algorithm given above, end conditions $\mathbf{b}_{12} = \mathbf{d}_1$, $\mathbf{b}_{13} = \mathbf{d}_2$ and $\mathbf{b}_{31} = \mathbf{d}_8$, $\mathbf{b}_{32} = \mathbf{d}_9$ were constructed by giving multiplicity 4 to u_3 and u_8.

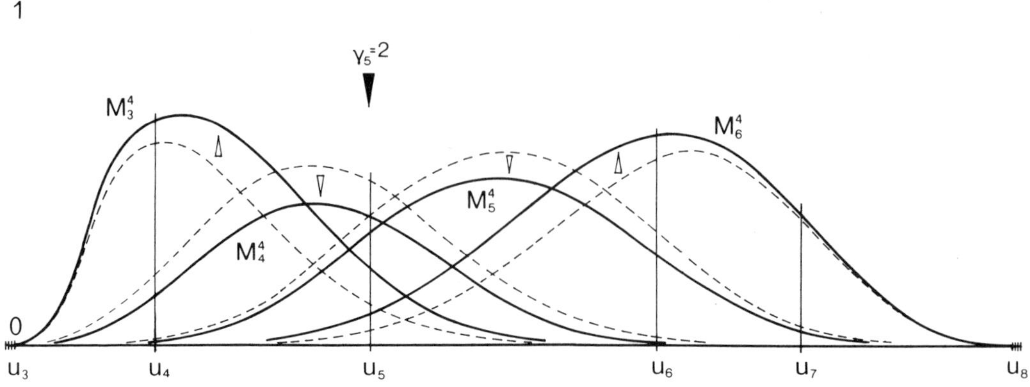

Figure 9. Quartic B-splines, example.

4. B-Splines. Any continuous composite piecewise polynomial curve of degree n can be written as

$$\mathbf{x}(u) = \Sigma_j \mathbf{b}_j B_j^n(u),$$

where the $B_j^n(u)$ are composed truncated Bernstein polynomials, corresponding to the Bézier points $\mathbf{b}_j$. For both types of curves considered above these Bézier points are constructed by repeated linear combinations from control points $\mathbf{d}_i$. Thus these curves can also be written as

$$\mathbf{x}(u) = \Sigma_i \mathbf{d}_i M_i^n(u)$$

where the $M_i^n(u)$ are "B-splines" corresponding to the $\mathbf{d}_i$. For example, in the case of $\Delta_{i+1}/\Delta_i = \delta$ and $\gamma_i = \gamma$, all i, such "B-splines" were considered by Barsky [1] and Barsky and DeRose [2] and called "β-B-splines." The ordinates of the Bézier points of these B-splines can easily be constructed from the identity

$$M_i^n(u) = \Sigma_j \delta_{i,j} M_j^n(u)$$

by the given algorithms above, where the Kronecker delta $\delta_{i,j}$ represents the 1D control polygon. For a curvature continuous composite cubic this was done by Boehm [5]. For a simplified torsion continuous composite quartic, $\alpha_i = \beta_i = \gamma_i$, some B-splines corresponding to Fig. 8 are shown in Fig. 9. The C^3 B-splines are dashed again. Obviously some quantity of these B-splines corresponding to $\mathbf{d}_4$ and $\mathbf{d}_5$ is pushed to their neighbors corresponding to $\mathbf{d}_3$ and $\mathbf{d}_6$.

Note that the $M_i^n(u)$ form a partition of unity. This follows immediately from the constructions above applied to $\Sigma_i 1 \cdot M_i(u)$ and the fact that the $B_j^n(u)$ form a partition of unity.

5. Surfaces. The existence of B-splines allows us to build up tensor product surfaces

$$\mathbf{x}(u, v) = \Sigma_i \Sigma_j \mathbf{d}_{i,j} M_j^n(v) M_i^n(u),$$

where the $\mathbf{d}_{i,j}$ are control points, and where $M_i^n(u)$ and $M_j^n(v)$ are the B-splines in the u- and v-direction, respectively, corresponding to different sets of knots and weight factors. These B-splines can be expressed in terms of composed truncated Bernstein polynomials $B_l^n(u)$ and $B_k^n(v)$, respectively. Thus such a surface can be written as

$$\mathbf{x}(u, v) = \Sigma_l \Sigma_k \mathbf{b}_{l,k} B_k^n(v) B_l^n(u).$$

It is easy to construct the Bézier points $\mathbf{b}_{l,k}$ of this surface by the use of the corresponding curve schemes above:

(i)   Determine the "mixed control points" $\mathbf{c}_{i,k}$ from the $\mathbf{d}_{i,j}$ by the use of the curve scheme in v-direction.

(ii)   Determine the Bézier points $\mathbf{b}_{l,k}$ from the $\mathbf{c}_{i,k}$ by the use of the curve scheme in u-direction.

From the well-known tensor product properties (see, e.g., Boehm, Farin and Kahmann [4]) follows immediately:

All curves $u = $ const as well as $v = $ const on the surface are curvature (and torsion) continuous. Moreover, any plane section of such a surface is curvature continuous (see Boehm [5] and the Appendix) and torsion continuous, because a plane curve always has torsion zero.

6.   Algebraic Continuity.   Sometimes (see, e.g., Barsky and DeRose [2], Herron [8]), a C^0 curve $\mathbf{x}(u)$ is called "smooth" if there exists a reparameterization $u = u(v)$, such that $\mathbf{x}(u(v))$ is C^r with respect to v, shortly written as C_v^r. C_v^r conditions at a point $u = u_0$ are easily derived using the chain rule, where " ′ " denotes differentiation with respect to v. Thus

$$\mathbf{x}'_+ = \mathbf{x}'_- \text{ is equivalent to } \dot{\mathbf{x}}_+ = \alpha \dot{\mathbf{x}}_-,$$

$$\mathbf{x}''_+ = \mathbf{x}''_- \text{ is equivalent to } \dot{\mathbf{x}}_+ = \alpha \ddot{\mathbf{x}}_- + v \dot{\mathbf{x}}_-,$$

etc. That is, algebraic C_v^r continuity coincides with geometric continuity, if $r \leq 2$. However,

$$\mathbf{x}'''_+ = \mathbf{x}'''_- \text{ is equivalent to } \dddot{\mathbf{x}}_+ = \alpha^3 \dddot{\mathbf{x}}_- + 3\alpha v \ddot{\mathbf{x}}_- + \mu \dot{\mathbf{x}}_-$$

only, which is more restrictive than (3d), because λ is restricted to $\lambda = 3v$ now, i.e., $\mathbf{b}_3$ is forced to lie on a straight line parallel to T.

7.   Smoothness of Higher Degree.   It is clear how to define higher degree geometric and algebraic smoothness. Continuing beyond (3d), geometric fourth degree smoothness is defined by

$$\mathbf{b}_4 = \mathbf{c}_4 + \beta_0 \mathbf{b}_0 + \cdots + \beta_3 \mathbf{b}_3, \quad \beta_0 + \cdots + \beta_3 = 0;$$

however, this is satisfied by all curves whose Bézier points $\mathbf{b}_0, \cdots, \mathbf{b}_3$ span $\mathbb{R}^3$. This is the general case.

However, fourth degree algebraic smoothness is derived from the fact that

$$\mathbf{x}''''_+ = \mathbf{x}''''_- \text{ is equivalent to } \ddddot{\mathbf{x}}_+ = \alpha^4 \ddddot{\mathbf{x}}_- + 6\alpha^2 v \dddot{\mathbf{x}}_- + (3v^2 + 4\alpha\mu)\ddot{\mathbf{x}}_- + \lambda \dot{\mathbf{x}}_-.$$

That is, $\mathbf{b}_4$ is also forced to a straight line parallel T.

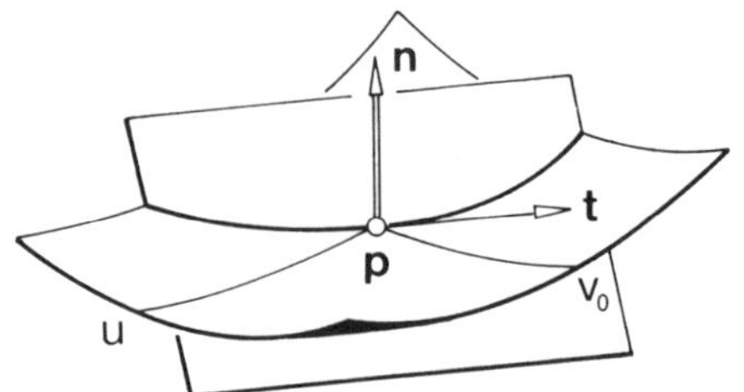

Figure 10.   Plane section of a surface.

Appendix.   Finally, we will add a short proof of the curvature property of the considered surfaces. Let "$-$" and "$+$" now denote left and right limits at a point $\mathbf{p} = \mathbf{x}(u, v_0)$ with respect to v, where $v = v_0$ is the knot line under consideration. One has

$$\mathbf{x}_{v+} = \mathbf{x}_{v-}, \quad \mathbf{x}_{uv+} = \mathbf{x}_{uv-};$$

however,

$$\mathbf{x}_{vv+} = \mathbf{x}_{vv-} + v_0 \mathbf{x}_{v-},$$

and thus for the coefficients E, F, G, of the first fundamental form (see, e.g., do Carmo [6]),

$$E_+ = \mathbf{x}_u \cdot \mathbf{x}_u = E_-, \ F_+ = \mathbf{x}_u \cdot \mathbf{x}_v = F_-, \ G_+ = \mathbf{x}_v \cdot \mathbf{x}_v = G_-.$$

Let $\mathbf{n} = (\mathbf{x}_u \wedge \mathbf{x}_v)/|\mathbf{x}_u \wedge \mathbf{x}_v|$ be the surface normal at $\mathbf{p}$. Since $\mathbf{n}_+ = \mathbf{n}_-$ and $\mathbf{n} \cdot \mathbf{x}_v = 0$ one gets for the coefficients e, f, g, of the second fundamental form (see, e.g., do Carmo [6])

$$e_+ = \mathbf{n} \cdot \mathbf{x}_{uu} = e_-, \quad f_+ = \mathbf{n} \cdot \mathbf{x}_{uv+} = f_-, \quad g_+ = \mathbf{n} \cdot \mathbf{x}_{vv+} = g_-.$$

Thus, both fundamental forms coincide at $\mathbf{p}$ as well as both second fundamental forms. As a consequence both normal curvatures in any tangent direction $\mathbf{t}$ coincide, as well as (by Meusnier's theorem, see, e.g., do Carmo [6]) the curvature of any section by a plane containing $\mathbf{t}$ (see Fig. 10).

REFERENCES

[1]   B. A. BARSKY, *The β-spline: A local representation based on shape parameters and on fundamental geometric measures*, Ph.D. dissertation, Univ. Utah, Salt Lake City, UT, 1981.

[2]   B. A. BARSKY AND T. D. DeROSE *Geometric continuity of parametric curves*, Report No. UCB/CSD 84/205, EECS, Univ. Calif. Berkeley, Berkeley, CA, 1985.

[3]   W. BOEHM, *Über die Konstruktion von B-spline–Kurven*, Computing, 18 (1977), pp. 161–166.

[4]   W. BOEHM, G. FARIN AND J. KAHMANN, *A survey on curve and surface methods in* CAGD, Computer Aided Geometric Design, 1 (1984), pp. 1–60.

[5]   W. BOEHM, *Curvature continuous curves and surfaces*, Computer Aided Geometric Design, 2 (1985), pp. 313–323.

[6]   M. P. DO CARMO, *Differential Geometry of Curves and Surfaces*, Prentice-Hall, Englewood Cliffs, NJ, 1976.

[7]   G. FARIN, *Some remarks on V^2-splines*, Computer Aided Geometric Design, 2 (1985), pp. 325–328.

[8]   G. HERRON, *Techniques for visual continuity*, this Volume, 1987.

[9]   G. M. NIELSON, *Some piecewise polynomial alternatives to splines under tension*, Computer Aided Geometric Design, R. E. Barnhill and R. F. Riesenfeld, eds., Academic Press, New York, 1974.

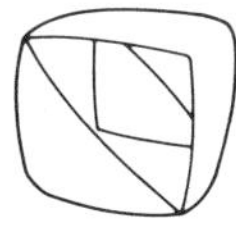

Energy Comparisons of Wilson–Fowler Splines with Other Interpolating Splines

FREDERICK N. FRITSCH

Abstract. The energy of the Wilson–Fowler spline through twenty data sets is compared with that of five other splines through the data. It is concluded that the WF-spline is sometimes better and sometimes worse than a parametric cubic spline. Areas for further research are indicated.

1. Introduction.

1.1. Background. The *Wilson–Fowler spline* (WF-spline) was introduced in the early 1960s (see Fowler and Wilson [4]) as a means for passing a smooth curve through a planar set of design points: (x_i, y_i), $i = 1, \cdots, n$. The WF-spline has been used in the APT N/C system (see IIT Research Institute [9]) ever since the TABCYL (tabulated cylinder) was introduced to allow point-defined curves.

This past decade has seen the development of many CAD/CAM systems, most of which contain some sort of spline entity. The most common of these is the B-spline curve, whose component functions are (usually cubic) B-splines (see de Boor [1]) in some parameter t. Such a spline will not, in general, coincide with a WF-spline through the same data points. Thus arises the need to compare WF-splines with other splines. The original intent of this study was to answer the question: "Is the WF-spline better than an ordinary parametric cubic spline?" (A negative answer might call for a rethinking of the way parts are defined by DoE and its contractors.)

1.2. Energy as a basis for comparison. The original cubic spline arose as a mathematical model for a draftsman's spline. Here a thin beam is constrained to pass through the data points and the location of its center line at equilibrium is sought. The *elastica* is the ideal spline, which minimizes the total energy

$$(1.1) \qquad E = \int_0^S [\kappa(s)]^2 \, ds,$$

185

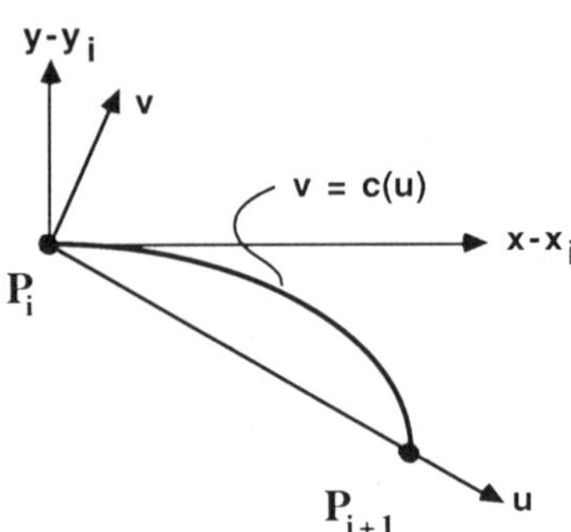

Figure 1.   Wilson–Fowler spline local coordinate system.

where s is arc length, S is the total length of the curve, and $\kappa(s)$ is the local curvature. If we regard the data as defining a function, $y = y(x)$, then

$$(1.2) \qquad ds = \sqrt{(1 + [y'(x)]^2)}\, dx,$$

and

$$(1.3) \qquad \kappa(x) = \frac{y''(x)}{(1 + [y'(x)]^2)^{3/2}},$$

so the energy integral (1.1) becomes

$$(1.4) \qquad E = \int_{x_1}^{x_n} \frac{[y''(x)]^2}{(1 + [y'(x)]^2)^{5/2}}\, dx.$$

If one makes the simplifying assumption that $[y'(x)]^2$ is everywhere so small relative to one that the denominator of (1.4) can be ignored, one obtains the "linearized" energy

$$(1.5) \qquad E_L = \int_{x_1}^{x_n} [y''(x)]^2\, dx.$$

It is well known (see de Boor [1, p. 63]) that the function that minimizes E_L subject to the constraint of passing through the data is the *natural cubic spline*. This function $y(x)$ is a cubic polynomial on each interval $[x_i, x_{i+1}]$, the pieces are joined so that y is continuous and has continuous first and second derivatives, and it satisfies the *free-end conditions:* $y''(x_1) = y''(x_n) = 0$. (One can similarly define cubic splines that satisfy *fixed-end conditions:* $y'(x_1) = d_1$; $y'(x_n) = d_n$.)

One of the motivations for developing the WF-spline was to obtain something closer to the true elastica by introducing a local coordinate system (see Fig. 1) on each segment, with the independent variable u in segment i running along the chord joining (x_i, y_i) with (x_{i+1}, y_{i+1}). In such a coordinate system, it was hoped, the cubic polynomial $v(u)$, representing the deviation from the chord, would have a value of

$$(1.6) \qquad E_i = \int_{u_i}^{u_{i+1}} [v''(u)]^2\, du$$

that is closer to the true energy (1.1) of the segment than if the original independent variable x had been retained. Thus it seems natural to use the true energy as the measure of comparison between WF- and other splines.

2. Comparison Splines.

2.1. Parametric piecewise cubics.   Let $0 = t_1 < t_2 < \cdots < t_n$ be preselected parameter values such that a parameterized curve $\mathbf{P}(t) = (x(t), y(t))$ passes through the ith data point at $t = t_i$:

$$(2.1) \qquad \mathbf{P}(t_i) \equiv (x(t_i), y(t_i)) = (x_i, y_i) \equiv \mathbf{P}_i, \qquad i = 1, \cdots, n.$$

Such a curve is a *parametric piecewise cubic curve* if each component function is a cubic polynomial in t on each subinterval $[t_i, t_{i+1}]$. Parametric piecewise cubics have the advantage of being invariant under linear coordinate changes (see de Boor [1, p. 319]). They also allow vertical tangents, which is impossible for ordinary piecewise cubic curves. It is shown in Fritsch [5] that a WF-spline is a parametric piecewise cubic in the *cumulative chord length parameterization*:

$$(2.2) \qquad u_1 = 0, \qquad u_{i+1} = u_i + L_i, \quad i = 1, \cdots, n-1,$$

where $L_i = \sqrt{[(x_{i+1} - x_i)^2 + (y_{i+1} - y_i)^2]}$ is the length of the ith chord. The component functions $x(u)$ and $y(u)$, however, turn out to be merely continuous. They have discontinuities in the first and second derivatives at the data points, even though the curve they describe has continuous tangent and curvature.

A *parametric cubic spline* (PC-spline) is a parametric piecewise cubic whose component functions are cubic splines, that is, $x(t)$ and $y(t)$ have continuous first and second derivatives. Fritsch [5] introduced a new parameterization $t = t(u)$ for a WF-spline so that the transformed component functions $(\hat{x}(t), \hat{y}(t)) = (x(u(t)), y(u(t)))$ have continuous derivatives with respect to t. This reparameterization takes the form

$$(2.3) \qquad t = t(u) = t_i + k_i(u - u_i), \qquad u \in [u_i, u_{i+1}],$$

where the u_i are as in (2.2),

$$(2.4) \qquad t_1 = 0, \qquad t_{i+1} = t_i + k_i(u_{i+1} - u_i), \quad i = 1, \cdots, n-1,$$

and the k_i are fixed constants chosen to make the derivatives continuous. Note that, due to the linear nature of (2.3), the result will be a parametric piecewise cubic in the new parameter t. It is not possible, in general, to make $\hat{x}$ and $\hat{y}$ have continuous second derivatives, so that a WF-spline is not a PC-spline.

Since arc length is not used as the parameter for the splines under study, the relations

$$(2.5) \qquad ds = \sqrt{([x'(t)]^2 + [y'(t)]^2)}\, dt,$$

and

$$(2.6) \qquad \kappa(t) = \pm \frac{x'(t)y''(t) - y'(t)x''(t)}{([x'(t)]^2 + [y'(t)]^2)^{3/2}},$$

must be used to compute the true energy (1.1) as

$$(2.7) \qquad E = \sum_{i=1}^{n-1} \int_{t_i}^{t_{i+1}} \frac{(x'y'' - y'x'')^2}{(x'^2 + y'^2)^{5/2}} \, dt.$$

2.2. v-splines. In order to allow for the application of local tension at the data points, Nielson [12] defined a v-*spline* to be the function $f(t)$ that interpolates given data (t_i, f_i), $i = 1, \cdots, n$, and minimizes the functional

$$(2.8) \qquad E_v[f] = \int_{t_1}^{t_n} [f''(t)]^2 \, dt + \sum_{i=1}^{n} v_i [f'(t_i)]^2,$$

where the v_i are fixed (nonnegative) *tension parameters*. The first term is the linearized energy (1.5) that is minimized by the cubic spline, so a v-spline will generally have only first derivative continuity. In fact, a v-spline is characterized by the *jump conditions*

$$(2.9) \qquad f''(t_{i^+}) - f''(t_{i^-}) = v_i f'(t_i),$$

from which the ordinary cubic spline is the special v-spline with $v_i = 0$, $i = 1, \cdots, n$.

A *parametric v-spline* is a parametric piecewise cubic, each of whose component functions is a v-spline (with the same v-values for each component). As demonstrated by Nielson [12], a parametric v-spline has continuous tangent and curvature. A PC-spline is a special parametric v-spline with $v_i = 0$, $i = 1, \cdots, n$. Fritsch [5] showed that a WF-spline is a parametric v-spline, provided one takes the jump conditions (2.9) as the defining relation and drops the restriction $v_i \geq 0$. Thus, all of the splines being considered here are v-splines, for a suitable choice of parameterization and v_i.

One may define a *uniformly shaped v-spline* to be one with $v_i = v$, $i = 1, \cdots, n$. Thus, a PC-spline is a special uniformly shaped v-spline with $v = 0$. It is of interest to compare the energy of the WF-spline and the PC-spline with that of the *optimal uniformly shaped v-spline* (OUSN-spline). This is the uniformly shaped v-spline that has the minimum value of the true energy E as computed from (2.7). (This v-spline was selected simply because it is computable via a univariate optimization algorithm and gives some indication of the improvement possible over a PC-spline. There is no reason to believe that choosing all v_i equal is a good idea, but we have not attempted minimizing over all possible v_i-values.)

3. The Tests.

3.1. The test data. The twenty data sets employed in these tests were as follows:

SIN1:   This is the sample data in Fowler and Wilson [4]; namely, $y_i = 2 \cdot \sin(x_i)$, for 28 uniformly spaced x-values in $[0, 3\pi/2]$. The values were given to six decimal places (which is more accurate than in the reference).

SIN2:   This is every third point from SIN1.

QCIR1:   This is a set of uniformly spaced points on the quarter circle with radius 1.5 and angles in $[0, 90°]$. The points were read in polar coordinates, at 5° increments, and converted to Cartesian coordinates internally.

QCIR2: This is every other point from QCIR1.

WRM:[1] This is a constructed data set used by W. R. Melvin [11] when he was testing his algorithm for computing a WF-spline.

SF1, SF2:[1] These are two constructed data sets used by S. K. Fletcher [2] as part of her spline testing procedure.

FNFk, $k = 1, \cdots, 5$:[1] This is a series of five test sets constructed by the author. (They were originally invented to visually demonstrate the derivative discontinuities of the WF-spline component functions, since most "reasonable" data sets yield derivative jumps smaller than 10^{-3}.)

RPN:[1] This data set was constructed by the author so that $(u_i + 7.99, x_i)$, with u_i given by (2.2), is the RPN14 data set of Fritsch and Carlson [6].

BMK1:[1] This is a set of data that has been used by LLNL as a benchmark for vendor-supplied splines. The points are taken from a pair of tangent ellipses and contain one inflection. They are given in polar coordinates at 2° increments.

BMK2:[1] This is a subset of BMK1, retaining its original character.

JJi, JMj: These are five actual design contours which are not available outside LLNL. All but the last are very similar to BMK1.

3.2. Parameterizations. Two parameterizations for splines have been mentioned above. The first is the cumulative chord length parameterization (2.2), which will be called the *natural parameterization*. The second is the parameterization (2.4) required to give the WF-spline component functions continuous derivatives, called the WF-*parameterization*. The five splines compared with the WF-spline in this study were the PC-spline and OUSN-spline using each of these parameterizations and the ordinary (nonparametric) cubic spline.

3.3 Boundary conditions. In order to completely determine a parametric v-spline, it is necessary to specify some sort of boundary conditions (BC). The BC employed in these tests were specified tangents $(x'(t_1), y'(t_1))$ and $(x'(t_n), y'(t_n))$. Since a WF-spline is determined simply by end-slopes, there is some arbitrariness in the choice of the magnitudes of the boundary tangent vectors. These tests used the formulas

$$(3.1) \qquad x'(u_1) = \frac{L_1}{[(x_2 - x_1) + S_1 \cdot (y_2 - y_1)]}, \qquad y'(u_1) = S_1 \cdot x'(u_1),$$

---

[1] These data are listed in the Appendix.

and their analogue at u_n, which arise naturally when one represents a WF-spline as a parametric piecewise cubic (see Fritsch and Springmeyer [7], Appendix A). In (3.1), L_1 and S_1 are the length and slope of the first chord. These BC are scaled appropriately for the transformation to the WF parameterization.

For the sine data, two different sets of BC were used. One is the "correct" boundary slopes: $S_1 = 2$, $S_n = 0$. The other is the "default" BC, namely that each boundary slope match that of the circle through the three end points. (Note that this is close to, but not the same as, the default BC of Fowler and Wilson [4, pp. 21–22].) For the other data sets, a single BC was chosen, making a total of 22 tests in all.

4. Test Results. The results of these tests, as run on a CRAY-1 using single precision arithmetic, are given in Tables 1 and 2. The default BC are indicated by

Table 1.   Results for WF-splines.

Data ID	n	BC	E_0	E_{WF}	arc length	k_{ratio}	$\nu_{\min}$	$\nu_{\max}$
SIN1	28	S2S0	3.7339	3.7337	7.9506	1.0075	−5.56	0.0
SIN1	28	Def.	3.6343	3.6338	7.9055	1.0075	−5.56	0.0
SIN2	10	S2S0	3.7361	3.7498	7.9115	1.0462	−14.44	0.0
SIN2	10	Def.	3.0508	3.0156	7.9023	1.0494	−14.37	0.0
QCIR1	19	N0S0	3.2624*	1.0472	2.3562	1.0000	−0.14	−0.14
QCIR2	10	N0S0	4.2967*	1.0472	2.3562	1.0000	−0.27	−0.27
WRM	4	Def.	—†	4.9150	2.4966	1.1945	−3.56	0.0
SF1	5	Def.	2.0696‡	0.7826	7.9263	1.1070	−3.16	−0.51
SF2	7	Def.	17.769‡	1.6048	6.6100	1.0429	−3.10	+0.10
FNF1	6	S1S1	1.5657‡	1.4088	13.738	1.0838	−5.63	−2.49
FNF2	6	S1S1	4.7422‡	1.6652	14.524	1.2505	−6.15	−2.70
FNF3	6	S1S1	45.099‡	2.6315	16.034	1.9302	−9.56	+0.61
FNF4	6	S1S1	245.19‡	3.2239	16.779	2.3239	−10.08	+1.80
FNF5	6	S1S1	—†	3.3066	16.881	2.3777	−10.13	+1.94
RPN	9	Def.	14.188§	5.2661	12.059	1.1551	−38.65	+0.05
BMK1	46	Def.	1.1056¶	0.5654	6.2650	1.0006	−0.26	−0.01
BMK2	11	Def.	3.6613¶	0.5641	6.2649	1.0097	−1.03	−0.04
JJ2	46	Def.	0.6278¶	0.5427	4.6008	1.0001	−0.07	−0.01
JJ3	46	Def.	1.2322¶	0.5451	4.6382	1.0002	−0.08	−0.01
JM1	46	Def.	1.1764¶	0.4049	6.4864	1.0003	−0.13	−0.00
JM2	46	Def.	1.0352¶	0.5449	6.7003	1.0008	−0.42	+0.00
JM3	85	Def.	0.2664	0.2664	16.298	1.0003	−0.27	0.0

* This is the "natural" spline. (Simulations of vertical slope yielded astronomical energy values.)

† Could not compute spline: two consecutive data points have same x-coordinate.

‡ The "natural" spline has smaller energy, but it is also greater than E_{WF}.

§ This is the "natural" spline. (Default BC gave energy ca. 1.7×10^7.)

¶ This is the "natural" spline. (Default BC gave energies in excess of 3000, due to a nearly vertical end slope.)

Table 2.   Energy comparisons.

Data ID	BC	E_{WF}	Natural parameterization			WF parameterization		
			ΔE for $v = 0$	ΔE for v_{opt}	v_{opt}	ΔE for $v = 0$	ΔE for v_{opt}	v_{opt}
SIN1	S2S0	3.7337	$-1.1E-4$	$-1.8E-4$	-1.15	$-1.2E-4$	$-1.9E-4$	-1.18
SIN1	Def.	3.6338	$-6.1E-5$	$-1.5E-4$	-1.28	$-8.5E-5$	$-1.7E-4$	-1.26
SIN2	S2S0	3.7498	$+2.5E-2$	$+1.6E-2$	-4.39	$+2.0E-2$	$+1.0E-2$	-4.41
SIN2	Def.	3.0156	$+2.9E-2$	$+1.5E-2$	-5.15	$+2.2E-2$	$+8.2E-3$	-5.04
QCIR1	N0S0	1.0472	$-3.0E-7$	$-5.0E-7$	-0.05			
QCIR2	N0S0	1.0472	$-4.9E-6$	$-7.8E-6$	-0.10			
WRM	Def.	4.9150	$-3.9E-2$	$-4.1E-2$	-0.73	$+1.5E-2$	$+1.4E-2$	$+0.64$
SF1	Def.	0.7826	$+7.2E-4$	$-4.9E-3$	-2.06	$+1.6E-3$	$-4.2E-4$	-1.21
SF2	Def.	1.6048	$-4.1E-4$	$-1.9E-3$	-1.41	$-6.7E-4$	$-2.8E-3$	-1.69
FNF1	S1S1	1.4088	$+4.5E-3$	$-6.5E-3$	-2.93	$-4.9E-3$	$-9.2E-3$	-1.75
FNF2	S1S1	1.6652	$-2.1E-2$	$-4.9E-2$	-4.03	$-3.6E-2$	$-3.6E-2$	-0.29
FNF3	S1S1	2.6315	$-5.6E-1$	$-6.5E-1$	-6.10	$-2.8E-1$	$-4.2E-1$	$+9.12$
FNF4	S1S1	3.2239	-1.0	-1.1	-6.62	$-4.1E-1$	$-8.0E-1$	$+17.0$
FNF5	S1S1	3.3066	-1.1	-1.2	-6.67	$-4.3E-1$	$-8.6E-1$	$+18.2$
RPN	Def.	5.2661	$-8.9E-4$	$-3.9E-3$	-3.73	$+1.8E-3$	$+1.5E-3$	-1.30
BMK1	Def.	0.5654	$-3.0E-8$	$-5.0E-8$	-0.05	$-^*$	$-^*$	$-^*$
BMK2	Def.	0.5641	$+6.0E-6$	$-9.9E-6$	-0.31	$-2.0E-7$	$-1.6E-5$	-0.30
JJ2	Def.	0.5427	$-5.0E-9$	$-8.0E-9$	-0.02	$-^*$	$-^*$	$-^*$
JJ3	Def.	0.5451	$-6.0E-9$	$-9.0E-9$	-0.02	$-^*$	$-^*$	$-^*$
JM1	Def.	0.4049	$-6.0E-9$	$-9.0E-9$	-0.03	$-^*$	$-^*$	$-^*$
JM2	Def.	0.5449	$+3.0E-8$	$+2.0E-11$	-0.07	$-4.0E-8$	$-7.0E-8$	-0.07
JM3	Def.	0.2664	$-4.0E-9$	$-7.0E-9$	-0.05	$-5.0E-9$	$-7.0E-9$	-0.05

* Results identical to those for natural parameterization to all digits given.

"Def." The notation "N0S0" means the curve has a zero normal (vertical slope) at (x_1, y_1) and a zero slope at (x_n, y_n). Similarly, "S2S0" indicates an initial slope of two and a final slope of zero. All reported energies were computed using a 50-panel Simpson's rule to approximate each of the $n-1$ integrals in (2.7). (Numerical tests indicate that the result is probably accurate to at least five decimal places.) The "optimal v," v_{opt}, was computed by subroutine LCLMIN of Hausman [8], with a requested final interval length of 10^{-3}. (Tests showed that the results are relatively insensitive to this convergence criterion.) Further details on the computations may be found in Fritsch and Springmeyer [7].

Table 1 contains details on the WF-spline through each of these data sets, with the indicated BC. Where possible, the ordinary cubic spline was also obtained, and its energy E_0 computed via (1.4). The other quantities in the table are the total arc length and energy of the curve; k_{ratio}, the ratio of the largest and smallest k_i in (2.4);[2] v_{min} and

---

[2] The departure of k_{ratio} from one is an indication of the amount of breakpoint rearrangement required to make the WF-spline component derivatives continuous.

v_{max}, the largest and smallest v_i-values, when represented as a v-spline in the WF parameterization.

Table 2 compares the WF-spline with the four other parametric v-splines discussed in § 3.2. The quantity $\Delta E = E - E_{WF}$ will be positive if, and only if, the WF-spline has the smaller energy. (Due to the probable accuracy of the integrals, ΔE is given only to one significant figure if it is smaller than 10^{-6} in magnitude.) The value of v_{opt} is also given for each of the OUSN-splines. In case $k_{ratio} = 1$, no reparameterization was necessary, and the last three columns have been left blank.

5. Observations and Conclusions.

5.1. The WF-spline usually has a smaller energy than the ordinary cubic spline, as predicted by Fowler [3]. The only exception is the SIN2 data with specified boundary slopes (see Table 1).

5.2. The majority of the v-values given in the tables are negative. This means that the curve is "looser" than the corresponding parametric cubic spline. It also means that the curve may not minimize functional (2.8). The fact that most values of v_{opt} are

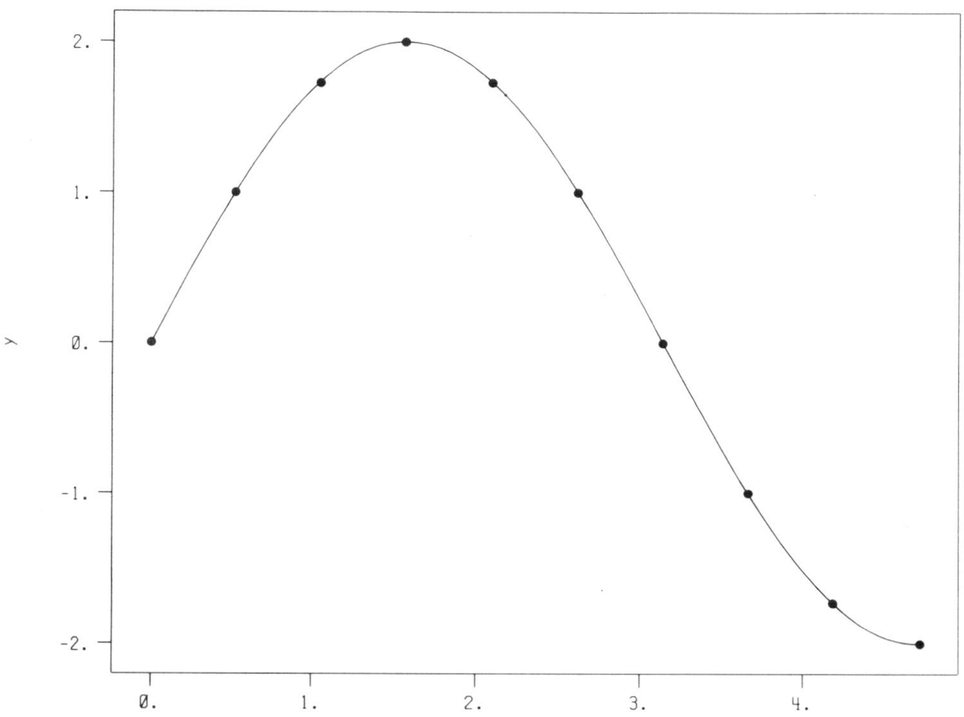

Figure 2. WF-spline through SIN2 data (default BC).

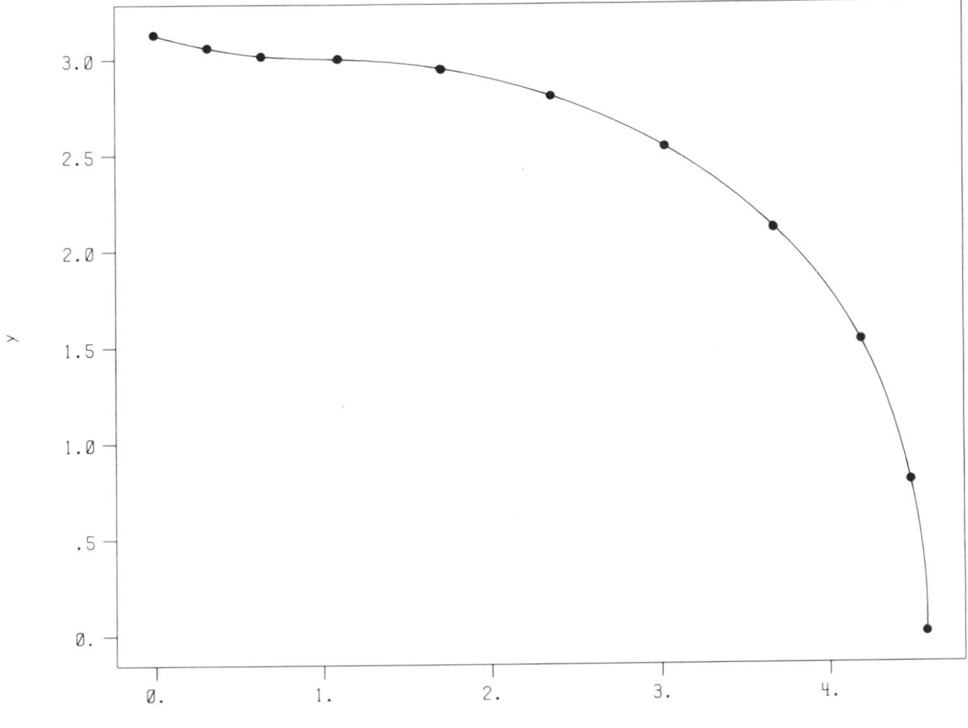

Figure 3. WF-spline through BMK2 data (default BC).

negative suggests that v-splines with negative (and unequal) v-values may provide a closer approximation to the true elastica, with its desirable shape properties, than standard cubic splines.

5.3. For one data set (SIN2) the WF-spline is better than both the PC-spline and the OUSN-spline[3] in either parameterization. For two sets (WRM and RPN) it is better than both in the WF parameterization, but worse than both in the natural parameterization. For one (SF1) it is better than PC but worse than OUSN in both parameterizations. For another (FNF1) it is better than the PC-spline only in the natural parameterization. It is clearly not possible to draw any general conclusion about the goodness of one of these parameterizations over the other. It should be noted that in all these cases except WRM and RPN, the five parametric splines are so similar as to be indistinguishable on the scale of a plot.

---

[3] This is possible, since the WF-spline is definitely *not* uniformly shaped.

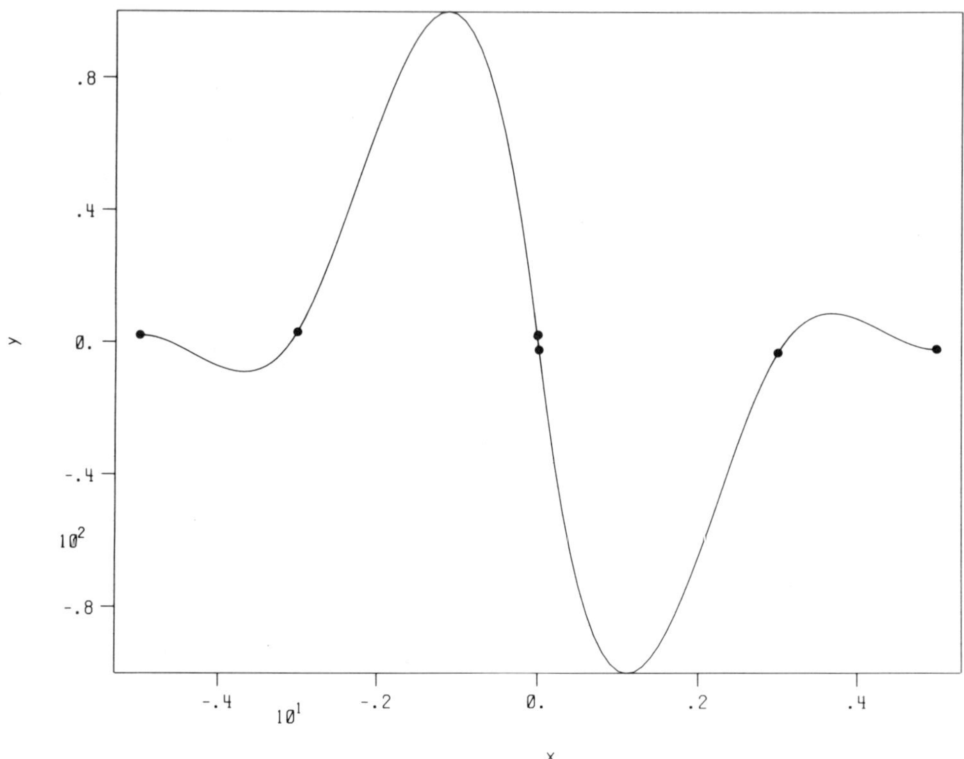

Figure 4. Ordinary cubic spline through FNF4 data (slope 1 at both ends). (Note the drastically different vertical scale than for Figs. 5–9.)

5.4. For "realistic" data sets (JJ*i*, JM*j*, BMK*k*), these results provide no basis for choosing one of these splines over another. The curves are indistinguishable and the WF-spline component functions have extremely small derivative discontinuities. See Figs. 2 and 3 for "typical" WF-splines.

5.5. On the other hand, there are some data sets (FNF3–FNF5) for which the WF-spline is much worse than the PC spline, and the natural parameterization is significantly better than the WF parameterization for the same method of choosing v. This is clearly illustrated by Figs. 4–9, in which the six splines through the FNF4 data set are seen to be clearly different.

5.6. The question posed in § 1.1 has not been completely answered. Conflicting interpretations of the results of these computations are possible:

(1)   WF-splines are as good as PC-splines for interpolating design data, so the present system may as well be left as is.

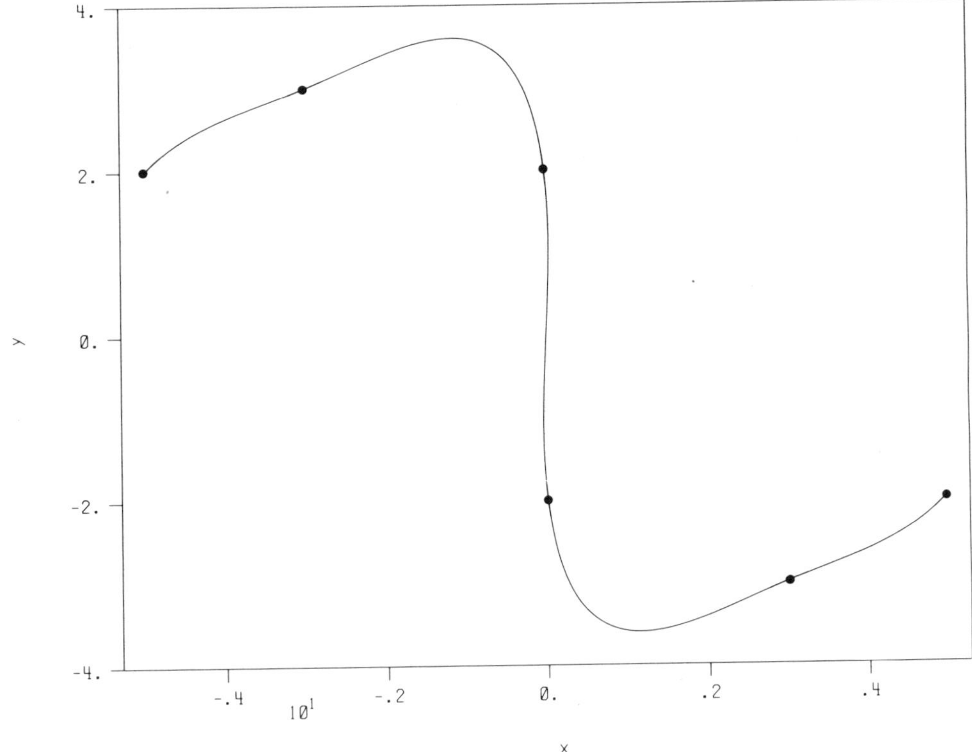

Figure 5. WF-spline through FNF4 data (slope 1 at both ends).

(2)   PC-splines are as good as or better than WF-splines. Since they are cheaper to compute and their mathematical properties are better established, perhaps it is time to consider changing from WF-splines to PC-splines.

It is clear that much more extensive testing on actual design data would be necessary to justify the expense of changing a working system.

6.   Open Questions.   It is clear that much more work is needed to supply the "why?" for the conclusions of the previous section and to continue progress toward better approximations to the true elastica. Some of the open questions include the following.

6.1.   Properties of WF-splines.   Why is the WF-spline good on the SIN2 data and so poor on the FNF data? For what types of data sets can the WF-spline be expected to be good?

The FNF data are really five representatives of a parameterized family of data sets (see Appendix), the parameter being μ, the magnitude of the abscissae of the third and

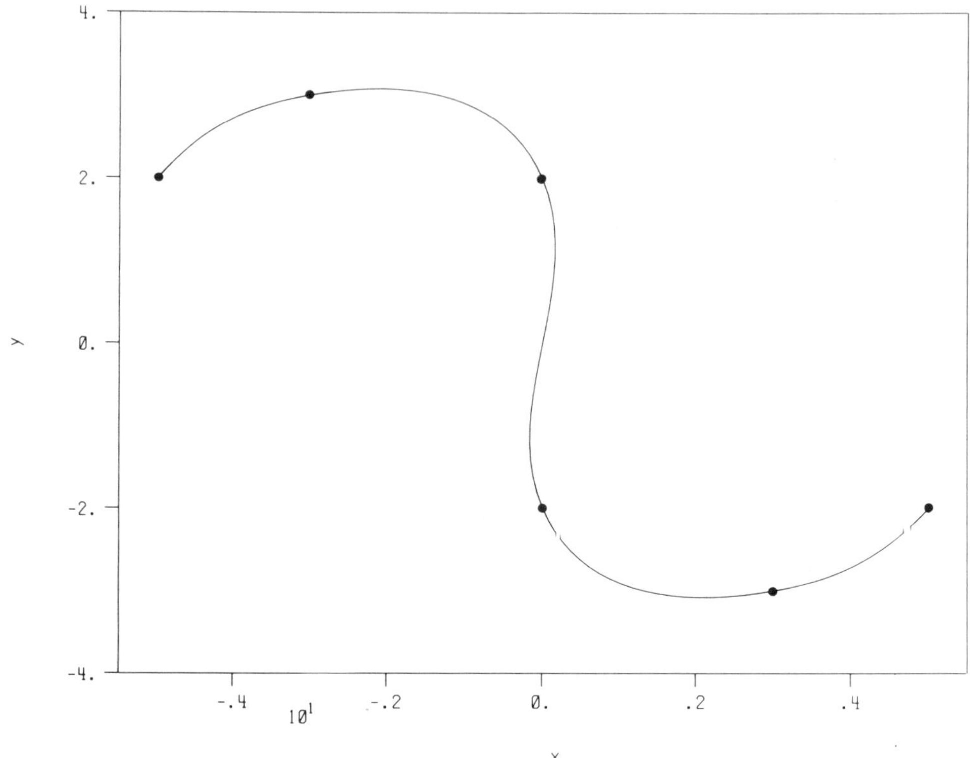

Figure 6. PC-spline through FNF4 data (slope 1 at both ends; natural parameterization).

fourth points. It is not evident from the data presented in Table 1, but there are really only two distinct v-values, due to the symmetry of the data. Their values, together with the values of μ and NIT, the number of iterations of the Melvin [11] algorithm required for computing the WF parameters, for the FNFk data sets are given in Table 3. Among the questions that suggest themselves are:

(a)   If the value of $\mu \in (0.5, 1.0)$ at which $v_2 = v_3$ were chosen, how would the energy of the WF-spline (which would be uniformly shaped) compare with that of the OUSN-spline?

Table 3

k	μ	v_2	v_3	NIT
1	1.0	−2.49	−5.63	3
2	0.5	−6.15	−2.70	4
3	0.1	−9.60	+0.61	6
4	0.01	−10.08	+1.80	15
5	0.0	−10.13	+1.94	9

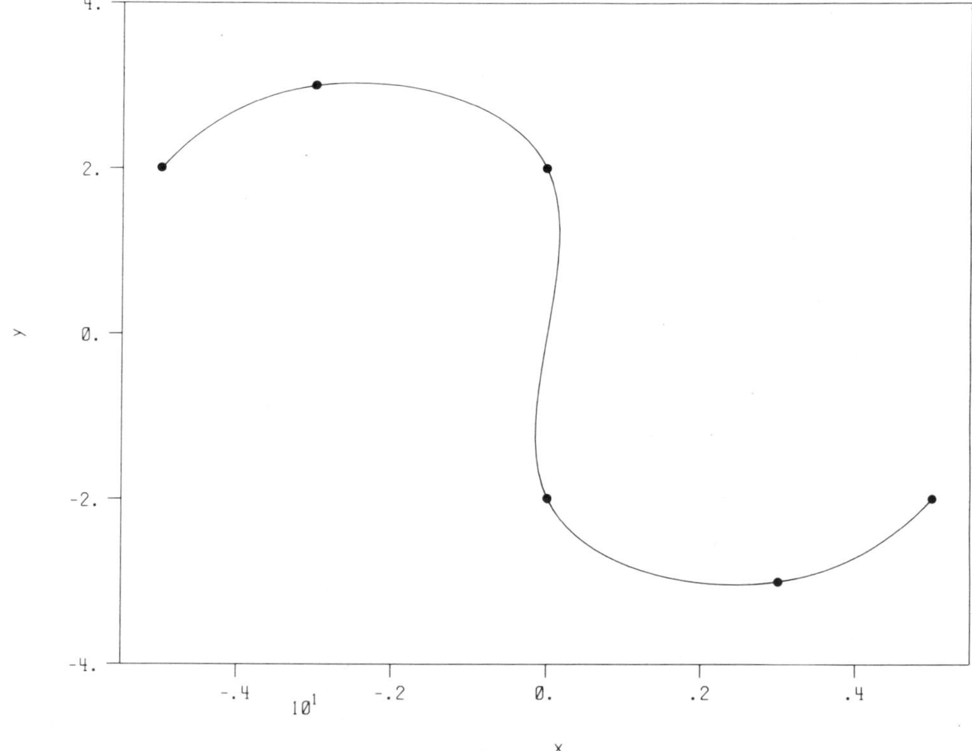

Figure 7. OUSN-spline through FNF4 data (slope 1 at both ends; natural parameterization).

(b)   Does anything special happen when $v_3 = 0$?

(c)   Why isn't NIT monotonic?

6.2.   Effect of parameterization and boundary conditions. Since such a dramatic change in energy and curve shape with parameterization was observed with the FNF data sets, it might be worth investigating "optimal parameterizations" for v-splines. For example, it may well be that neither parameterization considered here is "best" for SIN2 data. One possible approach is given by Marin [10].

The effect of the magnitudes of the boundary tangent vectors on the energy of the curve has not been investigated at all. This may well have biased the results of these tests in favor of the WF-spline and warrants further study.

6.3.   Optimal nonuniform v-splines. As noted in § 5.2, there is definite evidence that nonuniform v-splines, with both positive and negative v-values, may be better than any of the splines considered here. The author would like to encourage research into methods for choosing v-values which will provide better piecewise cubic approximations to the true elastica.

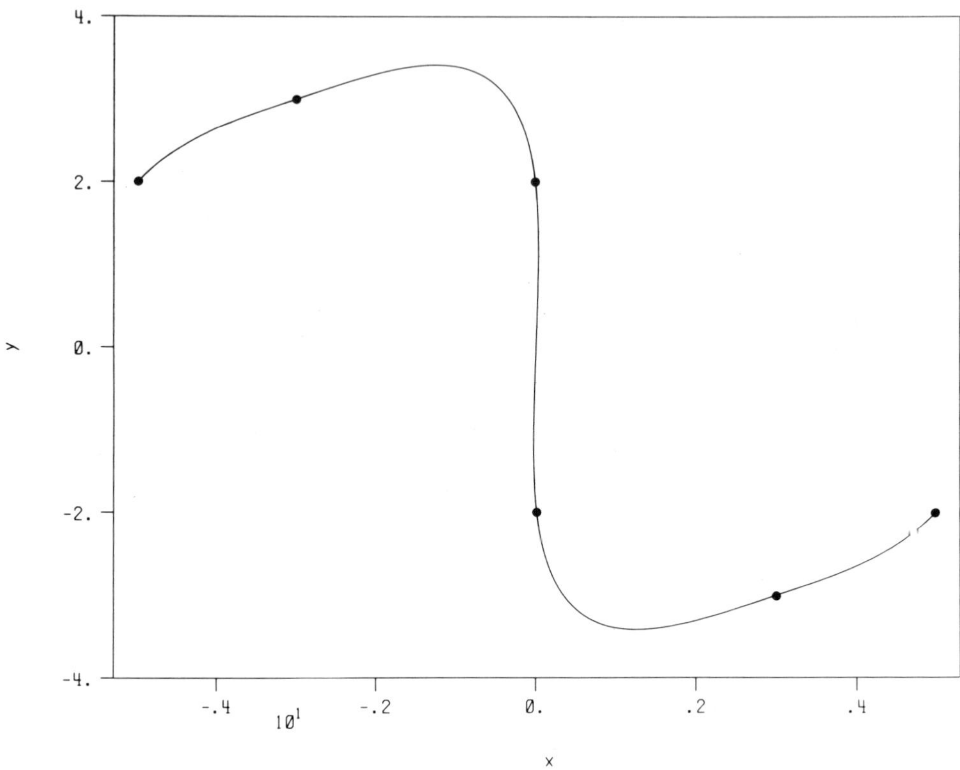

Figure 8.   PC-spline through FNF4 data (slope 1 at both ends; WF parameterization).

Appendix.   Listing of the Test Data.
WRM:

These four points are: $(0, 2)$, $(0.5, 1.5)$, $(1, 1)$, $(1, 0)$.

SF1:

These five points are: $(5, 0)$, $(4, 1)$, $(3, 4)$, $(1, 5)$, $(0, 5)$.

SF2:

These seven points are: $(-2.5, 1)$, $(-2, 2)$, $(-1, 2.5)$, $(0, 2.75)$, $(1, 2.5)$, $(2, 2)$, $(2.5, 1)$.

FNFk, $k = 1, \cdots, 5$:

These six points are: $(-5, 2)$, $(-3, 3)$, $(-\mu_k, 2)$, $(\mu_k, -2)$, $(3, -3)$, $(5, -2)$, where $\mu_1 = 1$, $\mu_2 = 0.5$, $\mu_3 = 0.1$, $\mu_4 = 0.01$, $\mu_5 = 0$.

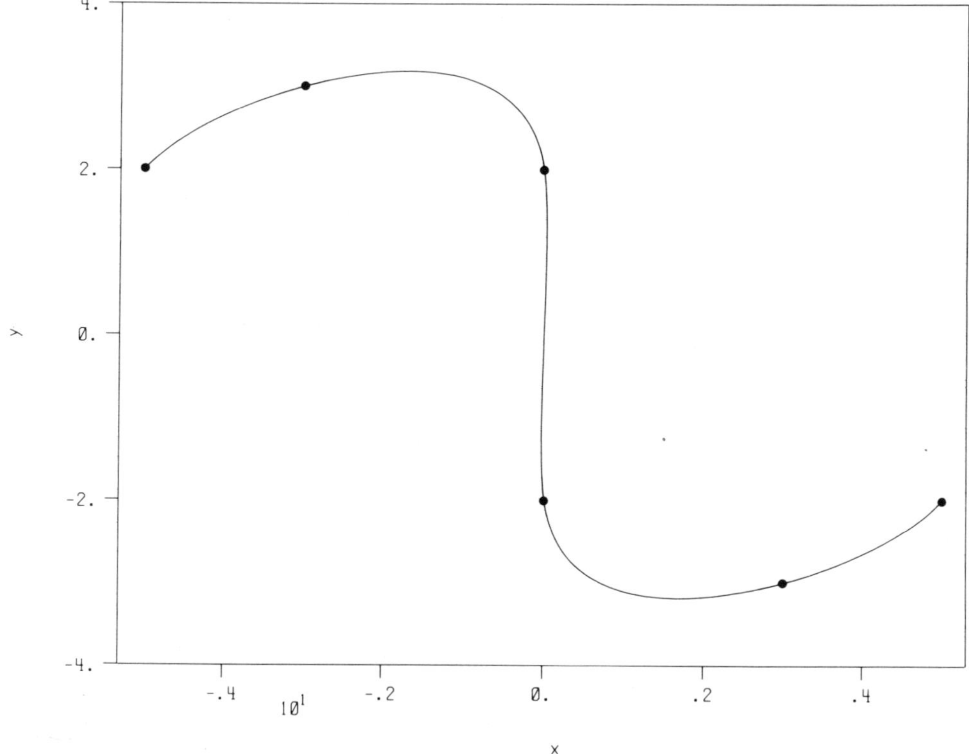

Figure 9. OUSN-spline through FNF4 data (slope 1 at both ends; WF parameterization).

RPN:

These nine points are:

i	x_i	y_i
1	0.0	0.0
2	0.000027	0.1
3	0.043722	0.189935
4	0.169183	0.684269
5	0.469428	1.084085
6	0.943740	1.728312
7	0.998636	3.727558
8	0.999919	6.727558
9	0.999994	11.727558

BMK1 (BMK2):

These data are listed below, in polar coordinates. (θ is in degrees.) Point numbers in parentheses indicate which points of BMK1 constitute BMK2.

i	θ_i	r_i	i	θ_i	r_i
1(1)	0.0	4.582575	24	46.0	3.774696
2	2.0	4.581565	25	48.0	3.717856
3	4.0	4.578537	26(6)	50.0	3.662133
4	6.0	4.573497	27	52.0	3.607701
5	8.0	4.566457	28	54.0	3.554702
6(2)	10.0	4.557429	29	56.0	3.503250
7	12.0	4.546433	30	58.0	3.453436
8	14.0	4.533491	31(7)	60.0	3.405328
9	16.0	4.516018	32	62.0	3.358977
10	18.0	4.491610	33	64.0	3.314416
11(3)	20.0	4.460876	34	66.0	3.271665
12	22.0	4.424473	35	68.0	3.230735
13	24.0	4.383080	36(8)	70.0	3.191622
14	26.0	4.337383	37	72.0	3.154470
15	28.0	4.288056	38	74.0	3.123414
16(4)	30.0	4.235752	39	76.0	3.099905
17	32.0	4.181087	40(9)	78.0	3.083516
18	34.0	4.124637	41	80.0	3.073963
19	36.0	4.066934	42	82.0	3.071082
20	38.0	4.008456	43(10)	84.0	3.074826
21(5)	40.0	3.949636	44	86.0	3.085257
22	42.0	3.890854	45	88.0	3.102553
23	44.0	3.832444	46(11)	90.0	3.127017

Acknowledgments. The author wishes to thank Greg Nielson, who pointed out to him the connection between β-splines and v-splines, Bill Gordon, who suggested the use of true energy as a measure of comparison, Becky Springmeyer, who wrote the program that generated the original results and produced the spline plots and Joe Janzen and John Martin who supplied him with realistic data for these tests. This work was performed under the auspices of the U.S. Department of Energy by Lawrence Livermore National Laboratory under contract W-7405-Eng-48. It was supported in part by the Applied Mathematics Research Program, Office of Energy Research.

REFERENCES

[1] C. DE BOOR, *A Practical Guide to Splines*, Springer-Verlag, New York, 1978.

[2] S. K. FLETCHER, CADCAM-001: *Spline transfer through* IGES, Report SAND 83-2131, Sandia National Laboratories, Albuquerque, NM, 1983.

[3] A. H. FOWLER, *Slopes for cubic spline fit: test curve*, internal memorandum, Union Carbide Nuclear Company, Oak Ridge, TN, 1961.

[4]   A. H. FOWLER AND C. W. WILSON, *Cubic spline, a curve fitting routine,* Y-12 Plant Report Y-1400 (Revision 1), Union Carbide Corporation, Oak Ridge, TN, 1966.

[5]   F. N. FRITSCH, *The Wilson–Fowler spline is a v-spline,* Preprint UCRL-93801, Lawrence Livermore National Laboratory, Livermore, CA, 1985. (Submitted to Computer Aided Geometric Design.)

[6]   F. N. FRITSCH AND R. E. CARLSON, *Monotone piecewise cubic interpolation,* SIAM J. Numer. Anal., 17 (1980), pp. 238–246.

[7]   F. N. FRITSCH AND R. R. SPRINGMEYER, *Software for plotting Wilson–Fowler and v-splines,* Computer Documentation Report UCID 20658, Lawrence Livermore National Laboratory, Livermore, CA, 1985.

[8]   R. F. HAUSMAN, JR., *A hybrid subroutine for function optimization on a line segment,* Computer Documentation Report UCID-30039, Lawrence Livermore Laboratory, Livermore, CA, 1972.

[9]   IIT Research Institute, APT *Part Programming,* McGraw-Hill, New York, 1967.

[10] S. P. MARIN, *An approach to data parametrization in parametric cubic spline interpolation problems,* J. Approx. Theory, 41 (1984), pp. 64–86.

[11] W. R. MELVIN, *Error analysis and uniqueness properties of the Wilson–Fowler spline,* Report LA-9178, Los Alamos National Laboratory, Los Alamos, NM, 1982.

[12] G. M. NIELSON, *Some piecewise polynomial alternatives to splines under tension,* in Computer Aided Geometric Design, R. E. Barnhill and R. F. Riesenfeld, eds., Academic Press, New York, 1974, pp. 209–235.

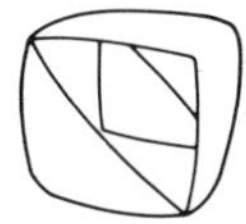

Assembling Triangular and Rectangular Patches and Multivariate Splines

THOMAS JENSEN

Abstract. Algorithms are developed for building complex, smooth surfaces from combinations of rectangular and triangular patches or combinations of rectangular and triangular splines. The criteria for smoothness are first presented and then applied in the development of tractable patches using Bernstein–Bézier bases. Subsequently, an algorithm for combining box spline surfaces of different topology is presented and used to construct spline surfaces on general, closed polyhedra.

Introduction. The number of parametric surface methods for the representation of geometry has grown steadily since the recognition of the need for such constructive methods for computer modeling and simulation in the 1960s. The reasons for the growth are simply the abstract richness of the subject and the fact that, as of yet, no single representation has proven to be sufficient for the variety of problems for which parametric surfaces are used. In fact, it is often the case that a single scheme is insufficient for a single object and that design would be easier, and perhaps qualitatively better, if more than one type of patch or surface could be used in its construction.

The assembly and constraint of vector-valued parametric patches to ensure smooth boundaries is a nontrivial problem. It requires precise specification and generalization of the definitions of smoothness (parametric continuity of the functions of adjacent elements has been shown to be insufficient for what might be called piecewise constructive elements, such as Bézier rectangular and triangular patches), as well as careful analysis of the particular elements themselves. Only recently have significant results been achieved for vector-valued parametric patches and then mostly for the assembly of patches of the same representation.

Constructions for matching individual, polynomial Bézier patches to others with the same topology (rectangles to rectangles, triangles to triangles) are presented in Farin [10] and Kahmann [15]. Herron [13] has addressed the problem of matching

203

rational, triangular patches to one another and his results are extensible, as will be seen. Piper [17] also has proposed a scheme for matching quartic Clough–Tocher elements. His interpolant is similar to what is proposed here for triangles, although the development is very different.

Charrot and Gregory [5] have addressed the issue of matching triangular and five-sided patches to rectangular patches. However, at this time, the author is unaware of any previous work matching box spline surfaces of different topologies.

The sequel consists of four sections. In the first, criteria for first and second order geometric continuity are proposed. In the second, a method for building arbitrary, mixed networks of individual, rectangular and triangular patches which satisfy the first order continuity criterion is developed and compared to the other existing single element methods. In the third section, a method for smoothly assembling box splines of different topology and order is presented. The paper concludes with a construction for defining multivariate box spline surfaces over general, closed polyhedra.

Throughout this paper, all equations, definitions, and figures are indexed sequentially by a single system.

1. Geometric Continuity. As in Herron, the following definition of geometric continuity of the first order formalizes the concept of "smooth" joins between the vector-valued images of a set of functions used piecewise to produce a larger surface.

DEFINITION 1.1. A surface S, composed of parametric patches is GC^1, geometrically smooth, if the surface is C^0 and there exists a unique tangent plane defined at each point of the surface.

Clearly, if a surface is C^1, it is GC^1 as well. Let

$$\{\mathbf{F}_i(r, s) = (f_{i1}(r, s), f_{i2}(r, s), f_{i3}(r, s))\}, \quad i = 1, 2$$

be two internally C^1 patches defined on the unit square and having the property that $\mathbf{F}_1(1, s) = \mathbf{F}_2(0, s)$. The union S of the two patches will be GC^1 if the unit normals of the two patches coincide on the common boundary or,

$$(1.2) \qquad \mathbf{N}_1 = \mathbf{N}_2 \qquad \text{where } \mathbf{N}_i = \frac{(\partial \mathbf{F}_i / \partial r \times \partial \mathbf{F}_i / \partial s)}{|(\partial \mathbf{F}_i / \partial r \times \partial \mathbf{F}_i / \partial s)|}.$$

Clearly, the r and s partials of each patch need not be of equal magnitude. In fact, the cross-boundary partials, the r partials in this example, need not be collinear. The four partials merely need to be coplanar.

As will be seen in § 2, this definition is useful as well as intuitive. We now turn to the definition of second order geometric continuity. Since the human eye is particularly adept at detecting discontinuities in the curvature of a surface, this is more than an abstract issue for applications such as automobile body design. Definition 1.1 equated GC^1 with a concise and general condition on the first partial derivatives; the following definitions from classical differential geometry provide a similar condition on the second partials, ensuring continuity of curvature (O'Neill [16]). They constitute a natural definition for GC^2.

DEFINITION 1.3. Let $\mathbf{v}$ be a tangent vector to a surface S at $\mathbf{p}$ and G be a real or vector-valued function on S. Then $\nabla_{\mathbf{v}} G(\mathbf{p})$, the co-variant derivative of G with respect to $\mathbf{v}$, is the common value of $(\partial/\partial t)(G(\mu(t)))(0)$ for all curves μ in S emanating at $\mathbf{p}$ which have initial velocity $\mathbf{v}$.

The co-variant derivative is particularly useful for this problem because it is independent of the parameterization of S, or more specifically, the parameterizations of the pieces of S. There are a variety of methods for computing it, even if S is defined implicitly. However, if S is defined parametrically by a function $\mathbf{F}$, the co-variant derivative can be evaluated directly. In that case, since $\mathbf{v}$ is a tangent vector to S at $\mathbf{p}$, there exist u_1 and u_2 such that $\mathbf{v} = u_1 \, \partial\mathbf{F}/\partial r + u_2 \, \partial\mathbf{F}/\partial s$. Letting $\mathbf{u} = (u_1, u_2)$, an explicit formula for a μ is $\mu(t) = \mathbf{F}(\mathbf{u}t + \mathbf{q})$, where $\mathbf{q}$ is the pre-image of $\mathbf{p}$.

DEFINITION 1.4.   Let $\mathbf{v}$ be a tangent vector to S and let $\mathbf{U}$ be the unit normal vector field on a neighborhood of $\mathbf{p}$ in S. The Shape Operator, $\text{Shape}_\mathbf{v} S(\mathbf{p})$, is defined by

$$(1.5) \qquad\qquad \text{Shape}_\mathbf{v} S(\mathbf{p}) = -\nabla_\mathbf{v} U(\mathbf{p}).$$

$\text{Shape}_\mathbf{v} S(\mathbf{p})$ is a symmetric linear operator on the two-dimensional space of the tangent plane of S at $\mathbf{p}$. Further, it has the property that if $\mathbf{v}$ is a unit tangent vector then $\text{Shape}_\mathbf{v} S(\mathbf{p})$ is the curvature of S in the v direction. The Shape Operator contains all the information about curvature of the surface of S in any direction.

DEFINITION 1.6.   A surface S is GC^2, has second order geometric smoothness, if and only if it is GC^1 and the Shape Operator, $\text{Shape}_\mathbf{v} S(\mathbf{p})$, is uniquely defined at each point.

In the constructions that follow, only GC^1 continuity is achieved. The author is unaware of any tractable GC^2 patches and this remains an open problem.

2. Constructing Mixed, GC^1 Networks of Triangular and Rectangular Patches.   In what follows, the implied domain of every rectangular patch is the unit square and the parameters corresponding to a point in that domain are its Cartesian coordinates. Triangular patches, in turn, are presumed to be defined over an equilateral triangle with sides of unit length; the parameters are the barycentric coordinates. The following properties of the barycentric coordinates of a triangle are stated without proof and can be found in a number of places in the literature (Barnhill [1]; Boehm [4]).

Let $\mathbf{J}_0, \mathbf{J}_1, \mathbf{J}_2$ denote the vertices of any triangle. Every point $\mathbf{p}$ inside the triangle can be uniquely described by a triple (b_0, b_1, b_2) where,

$$(2.1) \qquad\qquad \mathbf{p} = \sum_{i=0,2} b_i \mathbf{J}_i$$

and

$$(2.2) \qquad\qquad 1 = \sum_{i=0,2} b_i.$$

$$(2.3) \qquad b_i(\mathbf{J}_j) = \delta_{ij} \qquad \text{(the Kronecker delta function)}.$$

All boundary curves in this section are cubic Bézier curves. All cross-boundary derivatives are defined, as in Herron, as a product of two interpolants. One interpolates the magnitudes of the cross-boundary derivatives at the vertices of an

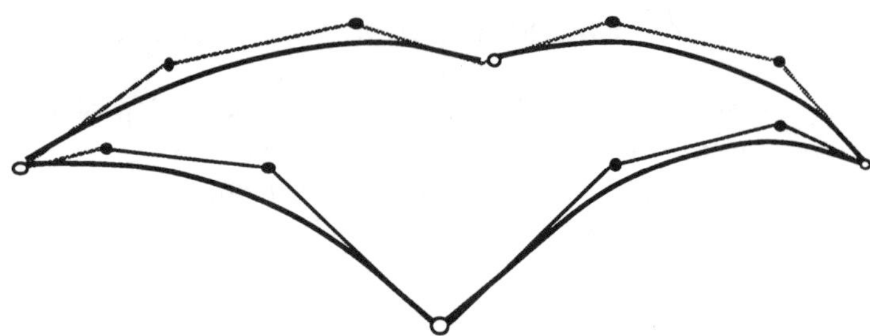

Figure 2.5.   Data defining rectangular patch.

edge; the other interpolates the directions of the derivatives at the vertices.

$$(2.4) \qquad \frac{\partial \mathbf{F}}{\partial \mathbf{D}} = \mathbf{R}(t) = q(t)\mathbf{Q}(t).$$

D, the direction of differentiation in the domain, is chosen to be dependent on the tangential derivative vector common to two adjacent patches and is not initially defined in terms of the parameterization of the patch. In particular, if **T** denotes a tangent derivative in the direction of an edge **e** at an end point of **e**, **D** is constructed so that **R** is perpendicular to **T**. The consequence of this minor, additional complexity is that the construction of a patch need not reference a neighbor's parameterization to ensure GC^1. The definition of **D** creates a common cross-boundary direction for differentiation which is independent of parameterization. Further, in the methods that follow in this section, only the corresponding $\mathbf{Q}(t)$'s, that is the direction terms of corresponding cross-boundary derivatives, are matched. This ensures the existence of a unique, common unit normal and so the conditions for GC^1 are met.

In contrast to Herron's implementation of these ideas, the cross-boundary derivatives are chosen to be cubic, rather than quadratic, and the triangular interpolant is polynomial. This increase in degree purchases fairer surfaces for both the rectangular and triangular elements.

We now refine these ideas and, in particular, turn to the problem of building suitable rectangular patches from cubic Bézier boundary curves, Fig. 2.5.

In order to be consistent with the development for triangles, we introduce the following notation for the domain and pertinent derivatives used to define the rectangular patch, Fig. 2.6.

$$(2.7) \qquad \begin{array}{l} \mathbf{K}_0 = (0,0), \quad \mathbf{K}_1 = (1,0), \quad \mathbf{K}_2 = (1,1), \quad \mathbf{K}_3 = (0,1), \\ \mathbf{e}_i = \mathbf{K}_j - \mathbf{K}_i, \qquad j = i + 1/\bmod 4. \end{array}$$

T_{ij} denotes the tangential derivative at vertex i in the direction towards vertex j. The rectangular patch itself, **F**, is represented as a convex combination, in the spirit of Brown and Little, Barnhill [1], of two tensor product Bézier patches. A nonstandard labeling of the Bézier control points, Fig. 2.6, is used for edge symmetry. The control points also have three subscripts instead of two: the first denotes the polynomial Bézier patch

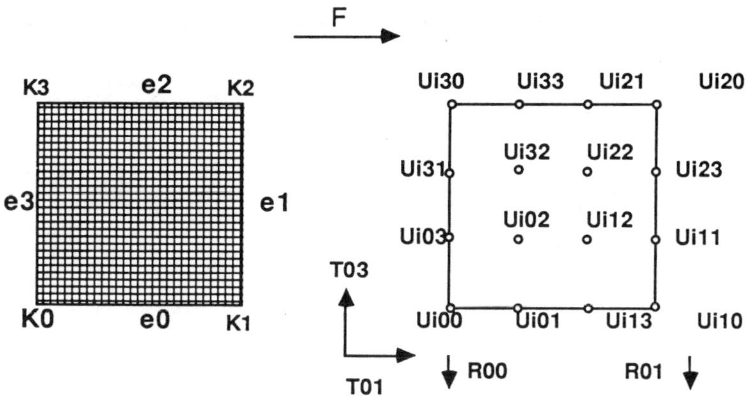

Figure 2.6.   Rectangular patch notation.

with which the control point is associated.

$$(2.8) \qquad \mathbf{F} = \frac{[r^2(1-r)^2\mathbf{F}_1 + s^2(1-s)^2\mathbf{F}_2]}{[r^2(1-r)^2 + s^2(1-s)^2]}$$

where

$$
\begin{aligned}
(2.9) \qquad \mathbf{F}_i = \sum_{j=0,3} & (1 - g_j(r,s))^3(1 - h_j(r,s))^3\mathbf{U}_{ij0} \\
& + 3(1 - g_j(r,s))^2 g_j(r,s)(1 - h_j(r,s))^3\mathbf{U}_{ij1} \\
& + 9(1 - g_j(r,s))^2 g_j(r,s)(1 - h_j(r,s))^2 h_j(r,s)\mathbf{U}_{ij2} \\
& + 3(1 - g_j(r,s))^3(1 - h_j(r,s))^2 h_j(r,s)\mathbf{U}_{ij3}
\end{aligned}
$$

and

$$g_0(r,s) = r, \quad g_1(r,s) = s, \quad g_2(r,s) = 1 - r, \quad g_3(r,s) = 1 - s,$$
$$h_0(r,s) = s, \quad h_1(r,s) = 1 - r, \quad h_2(r,s) = 1 - s, \quad h_3(r,s) = r.$$

The boundary control points of both of the tensor product polynomial patches, $\{\mathbf{U}_{ijk}\}$, $i = 1,2$, $j = 0, 3$, $k = 0, 1, 3$, are provided by design and coincide, $\mathbf{U}_{1jk} = \mathbf{U}_{2jk}$, $k = 0, 1, 3$. The interior control points, $\{\mathbf{U}_{ij2}\}$, are generated using the strategy described above. Let $\mathbf{R}_i(t)$ denote the cross-boundary derivative on $\mathbf{e}_i$, $t = g_i(r,s)$,

$$
\begin{aligned}
(2.10) \qquad \mathbf{R}_i(t) = \frac{\partial \mathbf{F}}{\partial \mathbf{D}_i} = & \left[\frac{(1-t)\mathbf{R}_{i0}}{|\mathbf{R}_{i0}|} + \frac{t\mathbf{R}_{i1}}{|\mathbf{R}_{i1}|} \right] \\
& \cdot [u_0(t)|\mathbf{R}_{i0}| + u_1(t)C_i(|\mathbf{R}_{i0}| + |\mathbf{R}_{i1}|)/2 + u_2(t)|\mathbf{R}_{i1}|]
\end{aligned}
$$

where

$$(2.11) \qquad u_0(t) = 2t^2 - 3t + 1, \quad u_1(t) = 4t - t^2, \quad u_2(t) = 2t^2 - t,$$

and $\{C_i > 0\}$, $i = 0, 2$ are free parameters controlling the interior shape of the surface.

The requirement that $\mathbf{R}_i(t)$ be perpendicular to the tangent derivatives at the end points of $\mathbf{e}$ yields the following conditions:

$$(2.12) \qquad \mathbf{R}_{i0} \cdot \mathbf{T}_{ij} = 0, \qquad \mathbf{R}_{i1} \cdot \mathbf{T}_{jk} = 0$$

where by simple differentiation $\mathbf{T}_{ij} = 3(\mathbf{U}_{i1} - \mathbf{U}_{i0})$ and $\mathbf{T}_{ji} = 3(\mathbf{U}_{j3} - \mathbf{U}_{j0})$. $\mathbf{D}_i$ changes along $\mathbf{e}_i$ and is defined by linear interpolation of its values at the end points of $\mathbf{e}_i$, which are in turn constrained by (2.12). In particular,

$$(2.13) \qquad \mathbf{D}_i = (1 - t)\mathbf{D}_{i0} + t\mathbf{D}_{i1}.$$

Without loss of generality we assume

$$(2.14) \qquad \mathbf{D}_{i0} \cdot \mathbf{e}_l = -1, \qquad \mathbf{D}_{i1} \cdot \mathbf{e}_j = 1.$$

Now all the tangent vectors to a surface are determined if the tangent vectors in two independent directions are known. If $\mathbf{u}$ and $\mathbf{v}$ are independent vectors in the domain and $\mathbf{w}$ is any other direction in the domain,

$$(2.15) \qquad \frac{\partial \mathbf{F}}{\partial \mathbf{w}} = \frac{\partial \mathbf{F}}{\partial \mathbf{u}} \frac{(\mathbf{v} \cdot \mathbf{w}^{\mathrm{perp}})}{(\mathbf{v} \cdot \mathbf{u}^{\mathrm{perp}})} + \frac{\partial \mathbf{F}}{\partial \mathbf{u}} \frac{(\mathbf{w} \cdot \mathbf{u}^{\mathrm{perp}})}{(\mathbf{v} \cdot \mathbf{u}^{\mathrm{perp}})}.$$

This can be applied to $\mathbf{R}_{i0}$ and $\mathbf{R}_{i1}$ and yields formulae for them in terms of the tangent edge derivatives. Substitution of these formulae into (2.12) and (2.14) yields two linear systems which in turn yield

$$(2.16) \qquad \begin{aligned} \mathbf{D}_{i0} &= -\frac{(\mathbf{T}_{ij} \cdot \mathbf{T}_{il})}{(\mathbf{T}_{ij} \cdot \mathbf{T}_{ij})} e_i - e_l, \\[2mm] \mathbf{D}_{i1} &= \frac{(\mathbf{T}_{jk} \cdot \mathbf{T}_{ji})}{(\mathbf{T}_{ji} \cdot \mathbf{T}_{ji})} e_i + e_j \end{aligned}$$

and

$$(2.17) \qquad \begin{aligned} \mathbf{R}_{i0} &= -\frac{(\mathbf{T}_{ij} \cdot \mathbf{T}_{il})}{(\mathbf{T}_{ij} \cdot \mathbf{T}_{ij})} T_{ij} + T_{il}, \\[2mm] \mathbf{R}_{i1} &= -\frac{(\mathbf{T}_{jk} \cdot \mathbf{T}_{ji})}{(\mathbf{T}_{ji} \cdot \mathbf{T}_{ji})} T_{ji} + T_{jk}. \end{aligned}$$

For $\mathbf{F}_1$ in (2.8), $\{\mathbf{U}_{1j2}\}$, $j = 0, 3$ are determined by evaluating (2.10) on $\mathbf{e}_0$ and $\mathbf{e}_2$ at $t = \frac{1}{3}$ and $t = \frac{2}{3}$, setting the result equal to the derivative of (2.9) in the direction $\mathbf{D}_i(t)$, using formula (2.15) again, and solving the two simple linear systems. This was done with a symbolic manipulation program, Macsyma, and the resulting formulae are linear combinations of control points.

Thus, $\mathbf{F}_1$ interpolates to the boundary curves, $\mathbf{R}_0$ and $\mathbf{R}_2$. $\{\mathbf{U}_{2j2}\}$ are determined analogously, using evaluation of the cross-boundary derivative $\mathbf{R}_i$ on $\mathbf{e}_1$ and $\mathbf{e}_3$. $\mathbf{F}_2$ interpolates to the boundary curves, $\mathbf{R}_1$ and $\mathbf{R}_3$. Thus, $\mathbf{F}$, the convex combination of $\mathbf{F}_1$

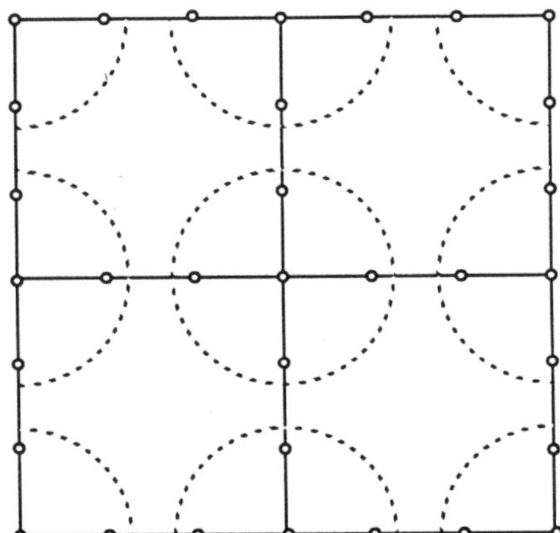

Figure 2.18.   Patches meet smoothly when each boundary control point set enclosed by a dotted curve resides in a single plane.

and $\mathbf{F}_2$, interpolates to both the boundary and all four cross-boundary derivatives. Adjacent patches join smoothly if they share a common boundary and if all the control points immediately adjacent to a vertex and the vertex itself are coplanar, Fig. 2.18.

The construction for triangles is very similar. Instead of a convex combination, a Clough–Tocher like macro-element of three quartic polynomial patches is used to obtain the necessary degrees of freedom, Fig. 2.19. The subpatches are represented as

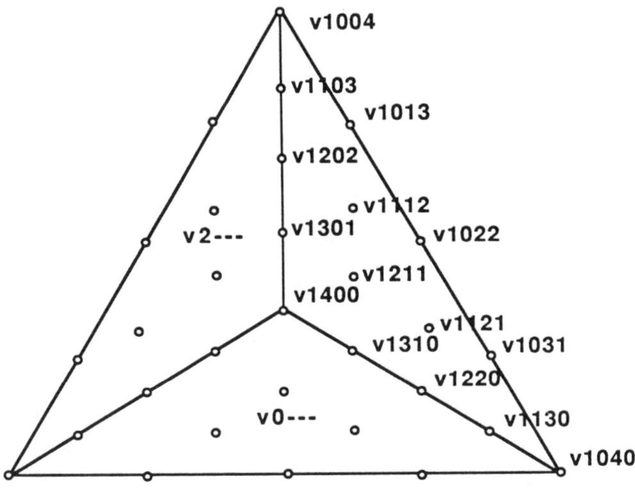

Figure 2.19.   Topology and control points for the three quartic polynomials defining the macro triangular patch.

quartic Bézier triangular patches,

$$(2.20) \qquad \mathbf{F}_i(b_1, b_2, b_3) = \sum_{j+k+l=n} B^n_{jkl}(b_1, b_2, b_3)\mathbf{V}_{ijkl}, \qquad i = 0, 2$$

where

$$(2.21) \qquad B^n_{jkl}(b_1, b_2, b_3) = \frac{n!\, b_1^j b_2^k b_3^l}{(j!\, k!\, l!)} \quad \text{and} \quad n = 4.$$

Note that the first subindex of the control point indicates the subtriangle to which it belongs. In the discussion that follows, substantial use will be made of degree elevation, i.e., representing an nth order polynomial as a polynomial of degree $n + 1$ (see Farin [11]). If

$$(2.22) \qquad \mathbf{F}_i = \sum_{j+k+l=n} B^n_{jkl}\mathbf{V}_{ijkl}, \quad \text{then } \mathbf{F}_i = \sum_{j+k+l=n+1} B^{n+1}_{jkl}\mathbf{P}_{ijkl} \quad \text{where}$$

$$\mathbf{P}_{ijkl} = (j\,\mathbf{V}_{ij-1kl} + k\mathbf{V}_{ijk-1l} + l\mathbf{V}_{ijkl-1})/n.$$

On edge $\mathbf{e}_i$, the cross-boundary derivative is again written as (2.10) and $\mathbf{D}_i$ has the form (2.13). Equations (2.14) are replaced with the constraint,

$$(2.23) \qquad \frac{\partial b_{i1}}{\partial \mathbf{D}_i} = 1$$

where b_{i1} is the barycentric coordinate corresponding to the interior vertex of the ith subpatch. The Gram–Schmidt orthogonalization process yields formulae analogous to (2.16) and (2.17) (see Herron [13]).

$$(2.24) \qquad \begin{aligned} \mathbf{D}_{i0} &= -\frac{(\mathbf{T}_{jk} \cdot \mathbf{T}_{ji})}{(\mathbf{T}_{jk} \cdot \mathbf{T}_{jk})} e_i - e_k, \\[2ex] \mathbf{D}_{i1} &= \frac{(\mathbf{T}_{ki} \cdot \mathbf{T}_{kj})}{(\mathbf{T}_{kj} \cdot \mathbf{T}_{ki})} e_i + e_j \end{aligned}$$

and

$$(2.25) \qquad \begin{aligned} \mathbf{R}_{i0} &= -\frac{(\mathbf{T}_{jk} \cdot \mathbf{T}_{ji})}{(\mathbf{T}_{jk} \cdot \mathbf{T}_{jk})} T_{jk} + T_{ji}, \\[2ex] \mathbf{R}_{i1} &= -\frac{(\mathbf{T}_{ki} \cdot \mathbf{T}_{kj})}{(\mathbf{T}_{kj} \cdot \mathbf{T}_{kj})} T_{kj} + T_{ki}. \end{aligned}$$

The linearity of the other barycentric coordinate functions and the explicit formula for $\mathbf{D}_i$, (2.13), can be used to derive their derivatives in the direction $\mathbf{D}_i$,

$$(2.26) \qquad \begin{aligned} \frac{\partial b_{i2}}{\partial \mathbf{D}_i} &= tY + \frac{(\mathbf{T}_{jk} \cdot \mathbf{T}_{ji})}{(\mathbf{T}_{jk} \cdot \mathbf{T}_{jk})} - 1, \\[2ex] \frac{\partial b_{i3}}{\partial \mathbf{D}_i} &= (1 - t)Y + \frac{(\mathbf{T}_{kj} \cdot \mathbf{T}_{ki})}{(\mathbf{T}_{kj} \cdot \mathbf{T}_{kj})} - 1 \end{aligned}$$

where

$$Y = -\frac{(\mathbf{T}_{jk} \cdot \mathbf{T}_{ji})}{(\mathbf{T}_{jk} \cdot \mathbf{T}_{jk})} - \frac{(\mathbf{T}_{kj} \cdot \mathbf{T}_{ki})}{(\mathbf{T}_{kj} \cdot \mathbf{T}_{kj})} + 1.$$

The boundary control points, $\{\mathbf{V}_{i0kl}\}$, $k + l = 4$, of the three subpatches are found by degree elevation of the cubic boundary curves. For $i = 0, 2$, $\mathbf{V}_{i130}$ and $\mathbf{V}_{i103}$ are determined by the requirement of internal C^1 continuity between subpatches. Each point must be the average of the three adjacent points on the boundary.

Since $\mathbf{F}_i$, (2.20), is quartic, $\partial \mathbf{F}_i / \partial b_{i1}$ is cubic. However, $\partial \mathbf{F}_i / \partial b_{i2}$ and $\partial \mathbf{F}_i / \partial b_{i3}$ restricted to $b_{i1} = 0$ are quadratic, since the boundary curves of the macro-element were restricted to be cubic (this can be seen by ordinary partial differentiation and evaluation). Hence,

$$(2.27) \qquad \frac{\partial \mathbf{F}_i}{\partial \mathbf{D}_i} = \frac{\partial \mathbf{F}_i}{\partial b_{i1}} \frac{\partial b_{i1}}{\partial \mathbf{D}_i} + \frac{\partial \mathbf{F}_i}{\partial b_{i2}} \frac{\partial b_{i2}}{\partial \mathbf{D}_i} + \frac{\partial \mathbf{F}_i}{\partial b_{i3}} \frac{\partial b_{i3}}{\partial \mathbf{D}_i}$$

is the sum of a cubic multiplied by a constant and two quadratics each multiplied by linear functions. $\partial \mathbf{F}_i / \partial \mathbf{D}_i$ is cubic. For each of the subpatches, $\mathbf{V}_{i1221}$ and $\mathbf{V}_{i112}$ are determined by this cross-boundary derivative. As with the rectangular patch, (2.10) is equated to (2.27) and evaluated at $\frac{1}{3}$ and $\frac{2}{3}$. The resulting two-dimensional linear system is solved for the two Bézier control points. Note that these two conditions plus the two implicit conditions matching the endpoints force the two cubics $\mathbf{R}_i$ and $\partial \mathbf{F}_i / \partial \mathbf{D}_i$ to coincide everywhere.

The internal boundaries between subpatches are constrained to be cubics and the remaining coefficients are found using the requirement of internal C^1 continuity. First, the internal boundaries are written as cubics with the tangent derivatives at the external vertices determined as described above. Then, the curves are degree elevated and the cross-boundary derivatives are evaluated twice on each side to generate conditions for the remaining points.

Once again, adjacent patches, whether rectangular or triangular, will join smoothly if they share a common boundary and if the control points about either of the two vertices and that vertex itself reside in a single plane, Fig. 2.28. This compares favorably to the smoothness conditions required by competitive elements. In Farin [10], for example, there are two conditions for geometric smoothness across a boundary between two triangular patches. First, adjacent triangles formed by degree elevated control points on a boundary and the original control points immediate to the boundary must form planar quadrilaterals. Second, the ratio of the areas of pairs of such adjacent triangles must be constant along the boundary.

The condition in Herron [13] is very similar to that proposed here. However, the patches described here are more flexible. Figure 2.29 shows a hemisphere modeled with four of the macro triangular elements. Figures 2.30 and 2.31 show four hemispheres with different values of C_i in (2.10). When $C_i = 1$, (2.10) for $R_i(t)$ reduces to Herron's original derivative function, which was a product of two linear interpolations. The flattening or dampening between adjacent elements that is apparent in the corresponding upper right-hand hemisphere of both figures is caused by linear interpolation of the direction vectors. When $C_i = 1.2$, the parabolic interpolation is

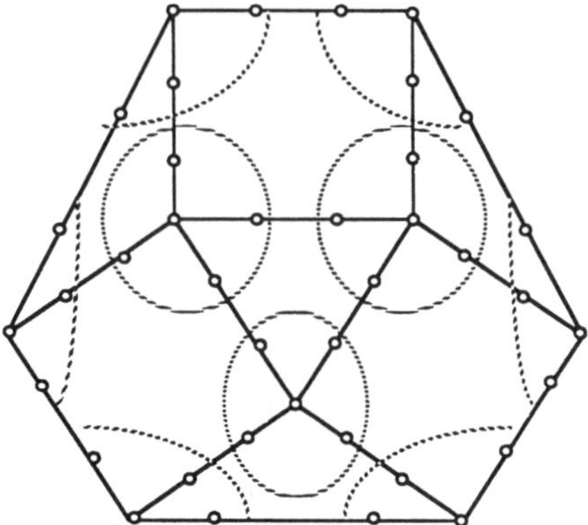

Figure 2.28.   All patches meet smoothly when each set of boundary control points enclosed by a dotted line is coplanar.

more nearly spherical and produces the more pleasing models in the lower left-hand corner of the diagrams. Figure 2.32 is a simple sailing ship modeled entirely with the triangular element.

The GC^1 triangular patch of Gregory and Charrot is based on Coons' blending functions, matches rectangular patches with arbitrary cubic boundaries and produces eighth order polynomials. The method is clearly competitive; for example, both it and the triangular method presented here require the same number of control points. Nevertheless, in design systems using subdivision for intersection calculations, three quartics are likely to be preferable to a single eighth order polynomial. Formulae for both of the elements presented here, in the form of Macsyma output, are available on request.

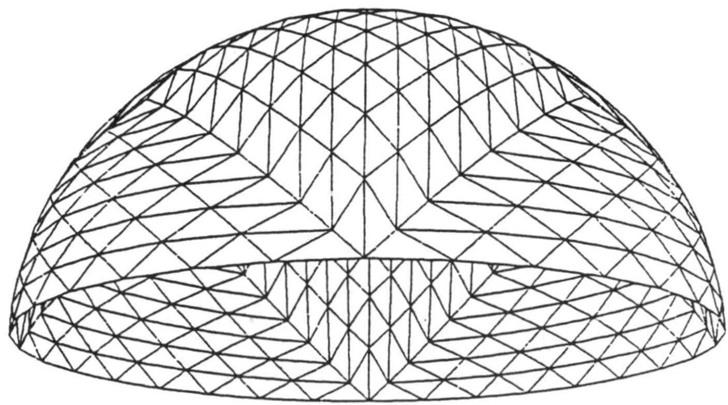

Figure 2.29.   A hemisphere modeled with four triangular elements.

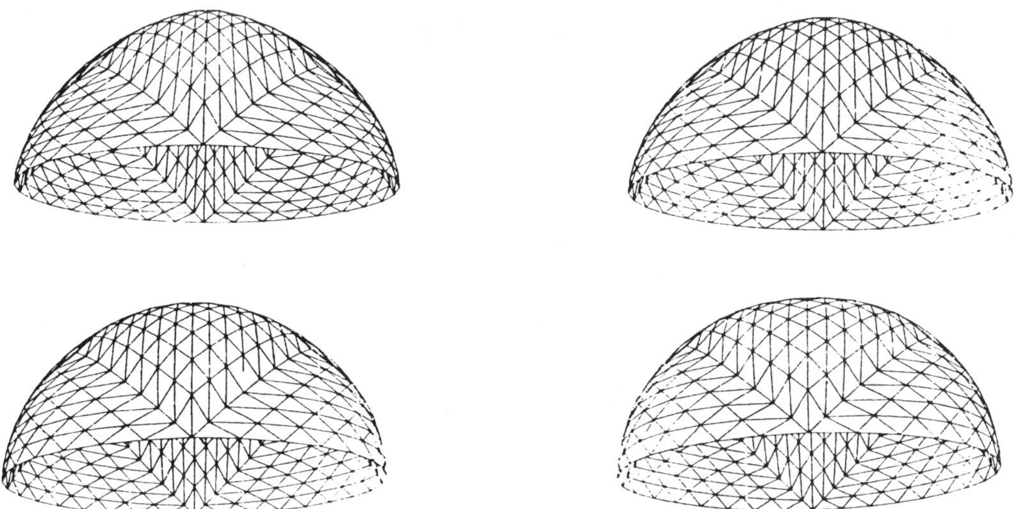

Figure 2.30. Four hemispheres modeled with the 3 quartic patch. The values for C_i left to right and top to bottom are 0.8, 1.0, 1.2 and 1.4.

3. Compositing Multivariate Box Splines.

3. Compositing Multivariate Box Splines. The cross-boundary derivative of a spline surface, like the surface itself, is a piecewise polynomial with an arbitrary number of segments and changes in shape and so assembling such surfaces with different intrinsic topologies requires a different approach from that of the previous section. The goal, however, remains the same: to allow construction of a single surface with both rectangular and triangular regions. We begin by defining box splines, the class of multivariate splines to be assembled.

DEFINITION 3.1. Let $B = \{y = a + \mu_1 a_1 + \cdots + \mu_m a_m \mid \mu_i$ in $[0, 1]$, a, a_i are linearly independent vectors in the affine space $A^m\}$. B is a "box" in A^m. a is called the box vertex.

DEFINITION 3.2 (de Boor and De Vore [8]). Let Π_1 be an affine map from A^m onto A^p, $p \le m$, which maps $\{a_i\}$, $i = 1, m$ onto $\{a_i\}$, $i = 1, p$. Let Π_2 be an affine map from A^p onto A^r, which maps a to b, $\{a_i\}$, $i = 1, p$ onto a set of distinct, not necessarily independent vectors $\{b_i\}$, $i = 1, r$. For any given point x in A^r, the set of all points y in A^m such that $\Pi_2\Pi_1 y = x$ forms a fiber X of the map $\Pi_2\Pi_1$. The value of the box spline function $M(B)$ at x is defined to be $\mathrm{vol}_{m-1}(X$ intersect $B)/\mathrm{vol}_{m-r}(U)$, where U is any fixed $(m - r)$-dimensional unit box parallel to X, Fig. 3.3.

Box splines generalize tensor products of univariate B-splines, admit triangular networks as well, and possess tractable subdivision formulae (Boehm [3]; Cohen et al. [6]; Dahmen and Micchelli [7]; Prautszch [18]). Paraphrasing Schoenberg, a box spline function can be characterized as the loss of intensity of parallel light beams passing through a box of an affine space A^m onto an affine space A^r. The shadow of B is the support of $M(B)$, Fig. 3.3.

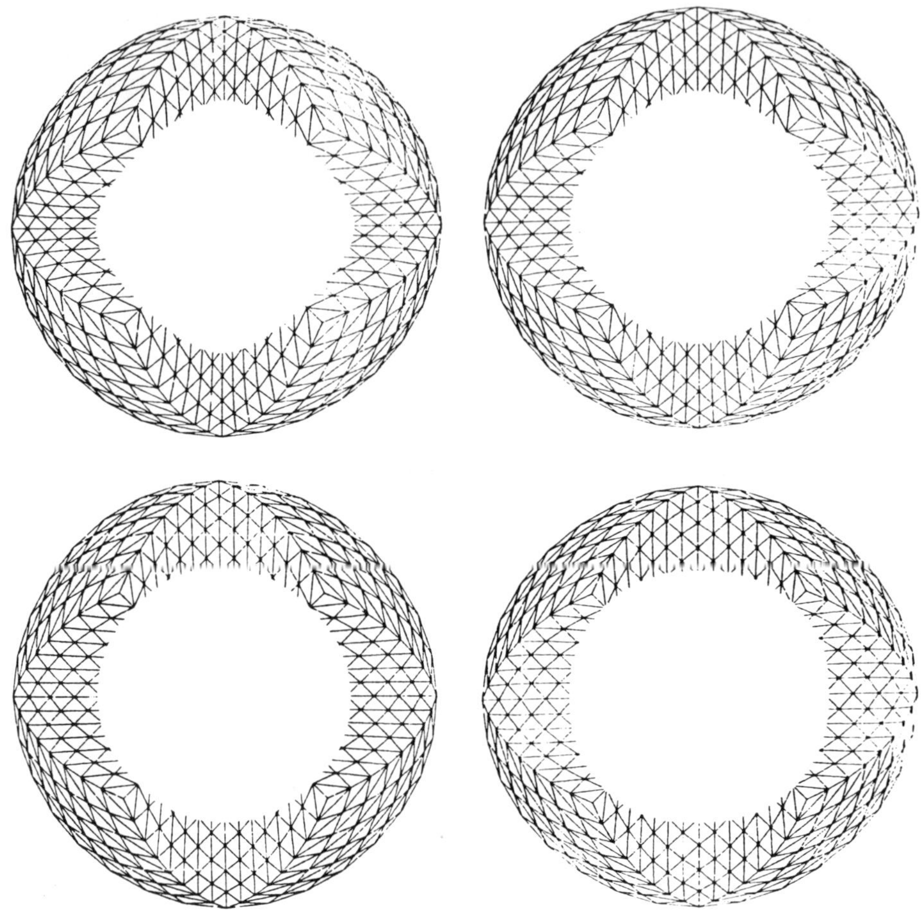

Figure 2.31.   The same four hemispheres, rotated and sectioned. The hemisphere at the lower left, with $c_i = 1.2$, is clearly superior.

The following further conditions are imposed to facilitate building surfaces from such basis functions and their assembly. b and $\{b_i\}$, $i = 1, r$ are restricted to coincide with the edges of a regular grid in A^r. The subscript i of a box spline denotes an integer multi-index defining b in terms of the grid vectors and hence specifies a unique box spline among a set of translates, where each box vertex corresponds to a vertex of the grid (see Boehm [4]). The superscript p denotes a multi-index $(p_1, p_2, \cdots, p_m)$, where p_j is the integer number of elements of $\{a_i\}$ which map to a_j under the affine map Π_1. Surfaces are defined by linear combinations of translates of a particular box spline, hence,

$$(3.4) \qquad\qquad S(x) = \sum d_i M_i^p(x).$$

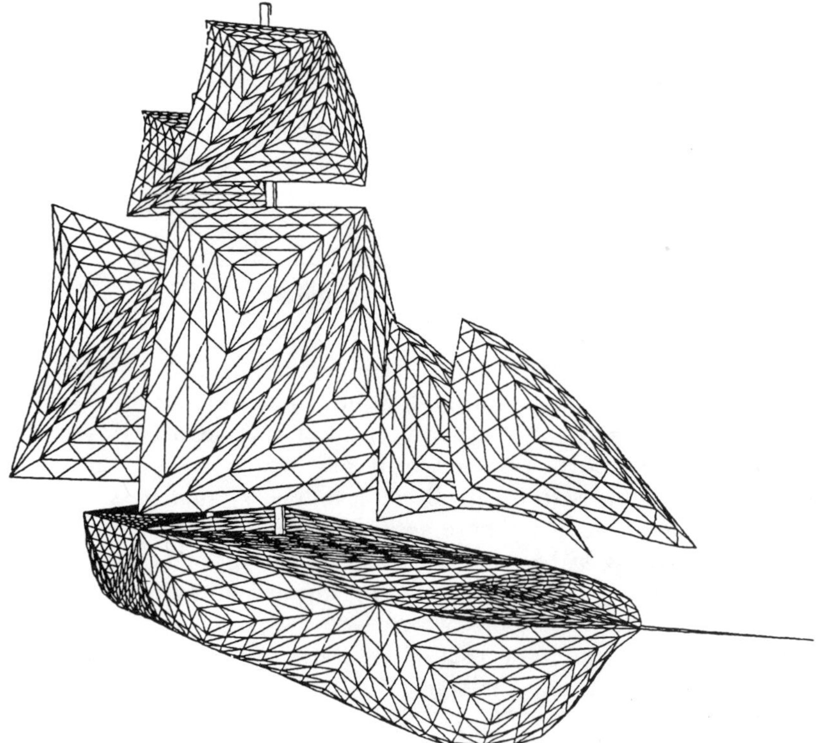

Figure 2.32.   A simple sailing ship.

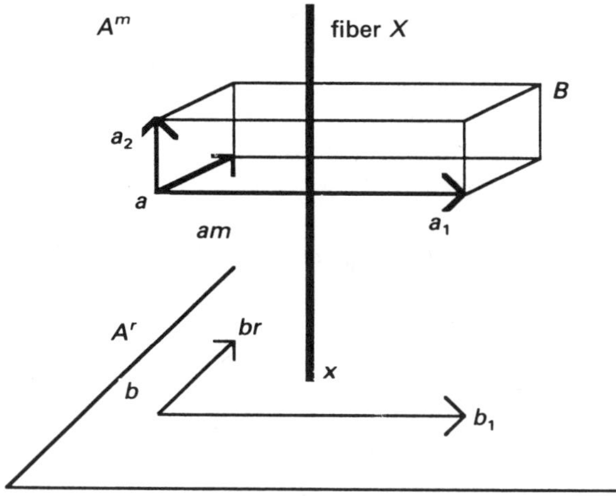

Figure 3.3.   Defining a box spline basis function.

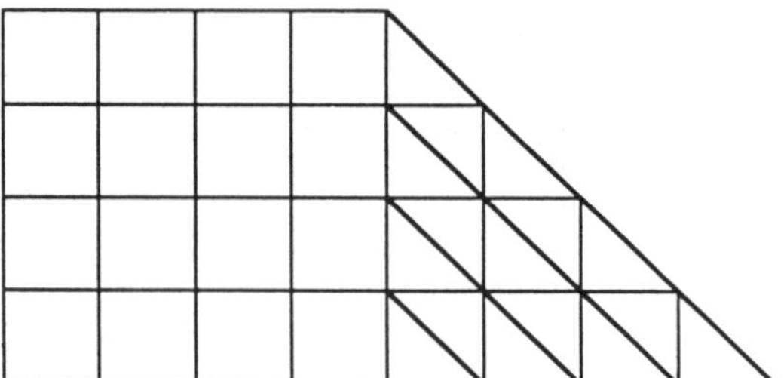

Figure 3.5. A simple composite network.

As with univariate splines, there exists a conceptually simple recursive algorithm for evaluation (Boehm [4]; de Boor and Hollig [9]), and $S(x)$ and $M_i^p(x)$ are of continuity class C^{m-s-1}, where $s = \min \{w \mid$ any w vectors of $\{b_i\}$ span $A^r\}$. Translates of a box spline do not necessarily form a basis, which may or may not pose problems in design.

The problem, again, is to create B-spline networks that possess both rectangular and triangular subnets, Fig. 3.5. There are at least two approaches to the problem. For example, one surface could be constrained to match another along a line of constant parameter; differentiating both surfaces along the line and equating the results would lead to sets of constraint equations which would ensure nth order continuity there. Another approach is to search for a set of box splines, $\{M'\}$, which would contain both the rectangular and the triangular spline in their linear span. That is the approach developed here.

A buffer domain is defined between the rectangular and triangular domains which is wide enough to separate all of the basis functions pertinent to both. In the buffer domain, the surface is represented as a linear combination of the elements of $\{M'\}$. Outside of the buffer domain, and at its edges, translates of the elements of $\{M'\}$ are consolidated in linear combinations to reproduce the original rectangular and triangular generating functions. The result is three independent domains with surfaces defined by three different generating sets, which, nonetheless, possess C^n joins, where n is the minimum degree of smoothness of the original rectangular and triangular splines. The method uses two basic properties of box spline basis functions: their global definition and their local support.

$\{M'\}$ is built by viewing the original spline surfaces as approximations and then taking the tensor product of the two. Let $\{M_{Ti}^p(x)\}$, $\{M_{Rj}^q(y)\}$ denote the box spline generating functions for the triangular and rectangular domains, respectively. $\{M_{Ti}^p(x)M_{Rj}^q(y)\}$ is a set of functions mapping A^{2r} into R^1. The natural restriction to A^r,

$$(3.6) \qquad\qquad\qquad C = \{M_{Ti}^p(x)M_{Rj}^q(x)\},$$

produces a set of generating functions with the following properties:

    (1) C is a partition of unity.

    (2) The linear span of C contains any polynomials that reside in both the span of $\{M_{Ti}^p(x)\}$ and the span of $\{M_{Rj}^q(x)\}$.

    (3) If x is the number of rectangular bases and y is the number of triangular bases covering the cardinality of C is at least xy. Many of these are 0.

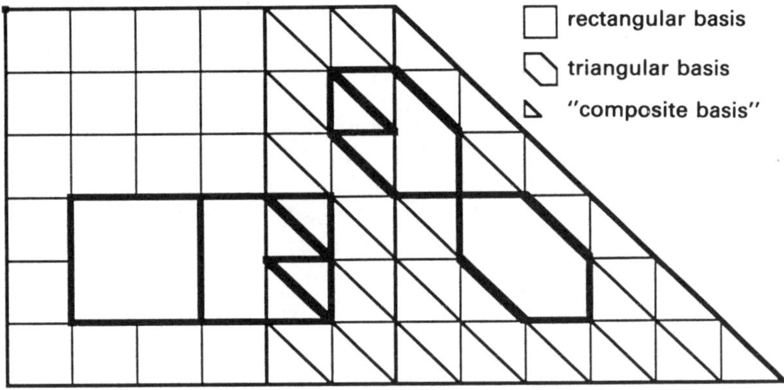

Figure 3.7.   A simple example of a composited box spline surface.

Figure 3.7 illustrates a simple example. The rectangle represents the support of the piecewise linear tensor product; the hexagon that of the linear finite element. The elements of C, the composite generators, are consolidated as shown to reproduce the primary box spline basis functions which straddle the buffer domain boundaries. The width of the buffer domain is the maximum of the widths of the primary box spline basis functions.

The number of nontrivial composite generating functions in the buffer domain can be large, particularly if the primary box splines are smoother and hence have larger support. However, just as the composite generating functions are grouped along the boundary to reproduce translates of the primary box splines, the remaining generating functions within the buffer domain can be collected to reproduce, as far as possible, translates of the box spline generator of one of the adjacent surfaces. This reduces the density of generating functions, and consequently control points, to that of the neighboring box spline surface.

Figures 3.8–3.10 (see color insert) show a C^2 bicubic tensor product spline composited with a C^2 quartic triangular spline. In Fig. 3.9, the portion of the surface corresponding to the buffer domain is drawn in yellow.

The surfaces were evaluated with the recursive algorithm of de Boor and Höllig, but could have been represented more efficiently by means of subdivision. An important issue that remains unanswered is the question of whether a tractable uniform basis or generating set exists for C.

4.   Composite Box Spline Surfaces Over Polyhedra.   We now apply the construction of the previous section to define spline surfaces over polyhedra. Let P be any closed polyhedron with faces $\{f_i\}$ and edges $\{e_{ij}\}$, Fig. 4.1. We assume that a box spline basis function and its translates $\{M^p_{ir}\}$, and a regular grid defined by tessellating vectors $\{b_{is}\}$ are associated with each f_i. The grids on f_i and f_j are constrained to agree on e_{ij}. The algorithm for constructing the surface has three stages.

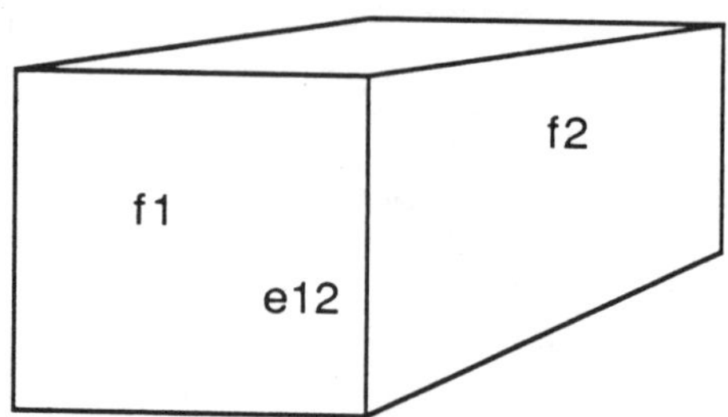

Figure 4.1.   A closed polyhedron.

First, P is chamfered along the edges e_{ij} to create buffer domains $\{c_{ij}\}$, Fig. 4.2. Ω_1 is chosen to satisfy:

$$(4.3) \qquad \Omega_1 = \arctan\left(-\cotan(\Omega_0) - \frac{d_j \cosec(\Omega_0)}{d_i}\right) \geq \frac{\pi}{2},$$

which implies

$$(4.4) \qquad d_{ij} = \frac{d_i}{\cos(\pi - \Omega_1)} = \frac{d_j}{\cos(\pi - \Omega_1)}.$$

(If $\Omega_0 = \pi/2$, $\Omega_1 = \arctan(-d_j/d_i)$.) This allows the box spline basis functions on f_i and f_j to be extended smoothly to c_{ij}.

Second, the modified polyhedron P is chamfered at vertices by the polygon whose vertices are the intersection of the various adjacent f_i and their adjacent c_{ij}, Fig. 4.5. By construction, the boundaries of c_{ijk} only meet other chamfers along edges, i.e., c_{ij}. c_{ijk} need not be planar if it has more than three vertices. If this is the case, an additional projection onto the plane of the first three is defined.

Finally, the chamfers of the modified polyhedron are used as buffer domains, as in § 3, and a single surface is defined. Let Π_{3ij} be the affine mapping from the Euclidean

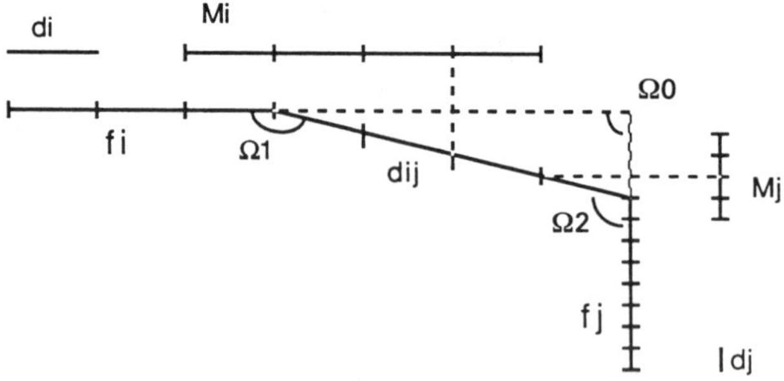

Figure 4.2.   Chamfering to create buffer domains.

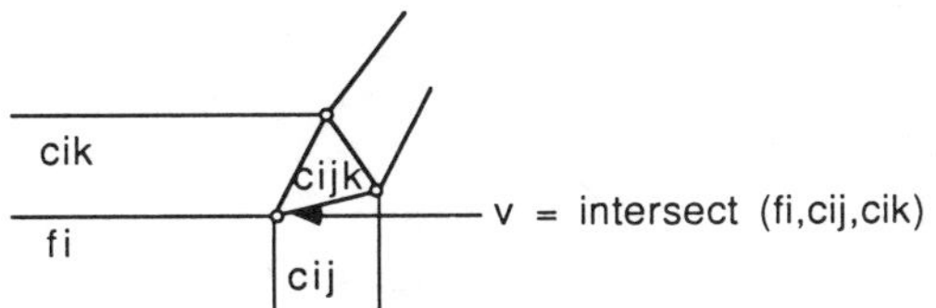

Figure 4.5.   Chamfered vertices.

space containing f_i to the Euclidean space containing c_{ij} defined by,

$$(4.6) \qquad \Pi_{3ij}(x) = [(x - s_{ik}) \cdot e_{ik}]e_{ik} + \frac{[(x - s_{ik}) \cdot e_{ik}^{\text{perp}}]}{\cos{(\pi - \Omega_1)}} e_{ik}^{\text{perp}} + t_c$$

where,

e_{ik} is the boundary in the affine space of f_i between f_i and c_{ij},

s_{ik} is the starting point of e_{ik},

t_{ik} is the terminating point of e_{ik},

s_{cl} is the point on c_{ij} corresponding to t_{ik}, and

t_{cl} is the point on c_{ij} corresponding to s_{ik}, Fig. 4.7.

The box spline basis functions and grids of f_i are projected onto c_{ij} by Π_{3ij}. M_{ir}^p, the box spline associated with f_i is defined by two affine projections, Π_{1i} and Π_{2i}, see § 3. The composition of affine maps is an affine map and M_{ir}^p is mapped to the box spline c_{ij} defined by Π_{1i} and $\Pi_{2i}\Pi_{3ij}$. The box splines thus projected are then composited by the method described in § 3. The same method is then applied again to create generating functions defined on the vertical domains from those of the edge domains and the primitive faces f_i which share vertices.

The generating functions in the buffer domains can be consolidated as in § 3.

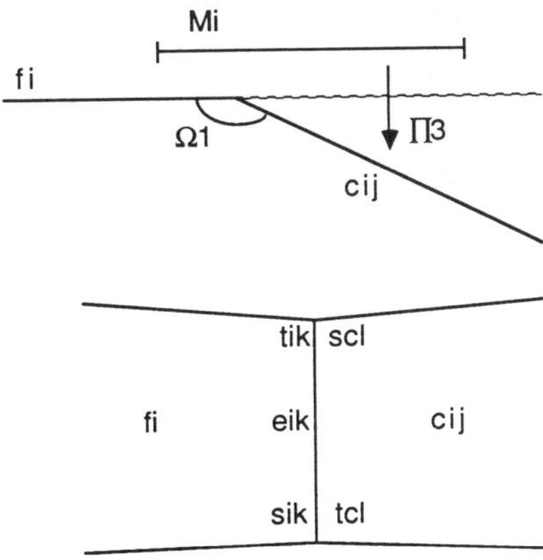

Figure 4.7.   The mapping Π_{3ij}.

5. Conclusion. Algorithms for assembling triangular and rectangular patches and splines have been developed. The methods for patches generalize and improve upon the innovative work of Herron. They are easily implemented.

The methods for spline surfaces are simpler analytically but much more complex computationally. Further, the degree of the surface in the buffer domains can be quite high. The method remains practical because the majority of the surface of the solid is covered by linear combinations of translates of box splines which can be of reasonably low degree.

REFERENCES

[1]   R. E. BARNHILL, *Representation and approximation of surfaces,* in Mathematical Software III, J. R. Rice, ed., Academic Press, New York, 1977.

[2]   W. BOEHM, G. FARIN, AND J. KAHMANN, *A survey of curve and surface methods in* CAGD, Computer Aided Geometric Design, 1 (1984), pp. 1–60.

[3]   W. BOEHM, *Subdividing multivariate splines,* Computer Aided Design, 15 (1983), pp. 345–352.

[4]   ———, *Calculating with box splines,* Computer Aided Geometric Design, 1 (1984), pp. 149–168.

[5]   P. CHARROT AND J. GREGORY, *A pentagonal surface patch for* CAGD, Computer Aided Geometric Design, 1 (1984), pp. 87–96.

[6]   E. COHEN, T. LYCHE AND R. RIESENFELD, *Discrete box splines and refinement algorithms,* Computer Aided Geometric Design, 1 (1984), pp. 131–148.

[7]   W. DAHMEN AND C. MICCHELLI, *Subdivision algorithms for the generation of box spline surfaces,* Computer Aided Geometric Design, 1 (1984), pp. 115–130.

[8]   C. DE BOOR AND R. DE VORE, *Approximation by smooth multivariate splines,* Trans. Amer. Math. Soc., 276 (1983), pp. 775–788.

[9]   C. DE BOOR AND K. HÖLLIG, *B-splines from parallelepipeds,* Mathematics Research Center Report #2320, Madison-Wisconsin, 1982.

[10] G. FARIN, *Smooth interpolation to scattered* 3D *data,* in Surfaces in Computer Aided Geometric Design, R. Barnhill and W. Boehm, eds., North-Holland, Amsterdam, 1983.

[11] ———, *Triangular Bernstein–Bézier patches,* Computer Aided Geometric Design, 1986, to appear.

[12] J. GREGORY, C^1 *rectangular and non-rectangular surface patches,* in Surfaces in Computer Aided Geometric Design, North-Holland, Amsterdam, 1983.

[13] G. HERRON, *Smooth closed surfaces with discrete triangular interpolants,* Computer Aided Geometric Design, 2 (1985), pp. 297–306.

[14] ———, *Techniques for visual continuity,* this Volume, 1987.

[15] J. KAHMANN, *Continuity of curvature between adjacent Bézier patches,* in Surfaces in Computer Aided Geometric Design, North-Holland, Amsterdam, 1983.

[16] B. O'NEILL, *Elementary Differential Geometry,* Academic Press, New York, 1966.

[17] B. PIPER, *Visually smooth interpolation with triangular Bézier patches,* this Volume, 1987.

[18] H. PRAUTSZCH, *Generalized subdivision and convergence,* Computer Aided Geometric Design, 2 (1985), pp. 69–75.

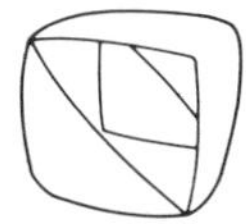

Visually Smooth Interpolation
with Triangular Bézier Patches

BRUCE R. PIPER

Abstract. We consider the problem of constructing two parametrically defined triangular Bézier patches that join in a visually smooth fashion and interpolate to given positions and tangent vectors at the endpoints of the common edge. Conditions for two such patches to join with tangent plane continuity are developed and used to solve this interpolation problem. A counterexample is given to demonstrate that, in general, cubic patches do not have enough degrees of freedom to solve the problem. The construction of a solution is presented for the quartic case. Using this solution, a scheme is developed to interpolate to positional and tangential data defined in a triangular mesh.

1. Introduction. To model a triangular mesh of three-dimensional positional and tangential data it is often desirable to develop parametrically defined triangular patch schemes which produce visually smooth surfaces. Visual smoothness between adjacent triangular patches will essentially be defined to mean tangent plane continuity. Since we want to be able to model surfaces with very general topology, it is preferable that we not view our surface as resulting from one map of a triangulated domain on $\mathbf{R}^2$ into $\mathbf{R}^3$. (With such a map we are, for instance, forced to introduce undesirable singularities in order to model sphere-like surfaces.)

We shall therefore view each patch as the image of one domain triangle. Thus, instead of considering one map from a triangulated domain onto our surface, we shall consider many maps of one domain triangle onto different patches of the surface. This approach also eliminates the difficulty of having to artificially construct a triangulated $\mathbf{R}^2$ domain for a triangular mesh of $\mathbf{R}^3$ data. Herron [8] used this idea in his scheme for constructing visually smooth closed surfaces.

In this paper we develop a scheme to interpolate to positional and tangential data in a triangular mesh using polynomial triangular patches. The polynomials are represented in Bernstein–Bézier form and we assume that the reader is familiar with the associated theory; see Farin [2], [4]. Previous work in this area has been done by

Farin [3] where an interpolation scheme was developed for this type of data. In his scheme it was necessary to enforce a relationship between the positions and tangents at neighboring vertices in the triangular mesh. The scheme we develop in the present paper does not require this relationship to hold. Jensen [9] also has proposed a scheme for matching quartic Clough–Tocher elements. His interpolant is similar to the scheme we propose, although the development is very different.

We proceed by first solving the problem of joining two patches together smoothly while interpolating to given positional and tangential data. Once this problem is solved, a straightforward modification of Farin's method [3] is used to create an interpolant to data defined in a triangular mesh. Thus, the majority of this paper is devoted to solving the problem for two patches.

In order to define this problem explicitly we use the following notation. Throughout this paper boldface letters denote vectors and nonboldface letters scalars. Let

$$(1.1) \qquad B_{i,j,k}^n(u, v, w) = \frac{n!}{i!\,j!\,k!} u^i v^j w^k, \quad i+j+k = n, \quad u+v+w = 1,$$

be the Bernstein polynomials of degree n where u, v and w are the barycentric coordinates; see Barnhill [1]. Also, let

$$(1.2) \qquad \mathbf{P}(u, v, w) = \sum_{i+j+k=n} B_{i,j,k}^n(u, v, w)\mathbf{p}_{i,j,k}$$

and

$$(1.3) \qquad \mathbf{Q}(u, v, w) = \sum_{i+j+k=n} B_{i,j,k}^n(u, v, w)\mathbf{q}_{i,j,k}$$

be adjacent triangular patches where $\mathbf{p}_{i,j,k}$ and $\mathbf{q}_{i,j,k}$ are the control points. For the cubic case this is illustrated in Fig. 1. Notice that $\mathbf{P}$ and $\mathbf{Q}$ are patches defined in terms of barycentric coordinates without reference to a specific domain. This is because the domain triangle for each patch may be chosen arbitrarily (e.g., these domain triangles may be chosen to be identical).

We consider the following problem. Assume we are given the control points

$$(1.4) \qquad \mathbf{p}_{n,0,0} = \mathbf{q}_{n,0,0}, \quad \mathbf{p}_{n-1,1,0} = \mathbf{q}_{n-1,1,0}, \quad \mathbf{p}_{n-1,0,1}, \quad \mathbf{q}_{n-1,0,1},$$

$$(1.5) \qquad \mathbf{p}_{0,n,0} = \mathbf{q}_{0,n,0}, \quad \mathbf{p}_{1,n-1,0} = \mathbf{q}_{1,n-1,0}, \quad \mathbf{p}_{0,n-1,1}, \quad \mathbf{q}_{0,n-1,1}.$$

For the tangent planes at the vertices to be well defined, the four points in (1.4) must be coplanar, distinct and $\mathbf{p}_{n-1,0,1}$ and $\mathbf{q}_{n-1,0,1}$ must lie on strictly opposite sides of the line through the points $\mathbf{p}_{n,0,0}$ and $\mathbf{p}_{n-1,1,0}$. A similar assumption is made on the points in (1.5). The problem is then to determine the remaining control points so as to create a visually smooth join between the two patches $\mathbf{P}$ and $\mathbf{Q}$. The main result of this paper is that it is necessary and sufficient that the degree n of the patches $\mathbf{P}$ and $\mathbf{Q}$ be greater than or equal to 4 in order for this problem to always have a solution.

In the next section we develop conditions for two adjacent Bézier patches to join in a visually smooth fashion. In § 3 a counterexample is presented to demonstrate that cubics are not always sufficient to solve the above problem. In § 4, we construct a

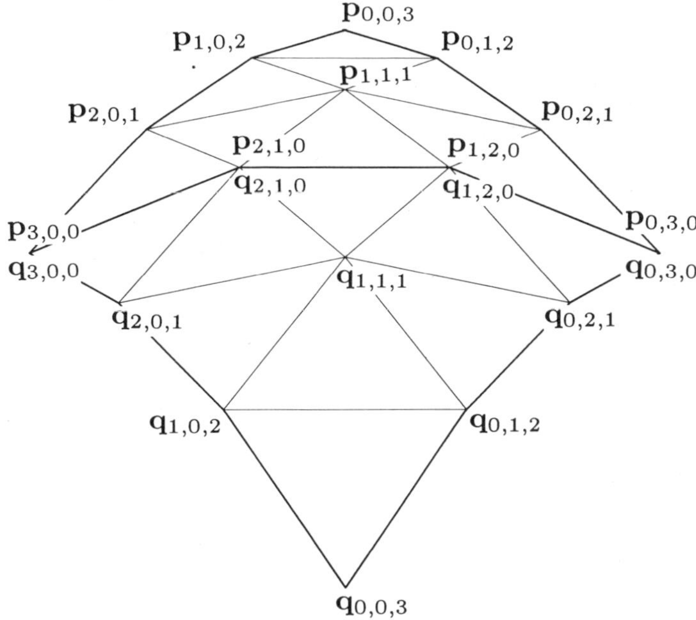

Figure 1. Control points for two neighboring Bézier patches.

solution to the problem using quartics. Using the results of § 4, an interpolation scheme for data in a triangular mesh is implemented in § 5 and pictorial examples of the surfaces it produces are presented in § 6.

2. Visual Continuity. Continuity between the patches $\mathbf{P}$ and $\mathbf{Q}$ as defined by (1.2) and (1.3) is ensured by insisting that the control points along the common edge be identical. For example, in Fig. 1 this means that

$$\mathbf{p}_{i,j,0} = \mathbf{q}_{i,j,0} \quad \forall i + j = 3.$$

Despite the fact that this condition implies the two patches are oriented in opposite directions, it simplifies the following mathematics. Therefore, we assume throughout that we are dealing with patches $\mathbf{P}$ and $\mathbf{Q}$ which join continuously along the edge defined by $w = 0$. This is a necessary requirement before we can consider tangent plane continuity.

First, we require a formula which enables us to find the tangent vectors at any point on a Bézier patch. A development of such a formula can be found in Farin [2] where, as here, a vector in terms of barycentric coordinates is a triple of coordinates that sum to zero. The tangent vector of the map $\mathbf{P}$ corresponding to the (domain) vector $\mathbf{a} = (a_1, a_2, a_3)$ is given by

$$(2.1) \quad \mathbf{D_a P}(u, v, w) = n \sum_{i+j+k=n-1} B_{i,j,k}^{n-1}(u, v, w)[a_1 \mathbf{p}_{i+1,j,k} + a_2 \mathbf{p}_{i,j+1,k} + a_3 \mathbf{p}_{i,j,k+1}]$$

where the operator $\mathbf{D_a}$ denotes the directional derivative in the direction of $\mathbf{a}$ scaled by the length of the corresponding domain vector. Note that this length is not the length of the vector $\mathbf{a}$ as written in barycentric coordinates but rather the length of the corresponding two-dimensional vector in the domain traingle. However, it is never necessary to calculate the length of this domain vector since the length of the tangent vector given by (2.1) is not important for our purposes.

Throughout this paper we assume that the maps $\mathbf{P}$ and $\mathbf{Q}$ have two independent directional derivatives at every point. This assumption implies that each patch has well defined tangent planes so that we only need to consider how the patches join together. The two patches $\mathbf{P}$ and $\mathbf{Q}$ form a visually smooth join if, at each point along the common edge, the tangent planes of each patch agree and the patches are situated in such a way that a cusp does not occur at the join.

The tangent plane of the patch $\mathbf{P}$ at a point along the common edge is spanned by the directional derivative of $\mathbf{P}$ taken in two different domain directions. Without loss of generality, we may choose these directions to be those corresponding to the vectors $\mathbf{a1} = (-1, 1, 0)$ and $\mathbf{a2} = (-1, 0, 1)$. The same directions are used for the patch $\mathbf{Q}$ and since we are assuming continuity between the patches, the tangent vectors $\mathbf{D_{a1}P}$ and $\mathbf{D_{a1}Q}$ are identical. Thus, the tangent planes of the two patches will agree if the tangent vectors $\mathbf{D_{a1}P}$, $\mathbf{D_{a2}P}$ and $\mathbf{D_{a2}Q}$ all lie in the same plane. To actually obtain a visually smooth join we must also ensure that cusps do not occur. This may be accomplished by requiring that the vectors $\mathbf{D_{a2}P}$ and $\mathbf{D_{a2}Q}$ lie on strictly opposite sides of the vector $\mathbf{D_{a1}P}$.

Using formula (2.1), it can be shown that

$$(2.2) \qquad \begin{aligned} \mathbf{D_{a1}P}(u, v, 0) &= \mathbf{D_{a1}Q}(u, v, 0) = n(\mathbf{J} - \mathbf{I}), \\ \mathbf{D_{a2}P}(u, v, 0) &= n(\mathbf{K} - \mathbf{I}) \quad \text{and} \quad \mathbf{D_{a2}Q}(u, v, 0) = n(\mathbf{L} - \mathbf{I}), \end{aligned}$$

where

$$\mathbf{I} = \sum_{i+j=n-1} B_{i,j,0}^{n-1}(u, v, 0)\mathbf{p}_{i+1,j,0},$$

$$\mathbf{J} = \sum_{i+j=n-1} B_{i,j,0}^{n-1}(u, v, 0)\mathbf{p}_{i,j+1,0},$$

$$\mathbf{K} = \sum_{i+j=n-1} B_{i,j,0}^{n-1}(u, v, 0)\mathbf{p}_{i,j,1},$$

$$\mathbf{L} = \sum_{i+j=n-1} B_{i,j,0}^{n-1}(u, v, 0)\mathbf{q}_{i,j,1}.$$

With this notation, a necessary condition for tangent plane continuity between the patches is that

$$(2.3) \qquad \operatorname{Det} \begin{bmatrix} 1 & \mathbf{I} \\ 1 & \mathbf{J} \\ 1 & \mathbf{K} \\ 1 & \mathbf{L} \end{bmatrix} = 0$$

for all u, v such that $u + v = 1$.

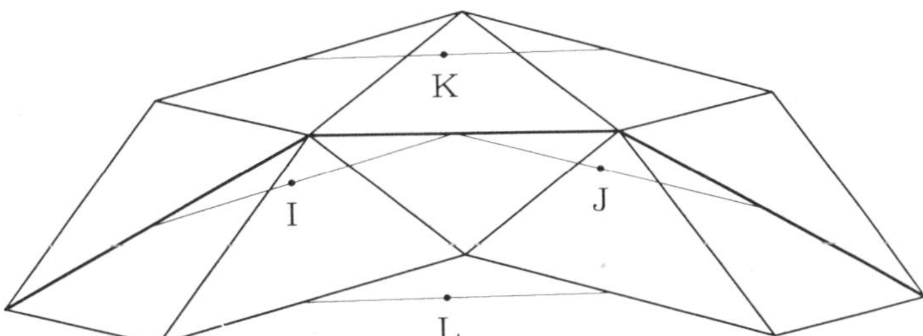

Figure 2.   Tangent plane continuity.

Since the value of a determinant is unchanged by elementary row operations, we may subtract the first row from each of the last three rows of the matrix in (2.3) to obtain scaled versions of the three vectors in (2.2). Hence, requiring (2.3) to hold is equivalent to requiring that the three vectors in (2.2) all lie on the same plane.

We may interpret (2.3) geometrically. To do this we consider the cubic case as illustrated in Fig. 2. Note that in this case $\mathbf{I}$, $\mathbf{J}$, $\mathbf{K}$ and $\mathbf{L}$ are all points determined by quadratic Bézier curves in the variables u and v. The points $\mathbf{I}$, $\mathbf{J}$ and $\mathbf{K}$ are on the tangent plane of $\mathbf{P}$ at u, v $(u + v = 1)$ and similarly the points $\mathbf{I}$, $\mathbf{J}$ and $\mathbf{L}$ are on the tangent plane of $\mathbf{Q}$ at u, v $(u + v = 1)$. If these four points are coplanar for all u, v such that $u + v = 1$ then the four rows of the matrix in (2.3) are dependent and hence the determinant is zero.

As stated above, (2.3) is only a necessary condition for tangent plane continuity. We still need to exclude the possibility of a cusp occurring at the join. This may be done by insisting that $\mathbf{K}$ and $\mathbf{L}$ lie on strictly opposite sides of the line through the points $\mathbf{I}$ and $\mathbf{J}$ for all u, v such that $u + v = 1$. In the next section we use (2.3) to show that cubics do not always provide sufficient freedom to solve the interpolation problem stated in the Introduction.

3.   The Cubic Case.   In view of the fact that in the functional case (i.e., where the data are related by a single function, $z = f(x, y)$) cubics are sufficient to solve the problem stated in the Introduction (Farin [3]), it is perhaps surprising that they are not always sufficient to solve the problem in the parametric case. Although for certain types of nonfunctional input data it might be that the problem can be solved with cubics, there exist cases where it cannot.

The input data for one such case is shown in Fig. 3. In this figure we have the control points

$$\mathbf{p}_{3,0,0} = \mathbf{q}_{3,0,0} = (0, 0, 0), \qquad \mathbf{p}_{2,1,0} = \mathbf{q}_{2,1,0} = (1, 0, 0),$$

$$\mathbf{p}_{1,2,0} = \mathbf{q}_{1,2,0} = (2, 0, 1), \qquad \mathbf{p}_{0,3,0} = \mathbf{q}_{0,3,0} = (4, 0, 1),$$

(3.1)

$$\mathbf{p}_{2,0,1} = (1, 1, 0), \qquad \mathbf{q}_{2,0,1} = (1, -1, 0),$$

$$\mathbf{p}_{0,2,1} = (3, 1, 1), \qquad \mathbf{q}_{0,2,1} = (3, -1, 1).$$

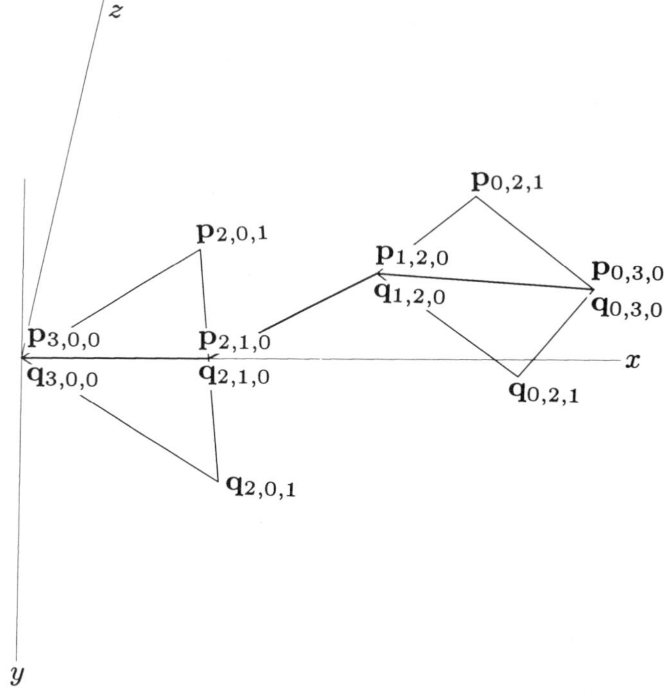

Figure 3.   Input data for which there does not exist a solution using cubics.

We claim that it is impossible to determine the control points $\mathbf{p}_{1,1,1}$ and $\mathbf{q}_{1,1,1}$ so as to satisfy (2.3). Since these are the only remaining control points which influence (2.3) and since (2.3) is a necessary condition for tangent plane continuity, establishing this claim proves that this is a case where cubics cannot be used to solve the problem stated in the Introduction.

In the present setting the left-hand side of (2.3) is a 6th degree polynomial in u and v. To write out this polynomial explicitly let

$$(3.2) \qquad (x_p, y_p, z_p) = \mathbf{p}_{1,1,1}, \qquad (x_q, y_q, z_q) = \mathbf{q}_{1,1,1}$$

be the x, y and z coefficients of the unknown control points. A straightforward calculation shows that the left-hand side of (2.3) for the data in (3.1) is

$$(3.3) \qquad 2uv[d_0 u^4 + d_1 4u^3 v + d_2 6u^2 v^2 + d_3 4uv^3 + d_4 v^4]$$

where $u + v = 1$ and

$$d_0 = (z_p + z_q) - 2,$$
$$d_1 = [(z_q y_p - z_p y_q) - (y_p - y_q) - (x_p + x_q) + (z_p + z_q) + 2]/2,$$
$$d_2 = [4(y_q x_p - y_p x_q) + 4(z_q y_p - z_p y_q) + 4(y_p - y_q) + 3(z_p + z_q) - 4]/6,$$
$$d_3 = [2(z_q y_p - z_p y_q) - (y_p - y_q) - (x_p + x_q) + (z_p + z_q) + 2]/2,$$
$$d_4 = 2(z_p + z_q) - 2.$$

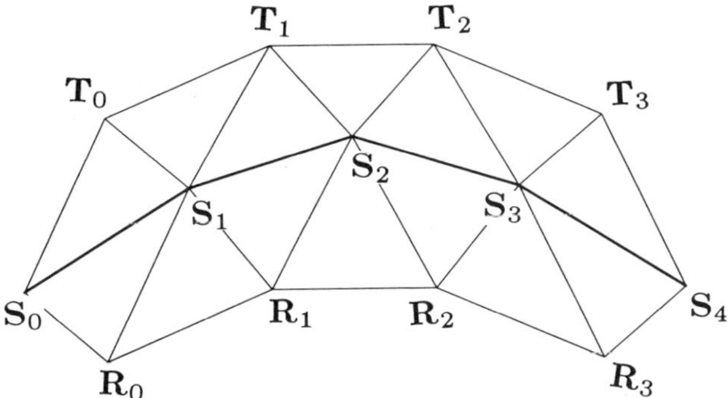

Figure 4. Bézier control points of two quartic patches that are relevant for tangent plane continuity.

Since uv is not identically zero and the one-dimensional Bernstein polynomials $((n!/i!\,j!)u^i v^j,\ i + j = 4)$ are independent, (3.3) vanishes for all u, v such that $u + v = 1$ if and only if all the d_i are identically zero. However, because it is not possible to pick z_p and z_q so as to make both d_0 and d_4 vanish, it follows that it is impossible to choose $\mathbf{p}_{1,1,1}$ and $\mathbf{q}_{1,1,1}$ so as to satisfy (2.3). Hence cubics cannot be used to solve the problem for this data set. In the next section we construct a solution using quartics.

4. The Quartic Case. In this section we prove the problem stated in the Introduction can always be solved with quartic patches. We first develop conditions on the unknown control points so that (2.3) holds and then show that these conditions can always be satisfied. To do this it is quite cumbersome to use (2.3) directly since the x, y and z components of the control points become interlaced. Therefore, in this section we use the equivalent condition that for each u, v $(u + v = 1)$ there exist scalars $E(u, v)$, $F(u, v)$, $G(u, v)$ and $H(u, v)$ not all zero, such that

$$(4.1) \qquad E(u, v)\mathbf{I} + F(u, v)\mathbf{J} + G(u, v)\mathbf{K} + H(u, v)\mathbf{L} = \mathbf{0},$$

and

$$(4.2) \qquad E(u, v) + F(u, v) + G(u, v) + H(u, v) = 0.$$

We choose $E(u, v)$, $F(u, v)$, $G(u, v)$, and $H(u, v)$ to be linear functions of u and v. In particular, we let

$$(4.3) \qquad \begin{aligned} E(u, v) &= e_1 u + e_2 v, & F(u, v) &= f_1 u + f_2 v, \\ G(u, v) &= g_1 u + g_2 v, & H(u, v) &= h_1 u + h_2 v. \end{aligned}$$

We proceed by first determining the coefficients in these linear expressions and then developing conditions on the unknown control points in terms of these coefficients.

To make the expansion of (4.1) more readable we introduce the notation illustrated in Fig. 4. In this figure we have

$$\begin{aligned} \mathbf{R}_i &= \mathbf{p}_{3-i,i,1}, & \mathbf{T}_i &= \mathbf{q}_{3-i,i,1}, & i &= 0, 1, 2, 3, \\ \mathbf{S}_i &= \mathbf{p}_{4-i,i,0} = \mathbf{q}_{4-i,i,0}, & & & i &= 0, 1, 2, 3, 4. \end{aligned}$$

Using this notation we may multiply out (4.1) to obtain

$$(4.4) \qquad \mathbf{C}_0 u^4 + \mathbf{C}_1 4u^3 v + \mathbf{C}_2 6u^2 v^2 + \mathbf{C}_3 4uv^3 + \mathbf{C}_4 v^4 = \mathbf{0},$$

where $u + v = 1$ and

$$\mathbf{C}_0 = e_1 \mathbf{S}_0 + f_1 \mathbf{S}_1 + g_1 \mathbf{R}_0 + h_1 \mathbf{T}_0,$$
$$\mathbf{C}_1 = [3(e_1 \mathbf{S}_1 + f_1 \mathbf{S}_2 + g_1 \mathbf{R}_1 + h_1 \mathbf{T}_1) + (e_2 \mathbf{S}_0 + f_2 \mathbf{S}_1 + g_2 \mathbf{R}_0 + h_2 \mathbf{T}_0)]/4,$$
$$\mathbf{C}_2 = [e_2 \mathbf{S}_1 + f_2 \mathbf{S}_2 + g_2 \mathbf{R}_1 + h_2 \mathbf{T}_1 + e_1 \mathbf{S}_2 + f_1 \mathbf{S}_3 + g_1 \mathbf{R}_2 + h_1 \mathbf{T}_2]/2,$$
$$\mathbf{C}_3 = [3(e_2 \mathbf{S}_2 + f_2 \mathbf{S}_3 + g_2 \mathbf{R}_2 + h_2 \mathbf{T}_2) + (e_1 \mathbf{S}_3 + f_1 \mathbf{S}_4 + g_1 \mathbf{R}_3 + h_1 \mathbf{T}_3)]/4,$$
$$\mathbf{C}_4 = e_2 \mathbf{S}_3 + f_2 \mathbf{S}_4 + g_2 \mathbf{R}_3 + h_2 \mathbf{T}_3.$$

Equation (4.4) is satisfied for all u, v such that $u + v = 1$ if and only if all the $\mathbf{C}_i$ are identically zero. Since the control points in the expression for $\mathbf{C}_0$ are all input data and are assumed to be coplanar, we can choose e_1, f_1, g_1 and h_1 (not all zero) so as to make $\mathbf{C}_0 = \mathbf{0}$ and so that $e_1 + f_1 + g_1 + h_1 = 0$. Recalling the restrictions we placed on the input data in (1.4) and (1.5) (i.e., that $\mathbf{R}_0$ and $\mathbf{T}_0$ are on strictly opposite sides of the line through $\mathbf{S}_0$ and $\mathbf{S}_1$) we see these scalars may be chosen so that $g_1 + h_1 = 1$. This normalization uniquely determines these scalars. We make an analogous choice for the scalars e_2, f_2, g_2 and h_2 so as to make $\mathbf{C}_4 = \mathbf{0}$ and so that $e_2 + f_2 + g_2 + h_2 = 0$. Thus we have chosen the scalars so that both $\mathbf{C}_0$ and $\mathbf{C}_4$ are zero and so that (4.2) is satisfied for all u, v such that $u + v = 1$.

Setting the remaining coefficients to zero gives us a linear system for the unknown control points $\mathbf{R}_1$, $\mathbf{R}_2$, $\mathbf{T}_1$, $\mathbf{T}_2$ and $\mathbf{S}_2$. To simplify the final form of the linear system we actually set $4\mathbf{C}_1 = 0$, $6\mathbf{C}_2 = 0$ and $4\mathbf{C}_3 = 0$. This system of three vector valued equations in five vector valued unknowns may be formally written as

$$(4.5) \qquad \mathbf{A}\vec{\mathbf{x}} = \vec{\mathbf{b}}$$

where

$$\mathbf{A} = \begin{bmatrix} 3g_1 & 3h_1 & 0 & 0 & 3f_1 \\ 3g_2 & 3h_2 & 3g_1 & 3h_1 & 3(f_2 + e_1) \\ 0 & 0 & 3g_2 & 3h_2 & 3e_2 \end{bmatrix},$$

$$\vec{\mathbf{x}} = \begin{bmatrix} \mathbf{R}_1 \\ \mathbf{T}_1 \\ \mathbf{R}_2 \\ \mathbf{T}_2 \\ \mathbf{S}_2 \end{bmatrix} \quad \text{and} \quad \vec{\mathbf{b}} = - \begin{bmatrix} e_2 \mathbf{S}_0 + f_2 \mathbf{S}_1 + g_2 \mathbf{R}_0 + h_2 \mathbf{T}_0 + 3e_1 \mathbf{S}_1 \\ 3(e_2 \mathbf{S}_1 + f_1 \mathbf{S}_3) \\ e_1 \mathbf{S}_3 + f_1 \mathbf{S}_4 + g_1 \mathbf{R}_3 + h_1 \mathbf{T}_3 + 3f_2 \mathbf{S}_3 \end{bmatrix}.$$

This system can be solved if the rows of the matrix $\mathbf{A}$ are independent. The only way the rows can be dependent is if $e_1 = e_2$, $f_1 = f_2$, $g_1 = g_2$ and $h_1 = h_2$. In this case we have $2(\mathbf{C}_1 + \mathbf{C}_3) = 3\mathbf{C}_2$ and so (4.5) reduces to a solvable two-by-five system. Hence a solution always exists.

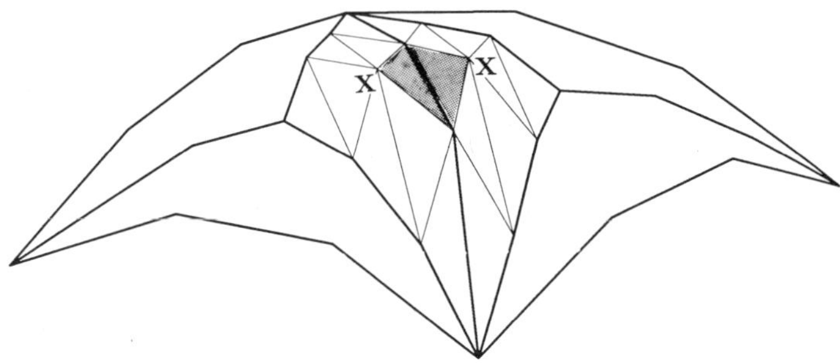

Figure 5. As a preprocessing stage we correct the points labeled by **X** to make the two shaded triangles coplanar.

Choosing the control points in the vector $\vec{x}$ so as to satisfy (4.5) produces a visually smooth join assuming the conditions to avoid cusps stated after formula (2.3) are met. In the next section we use (4.5) to construct an interpolation scheme.

5. An Interpolation Scheme. In order to interpolate to positional and tangential data defined in a triangular mesh, we use a Clough–Tocher-like scheme (see Strang and Fix [10]), where the mappings are quartic (instead of cubic) on each subtriangle. This idea was used by Farin [3] in his interpolation method.

In this section, we outline our interpolation scheme. The method proceeds by first determining a collection of candidate control points which define a piecewise cubic C^0 surface. This is then degree elevated and the resulting surface is modified so that the final result is visually smooth.

More precisely, suppose we are given positions and surface normals at each vertex of a triangular mesh. For each triangle we first define the control points $\mathbf{p}_{3,0,0}$, $\mathbf{p}_{0,3,0}$, $\mathbf{p}_{0,0,3}$ to be the vertices of the triangle. We then create $\mathbf{p}_{2,1,0}$ by the following procedure. The point $\mathbf{p}_{0,3,0}$ is projected onto the tangent plane at $\mathbf{p}_{3,0,0}$ and the vector from $\mathbf{p}_{3,0,0}$ to this projected point is scaled by an appropriate factor. (For the pictures in § 6, this factor was chosen to be one-third plus one-ninth the distance from $\mathbf{p}_{0,3,0}$ to the projected point.) The point $\mathbf{p}_{2,1,0}$ is then taken to be $\mathbf{p}_{3,0,0}$ plus this vector. The control points $\mathbf{p}_{2,0,1}$, $\mathbf{p}_{1,2,0}$, $\mathbf{p}_{0,2,1}$, $\mathbf{p}_{1,0,2}$, and $\mathbf{p}_{0,1,2}$ are chosen in an analogous fashion. This choice insures interpolation to the positions and surface normals at the vertices of the triangular mesh.

We define a candidate center control point of each triangle so that the resulting cubic patch will be a ruled surface if one of the sets of edge control points lies on a straight line (Farin [5]). This reduces unwanted undulations in the surface. The cubic triangular patches produced at this stage look good individually but often fit together at very sharp angles.

To mollify this effect, an additional preprocessing stage is implemented as follows. Each patch is subdivided at the centroid (see Goldman [7]) into three subtriangles. The center control points of each subtriangle are adjusted, if necessary, so that the two shaded triangles in Fig. 5 are coplanar. This is done by taking a weighted average of the two existing planes where the weights are chosen so as to favor the plane of the

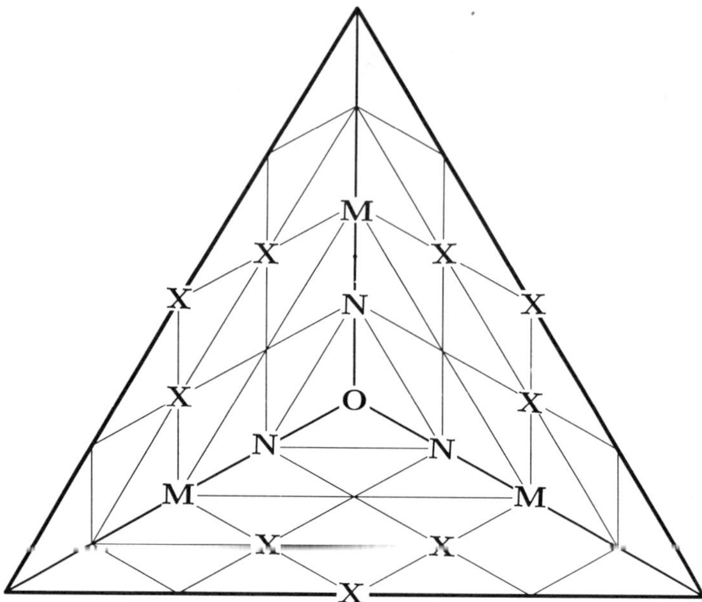

Figure 6.   The quartic Clough–Tocher split.

triangle with the smallest height. This choice helps to avoid "divots" in the surface due to "skinny" triangles. The control points on the interior edges are then modified to maintain smoothness between the subpatches of a traingle; see Farin [3].

Degree elevation (see Farin [3], [4]) is then performed on each subpatch of each patch in the triangular mesh. This gives us a collection of candidate control points for the surface we are constructing. These control points define a continuous piecewise quartic (actually cubic) surface that interpolates to the given positional and tangential data but, in general, does not satisfy the conditions for smoothness as given by (4.5).

For each patch boundary curve we modify the control points that occur as unknowns in (4.5). These are the points marked by **X** in Fig. 6. We seek a solution to (4.5) which minimizes the distances from the control points in the solution to the candidate control points. To do this, we first determine the matrix **A** and the right-hand side $\vec{\mathbf{b}}$. Then, letting $\vec{\mathbf{y}}$ be the vector corresponding to $\vec{\mathbf{x}}$ that contains the candidate control points, we solve

$$(5.1) \qquad\qquad \mathbf{A}\vec{\mathbf{e}} = \vec{\mathbf{b}} - \mathbf{A}\vec{\mathbf{y}}$$

in a least squares sense for each of the x, y, z components of the correction $\vec{\mathbf{e}}$. The corrected control points are then given by $\vec{\mathbf{e}} + \vec{\mathbf{y}}$.

These new control points define a surface with visually smooth joins between adjacent patches but, in general, only continuous joins across the interior edges formed by the Clough–Tocher split. In order to regain the smoothness across these edges, we modify the interior control points of each triangle. The control points to be modified

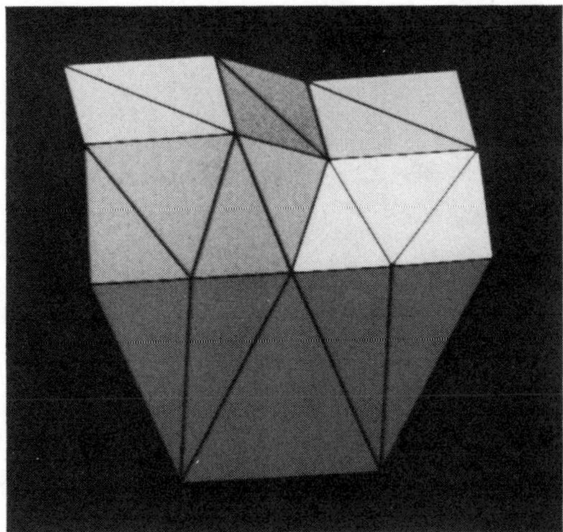

Figure 7. The triangulation of a set of positional data.

are labeled **M**, **N** and **O** in Fig. 6. We determine each of these control points as the centroid of the three that surround it. We first determine the **M**'s, then the **N**'s, and finally we determine **O**. This correction produces a parametrically C^1 join between interior subpatches of each triangle. Since this correction does not affect the tangent planes on the boundaries of the patch, the resulting surface interpolates to the input data and is visually smooth.

The above procedure produces a visually smooth surface for any set of candidate control points that interpolates to positions and normals at the vertices in a triangular

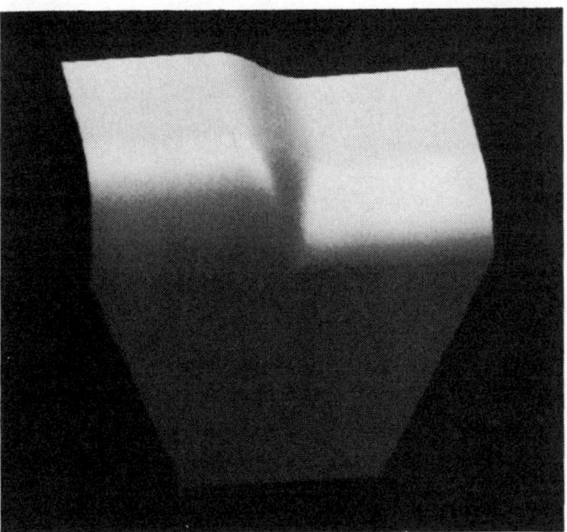

Figure 8. The interpolant applied to the data in Fig. 7.

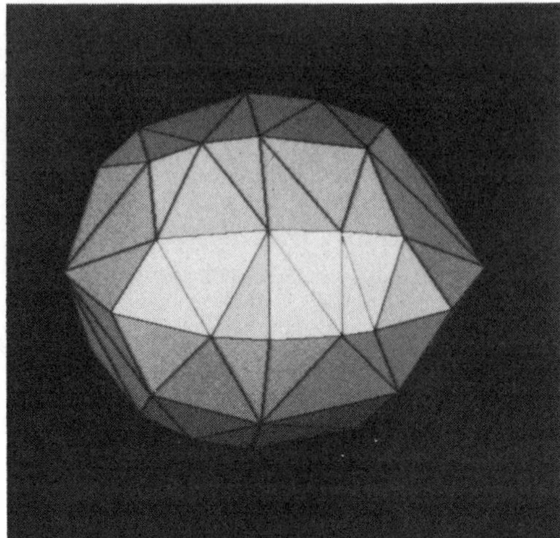

Figure 9. The triangulation of positional data on a closed surface.

mesh. Many choices (other than the ones made here) are possible for the remaining degrees of freedom in the set of candidate control points and further research is required to determine which choice produces the best surface.

It should be pointed out that in constructing the interpolant we have made no effort to ensure that the patches actually have well defined tangent planes or that they join without cusps. Although the restrictions we placed on the data in (1.4) and (1.5) ensure that we do not have cusps at the vertices of the triangular patches, they do not necessarily imply that the surface is free of cusps interior to the patches or on the edges

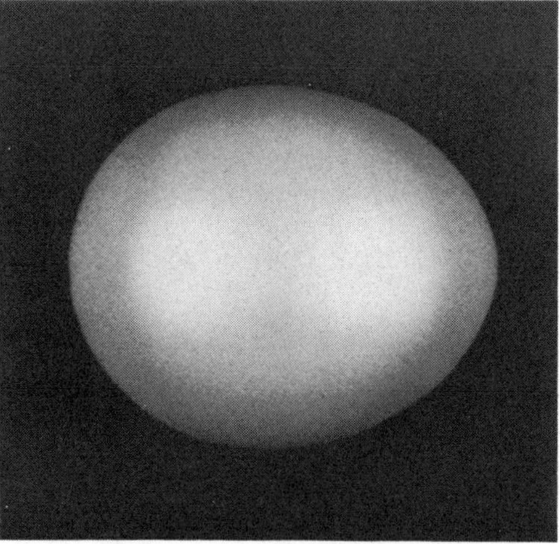

Figure 10. The interpolant applied to the data in Fig. 9.

of the patches. However, we believe that for many reasonable choices of the data this is the case. Further research is necessary to develop conditions on the input control points that guarantee that this interpolant is actually cusp-free.

6. Examples. In this section we present some examples of applying our interpolation scheme to a set of data defined in a triangular mesh. Such a data set is illustrated in Fig. 7. For this example, surface normals were user-specified so as to preserve selected planar regions. The resulting surface is shown in Fig. 8 where the picture was rendered with two point light sources and Phong shading; see Foley and Van Dam [6, pp. 577–584].

Figure 9 shows the triangulation of positional data on a closed surface. For this example, surface normals were generated at each point automatically. This was done in such a way so that surface normals of a sphere were reproduced if, at a particular data point, the surrounding data all lay on the same sphere. The resulting interpolant is shown in Fig. 10 where, as before, the picture was rendered with two point light sources and Phong shading.

Acknowledgments. This research was supported in part by the Department of Energy with contract DE-AC02-85ER12046 to the University of Utah and by a University of Utah Research Fellowship to the author. The author thanks Dr. A. J. Worsey for his help in writing this paper and Mark Watkins and Dianne Hansford for their help in making the pictures. Thanks also to Dr. G. Farin for his many useful comments on the content of this paper.

REFERENCES

[1] R. E. BARNHILL, *Representation and approximation of surfaces,* in Mathematical Software III, J. R. Rice, ed., Academic Press, New York, 1977, pp. 69–120.

[2] G. FARIN, *Bézier Polynomials over Triangles,* TR/91 Dept. Mathematics, Brunel Univ., Uxbridge, Middlesex, England, 1980.

[3] ———, *Smooth interpolation to scattered* 3D *data,* in Surfaces in Computer Aided Geometric Design, R. E. Barnhill and W. Boehm, eds., North-Holland, Amsterdam, 1983.

[4] ———, *Triangular Bernstein–Bézier patches,* Computed Aided Geometric Design, to appear.

[5] ———, private communication, 1985.

[6] J. D. FOLEY AND A. VAN DAM, *Fundamentals of Interactive Computer Graphics,* Addison-Wesley, Reading, MA, 1982.

[7] R. N. GOLDMAN, *Subdivision algorithms for Bézier triangles,* Computer Aided Design, 15 (1983), pp. 159–166.

[8] G. HERRON, *Smooth closed surfaces with discrete triangular interpolants,* Computer Aided Geometric Design, 2 (1985), pp. 297–306.

[9] T. JENSEN, *Assembling triangular and rectangular patches and multivariate splines,* this Volume, 1987.

[10] G. STRANG AND G. FIX, *An Analysis of the Finite Element Method,* Prentice-Hall, Englewood Cliffs, NJ, 1973.

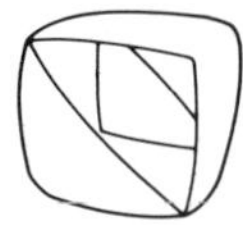

A Transfinite, Visually Continuous, Triangular Interpolant

GREGORY M. NIELSON

Abstract.   A new type of triangular surface patch is presented. Given three boundary curves and associated vectors representing outward surface normals there, a surface segment is defined over a triangle that matches this transfinite data on the boundary. A discretized version, called a six-parameter, G^1 patch, is developed and some examples are given.

1.   Introduction.   In this paper, we describe a new parametric surface patch which assumes arbitrary values and outward surface normals on all three edges of a triangular region. Because this interpolant matches data on the entire boundary of the triangle and not at a finite number of discrete points, we use the terminology of Gordon [7] and call it a transfinite interpolant. The C^1 interpolant for a rectangular domain originally described by Coons [5] is a transfinite interpolant, whereas the bicubic Hermite patch, which interpolates to position two partial derivative vectors and the cross partial derivative vectors at the four vertices, is a discrete interpolant. In fact, the bicubic Hermite patch is obtained from the Coons' patch by a process called "discretization" where the boundary functions required for the transfinite interpolant are provided by means of a univariate interpolation scheme along the boundary. In the case of the bicubic Hermite patch, both the position and cross-boundary tangents are taken as cubic Hermite interpolants. We should emphasize the importance of the generality provided by the transfinite interpolants. Even though these interpolants are usually not applied directly to transfinite boundary data, the fact that they can be applied to an extremely wide class of boundary information makes these interpolants useful in a wide variety of applications. For example, Nielson and Franke [11] used boundary data that consisted of linear combinations of exponential functions which resulted from a spline in tension curve network. This data was used to discretize the transfinite side-vertex interpolant described by Nielson [12].

The interpolants of this paper utilize outward surface normals rather than partial derivatives on the boundary. The property of a surface being C^1 or not can be altered by a reparameterization which does not affect the graph of the surface. Consequently, for surface design based upon joining surface patch segments, the conditions of C^1 continuity do not produce a surface that is any more or less visually continuous than one which is G^1 (i.e., a surface that has a continuously varying outward normal vector). The unit normal of a surface depends only upon the graph of the surface and not upon how the surface is represented. From an interactive design point of view, the concept of a surface normal is easy to visualize and it is usually quite easy to anticipate how changes in the normal will affect the surface. Except within the context of solid modeling, the subject of closed surfaces has not received much attention in CAGD. Herron [9], [10] gives an argument for visual continuity over C^1 continuity in the case of closed surfaces. Farin [6] exploits the Bernstein basis for polynomials to obtain geometric conditions that imply visual continuity. Barnhill [1] mentions both the concepts of visual continuity and the problems of closed surfaces. The interpolants introduced in this paper will allow for the representation of open or closed G^1 surfaces.

We now introduce some notation. Surface segments are vector valued functions

$$F(b_1, b_2, b_3) = \begin{pmatrix} x(b_1, b_2, b_3) \\ y(b_1, b_2, b_3) \\ z(b_1, b_2, b_3) \end{pmatrix}, \qquad (b_1, b_2, b_3) \in T$$

where the domain T is defined as

$$T = \{(b_1, b_2, b_3) : 0 \leq b_i \leq 1,\ i = 1, 2, 3;\ b_1 + b_2 + b_3 = 1\}.$$

The vertices of T are denoted by

$$V_1 = (1, 0, 0), \quad V_2 = (0, 1, 0), \quad V_3 = (0, 0, 1),$$

and the edges by

$$e_1 = \{(0, b_2, b_3) \in T\}, \quad e_2 = \{(b_1, 0, b_3) \in T\}, \quad e_3 = \{(b_1, b_2, 0) \in T\}.$$

Also, it is convenient to refer to the set of indices

$$I = \{(1, 2, 3), (2, 3, 1), (3, 1, 2)\}.$$

Often, we wish to refer to the parametric curves which are a result of restricting a function F defined over T to one of its edges. We use $F(e_i)$ to refer to this curve for the edge e_i and $F(\partial T)$ to collectively refer to all three.

Certain directional derivatives will also be of interest. The derivative taken in a direction parallel to an edge (say e_3),

$$(1) \qquad \frac{\partial F}{\partial e_3}(b_1, b_2, b_3) = F_{e_3}(b_1, b_2, b_3) = \lim_{\varepsilon \to 0} \frac{F(b_1 - \varepsilon, b_2 + \varepsilon, b_3) - F(b_1, b_2, b_3)}{\varepsilon},$$

is called a parallel-boundary tangent vector. The derivative taken in a direction perpendicular to an edge (say e_2),

$$(2) \qquad \frac{\partial F}{\partial \bar{e}_2}(b_1, b_2, b_3) = F_{\bar{e}_2}(b_1, b_2, b_3) = \lim_{\varepsilon \to 0} \frac{F(b_1 - \varepsilon, b_2 + 2\varepsilon, b_3 - \varepsilon) - F(b_1, b_2, b_3)}{\varepsilon},$$

is referred to as a cross-boundary tangent vector. The analogue for this derivative in the case of functions over triangular domains is often called a normal derivative since it is taken in a direction that is normal to an edge. We choose not to use this terminology here due to the potential confusion with the outward surface normal vector which is mentioned often in this paper. The cross product of cross-boundary derivatives and parallel-boundary derivatives yields this outward surface normal vector

$$(3) \qquad N[F](b_1, b_2, b_3) = \frac{F_{e_i} \times F_{\bar{e}_i}}{|F_{e_i} \times F_{\bar{e}_i}|}.$$

Based upon the previous notation, we now make a more formal statement of the properties of the interpolant we develop in this paper.

PROBLEM. *Let F be a parametric surface segment defined over T which is continuous and has a continuous outward normal vector $N[F]$. Assume that the cross-boundary and parallel-boundary tangent vectors are continuous on ∂T. Find $P[F]$ which is defined and continuous over T which has a continuous outward surface normal such that*

$$P[F](\partial T) = F(\partial T),$$

$$N[P[F]](\partial T) = N[F](\partial T).$$

2.   A Triangular G^1 Patch.   Many of the C^1 triangular interpolants for functions are based upon the combination of operators which are essentially univariate interpolation operators. For example, the BBG interpolant presented by Barnhill, Birkhoff and Gordon [3] is based upon univariate Hermite operators applied along lines parallel to the edges of a triangular domain. The side-vertex method (Nielson [12]) is based upon radial projectors which are similar operators applied along rays emanating from a vertex joining an opposing edge.

We would like to utilize these essentially univariate operators in the development of a G^1 patch. The univariate cubic Hermite operator produces a cubic interpolant to arbitrary end points and tangent vectors and has the representation

$$(4) \qquad H[V_0, V_1, V_0', V_1'](t) = H_0(t)V_0 + H_1(t)V_1 + \bar{H}_0(t)V_0' + \bar{H}_1(t)V_1'$$

where

$$H_0(t) = 1 - 3t^2 + 2t^3, \qquad H_1(t) = 3t^2 - 2t^3,$$
$$\bar{H}_0(t) = t - 2t^2 + t^3, \qquad \bar{H}_1(t) = t^3 - t^2, \quad 0 \le t \le 1$$

are the cardinal basis functions. For the present situation, we need an operator

$$g[V_0, V_1, N_0, N_1](t) = g(t)$$

with the properties

$$(5) \qquad g(0) = V_0, \quad g(1) = V_1, \quad (g'(0), N_0) = 0, \quad (g'(1), N_1) = 0.$$

If we take g to be the cubic Hermite operator, then we can see that the endpoint tangent vectors are not uniquely determined by the requirements of (5) and so we must impose some additional conditions. We require that $g'(0)$ be in the plane containing N_0

and $V_1 - V_0$ and that $g'(1)$ be in the plane of N_1 and $V_1 - V_0$. These conditions imply that

$$g'(0) = a[(V_1 - V_0) - (N_0, V_1 - V_0)N_0] = aN_0 \times (V_1 - V_0) \times N_0,$$
$$g'(1) = b[(V_1 - V_0) - (N_1, V_1 - V_0)N_1] = bN_1 \times (V_1 - V_0) \times N_1,$$

for arbitrary constants a and b. In order for the normals of the surfaces based upon this operator to be well defined, it is important that these tangent vectors do not have zero length. Therefore, we must assume that neither N_0 nor N_1 is parallel to $V_1 - V_0$. For our use of this operator, we have not found this to be a severe restriction. This is due in part to the fact that a chord $(V_1 - V_0)$ is an approximation to a tangent of the surface and therefore tends to be perpendicular to the surface normal and not parallel to it. Based upon the above assumptions and conditions we define

$$g[V_0, V_1, N_0, N_1](t) = g_{\alpha,\beta}[V_0, V_1, N_0, N_1](t) = g(t)$$

(6)
$$= H_0(t)V_0 + H_1(t)V_1 + \alpha \bar{H}_0(t) \frac{N_0 \times (V_1 - V_0) \times N_0}{|N_0 \times (V_1 - V_0) \times N_0|}$$

$$+ \beta \bar{H}_1(t) \frac{N_1 \times (V_1 - V_0) \times N_1}{|N_1 \times (V_1 - V_0) \times N_1|}$$

where H_0, H_1, $\bar{H}_0$ and $\bar{H}_1$ are the cardinal Hermite basis functions defined above in equation (4) and α and β are arbitrary nonzero constants. These values, α and β, can be used much like tension parameters to affect the shape of $g(t)$, but for the most part, we will take them to both be unity.

We now describe our transfinite interpolant. It is based upon the ideas of the side-vertex method and the previously defined operator g. The arguments of g are a vertex and the point on the opposing edge where a ray emanating from the vertex and passing through the point (b_1, b_2, b_3) intersect. These boundary points have the representation (cf. Fig. 1.):

$$\frac{b_j V_j + b_k V_k}{1 - b_i}, \qquad (i, j, k) \in I.$$

THEOREM 1.   *Let*

$$G_i[F](b_1, b_2, b_3) = g\left[F(V_i), F\left(\frac{b_j V_j + b_k V_k}{1 - b_i}\right), N[F](V_i), N[F]\left(\frac{b_j V_j + b_k V_k}{1 - b_i}\right)\right](1 - b_i),$$

then

(7)
$$G[F] = W_1 G_1[F] + W_2 G_2[F] + W_3 G_3[F]$$

where

$$W_i = \frac{b_j^2 b_k^2}{b_1^2 b_2^2 + b_2^2 b_3^2 + b_3^2 b_1^2}, \qquad (i, j, k) \in I$$

satisfies the interpolation conditions

$$G[F](\partial T) = F(\partial T), \qquad N[G[F]](\partial T) = N[F](\partial T).$$

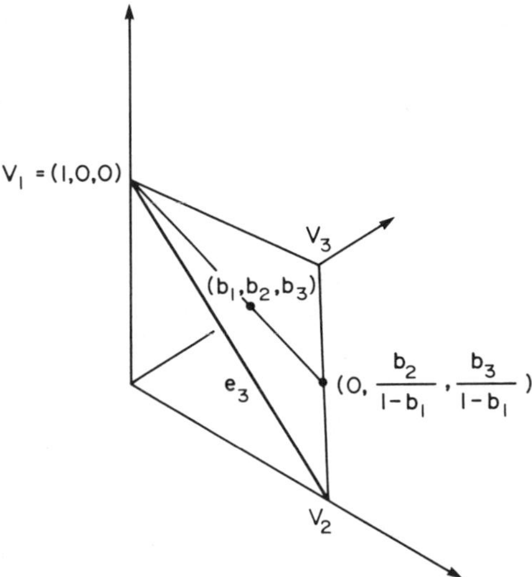

Figure 1.   The domain T.

Proof.   Based upon the properties of the operator g defined above, it is easy to see that G_i interpolates to F on e_i and since

$$W_i(e_j) = \delta_{ij}, \qquad i, j = 1, 2, 3$$

we have that G interpolates to F on the entire boundary of T.

Also, using the fact that

$$\frac{\partial W_i}{\partial \bar{e}_i}(e_i) = 0, \qquad i = 1, 2, 3$$

we have that the cross-boundary derivative of G is equal to the cross-boundary derivative of G_i on the edge e_i. That is

$$\frac{\partial G}{\partial \bar{e}_i}(e_i) = \frac{\partial G_i}{\partial \bar{e}_i}(e_i), \qquad i = 1, 2, 3.$$

Since both G and G_i interpolate to F on the edge e_i we know that they have the same parallel-tangent vector on this edge. By forming the cross-product of this tangent vector and the cross-boundary tangent vector, we may conclude that

$$N[G_i](e_i) = N[G](e_i), \qquad i = 1, 2, 3.$$

Again, based upon the definition of g and the fact that G_i interpolates to F on e_i we have that

$$N[G_i](e_i) = N[F](e_i), \qquad i = 1, 2, 3$$

and this completes the argument. $\square$

3. A Six-Parameter, G^1 Patch. In this section, we describe a discretization of the transfinite patch of the previous section. In the case of C^1 triangular interpolants for function data, a common approach to discretization is to take the function values along an edge as cubic *Hermite* interpolants and the cross-boundary normal derivatives as *linear* interpolants. In this way, the boundary data and consequently the entire triangular approximation depends only upon the nine values at the vertices, namely, the value of the function and its two first order partial derivatives at each of the vertices of the triangular domain. Since any two of these triangular interpolants joined along a common edge will yield a C^1 function, this type of interpolant is often called a nine-parameter, C^1 interpolant (element). In a similar manner we wish to obtain a six-parameter, G^1 interpolant. This type of interpolant will depend solely upon the position and outward surface normal at the three vertices of T. We denote these values by

$$\begin{aligned}
F_1 &= F(1, 0, 0), & N_1 &= N[F](1, 0, 0),\\
F_2 &= F(0, 1, 0), & N_2 &= N[F](0, 1, 0),\\
F_3 &= F(0, 0, 1), & N_3 &= N[F](0, 0, 1).
\end{aligned}$$

For each of the three boundary position curves, we will use the operator g defined in (6). Thus, we have

$$\begin{aligned}
F(1 - t, t, 0) &= g[F_1, F_2, N_1, N_2](t),\\
F(0, 1 - t, t) &= g[F_2, F_3, N_2, N_3](t),\\
F(t, 0, 1 - t) &= g[F_3, F_1, N_3, N_1](t).
\end{aligned}$$

Based upon our comments above concerning linearly varying cross-boundary derivatives for C^1 function interpolants, we may attempt to use a linearly varying outward surface normal. In general, this will not be consistent with the boundary position curves we have already selected since the parallel tangent vectors on an edge must be perpendicular to the outward surface normal. In order to overcome this potential inconsistency, we take the boundary surface normals as:

$$N[F](1 - t, t, 0) = N_3(t) = \frac{\dfrac{\partial F}{\partial e_3}(t) \times \left[(1 - t)N_1 \times \dfrac{\partial F}{\partial e_3}(0) + tN_2 \times \dfrac{\partial F}{\partial e_3}(1)\right]}{\left|\dfrac{\partial F}{\partial e_3}(t) \times \left[(1 - t)N_1 \times \dfrac{\partial F}{\partial e_3}(0) + tN_2 \times \dfrac{\partial F}{\partial e_3}(1)\right]\right|},$$

$$(8) \quad N[F](0, 1 - t, t) = N_1(t) = \frac{\dfrac{\partial F}{\partial e_1}(t) \times \left[(1 - t)N_2 \times \dfrac{\partial F}{\partial e_1}(0) + tN_3 \times \dfrac{\partial F}{\partial e_1}(1)\right]}{\left|\dfrac{\partial F}{\partial e_1}(t) \times \left[(1 - t)N_2 \times \dfrac{\partial F}{\partial e_1}(0) + tN_3 \times \dfrac{\partial F}{\partial e_1}(1)\right]\right|},$$

$$N[F](t, 0, 1 - t) = N_2(t) = \frac{\dfrac{\partial F}{\partial e_2}(t) \times \left[(1 - t)N_3 \times \dfrac{\partial F}{\partial e_2}(0) + tN_1 \times \dfrac{\partial F}{\partial e_2}(1)\right]}{\left|\dfrac{\partial F}{\partial e_2}(t) \times \left[(1 - t)N_3 + \dfrac{\partial F}{\partial e_2}(0) + tN_1 \times \dfrac{\partial F}{\partial e_2}(1)\right]\right|}.$$

It is easy to verify that

$$
\begin{aligned}
N_1(0) &= N_2, & N_1(1) &= N_3, \\
N_2(0) &= N_3, & N_2(1) &= N_1, \\
N_3(0) &= N_1, & N_3(1) &= N_2,
\end{aligned}
$$

(9)

and since these boundary surface normals are uniquely determined by data at endpoints of the edges only, any two patches which are joined along a common edge and which have been discretized in this fashion, will have a common surface normal along this edge.

The above comments assume that the normalizing values of equations (8) are never zero. While we have never observed this to be the case for any of our applications, it is theoretically possible that these surface normals are not well defined. For our results to be valid, we must assume that these vectors never degenerate to zero. Possibly there are better methods for discretizing boundary normals that would not necessitate this assumption and we plan to pursue this in the future. In some ways, the present method of discretizing is the simplest possible. The surface normal must be perpendicular to $\partial F/\partial e_i$ and so it can be represented as the cross-product of $\partial F/\partial e_i$ and some other vector and we have chosen this vector to be a linear combination of the two vectors necessary for the conditions of (9) to hold.

Based upon the above choices for the boundary data we may write down the discretized form of each of the three radial projectors

$$
(10) \quad g_i(b_1, b_2, b_3) = g\left[F_i, g[F_j, F_k, N_j, N_k]\left(\frac{b_j V_j + b_k V_k}{1 - b_i}\right), N_i, N_i\left(\frac{b_j V_j + b_k V_k}{1 - b_i}\right)\right](1 - b_i)
$$

and note that g_i interpolates to the boundary position data on all three edges and to boundary outward surface normals on the edge e_i. This last observation allows us to use some simpler weight functions than those used previously for the transfinite interpolant.

THEOREM 2. *If A_1, A_2, A_3 are three operators with the properties that*

$$
\begin{aligned}
A_i[F](\partial T) &= F(\partial T), \qquad i = 1, 2, 3, \\
N[A_i[F]](e_i) &= N[F](e_i),
\end{aligned}
$$

then

$$
A[F] = \frac{b_2 b_3 A_1[F] + b_1 b_3 A_2[F] + b_1 b_2 A_3[F]}{b_2 b_3 + b_1 b_3 + b_1 b_2}
$$

interpolates to F and its outward surface normal to the entire boundary of T.

Proof. Since the weight functions sum to unity, it is clear that

$$
A[F](e_i) = F(e_i), \qquad i = 1, 2, 3.
$$

If we let

$$
\beta_i = \frac{b_j b_k}{b_1 b_2 + b_1 b_3 + b_2 b_3}, \qquad (i, j, k) \in I
$$

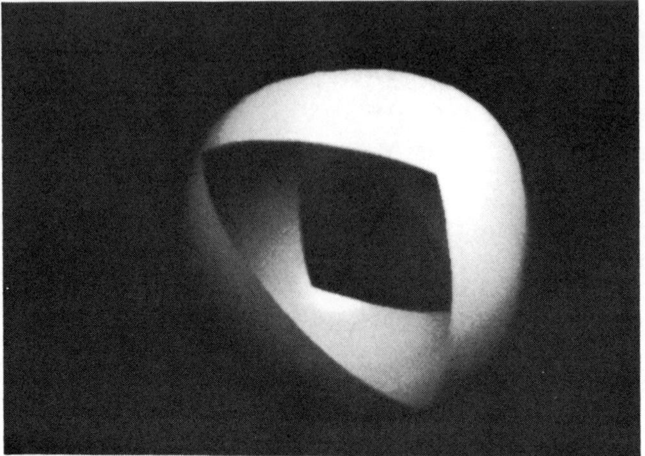

Figure 2a.   A closed surface with two patches not displayed.

Figure 2b.   A closed surface with tension, $\alpha = \beta = 0.25$.

then we have that

$$\frac{\partial A[f]}{\partial \bar{e}_i}(e_i) = \beta_1(e_i)\frac{\partial A_1[F]}{\partial \bar{e}_i}(e_i) + A_1[F](e_i)\frac{\partial \beta_1}{\partial \bar{e}_i}(e_i)$$
$$+ \beta_2(e_i)\frac{\partial A_2[F]}{\partial \bar{e}_i}(e_i) + A_2[F](e_i)\frac{\partial \beta_2}{\partial \bar{e}_i}(e_i)$$
$$+ \beta_3(e_i)\frac{\partial A_3[F]}{\partial \bar{e}_i}(e_i) + A_3[F](e_i)\frac{\partial \beta_3}{\partial \bar{e}_i}(e_i).$$

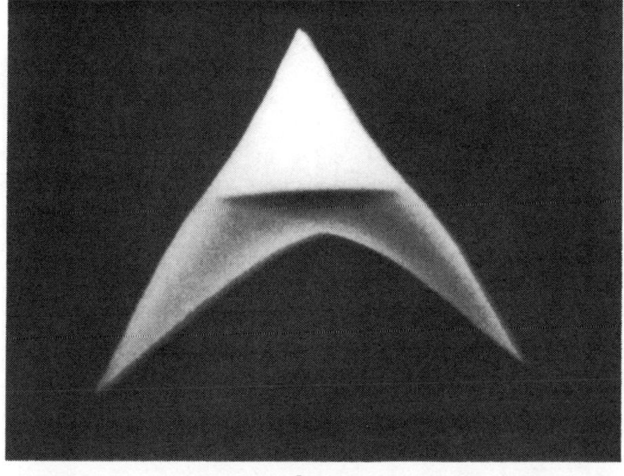

3a

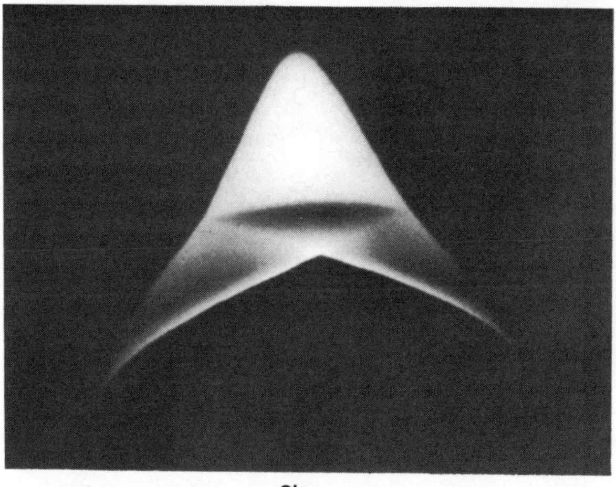

3b

Figure 3. The surface illustrated in Fig. 3a has average normals and the one in Fig. 3b has user specified normals that are only slightly different.

Since $\beta(e_j) = \delta_{ij}$ and $A_i[F](\partial T) = F(\partial T)$, this reduces to

$$\frac{\partial A[F]}{\partial \bar{e}_i}(e_i) = \frac{\partial A_i[F]}{\partial \bar{e}_i}(e_i) + F(e_i)\left[\frac{\partial \beta_1}{\partial \bar{e}_i}(e_i) + \frac{\partial \beta_2}{\partial \bar{e}_i}(e_i) + \frac{\partial \beta_3}{\partial \bar{e}_i}(e_i)\right].$$

Again we use the fact that $\beta_1 + \beta_2 + \beta_3 = 1$ to conclude that this last term is zero. The remainder of the argument follows that of Theorem 1. $\square$

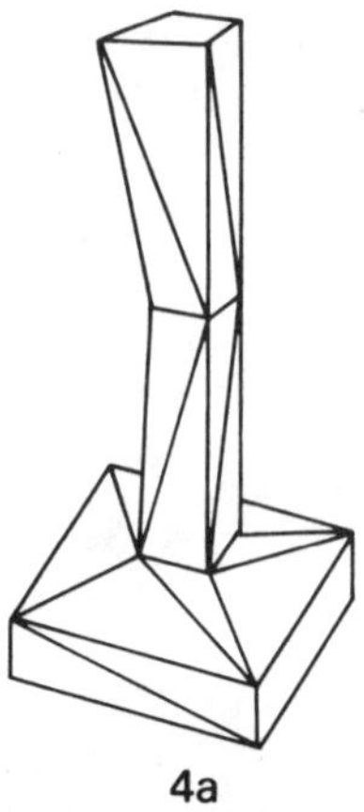

4a

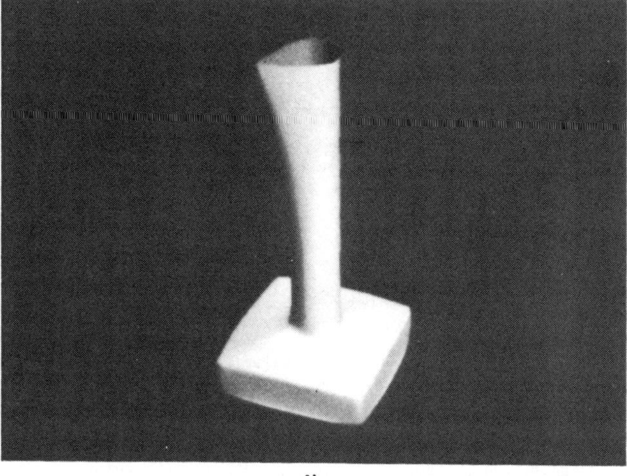

4b

Figure 4. A G^1 surface comprised of triangular patches (Fig. 4b). The data points are shown in Fig. 4a.

With this in mind, we define our six-parameter, G^1 patch to be

$$\bar{G} = \bar{G}[F_1, F_2, F_3, N_1, N_2, N_3]$$
$$= \beta_1 g_1 + \beta_2 g_2 + \beta_3 g_3$$

where g_i is given by (10).

We now present a few examples of this G^1 patch. The first is based upon the eight vertices of a unit cube and illustrates the possibility of defining a closed G^1 surface. There are twelve triangular patches, two per face of the cube, and the unit normals are taken as

$$\left(\pm \frac{1}{\sqrt{3}}, \pm \frac{1}{\sqrt{3}}, \pm \frac{1}{\sqrt{3}} \right).$$

A picture of a near spherical blob is not very interesting or revealing so we have omitted a couple of the triangular patches in the photograph of Fig. 2a. In Fig. 2b, we show the effect of the tension parameters α and β. The data is the same as that for the object of Fig. 2a except that $\alpha = \beta = 0.25$.

The next example shows the effect of varying the surface normals. The photographs of Fig. 3 use the vertices $V_1 = (1, 0, 0)$, $V_2 = (0, 1, 0)$, $V_3 = (0, 0, 1)$, $V_4 = (\frac{1}{2}, 0, \frac{1}{2})$, $V_5 = (\frac{1}{2}, \frac{1}{2}, \frac{1}{2})$ and $V_6 = (0, \frac{1}{2}, \frac{1}{2})$. Figure 3a uses normals for a vertex obtained by averaging the normals of all triangles that involve this vertex. Figure 3b uses slightly different values:

$$N_1 = (1, 0, 0), \quad N_2 = (0, 1, 0), \quad N_3 = (0, 0, 1),$$

$$N_4 = \left(\frac{\sqrt{2}}{2}, 0, \frac{\sqrt{2}}{2}\right), \quad N_5 = \left(\frac{\sqrt{3}}{3}, \frac{\sqrt{3}}{3}, \frac{\sqrt{3}}{3}\right), \quad N_6 = \left(0, \frac{\sqrt{2}}{2}, \frac{\sqrt{2}}{2}\right).$$

This example also serves to point out an obvious but quite important point. Even a simple object requires a lot of surface normals to be specified. While it is useful for selective modification to be able to specify normals, there must be some automatic way to select default values. The technique of local averages we mentioned above does not work very well in general. This is not surprising since a similar technique for scattered data interpolation does not work well either.

In Fig. 4b we show a G^1 surface based on the data of Fig. 4a. The normals for this example are taken from a rectangular v-spline (cf. Nielson [13]). We simply replaced the rectangular Hermite patch with two G^1 patches. The surface is essentially the same and so there is nothing really gained for this type of gridded data. The main purpose for including this example is to illustrate the potential of G^1 patches once a reasonable technique for selecting surface normals is developed.

Acknowledgments.   We appreciate the support of the U.S. Office of Naval Research. We thank Dick Franke for the time and effort he spent on this research. We also wish to thank Gerald Farin for the work involved in creating this reference text and for his contributions to this particular paper.

REFERENCES

[1]   R. E. BARNHILL, *Surfaces in computer aided geometric design: A Survey with new results*, Computer Aided Geometric Design, 2 (1985), pp. 1–17.

[2]   ———, *Representation and approximation of surfaces*, in Mathematical Software III, J. R. Rice, ed., Academic Press, New York, 1977, pp. 69–120.

[3]   R. E. BARNHILL, G. BIRKHOFF AND W. J. GORDON, *Smooth interpolation in triangles*, J. Approx. Theory, 8 (1973), pp. 114–128.

[4]   W. BOEHM, G. FARIN AND J. KAHMANN, *A survey of curve and surface methods in* CAGD, Computer Aided Geometric Design, 1 (1984), pp. 1–60.

[5]   S.A. COONS, *Surfaces for computer aided design of space forms*, Dept. Mechanical Engineering, MIT, 1964, revised 1967.

[6]   G. FARIN, *A construction for the visual C^1 continuity of polynomial surface patches*, Comput. Graphics Image Processing, 20 (1982), pp. 272–282.

[7]   W. J. GORDON, *Blending-function methods of bivariate and multivariate interpolation and approximation*, SIAM J. Numer. Anal., 8 (1971), pp. 158–177.

[8]   J. A. GREGORY, C^1 *rectangular and non-rectangular surface patches,* in Surfaces in Computer Aided Geometric Design, R. E. Barnhill and W. Boehm, eds., North-Holland, Amsterdam, 1983.

[9]   G. HERRON, *Smooth closed surfaces with discrete triangular interpolants,* Computer Aided Geometric Design, 2 (1985), pp. 297–306.

[10]   ———, *Techniques for visual continuity,* this Volume, 1987.

[11]   G. M. NIELSON AND R. FRANKE, *A method for construction of surfaces with tension,* in Surfaces, R. E. Barnhill and G. M. Nielson, eds., Rocky Mountain Mathematics Consortium, Tempe, AZ, 1984.

[12]   G. M. NIELSON, *The side-vertex method for interpolation in triangles,* J. Approx. Theory, 25 (1979), pp. 318–336.

[13]   ———, *Rectangular v-splines,* IEEE Comput. Graphics and Applications, 6 (1986), pp. 35–40.

Shape Constrained Curve
and Surface Fitting

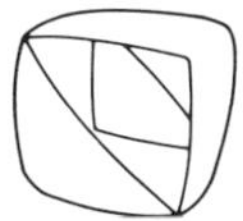

KEN W. BOSWORTH

Abstract. A general shape constrained approximation problem is recast into a finite-dimensional semi-infinite convex programming problem with special geometric properties. An algorithm for the solution of the numerical problem is briefly sketched, and several standard examples of shape constrained approximation problems are presented and solved.

1. Introduction. The following problem, (P), often arises, albeit sometimes in disguised form, in engineering modeling processes and numerical analysis:

GIVEN: (1) Data, $f(p)$, on a compact set $I \subset R^k$.

(2) A finite-dimensional subspace, V, of continuous functions defined on I;

$$V := \lim \{f_1, f_2, \cdots, f_n\}.$$

(P)

(3) A norm (or seminorm), $\|\cdot\|$, defined on the linear space gotten by adjoining f to V.

FIND: The best approximation, g_{ba}, to f from the space V; i.e., solve:

$$\|f - g_{ba}\| = \min_{g \in V} \|f - g\| \quad \text{for } g_{ba} \in V.$$

More often than not, one desires g_{ba} to possess additional attributes or qualities. In some instances, g_{ba} should *inherit* structure from f, e.g., g_{ba} should mimic f in "shape" as closely as possible (f concave up (down) $\Rightarrow g_{ba}$ concave up (down)), or g_{ba} should agree with f at certain prescribed points in I, etc. In other situations, g_{ba} should *not resemble* the data f too closely, e.g., if f is known to be noisy, g_{ba} should fit f "well," but not "wiggle" or "ripple" excessively. In such instances we impose *external* shape conditions on g_{ba}. From either end of the spectrum,

<table>
<tr><td align="center">Inheriting structure from
the data f</td><td align="center">$\Leftrightarrow$</td><td align="center">Divorcing g_{ba} from f's
undesirable qualities,</td></tr>
</table>

one casts the same abstract problem. In the formulation of problem (P) one adds a fourth "given":

(4) A pointwise indexed set of linear (shape) constraints of the form:

$$a(p) \leq L(p;f,g) \leq b(p), \quad p \in \tilde{I} \subset R^k, \quad \tilde{I} \text{ compact.}$$

Here, $L(p;f,g)$ is a linear functional operating on both f and $g \in V$, or on g alone. In the revised problem (P), the best approximation, g_{ba}, must now come from the class of those g's in V satisfying (4), i.e., satisfying the given collection of shape constraints.[1]

We cite here a few specific examples of (P):

Example 1. Synthesis. Inheriting structure. The data f on $[a, b]$ is a draftsman's curve, drawn between known points of inflection $\{x_i, y_i\}_{i=1}^m$. Δ is a partition of $[a, b]$, over which is defined the linear space of splines of fixed degree k, $V = S_\Delta^k$. One seeks a mathematical representation of f, $g_{ba} \in S_\Delta^k$, which yields: minimize $\|f - g\|_\infty$ over all $g \in S_\Delta^k$ subject to $g(x_i) = y_i$, $i = 1, \cdots, m$, and $g''(x) \geq 0$ for all x such that $f''(x) \geq 0$, and $g''(x) \leq 0$ for all x such that $f''(x) \leq 0$.

Example 2. Synthesis. Creating structure. Δ is a triangulation of $I \subset R^2$, over which is defined a linear space $S_\Delta^{k,m}$ of bivariate splines of fixed degree k and global smoothness m. Data f is given at *scattered* locations $D \subset I$. One seeks $g_{ba} \in S_\Delta^{k,m}$ which yields: mimimize $\|f - g\|_{\infty, D}$ over all $g \in S_\Delta^{k,m}$ subject to $|g_x| \leq M$ and $|g_y| \leq M$ on $I \backslash D$. Here, $\|f\|_{\infty, D} := \sup_{p \in D} |f(p)|$ is a seminorm (reflecting the sparsity of data), and the bounds on the derivatives attempt to control the approximation g in regions of sparse data.

Example 3. Reconstruction. $I = [0, b]$, $b \gg 1$. Data f is experimental data, known to be uniformly noisy measurements of a distribution g which satisfies: $\int_0^b g \, dx = 1$, $g(x) \geq 0$, $g'(x) \geq 0$, $g(0) = 0$, and $g(b) = 1$. Also, g is in $V := \lin \{\exp [k_i(x - x_i)]\}_{i=1}^n$; k_i, and x_i known. Then one seeks $g_{ba} \in V$ yielding: minimize $\|f - g\|_1$ over all $g \in V$ subject to g, meeting the above constraints.

2. Transcription of Problem (P). In this section we state a transcription of problem (P) into an "equivalent" optimization problem (P_T); this transcription is basically due to A. Haar [9]. Problem (P) is equivalent, in a sense to be made precise shortly, to the following semi-infinite convex program (P_T):

$$(P_T) \qquad \begin{aligned} &\text{maximize } \bar{e}_{n+1} \cdot \bar{x} \\ &\text{subject to } \bar{x} \in K \cap C. \end{aligned}$$

Problem (P_T) appears to be a linear program in the variables $\bar{x}$, but is a semi-infinite convex program by virtue of the feasible region being defined by the intersection of a convex symmetric "gauge body":

$$K := \{\bar{x} \in R^{n+1} \mid \|\bar{f}(p) \cdot \bar{x}\| \leq 1\},$$

---

[1] Here, $a(p)$ and $b(p)$ are extended real valued functions. In addition, $a(p)$ is continuous on $\tilde{I} \backslash \{a^{-1}(-\infty)\}$ and $b(p)$ is continuous on $\tilde{I} \backslash \{b^{-1}(\infty)\}$.

with a *"cone of constraints"*:

$$C := \{\bar{x} \in R^{n+1} \mid \bar{n}_+(p) \cdot \bar{x} \le 0, \bar{n}_-(p) \cdot \bar{x} \le 0, \forall p \in \tilde{I}\}$$
$$\cap \{\bar{x} \in R^{n+1} \mid \bar{e}_{n+1} \cdot \bar{x} \ge 0\},$$

with vertex at $\bar{O}_{n+1}$.

Here

$$\bar{x} = (x_1, \cdots, x_n, x_{n+1}), \qquad \bar{e}_{n+1} = (0, \cdots, 0, 1),$$
$$\bar{f}(p) = (f_1(p), \cdots, f_n(p), f(p)),$$
$$\bar{n}_+(p) = (\beta_1, \cdots, \beta_n, a(p) - \beta_{n+1}),$$
$$\bar{n}_-(p) = (-\beta_1, \cdots, -\beta_n, \beta_{n+1} - b(p)).^2$$

The dot product $\bar{u} \cdot \bar{v}$ is the standard Euclidean inner product in R^{n+1} (recall that n is the dimension of V). The vectors $\bar{n}_+(p)$ and $\bar{n}_-(p)$ were obtained by observing that the action of any linear functional L on lin $\{f_1, \cdots, f_n, f\}$ must have the representation:

$$L\left(p; f, \sum_{i=1}^{n} \alpha_i f_i\right) = \sum_{i=1}^{n} \alpha_i \beta_i(p) + \beta_{n+1}(p),$$

where

$$\beta_i(p) = L(p; 0, f_i), \qquad i = 1, \cdots, n,$$

and

$$\beta_{n+1}(p) = L(p; f, 0).$$

In (P_T) the variable x_{n+1} is sought to be maximized, for generally: $\max_{\bar{x} \in K} x_{n+1} = \min_{g \in V} \|f - g\|$. The cone C merely expresses the fact that the best approximation must also meet the shape constraints being posed in (P). The full equivalence of (P) and (P_T) for *any norm* is as follows (see Bosworth [1]):

(1) If $\bar{v} = (v_1, \cdots, v_{n+1})$ solves (P_T) *and* $\bar{e}_{n+1} \cdot \bar{v} > 0$, then with $\alpha_i = -v_i/v_{n+1}$, $i = 1, \cdots, n$, $g_{ba} := \sum_{i=1}^{n} \alpha_i f_i$ solves (P). Also, $1/v_{n+1} = \|f - g_{ba}\| = \min \|f - g\|$ over all $g \in V$ meeting the shape constraints. $\bar{v}$ and hence g_{ba} may not be unique.

(2) If $\bar{v}$ solves (P_T) but $v_{n+1} = 0$, then the constraints imposed on (P) *admit no candidates from the space V*, i.e., (P) is overconstrained.

(3) If (P_T) is unbounded in the $\bar{v}$-direction, with $v_{n+1} > 0$, then (P) is solved by $g_{ba} := \sum_{i=1}^{n} (-v_i/v_{n+1}) f_i$ with $\|f - g_{ba}\| = 0$. Thus, f is in V, f has representation g_{ba}, and g_{ba} meets all the posed shape constraints.

3. Solution of the Transcribed Problem (P_T). Problem (P_T) may be attacked by one of the semi-infinite versions of the Simplex Algorithm as described in Glashoff and Gustafson [7] or in Roleff [10], or by means of a cutting plane algorithm for convex programs as adapted to semi-infinite programs such as in Gribik [8]. A new algorithm has been developed by this author for handling (P_T) solely. The key idea to the

---

[2] For $p \in a^{-1}(-\infty)$, set $\bar{n}_+(p) := \bar{O}_{n+1}$. For $p \in b^{-1}(\infty)$, set $\bar{n}_-(p) := \bar{O}_{n+1}$.

algorithm is given below; one may refer to Bosworth [1], [2] for details and a FORTRAN 77 implementation of the algorithm.

The Tunnelling Algorithm, as it is called, is based on the special geometric structure of the convex program (P_T). One solves this program (involving a possibly infinite number of linear constraints) via an iterative process which selects (via heuristics based on the geometric structure of (P_T)) and solves at each stage a "minimal linear subprogram" of (P_T). These subprograms are *exchanged* as the iterations progress, rather than retained and augmented together (thus the linear program remains a manageable, in fact, constant, size). The result that one obtains is that the linear programs "converge" in some sense to a "linear program" representing $K \cap C$ at an optimal solution of (P_T), that is, representing the Kuhn–Tucker Optimality relations at the solution $\bar{v}$ to (P_T). See Bosworth [3] for the concise discussion of this exchange process and its relation to existing alternate methods of solution of (P), e.g., Remes-type algorithms, etc.

The Tunnelling Algorithm has been tested on a large variety of problems. The dimension of the approximating space has been allowed to reach 49 in unconstrained problems and 36 in constrained problems, a realistic situation in multivariate spline approximation problems; and various families of approximating functions have been used, e.g., uni- and multivariate B-splines, tensor products of univariate splines, polynomials, trigonometric functions, and exponential functions. In particular, the approximating space V is not restricted to be a Haar subspace. The successful incorporation of shape constraints has been tested in problems where it was known, a priori, that the cone of constraints C was not "singular," a concept not rigorously defined in this paper. Let it suffice now to say a "nonsingular" cone of constraints C corresponds to an imposition of shape constraints which does not "essentially" remove any degrees of freedom in the approximating space. (The algorithm can handle situations in which degrees of freedom have been removed in a *predictable* manner, e.g., via known, independent interpolation constraints. It is only when shape constraints "gang up" and collectively act as interpolation constraints that troubles arise.) The algorithm as presently coded tests for singularity at the very onset of a run, and if detected, aborts the run. See Bosworth [1] for a catalogue of some early test cases run by the Tunnelling Algorithm, and comparisons in efficiency (and success) with other existing methods. Preliminary results show a favorable trend of success, under the above-mentioned restriction.

We turn now to the presentation and solution of several concrete test problems illustrating the various situations in which (P) arises.

4. Numerical Results. In each example presented in this section, the type of shape constrained problem at hand will be stated (all problems are solved in the uniform norm), the approximating space of functions, V, will be briefly noted (the exact form of the basis functions will not generally be given, as most workers in this field will be familiar with minor variants of these function spaces already), the data, f, will be briefly described and presented graphically, and the best approximations to the data *both* in the absence of shape constraints *and* in the presence of shape constraints will be graphically presented. Notable differences in the behavior of these two solutions will be pointed out when necessary.

Example 1. *Monotone approximation to noisy data, known to have underlying monotone structure.* The approximating space is chosen to be the quadratic B-splines on $[-1, 1]$ with equally spaced knot sequence $\{-2.0, -1.5, -1.0, \cdots, 1.0, 1.5, 2.0\}$; dim $V = 6$. See de Boor [4] for details. Figure 1a is the data f, given at 201 points, Fig. 1b the best unconstrained spline approximation to f, and Fig. 1c the best monotone spline approximation to f. Note the obvious result: imposing monotonicity increases the uniform norm of the error in the fit.

Example 2. *Convex approximation to noisy data, known to have underlying convex structure.* The approximating space is that of the previous example, as is the data f, shown in Fig. 1a. Appearing in Fig. 1b is the best unconstrained approximation to f, and is clearly not convex. In Fig. 2a the best convex spline approximation to f is shown. Note the discontinuous second derivative in the approximation. A more reasonable space V is the *cubic* B-spline space on $[-1, 1]$ with equally spaced knot sequence $\{-2.5, -2.0, \cdots, 2.0, 2.5\}$ and dim $V = 7$. Figure 2b shows the best convex cubic spline fit to the data. In this instance, the imposition of continuity of second derivatives causes the best cubic (convex) approximation to suffer a larger uniform error than does the quadratic (convex) spline.

It should be noted in both Examples 1 and 2 in this section that the data set is much larger than the requisite set of data for performing *shape constrained interpolation* from the spline spaces. One is referred to the work of Fritsch and Carlson [6], and Fritsch and Butland [5], concerning monotonicity preserving interpolation in spline spaces. Another note is in order: if by accident an interpolant to the data exists in the spline space, and it preserves the shape of the data, then this will be the best (constrained) approximant selected by the method of this paper.

Example 3. *Bounded derivative constraints.* The approximating space is the tensor product space W on $[-1, 1] \times [-1, 1]$ gotten from taking $W = V \otimes V$, V as in Example 1 in this section; dim $W = 36$. Figure 3a shows the data f, actual topographic contour data from near Lake Placid, New York. To try to preserve the essential features of the data (the locations of the ponds, ridges, mountaintops, etc.) the partial derivatives of the best approximation were constrained to be nearly equal to the *estimated* partial derivatives of the data f (obtained by central difference quotients) in these locations of critical behavior. (The amount of constraint exercised is a design parameter, but should not be any smaller than the estimated error in the derivative approximations.) Figure 3b shows the best tensor product approximation to the data, and Fig. 3c shows the best derivative constrained approximation to the data. The constrained approximation appears very similar to the unconstrained approximation, the only notable differences being the precise preservation in the constrained situation of the flat foreground (a lake) and the locations of the mountain summits. (These differences are evident only upon direct computation of the actual partial derivatives of the approximants in these locations.)

Example 4. *Second derivative constraints.* In this example, the approximating space Z on $[-1, 1] \times [-1, 1]$ is gotten from the linear span of the tensor product of $\{1, \cos(\pi x/2), \sin(\pi x/2)\}$ with itself, and the bivariate polynomials $\{x, y, xy, x^2, y^2\}$; dim $Z = 14$. The data is a set of measurements, with uniform unknown error, of the deflection f of a thin plate obeying the partial differential equation $\Delta f = p$ on $I = (-1, 1) \times (-1, 1)$ and satisfying the boundary conditions $f|_{\partial I} = h$. Here, the load p

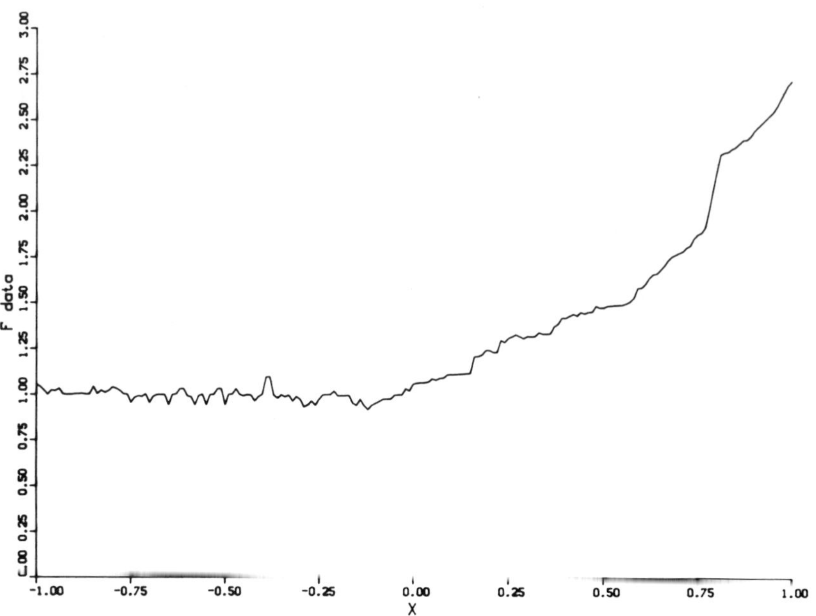

Figure 1a. Noisy data f with underlying monotone and convex structure. 201 data points on $[-1, 1]$.

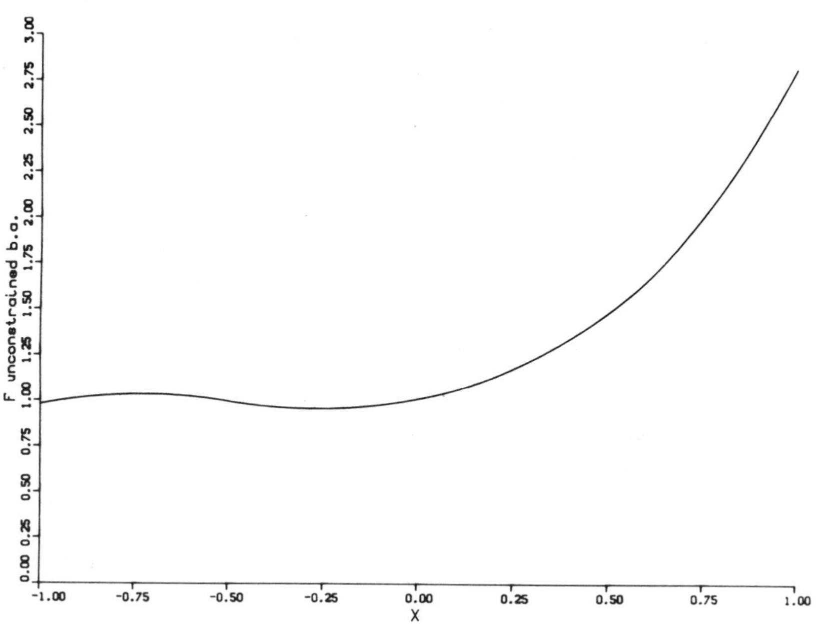

Figure 1b. The quadratic spline best uniform approximation to the data in Fig. 1a. The spline's derivatives are not constrained. Maximum error: 0.134.

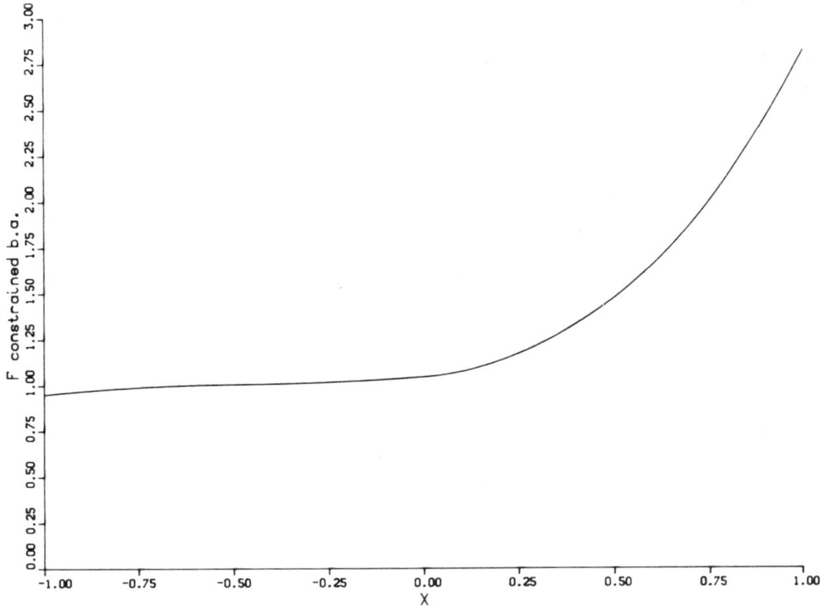

Figure 1c. The quadratic spline best uniform approximation to the data in Fig. 1a under monotonicity constraints. Maximum error: 0.144.

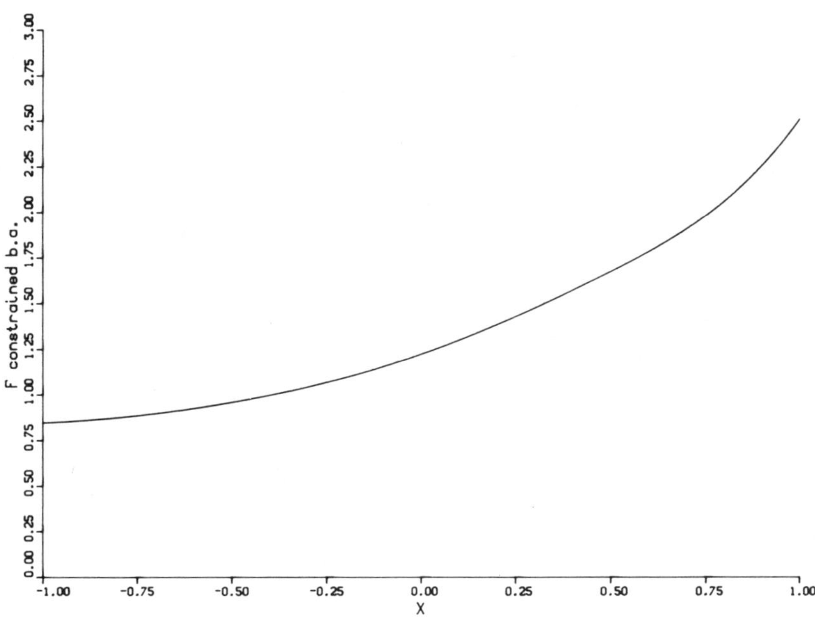

Figure 2a. The quadratic spline best uniform approximation to the data in Fig. 1a under convexity constraints. Maximum error: 0.169.

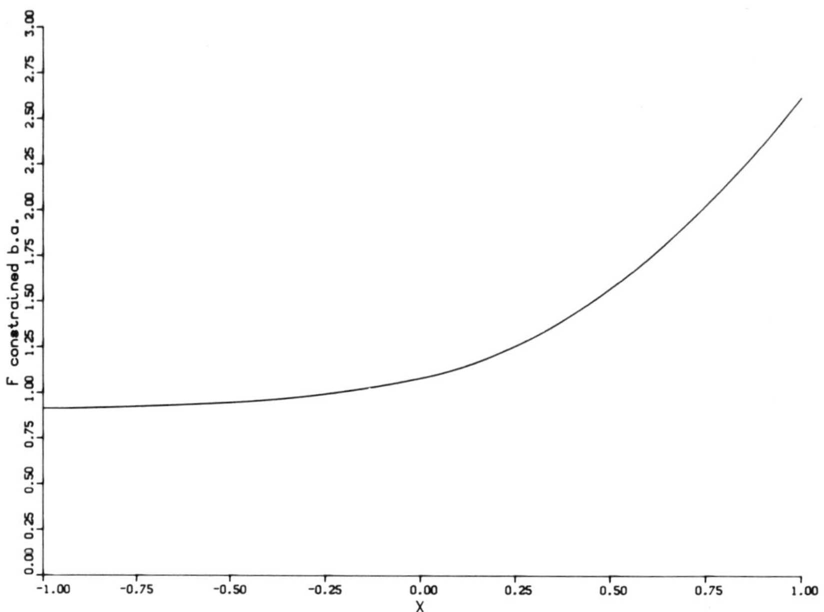

Figure 2b. The cubic spline best uniform approximation to the data in Fig. 1a under convexity constraints. Maximum error: 0.237.

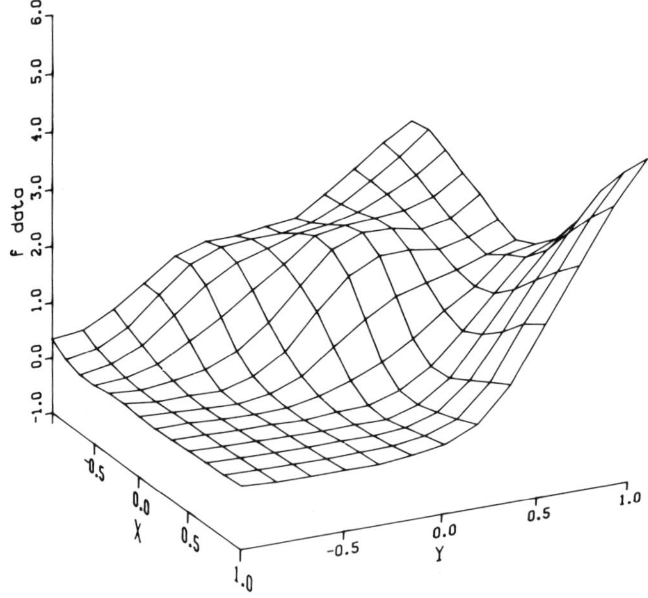

Figure 3a. Topographic data taken from near Lake Placid, New York. Elevations in hundreds of feet. 13×13 data grid.

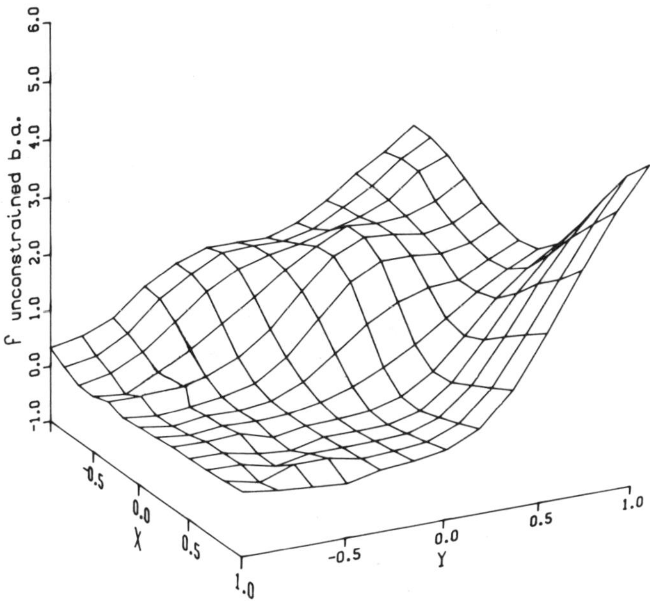

Figure 3b. Best tensor product spline uniform approximation to the data in Fig. 3a. No derivative constraints.

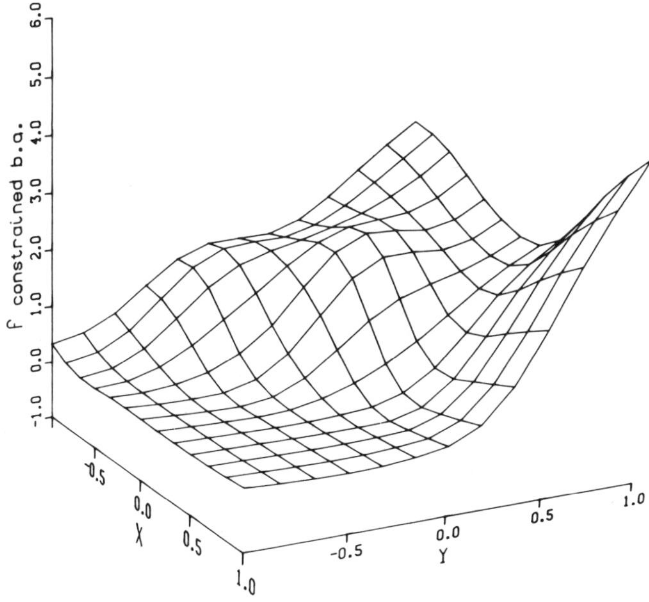

Figure 3c. Best tensor product spline uniform approximation to the data in Fig. 3a under bounded partial derivative constraints.

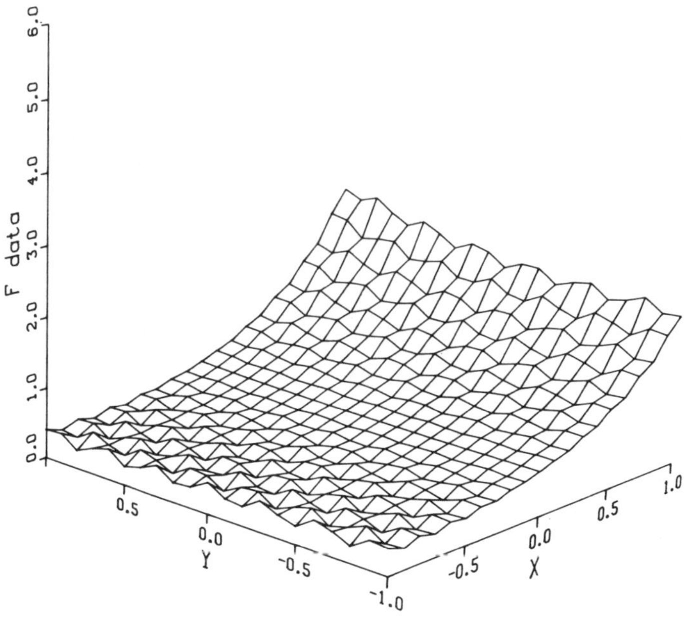

Figure 4a. Deflection f of a thin plate, with noise: $f = f_{\text{smooth}} + f_{\text{noise}}$. 21×21 data grid.

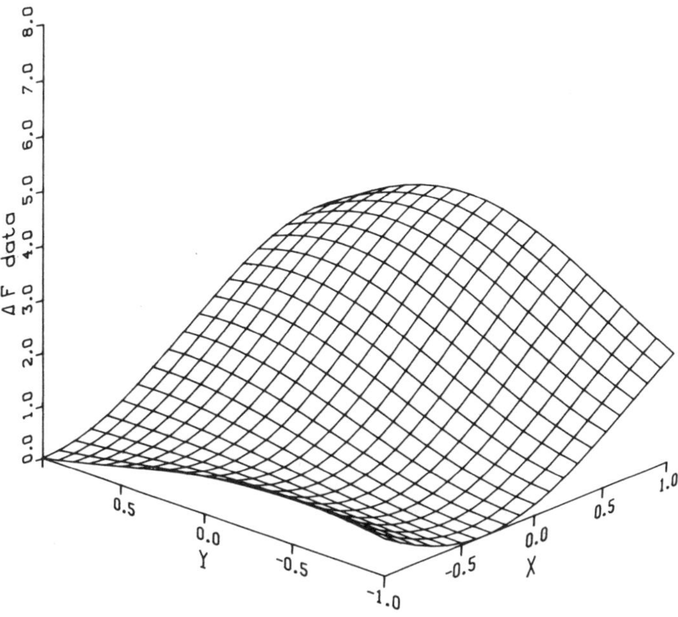

Figure 4b. The underlying distributed load $p := \Delta f_{\text{smooth}}$.

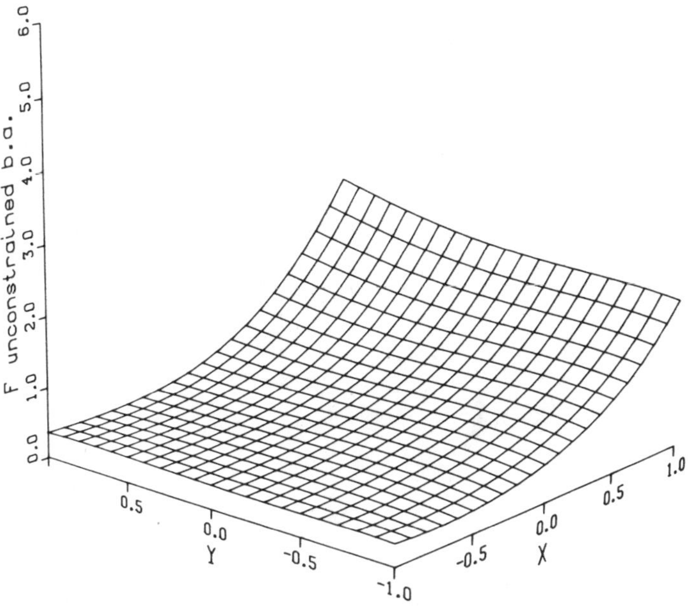

Figure 4c. Best uniform approximation to the data in Fig. 4a from the space Z. No constraints imposed.

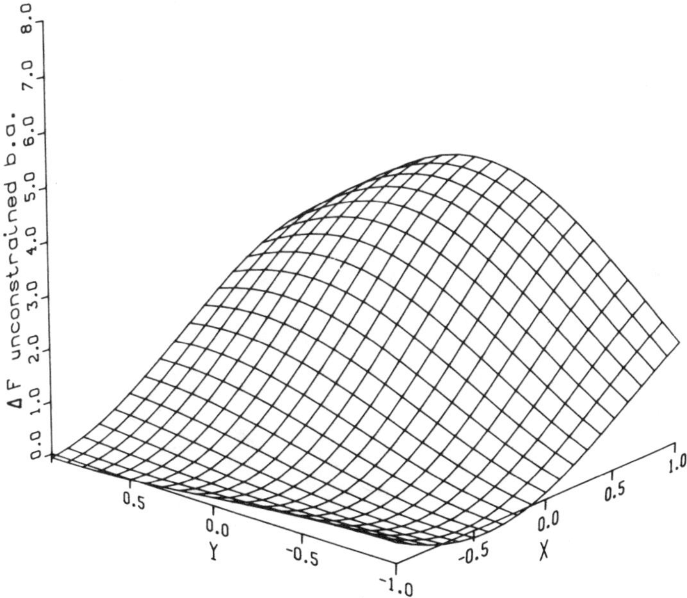

Figure 4d. Laplacian of the approximant in Fig. 4c. Note the regions of slight negativity of the Laplacian.

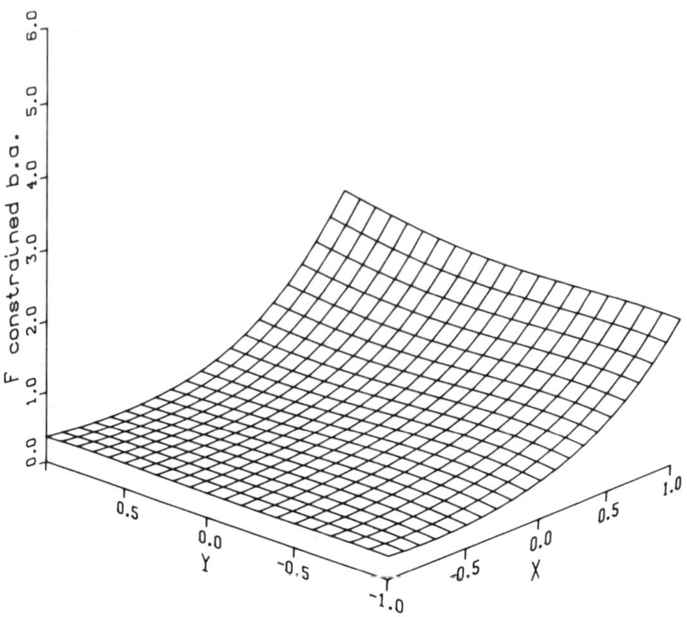

Figure 4e.   Best uniform approximation to the data in Fig. 4a from the space Z, with nonnegative Laplacian constraint imposed.

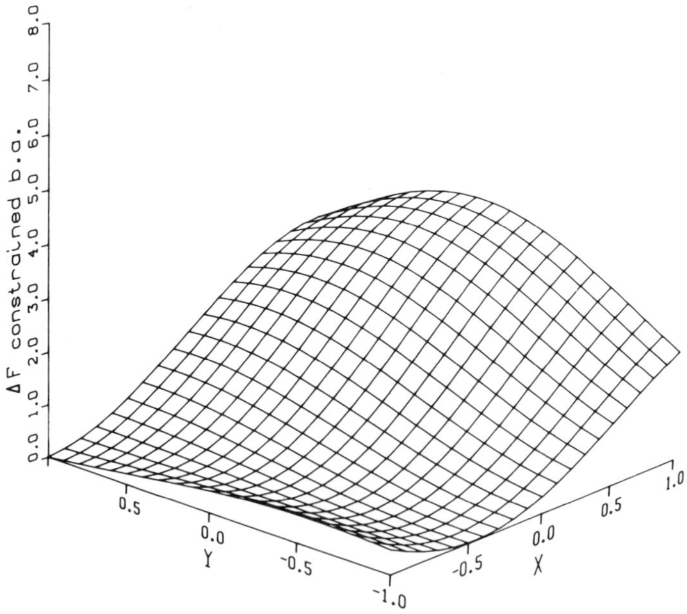

Figure 4f.   Laplacian of the approximant in Fig. 4e. Note the nonnegativity of the Laplacian over the entire region I.

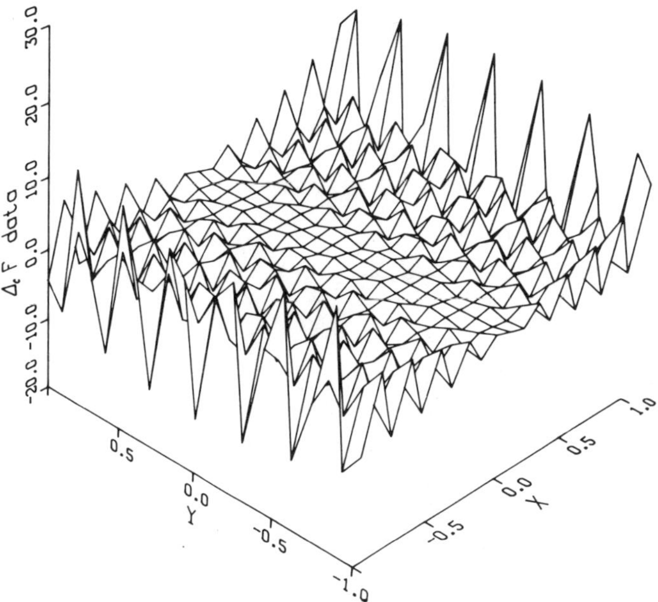

Figure 4g. The "approximate" Laplacian of f as calculated from central difference quotients, applied to the noisy f. Note the severe impact of the noise, f_{noise}, on these calculations.

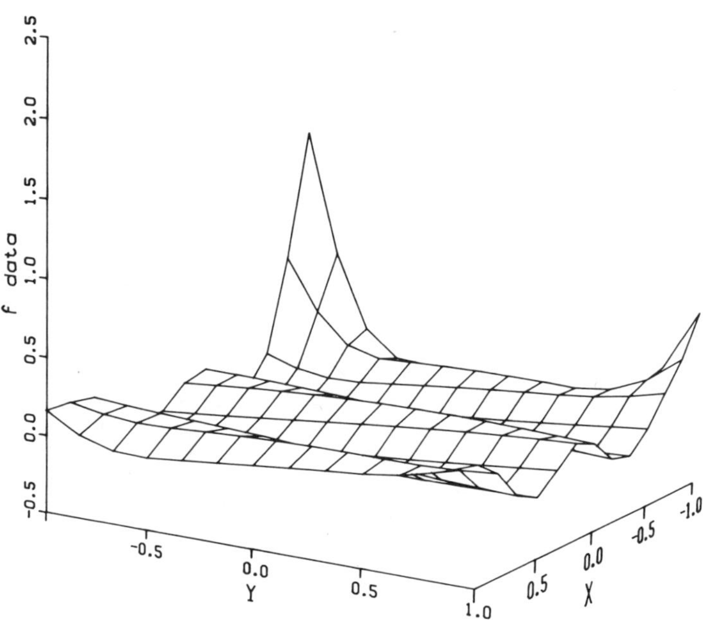

Figure 5a. Deflection f on a thin plate, with noise. 13×13 data grid.

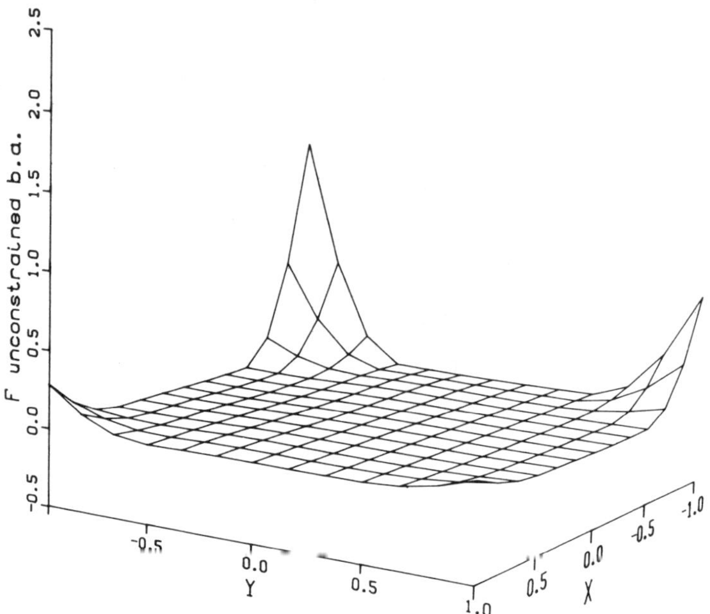

Figure 5b.   Best tensor product spline uniform approximation to the data in Fig. 5a, with no constraints imposed.

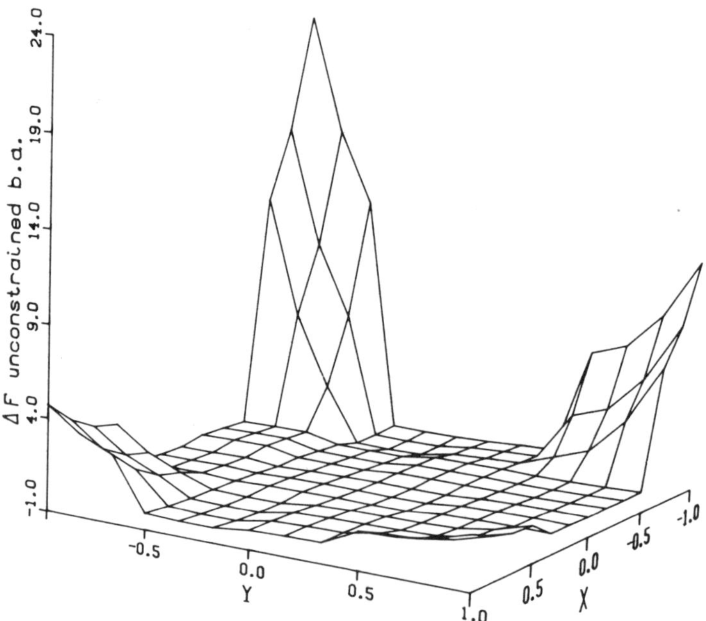

Figure 5c.   Laplacian of the approximant in Fig. 5b. Note the nonzero values of the Laplacian in the cross-shaped interior region.

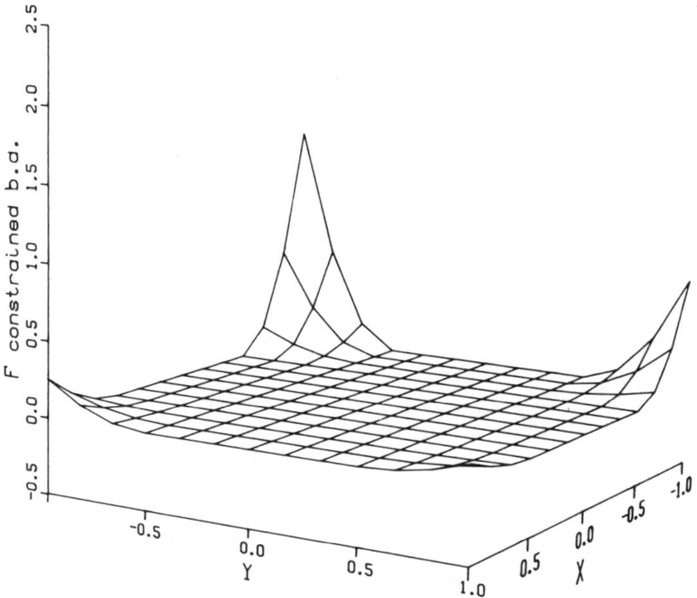

Figure 5d.   Best tensor product spline uniform approximation to the data in Fig. 5a, with nonnegative Laplacian constraint imposed.

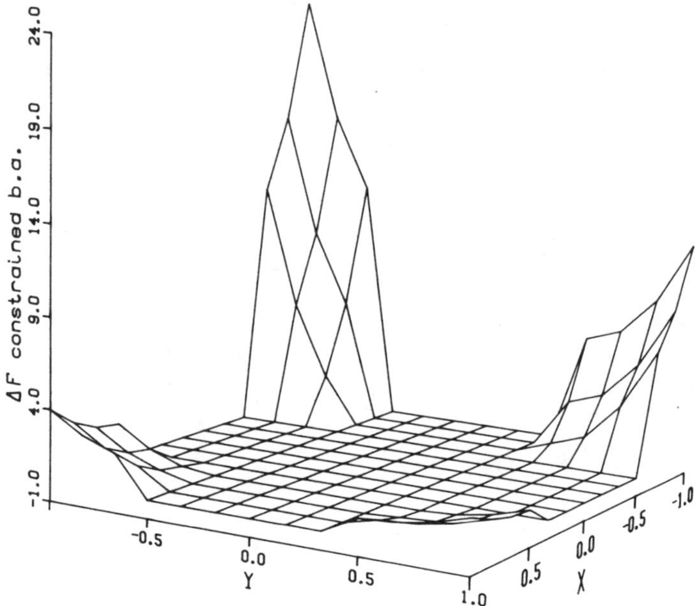

Figure 5e.   Laplacian of the approximant in Fig. 5d. Note the zero values of the Laplacian on the cross-shaped interior region. This approximate load agrees, in essence, with the actual load p.

is unknown but for the property $p \geq 0$ on I. The boundary values are assumed known. One approximates f over I from the space Z under the posed constraint $\Delta f \geq 0$. The resulting best constrained approximation g_{ba} can then be used to find a good estimate of the applied load p by setting $p = \Delta g_{ba}$. We chose the space Z since it is known to contain "superharmonic" functions. (Any space with this property would suffice.) Figure 4a shows the measured (noisy) deflection f caused by the distributed load p shown in Fig. 4b. The best unconstrained approximation is shown in Fig. 4c and the resulting load calculated from it in Fig. 4d. Note the negative values of the estimated load appearing at some locations in I. The best superharmonic approximation to the data f is given in Fig. 4e and the resulting load calculated from it in Fig. 4f. Shown in Fig. 4g is the load p as calculated from the *noisy* data f using centered differences. Note the rather poor reproduction obtained by this method, and the strikingly smooth results obtained by the constrained approximation method. One would in practice use a more natural norm in this type of problem, e.g., a Sobolev norm; however, the basic algorithm is indifferent to the norm used, the uniform norm being chosen here for simplicity of preparation of examples.

Example 5. Superharmonic approximation, using basis elements with local support. In this example, the space W is the tensor product space appearing in Example 3 in this section. The problem in this example is as in Example 4, with data f shown in Fig. 5a representing noisy measurements of the deflection caused by a load p, not shown here. The best approximation in the unconstrained case is shown in Fig. 5b, and the corresponding load calculated from it in Fig. 5c. The best approximation to the data under the superharmonicity constraint is shown in Fig. 5d. The approximate load calculated from the best constrained approximation is shown in Fig. 5e. The Laplacian of the constrained approximant reproduces the nonnegative load p, whereas the Laplacian of the unconstrained approximant fails to remain positive in the cross-shaped region where the load was actually zero. Note that the use of a locally supported basis allows the computation of approximations with nonharmonic properties (flat spots, discontinuous second derivatives, etc.), a desirable property in some instances.

5. Concluding Remarks. It is hoped that the above presented optimization viewpoint of shape constrained approximation problems may aid in the analysis of practical problems of CAGD by providing an alternate geometric framework in which to work (constrained approximation vs. interpolation). The work in this "interdiscipline" is incomplete at this time, but is proceeding along the following lines:

(1) Elimination of labor-intensive preprocessing of problems on the part of the user. Various standard types of shape constraints, choice of approximating space, and choice of norm should be provided as options in the use of the Tunnelling Algorithm program.

(2) An automatic means of satisfactorily handling so-called singular problems needs to be developed. In this context, singular problems are not necessarily unsolvable, but rather solvable with a much higher degree of preprogram user intervention.

(3) Incorporation of nonlinear pointwise shape constraints into (P_T) in a manner allowing attack by the Tunnelling Algorithm.

(4) Incorporation of global nonlinear shape constraints into (P_T). For example, bounds may be desired on the global Gaussian curvature of a surface. Once (3) is possible, the incorporation of (4) should follow.

(5) Determining the suitability of the approximating space V with respect to the shape constraints being posed. For example, in multivariate scattered data approximation via triangular splines, the choice of the triangulation affects the degree of flexibility available to meet shape "taming" constraints.

(6) Determination of suitable constraints with respect to a given approximating space. Guidelines (5) and (6) together will give workers in the field a catalog of suitable pairs of spaces and constraint types.

REFERENCES

[1] K. W. BOSWORTH, *A general method for the computation of best uniform norm approximations*, Ph.D. thesis, Rensselaer Polytechnic Institute, Troy, NY, 1984.

[2] ———, *The Tunnelling Algorithm: A general method for the computation of constrained uniform norm approximations*, U.S.U. Research Report 1985/29, Utah State Univ., Logan, UT, 1985.

[3] ———, *Multiple Exchangelike Algorithms for Best Constrained Approximation*, submitted to the Proc. of the 5th International Symposium on Approximation Theory, Academic Press, New York, 1986.

[4] C. DE BOOR, *A practical guide to splines*, in Applied Mathematical Sciences, Vol. 27, Springer-Verlag, New York, 1978.

[5] F. N. FRITSCH AND J. BUTLAND, *A method for constructing local monotone piecewise cubic interpolants*, SIAM J. Sci. Statist. Comput., 5 (1984), pp. 300–304.

[6] F. N. FRITSCH AND R. E. CARLSON, *Monotone piecewise cubic interpolation*, SIAM J. Numer. Anal., 17 (1980), pp. 238–246.

[7] K. GLASHOFF AND S. GUSTAFSON, *Linear optimization and approximation*, in Applied Mathematical Sciences, Vol. 45, Springer-Verlag, New York, 1983.

[8] P. R. GRIBIK, *A central cutting-plane algorithm for semi-infinite programming problems*, in Semi-Infinite Programming, Lecture Notes in Control and Information Sciences, Vol. 15, Springer-Verlag, New York, 1979, pp. 66–82.

[9] A. HAAR, *Uber die Minkowskische Geometrie und die Annäherung an stetige Funktionen*, Math. Ann., 78 (1918), pp. 294–311.

[10] K. ROLEFF, *A stable multiple exchange algorithm for linear SIP*, in Semi-Infinite Programming, Lecture Notes in Control and Information Sciences, Vol. 15, Springer-Verlag, New York, 1979, pp. 83–96.

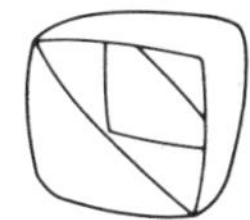

Shape Control of Curves
and Surfaces through
Constrained Optimization

ALAN K. JONES

Abstract. Sufficient conditions for the convexity of a parametric rational spline curve can be written as quadratic inequality constraints relating the curve's B-spline coefficients. Thus we can view the construction of a convex curve as a constrained optimization problem. The same technique can be applied to tensor product spline surfaces to ensure simultaneous convexity of an entire family of constant parameter curves.

Introduction. The purpose of this paper is to summarize some recent results on the control of convexity behavior in parametric curves and to describe one way in which these techniques can be extended to provide useful control over the shape of parametric surfaces.

Section 1 is an overview of the techniques described in Ferguson and Jones [3], [4], in which parametric spline curves with desired convexity behavior are obtained as solutions to constrained optimization problems. Sections 2 and 3 show that surface shape constraints that arise in certain practical engineering situations can be viewed as the application of curve convexity control to an entire family of isoparameter curves simultaneously. Section 4 shows how tensor product spline surfaces constructed by parametric blending of spline curves may violate these surface shape constraints, even when the curves to be blended appear well behaved. Section 5 introduces a set of nonlinear inequality constraints relating the spline coefficients of a surface which guarantee good surface shape behavior in the sense of §3. We conclude with a computational example showing how numerical constrained optimization can be used to generate surfaces satisfying these inequality constraints.

Throughout §§ 4 and 5, a simple model case consisting of a single, isolated bicubic patch is used as a running example. Certain computational details are supplied only for this model case. Considerations that arise when the techniques are applied to more general tensor product spline surfaces will be presented in Ferguson, Frank and Jones [2].

1. Shape Control for Parametric Curves.   In this section we show how to modify a parametric rational spline curve so that a specified part of that curve becomes convex. We assume that the curve is to be parameterized by a function $\mathbf{f}:[0, 1] \to \mathbf{R}^3$, and that $\mathbf{f}$ is expanded in terms of some fixed B-spline basis $\{b_j\}$ as $\mathbf{f}(t) = (\sum w_j \mathbf{P}_j b_j(t))/(\sum w_j b_j(t))$, where all of the weights w_j are strictly positive. Since the B-splines form a partition of unity, the special case when $w_j = 1$ for all j can be written simply as $\mathbf{f}(t) = \sum \mathbf{P}_j b_j(t)$, and most practical applications have been of this "nonrational" case.

Since "convexity" is not a useful concept for general space curves, we assume that a unit vector $\mathbf{N}$ has been specified, defining a fixed planar projection in which the curve is to be controlled. Our principal tool will be the inequalities introduced in the following definition.

DEFINITION 1.1.   Data for a *polygon convexity constraint* on the rational spline curve $\mathbf{f}$ consist of the following:

(1)   a subinterval $[a, b]$ of the parameter domain of $\mathbf{f}$,

(2)   a unit vector $\mathbf{N}$.

Let J and K be such that $b_J, b_{J+1}, \cdots, b_K$ are the B-splines which do not vanish uniformly on $[a, b]$. Then the curve $\mathbf{f}$ is said to satisfy the constraint if

$$(\mathbf{P}_{j+1} - \mathbf{P}_j) \times (\mathbf{P}_{k+1} - \mathbf{P}_k) \cdot \mathbf{N} > 0 \quad \text{whenever } J \leq j < k < K.$$

If $\mathbf{f}$ satisfies such a constraint, then the projection onto any plane perpendicular to $\mathbf{N}$ of the polygon defined by connecting $\mathbf{P}_J, \mathbf{P}_{J+1}, \cdots, \mathbf{P}_K$ must be convex in the sense that it crosses no fixed line more than twice. This, in turn implies that the planar projection of $\mathbf{f}([a, b])$ is convex in the same sense. The argument relies on the following well-known corollary of the variation diminishing property of B-splines.

PROPOSITION 1.1.   *Let* $\mathbf{f}(t) = (\sum w_j \mathbf{P}_j b_j(t))/(\sum w_j b_j(t))$ *be a rational spline curve in the plane, with all weights* $w_j > 0$. *Let* L *be a fixed line in the plane. Then the curve* $\mathbf{f}$ *crosses the line* L *no more often than does the spline polygon defined by the vertices* $\mathbf{P}_j$.

Proof.   Let $\mathbf{U}$ be a vector perpendicular to L, and c the constant such that, if $A(\mathbf{x}) = \mathbf{U} \cdot \mathbf{x} + c$, then $A(\mathbf{x}) = 0$ when $\mathbf{x}$ is on the line L, $A(\mathbf{x}) > 0$ when $\mathbf{x}$ is on one side of L, and $A(\mathbf{x}) < 0$ when $\mathbf{x}$ is on the other. Write

$$A(\mathbf{f}(t)) = \mathbf{U} \cdot \left(\sum w_j \mathbf{P}_j b_j(t)\right) \Big/ \left(\sum w_j b_j(t)\right) + c\left(\sum w_j b_j(t)\right) \Big/ \left(\sum w_j b_j(t)\right)$$

$$= \left(\sum w_j(\mathbf{U} \cdot \mathbf{P}_j + c)b_j(t)\right) \Big/ \left(\sum w_j b_j(t)\right)$$

$$= \sum W_j(t)A(\mathbf{P}_j)b_j(t),$$

where the factors $W_j(t) = w_j/(\sum w_j b_j(t))$ are all positive. But by Corollary XI.4 of de Boor [1], $A(\mathbf{f}(t))$ crosses zero no more often than the sequence of coefficients $W_j A(\mathbf{P}_j)$, and the result follows. $\square$

Several remarks should be made at this point. Firstly, note that Definition 1.1 is phrased in terms of strict inequalities. It is possible to replace the " $>$ " conditions with " $\geq$ " but care is needed to handle the degeneracies that may occur. See Ferguson and Jones [3], [4] for details. Furthermore, the " $\geq$ " conditions do not generalize nicely to surfaces. The only restriction imposed by the strict inequalities is that constrained regions of $\mathbf{f}$ may not contain embedded straight line segments. However, if it is known that a certain part of a curve should be a straight line, this can, and should, be phrased as a collection of linear equality relations on the B-spline coefficients, which will be far more tractable than the quadratic inequality conditions of Definition 1.1.

Secondly, the choice of orientation of $\mathbf{N}$ is significant. Together with the orientation of the curve as determined by its parameterization $\mathbf{f}$, $\mathbf{N}$ determine the "sense" of convexity of the curve. This is analogous to the distinction for scalar valued functions between "convex up," $y''(x) > 0$, and "convex down," $y''(x) < 0$.

Thirdly, the choice of interval $[a, b]$ is also important. It cannot be made too large, since the polygon, and hence the curve, is prevented from turning through more than a half-circle while in the interval. It is generally necessary to define a sequence of local polygon constraints, and piece them together to give local convexity for the curve. This has the additional advantage of reducing the total number of constraint inequalities, which, for any given constraint, grows quadratically in the number of B-splines involved, $(K - J + 1)$.

If $t_1, t_2, \cdots, t_N$ are the distinct knots of $\mathbf{f}$, then the recommended procedure is to define a separate constraint for each of the overlapping intervals $[t_1, t_3], \cdots, [t_{N-2}, t_N]$. Note that, unless we can guarantee by some other means that the curve has a well defined tangent vector at each t_i, it is not safe to constrain only the nonoverlapping intervals $[t_1, t_2], \cdots, [t_{N-1}, t_N]$, since this does not prevent cusps at the knots.

The sufficient conditions outlined above are not vacuous. For example, in Jones [6], cubic spline curves satisfying these constraints are constructed which interpolate prespecified position and derivative data. The procedure requires enough knots so that value, first and second derivative data can be specified independently at each interpolation point, and involves forcing the curvatures at the interpolation points to be "sufficiently small." The result is that curvature "bunches up" in between the interpolation points, and in the limit the resulting curve resembles a string of linear arcs. This behavior is similar to that of tension-parameter methods, of which Schweikert [8] and Nielson [7] are typical. In many practical cases it is known that much better behaved convex interpolants exist. The problem is that the geometry of the "feasible region," the subset of spline function space consisting of splines satisfying the convex polygon constraints, is very complicated, and these good functions seem to lie in regions of the feasible set not accessible to such simple methods.

A different strategy was adopted in Ferguson and Jones [3], [4]. The idea is to start with an "initial guess" interpolant $\mathbf{f}^0(t) = (\sum w_j \mathbf{P}_j^0 b_j(t))/(\sum w_j b_j(t))$ which is smooth, but may have extraneous inflections. The function $\mathbf{f}^0$ might be, for example, a natural C^2 cubic spline interpolant to a set of position and (optional) derivative data. We then search for a small perturbation of the spline coefficients of $\mathbf{f}^0$ which satisfies simultaneously the polygon convexity constraints and the necessary interpolation constraints. The weights w_j are left unchanged.

This search can be viewed as a constrained optimization problem, where the objective function is, for example,

$$F(\mathbf{P}_1, \mathbf{P}_2, \cdots) = \sum |\mathbf{P}_j - \mathbf{P}_j^0|^2,$$

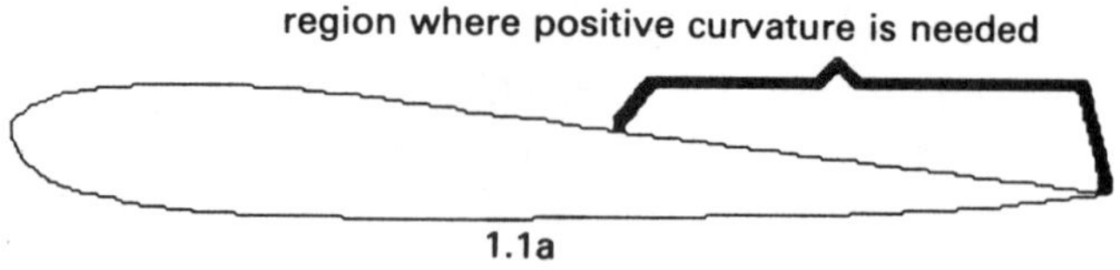

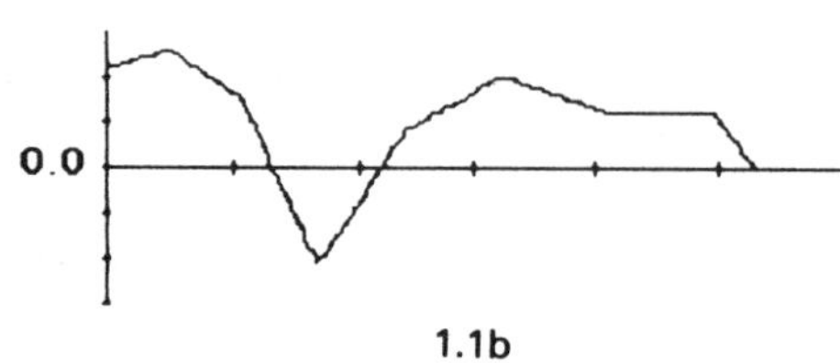

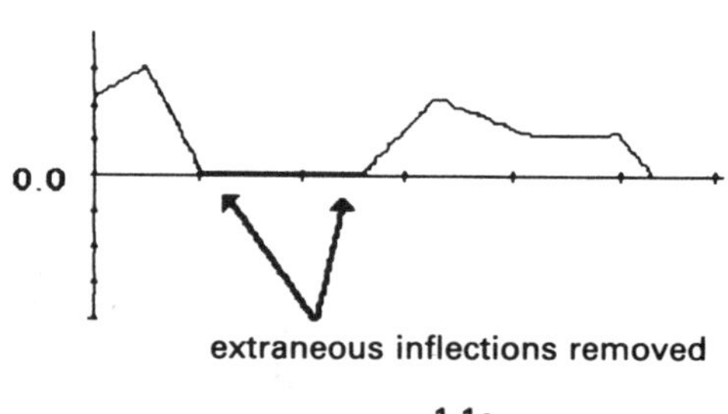

Figure 1.1.   Shape control for a parametric curve. 1.1a Hypothetical airfoil. 1.1b Curvature of the region before shape control. 1.1c Curvature of the region after slope control.

and the starting point is $(\mathbf{P}_1^0, \mathbf{P}_2^0, \cdots)$. The polygon convexity conditions of Definition 1.1 represent quadratic inequality constraints on the variables $\mathbf{P}_j$, while the interpolation conditions are linear equality constraints. By adding additional knots to the initial guess curve $\mathbf{f}^0$ as necessary, we guarantee that a feasible choice of the spline coefficients $(\mathbf{P}_1, \mathbf{P}_2, \cdots)$ exists. By our choice of the objective function F we ensure that any feasible local minimum of F will be an acceptable curve, even though the problem posed may have no unique global minimum.

Given the complicated geometry of the feasible region, finding such a feasible local minimum can be a severe test of library optimization software. However the routine NPSOL described in Gill et al. [5] seems to do quite well. Figure 1.1 is typical of real applications. This shows an interval along a hypothetical airfoil curve where the design engineer has specified "positive curvature," in the sense that $\mathbf{f}' \times \mathbf{f}'' \cdot \mathbf{N} > 0$, where the curve is being traced counterclockwise, and $\mathbf{N}$ is a unit vector pointing into the page. The first curvature plot shows the behavior in this interval of the curve the engineer originally constructed. Note the region of negative curvature and the two associated inflections. A set of polygon convexity constraints was defined covering the specified interval, following the procedure recommended above. The result, as seen on the second curvature plot, was removal of the undesired region of negative curvature and its associated inflections.

2. Surface Analysis in Engineering Practice. Our discussion will be aimed at the problem of constructing a surface which interpolates a single family of noninter-secting curves. For example, if the surface involved is the wing of an airplane, it may be defined by a family of airfoil cross-sections. The curves are given as parametric polynomial splines, with a common degree and knot set, and the surface is to be a parametric tensor product polynomial spline.

We assume that a parameterization has been established, by means of which each input curve can be regarded as a constant parameter line of the eventual surface, and write the jth such curve as $\mathbf{g}(- , v_j)$. The problem is to extend the map $\mathbf{g}$ to the entire (u, v) parameter domain, by parametric blending between the input curves. In a typical surface fitting application, great care will have been taken to control the convexity properties of the individual curves. Precise blending procedures, for example ruled surfaces, will have been specified by the user in some parts of the surface. In other regions, however, only "spline blending" will be specified, and the details will be left to the system, trusting that the resulting shape behavior will be acceptable. The question is, what does the word "acceptable" mean in this context?

The analysis of surface shape is far more difficult than the corresponding problem for curves. Engineers have had to face such problems since long before the development of modern three-dimensional graphics. Frequently they have resorted to extracting families of curves, preferably planar, from a surface, and analyzing them instead. In some cases, this procedure reflects the realities of the engineering problem involved. For example, the aerodynamic behavior of a wing may be largely determined by the properties of planar sections taken parallel to the expected airflow. In other applications, it is just traditional practice.

In any case, no new computerized method is likely to be accepted by practitioners unless it reproduces as closely as possible the sorts of control they have become accustomed to under previous systems. Therefore, it is instructive to examine a system which has been in active use for some years in working engineering groups, to see what these people are likely to feel that they need. We will describe the application of one such system to a wing-like surface.

The first set of inputs is a collection of "control curves" which run longitudinally down the wing, i.e., transverse to the air flow direction. Each control curve is specified in two planar views, the top view (XY plane) and the rear view (YZ plane). In each view, a control curve is represented as a string of straight line segments, circular arcs, and splines, with user-defined slope matching conditions at the joins. Figure 2.1 is a highly simplified hypothetical example.

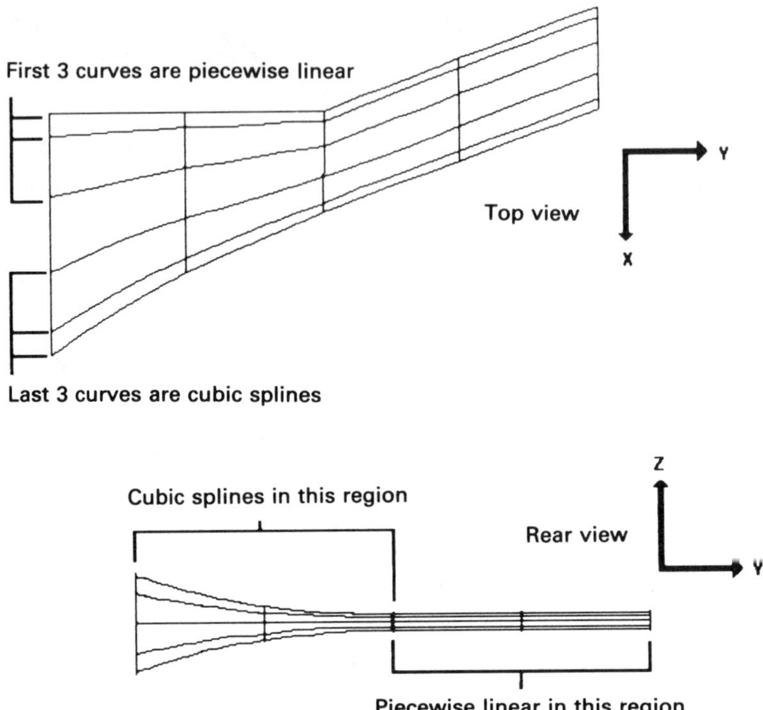

Figure 2.1.   An existing system for surface construction uses control curves specified in two plane views.

The second set of inputs is a collection of "blending rules," each of which is effective between a specified range of control curves, and over a specified range of y coordinates. A typical rule might specify that, along each y value between y_0 and y_1, the surface is to be obtained by piecewise linear blending between the first three control curves, and spline blending between the rest. See Fig. 2.2.

The essential independence of the two plane views extends to such basic features as derivative continuity along the control curves. Users frequently insist, for example, that they wish to produce the situation shown in Fig. 2.1, where some control curves have discontinuities in dx/dy at points where dz/dy is continuous. This is at first glance a surprising procedure, since it actually destroys geometric C^1 continuity. The surface normal must be discontinuous at such points, unless they are local extrema of z. However, even experienced users of the system were often unaware of this effect, because surface normal continuity was not the behavior of interest. What mattered to these practitioners was providing smooth projections of the control curves on the YZ plane.

3. Surface Shape Control Based on Isoparameter Curves. As the foregoing demonstrates, practical engineering is not always directly concerned with three-

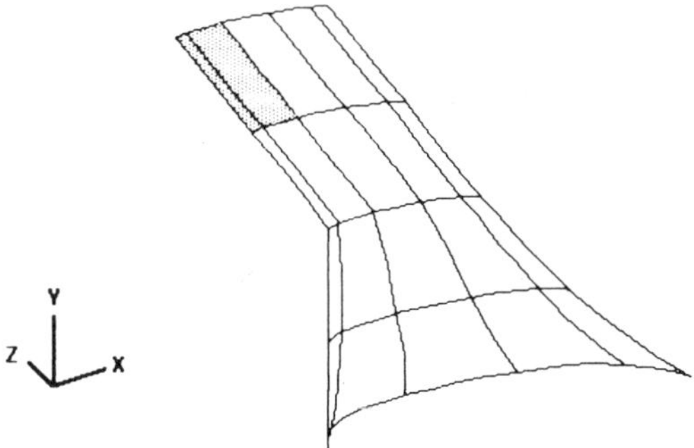

Figure 2.2. Rules for blending between control curves in an existing surface geometry system. ▦ ruled surface; □ cubic blending.

dimensional surface geometry. Control of a family of constant parameter lines in one or more plane views may be quite satisfactory. Therefore, we extend Definition 1.1 to surfaces as follows.

DEFINITION 3.1.   Data for an *isoparameter convexity constraint* on a (rational or "nonrational") tensor product spline surface $\mathbf{g}$ consist of the following:

(1)   a rectangular region $[a, b] \times [c, d]$ in the parameter domain of $\mathbf{g}$,

(2)   a unit vector $\mathbf{N}$.

The surface $\mathbf{g}$ is said to satisfy the isoparameter convexity constraint if, for each v in $[c, d]$, the isoparameter curve $\mathbf{g}(-, v)$ satisfies a polygon convexity constraint with data $[a, b]$ and $\mathbf{N}$.

Clearly, if a surface $\mathbf{g}$ is obtained by blending curves $\mathbf{g}(-, v_j)$ as in § 2, then a necessary condition for $\mathbf{g}$ to satisfy such a constraint is that each input curve $\mathbf{g}(-, v_j)$ must itself satisfy the associated polygon convexity constraint. In the next section, we show that this condition is not, unfortunately, sufficient.

4.   How Shape Control Fails in Practice.   Given an isoparameter convexity constraint and a set of curves $\mathbf{g}(-, v_j)$ which satisfy the associated polygon convexity constraint, how can a tensor product surface which interpolates these curves fail to satisfy the isoparameter convexity constraint? One possible cause of bad behavior is the generalization to two dimensions of well-known problems with spline curve interpolation. In Fig. 4.1, for example, a cubic spline blending of a family of ellipses has resulted in a self-intersecting surface. This suggests that something more than uncontrolled spline blending is needed.

In this case it is possible to apply the techniques of § 1 to prevent wiggles in the cross-interpolant. However, in general, imposing control in the v direction is not the way to insure good behavior in the u direction. We shall now show that even the strong shape control provided by piecewise linear blending in the v direction does not do so.

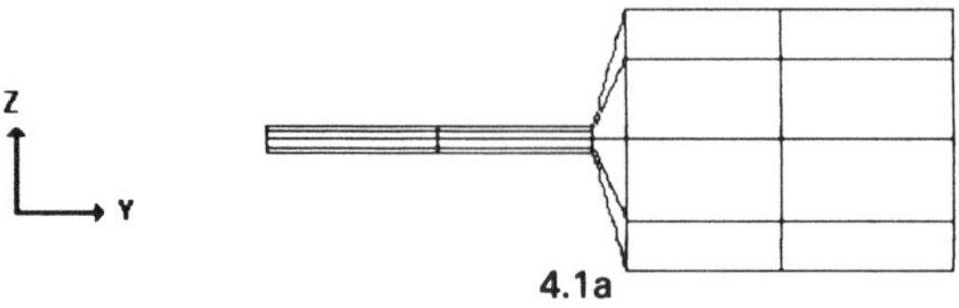

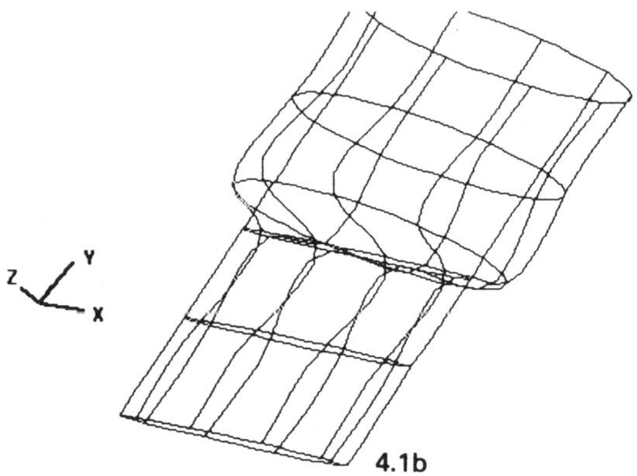

Figure 4.1.   A failure of surface shape control with cubic blending. 4.1a Side view of 6 radically different ellipses. 4.1b Naive cubic spline blending gives self-intersecting surface.

This amounts to constructing an example to demonstrate that the space of convex parametric curves is not itself convex. A corollary is that convexity for parametric curves cannot be phrased as a linear inequality constraint. This is why enforcing convexity of parametric curves is fundamentally more difficult than is the case with explicit curves $y = y(x)$, where convexity is equivalent to the condition $y''(x) \geq 0$ for all x.

The simple model we will analyze consists of linear blending between two cubic spline curves, lying in parallel planes. Each spline will use the knot set $\{0, 0, 0, 0, 1, 1, 1, 1\}$, and thus have only a single span. Figure 4.2 shows a particularly instructive example in which the curve $\mathbf{g}(-, 0)$ has coefficients

$$\mathbf{P}_{00} = (0, 18, 0), \quad \mathbf{P}_{10} = (6, 12, 0), \quad \mathbf{P}_{20} = (14, 14, 0), \quad \mathbf{P}_{30} = (18, 16, 0),$$

while $\mathbf{g}(-, 1)$ has coefficients

$$\mathbf{P}_{01} = (0, 12, 20), \quad \mathbf{P}_{11} = (2, 8, 20), \quad \mathbf{P}_{21}(4, 6, 20), \quad \mathbf{P}_{31} = (12, 0, 20).$$

(Only the spline polygons appear in Fig. 4.2. The cubic curves are shown in Fig. 5.1.)

We write the edges of the polygons as $\mathbf{D}_{jr} = \mathbf{P}_{j+1r} - \mathbf{P}_{jr}$, for $j = 0, 1, 2$ and $r = 0, 1$. The curve at level v of the linear blend between the two spline curves is itself a cubic

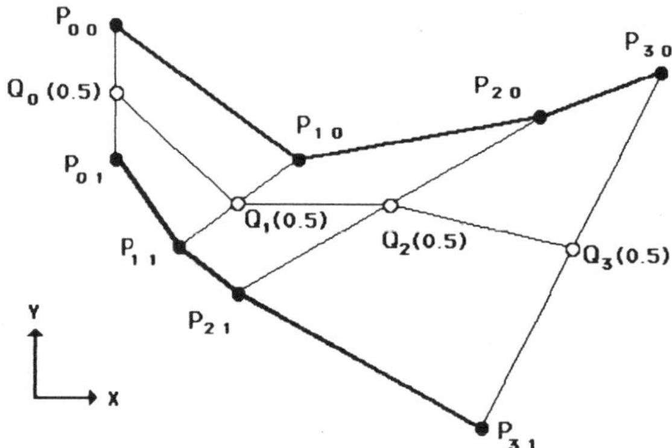

Figure 4.2   A failure of surface shape control with linear blending.

spline with the same knot set, with coefficients $\mathbf{Q}_j(v) = v\mathbf{P}_{j1} + (1-v)\mathbf{P}_{j0}$, and spline polygon edges $\mathbf{E}_j(v) = v\mathbf{D}_{j1} + (1-v)\mathbf{D}_{j0}$.

We now define an isoparameter shape constraint as in Definition 3.1, taking for $\mathbf{N}$ a common unit normal of the two parallel planes, in this case $(0, 0, 1)$, and for $[a, b]$ the entire parameter domain $[0, 1]$. In the example, the $v = 0$ spline polygon is convex "out," wrapping counterclockwise around the normal pointing out of the page. Analytically, we see this because the vector triple products $\mathbf{D}_{00} \times \mathbf{D}_{10} \cdot \mathbf{N} = 60$, $\mathbf{D}_{10} \times \mathbf{D}_{20} \cdot \mathbf{N} = 8$, and $\mathbf{D}_{00} \times \mathbf{D}_{20} \cdot \mathbf{N} = 36$ are all positive. Similarly for the spline polygon at $v = 1$, since $\mathbf{D}_{01} \times \mathbf{D}_{11} \cdot \mathbf{N} = 4$, $\mathbf{D}_{11} \times \mathbf{D}_{21} \cdot \mathbf{N} = 4$, and $\mathbf{D}_{01} \times \mathbf{D}_{21} \cdot \mathbf{N} = 20$ are also positive.

On the other hand, the spline polygon at the $v = 0.5$ level is not convex, as shown by the fact that $\mathbf{E}_0(0.5) \times \mathbf{E}_1(0.5) \cdot \mathbf{N} = 25$ and $\mathbf{E}_1(0.5) \times \mathbf{E}_2(0.5) \cdot \mathbf{N} = -10$ are of opposite signs. (In fact, the curvature plot in Fig. 5.1 shows that the cubic spline curve $\mathbf{g}(-, 0.5)$ is also not convex, although the inflection is much harder to detect visually.) How is it that $\mathbf{E}_1(v) \times \mathbf{E}_2(v) \cdot \mathbf{N}$ changes sign in the interval $[0, 1]$, and how could this have been prevented?

Expand the triple product as

$$\mathbf{E}_1(v) \times \mathbf{E}_2(v) \cdot \mathbf{N} = v^2[\mathbf{D}_{11} \times \mathbf{D}_{21} \cdot \mathbf{N}] + v(1-v)[\mathbf{D}_{11} \times \mathbf{D}_{20} \cdot \mathbf{N}]$$
$$+ v(1-v)[\mathbf{D}_{10} \times \mathbf{D}_{21} \cdot \mathbf{N}] + (1-v)^2[\mathbf{D}_{10} \times \mathbf{D}_{20} \cdot \mathbf{N}].$$

Clearly, it is enough for all four terms on the right to be positive. We formalize this observation as follows.

PROPOSITION 4.1.   *In the situation of Fig. 4.2, the isoparameter convexity constraint is satisfied if $\mathbf{D}_{jr} \times \mathbf{D}_{ks} \cdot \mathbf{N} > 0$ for $j, k = 0, 1, 2$ and $r, s = 0, 1$.*

Generalization to higher order splines is illustrated in the following.

PROPOSITION 4.2.   *Let a tensor product spline surface be given by $\mathbf{g}(u, v) = \sum \mathbf{P}_{jr} b_j^m(u) b_r^n(v)$, where $\{b_j^m\}$ and $\{b_r^n\}$ are B-spline bases of orders m and n in the u and v parameter directions, respectively. Let an isoparameter convexity constraint be given as in Definition 3.1. Let J, K, R, S be such that $b_J^m, b_{J+1}^m, \cdots, b_K^m$ are the*

B-splines in u which do not vanish uniformly on $[a, b]$, while b_R^n, b_{R+1}^n, $\cdots$, b_S^n are the B-splines in v which do not vanish uniformly on $[c, d]$. Then, writing $\mathbf{D}_{jr} = \mathbf{P}_{j+1r} - \mathbf{P}_{jr}$ as before, the isoparameter convexity constraint is satisfied if

$$\mathbf{D}_{jr} \times \mathbf{D}_{ks} \cdot \mathbf{N} > 0 \quad \text{for } J \le j < k < K \text{ and } R \le r, s \le S.$$

Proof. For any fixed v, we can write $\mathbf{g}(u, v) = \sum \mathbf{Q}_j(v) b_j^m(u)$, where $\mathbf{Q}_j(v) = \sum \mathbf{P}_{jr} b_r^n(v)$. Then the isoparameter convexity constraint is satisfied if, for every v in $[c, d]$, we have

$$(\mathbf{Q}_{j+1}(v) - \mathbf{Q}_j(v)) \times (\mathbf{Q}_{k+1}(v) - \mathbf{Q}_k(v)) \cdot \mathbf{N} > 0 \quad \text{for } J \le j < k < K.$$

The triple product can be expanded as

$$\left[\sum b_r^n(v)(\mathbf{P}_{j+1r} - \mathbf{P}_{jr}) \right] \times \left[\sum b_s^n(v)(\mathbf{P}_{k+1s} - \mathbf{P}_{ks}) \right] \cdot \mathbf{N} = \sum b_r^n(v) b_s^n(v) \mathbf{D}_{jr} \times \mathbf{D}_{ks} \cdot \mathbf{N},$$

with the sum taken over r and s. Since, for each v, all the B-spline products $b_r^n(v) b_s^n(v)$ are nonnegative, and at least one is nonzero, the result follows. $\square$

A similar but lengthier calculation establishes the following analogous result for rational spline surfaces.

PROPOSITION 4.3.   *With notation as in Proposition 4.2, let*

$$\mathbf{g}(u, v) = \left(\sum w_{jr} \mathbf{P}_{jr} b_j^m(u) b_r^n(v) \right) \Big/ \left(\sum w_{jr} b_j^m(u) b_r^n(v) \right),$$

where $w_{jr} > 0$ for all j, r. Then the isoparameter convexity constraint is satisfied if

$$(\mathbf{P}_{j+1r} - \mathbf{P}_{js}) \times (\mathbf{P}_{k+1p} - \mathbf{P}_{kq}) \cdot \mathbf{N} > 0 \quad \text{for } J \le j < k < K \text{ and } R \le p, q, r, s \le S.$$

It is useful in the above to distinguish between the "diagonal" constraints, i.e., those for which $r = s$ in the case of Proposition 4.2, or $p = q = r = s$ in the case of Proposition 4.3, and the "off-diagonal" or "cross-term" constraints. The diagonal inequalities enforce convexity along rows of the B-spline control net, while the cross-terms impose compatibility conditions on nearby rows.

In the example of Fig. 4.2, the diagonal terms are all positive by hypothesis, and convexity can fail only if the sum of the cross-terms is large and negative. This in turn can occur only if two things happen simultaneously, as illustrated in Fig. 4.2. First, there must be a substantial "twist" between the two curves, in order to violate the hypothesis of Proposition 4.1. Second, there must be a mismatch in the relative lengths of corresponding legs of the two polygons, e.g., $|\mathbf{D}_{10}| \gg |\mathbf{D}_{11}|$ and $|\mathbf{D}_{20}| \ll |\mathbf{D}_{21}|$, indicating large out-of-phase variations in parametric velocity (in the plane view defined by $\mathbf{N}$) between the two curves. This phenomenon can occur when the parameter values assigned along the input data curves are based on compromises between a large collection of parallel curves, or are influenced by behavior invisible in the given plane view.

In many cases that occur in practice, for example in scenarios like that of Fig. 2.2, neither problem arises. Thus, at least for a piecewise bilinear surface, the cross-terms would not be large enough to matter. Limited experimental evidence suggests that the same may be true for higher order interpolants. For example, examination of a C^1 bicubic surface fit by current production methods to the data in Fig. 2.2 shows that the magnitudes of negative triple products in the intermediate v-level polygons are in all cases bounded above by those of negative diagonal terms. In cases like this, controlling the rows of B-spline coefficients parallel to the view plane is important, but preventing every possible mismatch between adjacent rows, as is required by the hypothesis of Proposition 4.2, apparently is not.

5. Guaranteeing Shape Control through Constrained Optimization. Proposition 4.2 provides us with tools to construct a surface satisfying one or more isoparameter convexity constraints by the method described for curves in § 1. Just as before, we start with an "initial guess" surface $\mathbf{g}^0(u, v) = \sum \mathbf{P}_{jr}^0 b_j^m(u) b_r^n(v)$ generated by any desired technique, for example natural spline interpolation. Define an objective function $F(\mathbf{P}_{00}, \cdots) = \sum |\mathbf{P}_{jr} - \mathbf{P}_{jr}^0|^2$. Then, starting from the initial point $(\mathbf{P}_{00}^0, \cdots)$, search for a minimum of $F(\mathbf{P}_{00}, \cdots)$ subject to the quadratic inequality constraints generated according to Proposition 4.2, together with any desired interpolation constraints. The resulting surface $\mathbf{g}(u, v) = \sum \mathbf{P}_{jr} b_j^m(u) b_r^n(v)$ will then satisfy the desired isoparameter convexity constraints.

When can we be sure a priori such a solution exists? For the case of blending between two input curves, Proposition 4.2 itself provides some assistance. Specializing to the case where the B-splines $\{b_r^0\}$ are linear provides a set of conditions that involve only the coefficients of the input curves. If these compatibility conditions are satisfied, then we know that the surface defined by linear blending between the input curves satisfies the isoparameter constraints, so the feasible region is not empty. Techniques similar to those of Jones [6] can establish similar conditions for the existence of smooth blends satisfying the conditions of Proposition 4.2 in multicurve cases. However, much work remains to be done in this area.

As was pointed out in the previous section, the sufficient conditions given by Proposition 4.2 can be unnecessarily restrictive. (Indeed, the polygons of Fig. 4.2 are a carefully constructed example of this, as will be shown below.) What appears to be a better choice is motivated by the following observation. Rearranging terms in the cross-product expansion of § 4, we have

$$\mathbf{E}_1(v) \times \mathbf{E}_2(v) \cdot \mathbf{N} = q(v) = v^2[\mathbf{D}_{11} \times \mathbf{D}_{21} \cdot \mathbf{N}]$$
$$+ 2v(1 - v)[(\mathbf{D}_{11} \times \mathbf{D}_{20} \cdot \mathbf{N} + \mathbf{D}_{10} \times \mathbf{D}_{21} \cdot \mathbf{N})/2]$$
$$+ (1 - v)^2[\mathbf{D}_{10} \times \mathbf{D}_{20} \cdot \mathbf{N}]$$

where the expressions in brackets are just the Bézier coefficients of the scalar-valued quadratic q over the unit interval. A weaker sufficient (but still not necessary) condition for positivity of q is, therefore, that each of these expressions be positive.

Similar B-spline expansions can be derived for $q(v)$ in the case where $\mathbf{g}$ is any tensor product spline surface, once we know how to write products of nth order B-splines as sums of B-splines of order $(2n - 1)$. That is, we need coefficients A_{ijk}^n such that

$$b_i^n(v) b_j^n(v) = \sum A_{ijk}^n b_k^{(2n-1)}(v).$$

For this to be possible, the knot set for the B-spline basis $\{b_k^{(2n-1)}\}$ must be chosen with some care. For example, if t is a knot of multiplicity k for basis $\{b_i^n\}$, then derivatives of the products $b_i^n(v)b_j^n(v)$ of order greater than $(n-k-1)$ may be discontinuous, so t must appear as a knot of multiplicity at least $(k+n-1)$ in the knot set for the basis $\{b_k^{(2n-1)}\}$.

We can now sharpen Proposition 4.2 as follows.

PROPOSITION 5.1. *Let the hypothesis and notation be as in Proposition 4.2. Let* $\{\sum b_t^{(2n-1)}\}$ *be a B-spline basis chosen as above, and let* T, U *be such that* $b_T^{(2n-1)}$, $b_{T+1}^{(2n-1)}$, $\cdots$, $b_U^{(2n-1)}$ *are the B-splines that do not vanish uniformly on* $[c, d]$. *Then the isoparameter convexity constraint is satisfied if*

$$\sum A_{rst}^n \mathbf{D}_{jr} \times \mathbf{D}_{ks} \cdot \mathbf{N} > 0, \quad for\ J \le j < k < K,\ and\ T \le t \le U.$$

(*Each sum is taken over* $R \le r$, $s \le S$.)

Proof. As in the proof of Proposition 4.2, we need to show for each j and k that

$$\sum b_r^n(v)b_s^n(v)\mathbf{D}_{jr} \times \mathbf{D}_{ks} \cdot \mathbf{N} > 0 \quad \text{where the sum is over } r \text{ and } s.$$

But we can rewrite the left-hand side as

$$\sum \left(\sum A_{rst}^n \mathbf{D}_{jr} \times \mathbf{D}_{ks} \cdot \mathbf{N}\right) b_t^{(2n-1)}(v),$$

where the inner sum is over r and s, and the outer sum is over t.   □

Proposition 4.3 can be sharpened similarly to provide an analogous criterion for rational spline surfaces. See Ferguson, Frank and Jones [2] for more details.

The coefficients A_{ijk}^n can be calculated for any knot set using the de Boor–Fix dual functionals described in Lemma IX.1 of de Boor [1]. One calculates functionals dual to the basis $\{b_k^{(2n-1)}\}$ and applies them to the B-spline products $b_i^n b_j^n$. Computational details will be presented in Ferguson, Frank and Jones [2]. However, in the two-curve case, where the bases $\{b_j^n\}$ and $\{b_k^{(2n-1)}\}$ reduce to the Bézier bases of the appropriate order, the calculation is particularly simple, and can be carried out by hand.

We will illustrate this point with a simple computational example. For the case of cubic blending between two single-span cubic spline curves, the resulting surface will be a bicubic patch. Writing binomial coefficients as $B_{nk} = n!/k!\,(n-k)!$, the Bézier bases involved are $\{B_{3i}v^i(1-v)^{(3-i)}:i=0,\cdots,3\}$ and $\{B_{6k}v^k(1-v)^{(6-k)}:k=0,\cdots,6\}$. By inspection, $A_{ijk}^n=0$ unless $i+j=k$. The distinct nonzero values are $A_{000}^4=1$, $A_{011}^4=1/2$, $A_{022}^4=1/5$, $A_{033}^4=1/20$, $A_{112}^4=3/5$, and $A_{123}^4=9/20$. The rest of the table is filled out using the symmetries $A_{ijk}^n=A_{jik}^n$ and $A_{ijk}^n=A_{(n-1-i)(n-1-j)(2n-2-k)}^n$.

The inequalities of Proposition 4.2 become in this case $\sum A_{rst}^4 \mathbf{D}_{jr} \times \mathbf{D}_{ks} \cdot \mathbf{N} > 0$ for $0 \le j < k < 3$ and $0 \le t \le 6$. The inequalities for $t=0$ and $t=6$ involve only the two input data curves, which are to be held fixed. The remaining five are conveniently written as

$$F_{01}^{jk} > 0, \qquad\qquad 2F_{13}^{jk} + 3F_{22}^{jk} > 0,$$
$$2F_{02}^{jk} + 3F_{11}^{jk} > 0, \qquad F_{23}^{jk} > 0,$$
$$F_{03}^{jk} + 9F_{12}^{jk} > 0,$$

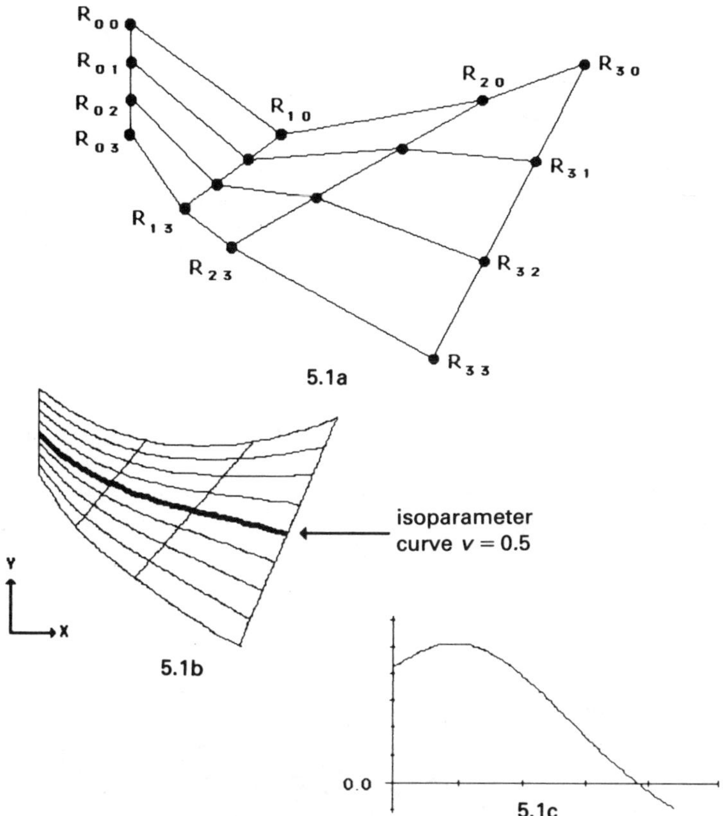

Figure 5.1.   A bicubic patch without shape control. 5.1a The B-spline control net. 5.1b The bicubic surface. 5.1c Curvature plot of the isoparameter curve $v = 0.5$.

where we set $F^{jk}_{pq} = (\mathbf{D}_{jp} \times \mathbf{D}_{kq} + \mathbf{D}_{jq} \times \mathbf{D}_{kp}) \cdot \mathbf{N}$. Since there are three allowed combinations of j and k, we see that imposing an isoparameter convexity constraint on an isolated bicubic patch while holding two boundary curves fixed involves a system of 15 quadratic inequalities.

Recall that Fig. 4.2 shows the control net of a spline surface which is cubic in the u direction and linear in the v direction. Figure 5.1 shows the control net of exactly the same surface, expressed as a bicubic patch with coefficients $\mathbf{R}_{js}$. In terms of the coefficients $\mathbf{P}_{jr}$ of Fig. 4.2, we have set

$$\mathbf{R}_{j0} = \mathbf{P}_{j0},$$
$$\mathbf{R}_{j1} = \frac{2\mathbf{P}_{j0} + \mathbf{P}_{j1}}{3},$$
$$\mathbf{R}_{j2} = \frac{\mathbf{P}_{j0} + 2\mathbf{P}_{j1}}{3}, \qquad \text{for } j = 0, \cdots, 3.$$
$$\mathbf{R}_{j3} = \mathbf{P}_{j1},$$

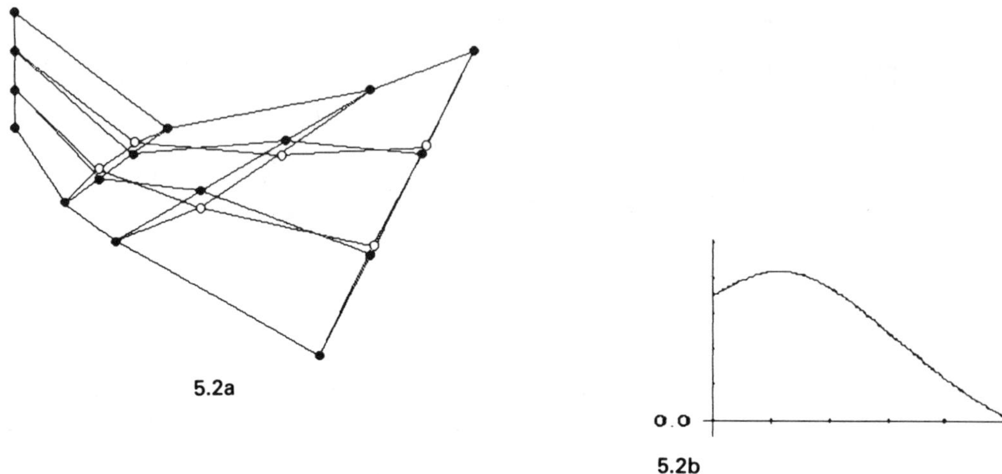

Figure 5.2.   Modifying the shape of a bicubic patch. 5.2a B-spline control nets before and after modification. ● original control points; ○ modified control points. 5.2b Curvature plot of the isoparameter curve $v = 0.5$ in the modified patch.

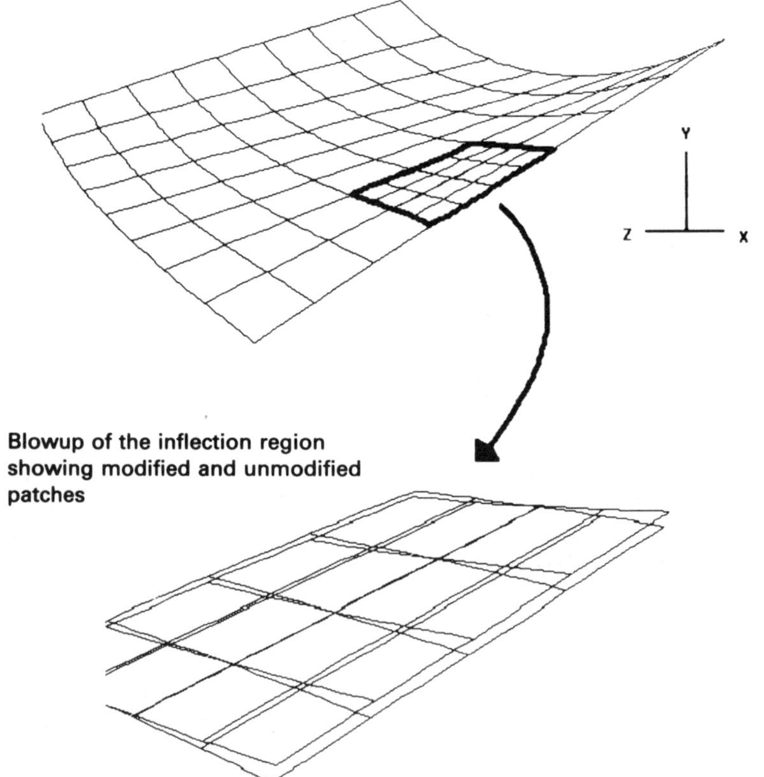

Figure 5.3.   Rotated view of the modified bicubic patch.

If the first and last rows are held fixed, i.e., if we force interpolation of the original curves at $v = 0$ and $v = 1$, then the inequalities of Proposition 4.2 cannot be satisfied, since, by construction, $(\mathbf{R}_{20} - \mathbf{R}_{10}) \times (\mathbf{R}_{33} - \mathbf{R}_{23}) \cdot \mathbf{N} = -64 < 0$, and no choice of the coefficients $\mathbf{R}_{js}$ for $s = 1, 2$ can change that fact. However, the weaker inequalities of Proposition 5.1, as derived above, can be satisfied in this case. Using the mean square perturbation functional F described at the beginning of this section, and the subroutine NPSOL documented in Gill et al. [5], a feasible choice of the coefficients $\mathbf{R}_{js}$ was easily found. The modified control net and the surface it determines are shown in Figs. 5.2 and 5.3.

REFERENCES

[1] C. DE BOOR, *A Practical Guide to Splines*, Springer-Verlag, New York, 1978.
[2] D. R. FERGUSON, P. FRANK AND A. K. JONES, *Designing surfaces using constrained optimization methods*, 1987, in preparation.
[3] D. R. FERGUSON AND A. K. JONES, *Parametric Curve Fitting with Shape Control*, Report ETA-TR-20, Engineering Technology Applications Div., Boeing Computer Services Co., Seattle, WA, 1984.
[4] ———, *Parametric curve fitting with shape control*, SIAM J. Sci. Statist. Comput., submitted.
[5] P. E. GILL, W. MURRAY, M. SAUNDERS AND M. H. WRIGHT, *User's Guide for* SOL/NPSOL, Report SOL 83-12, Dept. Operations Research, Stanford Univ., Stanford, CA, 1983.
[6] A. K. JONES, *An algorithm for convex parametric splines*, Report ETA-TR-29, Engineering Technology Applications Div., Boeing Computer Services Co., Seattle, WA, 1982.
[7] G. M. NIELSON, *Some piecewise polynomial alternatives to splines under tension*, in Computer Aided Geometric Design, R. E. Barnhill and R. F. Riesenfeld, eds., Academic Press, New York, 1974.
[8] D. G. SCHWEIKERT, *An interpolation curve using splines in tension*, J. Math. Physics, 45 (1966), pp. 312–317.

Interpolation to Arbitrary Data on a Surface

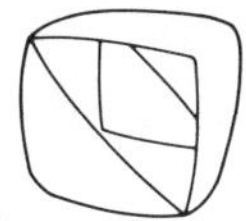

ROBERT E. BARNHILL, BRUCE R. PIPER AND KIM L. RESCORLA

Abstract. Two methods of interpolation to arbitrarily located data on surfaces are presented. The constructive mathematical methods are illustrated by means of color computer graphics for an aerodynamic application and for interpolation on the sphere.

1. Introduction. A problem that occurs frequently in science and engineering is the representation of information associated with points that are arbitrarily located on a surface in three-dimensional space. More precisely, the problem is to define a real valued map from a surface in three dimensions that interpolates to a given set of data.

Applications of such methods include the following:
(1) pressure on the surface of a wing,
(2) temperature on the surface of an internal human organ,
(3) electrical activity on the surface of the human brain.

A NASA-Ames problem of this type is the representation of pressure on an oblique wing, given the pressures at certain points on the wing. Building a pressure "surface" is necessary in order to evaluate the pressure at an arbitrary point on the wing. Barnhill, Piper and Stead [4] obtained preliminary results on this problem. This problem is an example of creating a "surface on a surface" (Barnhill [2]). The methods used in this research can be applied to other multidimensional problems. The Utah Math Computer Aided Geometric Design (CAGD) Group plans to apply these and related algorithms to the problem of mapping the human brain and its neuroscience functions in the international, multidisciplinary BRAINMAP Project.

A difficult task involved in such problems is the display of four-dimensional information. This rendering problem is addressed in the article by Barnhill, Makatura and Stead [6], which contains several color computer graphics illustrations including aerodynamic applications.

The organization of this paper is as follows: a solution tailored for the oblique wing problem is presented, followed by a solution to the more general problem of interpolation to arbitrarily located data on a convex surface. These constructive mathematical solutions have been implemented on color computer graphics equipment and illustrations of the results are shown for two examples: interpolation to oblique wing data and interpolation to data given on the sphere.

2. Modeling Pressure on a Wing. In the oblique wing project at NASA-Ames, the pressure is measured at locations ("plug taps") on the wing in wind tunnel experiments. The goal is to estimate the pressure at an arbitrary point on the wing. A complication is that the taps are unrelated to the points which define the geometry of the wing.

We propose the following algorithm:

(1) Geometry step:

(a) Domain of influence: Consider the top and bottom of the wing separately. The procedure to follow is performed on the top and bottom of the wing separately.

(b) Representation of wing: Approximate the top surface of the wing by a C^1 piecewise bicubic surface.

(2) Pressures at wing geometry points:

Determine the pressures at the defining "wing geometry points," that is, the positions that determine the mapping defined above.

(3) Pressures at arbitrary points on the wing:

Use another piecewise bicubic map that interpolates to the pressures at the wing geometry points to determine the pressure at an arbitrary point of the wing.

The details of these steps comprise the remainder of this section.

2.1. Representation of wing. The wing geometry points are points in R^3 which define the surface of the wing. They have a gridded structure and are of the form

$$f^{i,j}, \quad i = 0, \cdots, 14, \quad j = 0, \cdots, 16.$$

We estimate the derivative data needed for piecewise bicubic interpolation and call the overall interpolant f:

$$f : D = [0, 14] \times [0, 16] \to \mathbf{R}^3.$$

Thus the $f^{i,j}$ are the images of the gridpoints in D under the mapping f. That is, the top of the wing is the image of the rectangle D under a C^1 map which takes $\{(0, v) : 0 \le v \le 16\}$ onto the leading edge, $\{(14, v) : 0 \le v \le 16\}$ onto the trailing edge and the other two sides of D onto single points with u-tangents and uv-derivatives zero along these degenerate edges, see Fig. 1. Piecewise bicubic interpolation is used instead of tensor product spline interpolation in order to localize the interpolant and to more nearly maintain the wing's convexity.

2.2. Pressures at wing geometry points. The given data are of the form $(x_i; p_i)$ where the x_i are the three-dimensional data sites (plug taps) and the pressures p_i are measured. For each wing geometry point $f^{i,j}$, a corresponding pressure $p^{i,j}$ is calculated

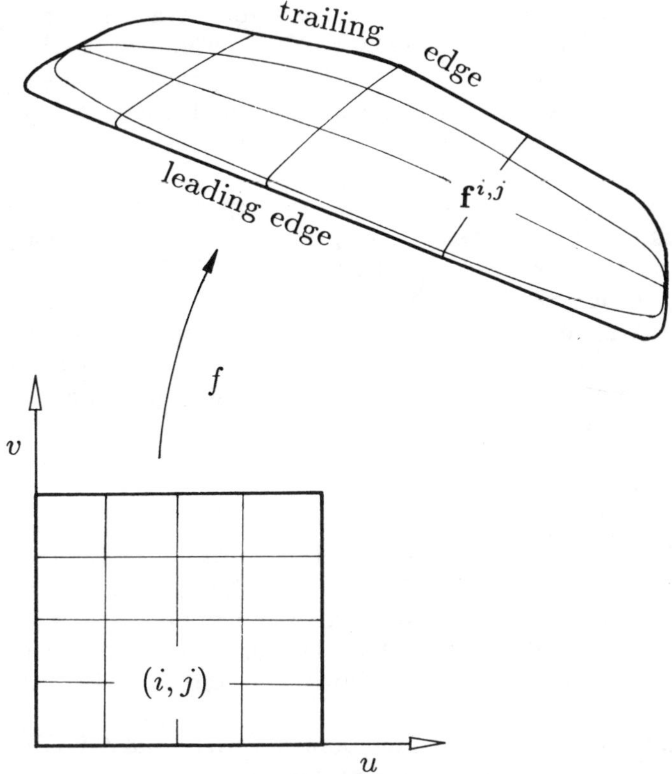

Figure 1.   The piecewise bicubic representation of the wing.

as follows. Form the hyperbolic multiquadric (Hardy [10]) M which is of the form

$$M = M(x) = \sum_{k=1}^{n} c_k \{d_k^2(x) + R\}^{1/2},$$

where $d_k(x)$ is the three-dimensional Euclidean distance from x to x_k, R is a specified positive number, the sum is over the n plug taps closest to x, and the numbers c_k are the solution of the $n \times n$ interpolation system $M(x_k) = p_k$. Thus the multiquadric is a function of $i, j: M = M_{i,j}$. The desired pressure at the wing geometry point is simply its image under its multiquadric M:

$$p^{i,j} = M_{i,j}(f^{i,j}).$$

In the examples below, $R = 0.1$ and $n = 30$.

We observe that the plug taps may not be on the geometric wing constructed as the piecewise bicubic image of D. However, since the location of the plug taps enters our solution only via three-dimensional distances, this possible complication does not matter.

2.3. Pressures at arbitrary points on the wing. We define the pressure at an arbitrary point x on the wing by $p = p(x) = g(f^{-1}(x))$ where g is a C^1 piecewise bicubic interpolant to the pressures $p^{i,j}$ at the wing geometry points. However, in practice, we need not deal with inverse functions: we evaluate and display the piecewise bicubic interpolant g to the pressures $p^{i,j}$ along isoparametric lines in D.

2.4. Example. To render the pressure we display surfaces which are "painted" by the interpolant. That is, we define a map from the range of pressure values to the spectrum of colors and paint each point on the surface the color which is associated with the pressure value given by the interpolant.

The pressures computed for wind tunnel data are displayed in Figs. 2a and 2b (see color insert) for the top and bottom of the wing, respectively. Points on the wing with low pressures are colored blue and points with high pressures are colored red, according to the color key shown.

All of the color pictures were rendered on an ADAGE 3000 using Phong shading (Foley and Van Dam [8]).

3. Interpolation to Scattered Data on a Convex Surface. The wings encountered in the NASA project are locally convex. Thus, in this section we present a solution to the following problem: given positional data on a convex "domain surface" S, where S is not explicitly known, and a real number associated with each position, interpolate to this information in a way that takes account of both the structure and shape characteristics of S.

As before, let $\mathbf{x}_i$ be the points on the domain surface and p_i be the associated real values. We construct a mapping P from S to the real line. To do this we actually create the map P from $\mathbf{R}^3$ to $\mathbf{R}^1$ which is then evaluated at points on the domain surface S. It is done in this fashion because we do not (in this method) assume we have an explicit form of the surface S.

More explicitly, we use a localized inverse distance weighting method (see Shepard [13]; Franke and Nielson [9]; Barnhill, Dube and Little [3]; and Stead [14]) of the form

$$P(\mathbf{x}) = \sum_{i=1,n} \frac{l_i(x)q_i(x)}{d_i(\mathbf{x})^2} \Big/ \sum_{i=1,n} \frac{1}{d_i(\mathbf{x})^2}, \qquad \mathbf{x} \in \mathbf{R}^3$$

where l_i are the "weight" functions, q_i are the "nodal" functions which interpolate to p_i, and d_i are "distance" functions. The l_i are constructed so as to capture the structure of the surface by suitably localizing the interpolant in $\mathbf{R}^3$. This allows us to use both the q_i and the d_i to take into account the (approximated) shape of the surface.

To actually construct these functions involves several independent steps. We now present an outline of these steps.

(1) *Tessellation.* We must create a structure on the surface to determine which points are neighbors of a given data site $\mathbf{x}_i$. This is done by a Dirichlet tessellation in $\mathbf{R}^3$ of the given domain points plus some additional "barrier" points which are inserted to separate the different "pieces" of the surface. This gives us the structure of the surface at a coarse level.

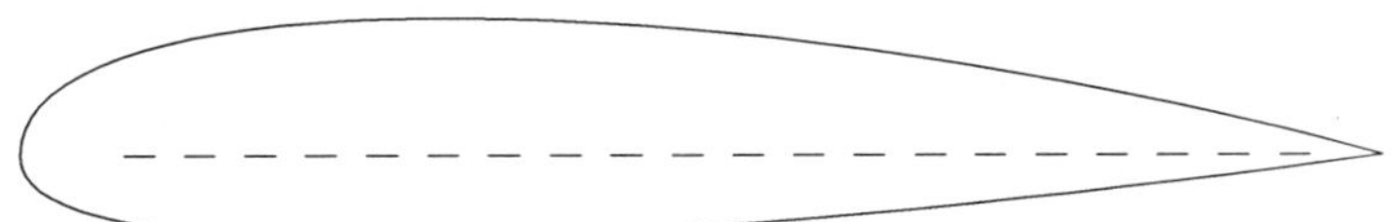

Figure 3.   The cross-section of a surface (solid line) with inserted barrier (dashed line).

(2) *Weight functions.*   Weight functions l_i are constructed on $\mathbf{R}^3$ which have local support relative to the neighbor structure we created on the points. These give us a tool for separating different regions of the surface in $\mathbf{R}^3$.

(3) *Shape approximation.*   To approximate the shape of the convex domain surface, we generate a sphere at each $\mathbf{x}_i$ that passes through $\mathbf{x}_i$ and approximates in a least squares sense the neighboring domain data points. This gives us, for each point $\mathbf{x}_i$, an approximate local shape of the domain surface which is used in constructing the distance and nodal functions.

(4) *Distance functions.*   We use the sphere generated at $\mathbf{x}_i$ in step 3 in defining the distance function d_i. This distance function serves to measure the approximate geodesic distance along the surface from the point $\mathbf{x}_i$ to any other point on the surface. It is defined over all of $\mathbf{R}^3$ by projecting a point in $\mathbf{R}^3$ onto the sphere at $\mathbf{x}_i$ and measuring the geodesic distance along the sphere from the projection to $\mathbf{x}_i$.

(5) *Nodal functions.*   We then create nodal functions q_i at $\mathbf{x}_i$ that interpolate to the value p_i at $\mathbf{x}_i$. These functions map $\mathbf{R}^3$ to $\mathbf{R}^1$ and approximate in a least squares sense the values p_j at the neighboring points $\mathbf{x}_j$. They take into account the nature of the domain surface in that we use the distance functions estimated in step 4 in defining weights for the least squares approximation.

We omit further details of steps 3–5 and concentrate on the more general aspects of the tessellation and the construction of weight functions in the next two sections.

3.1.   Tessellation.   In order to define a local mapping it is first necessary to get some idea of which domain data points are neighbors on the surface. We consider some user specified barriers which separate the different pieces of the surface. For example to define the surface given by the profile in Fig. 3 we may insert a barrier which consists of the plane whose profile is given by the dotted line. In general these barriers must be user specified if no other information (other than points on the domain surface) is given. The types of possible barriers can include finite or infinite planes and lines as well as points and circles. As an additional option, the user may specify that only selected data points are to be affected by a given barrier.

The barriers are incorporated by choosing a set of barrier points which lie on the barrier and are adequate to represent that barrier for the given data point set. These are obtained by perpendicularly projecting each of the given domain data points onto the barrier (line, plane or circle) and including these (after checking for duplicity) in the barrier point list.

We then concatenate the barrier point list with the given domain point list and form the $\mathbf{R}^3$ Dirichlet tessellation (Bowyer [7]) of the entire point list. The neighbors of

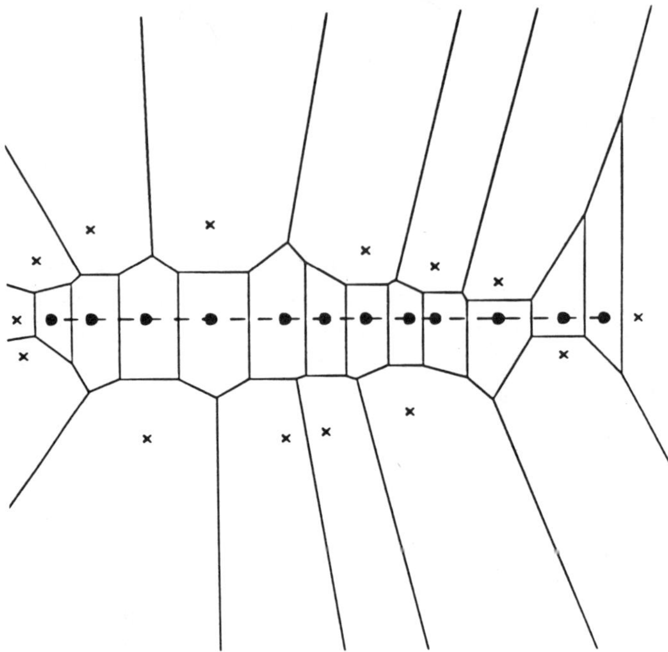

Figure 4. The Dirichlet tessellation of the data points (marked by crosses) and barrier points (marked by circles). The dashed line represents the barrier. Each barrier point is a perpendicular projection of a data point onto the barrier. The neighbors of any given data point include several barrier points and two other data points thus determining a closed curve.

a given point $\mathbf{x}_i$ as specified by the Dirichlet tessellation fall into two groups: (1) the barrier point neighbors; and (2) the nonbarrier point (given data points) neighbors. The nonbarrier point neighbors form the structure of the domain surface.

In many cases this structure is actually a triangulation which provides a piecewise linear approximation to the surface. We intend to pursue this method of triangulating points defined on a surface in future research.

This algorithm is illustrated by the two-dimensional tessellation in Fig. 4.

3.2. Weight functions. To incorporate the neighbor information constructed at step 1 in our map P we construct weight functions l_i for each data point $\mathbf{x}_i$. These functions map $\mathbf{R}^3$ to $\mathbf{R}^1$ and have support in a localized region about the point $\mathbf{x}_i$.

Each function l_i is the product of functions ll_j where the index j runs over the set $\mathbf{J}$ of the indices of all neighbors (barrier and nonbarrier) of $\mathbf{x}_i$.

$$l_i = \prod_{j \in \mathbf{J}} ll_j.$$

For each neighbor j, ll_j is a function whose contours are planes parallel to that of the

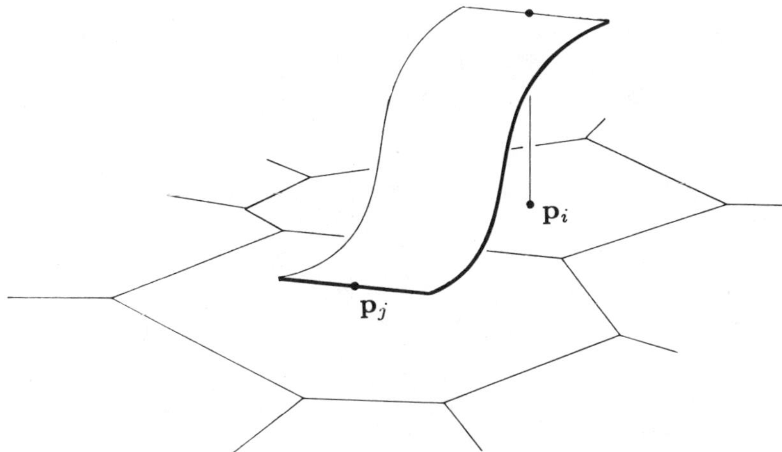

Figure 5.   The picture of a particular ll_j in the bivariate case.

plane which is perpendicular to the vector from $\mathbf{x}_i$ to $\mathbf{x}_j$. If we call this vector $\mathbf{v}$ we may write

$$ll_j(\mathbf{x}) = \begin{cases} 1 & \text{if } t < 0, \\ H_0(t) & \text{if } 0 \le t \le 1, \\ 0 & \text{if } t > 1, \end{cases}$$

where

$$t = \frac{((\mathbf{x} - \mathbf{x}_i), \mathbf{v})}{(\mathbf{v}, \mathbf{v})}$$

and H_0 is the cubic Hermite basis function which is one at $t = 0$ and zero at $t = 1$ and has zero first derivatives at $t = 0$ and $t = 1$. The weight function ll_j in the bivariate case is the ruled surface illustrated in Fig. 5.

The function l_i has support which is the intersection of the supports of the neighboring ll_j. The support of l_i is always convex. It is illustrated for the bivariate case in Fig. 6 by the shaded convex region.

The functions l_i provide local support weighting functions whose supports capture the neighbor structure as given by the Dirichlet tessellation. Notice that the choice of barrier points (in step 1) guarantees that if two points are separated by a planar barrier, the supports of their corresponding weighting functions are disjoint. This is illustrated in Fig. 7. Thus the barrier points work in conjunction with these functions to provide the appropriate local support.

3.3.   Example.   This method interpolates to arbitrarily located data on a convex surface. As an example, we take 16 points arbitrarily located on a sphere. We assume that a data value on one side of the sphere should not influence the interpolant on the

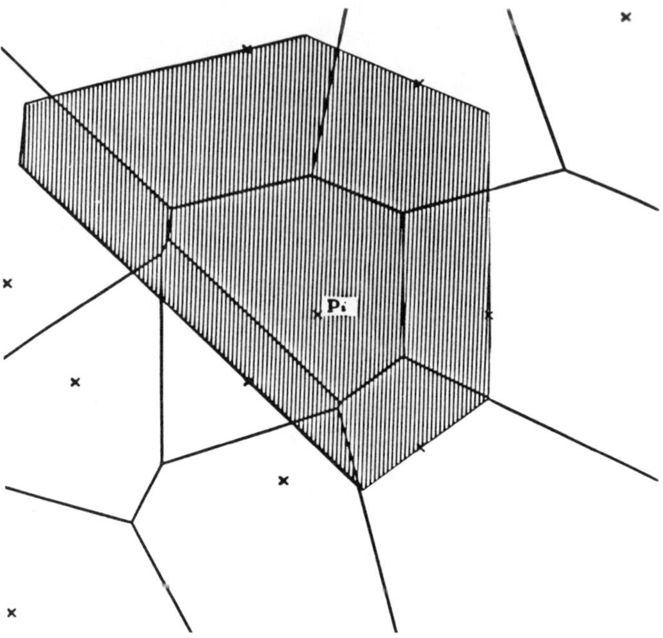

Figure 6. The support of the function l_i (shaded region). The crosses show the data sites and the solid lines show the Dirichlet tessellation.

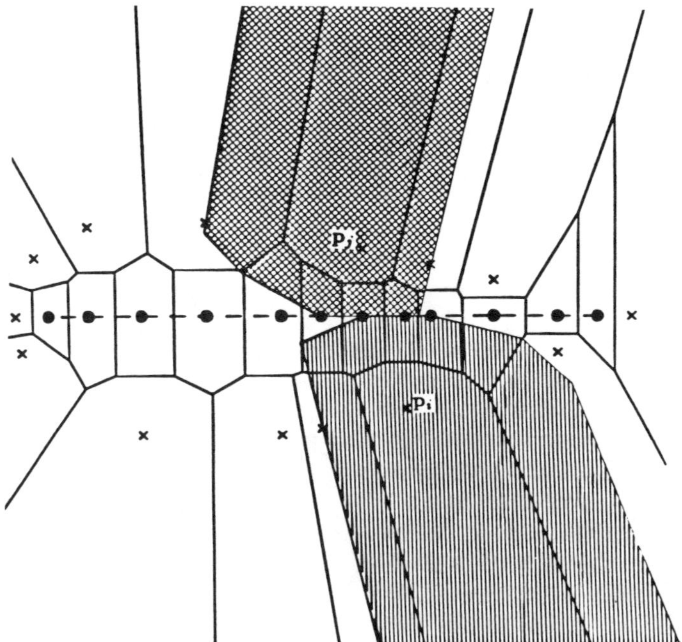

Figure 7. The supports of the functions l_i (shaded region) and l_j (cross hatched region). Since the data points $\mathbf{p}_i$ and $\mathbf{p}_j$ are on opposite sides of the barrier, their regions of support are disjoint.

opposite side of the sphere. Thus, the appropriate barrier is simply the point at the center of the sphere. Figure 8a (see color insert) shows data defined on the sphere and Fig. 8b shows the interpolant to this data set.

Acknowledgments.   The continuing fruitful research collaboration with Dr. Sarah E. Stead is enthusiastically acknowledged. This research was supported by the Department of Energy with contract DE-AC02-85ER12046 and NASA-Ames Interchange Number NCA2-1R821-401 with the University of Utah and by a University of Utah Research Fellowship to B. R. Piper. Thanks to G. Farin for his many useful comments on this manuscript.

REFERENCES

[1]   R. E. BARNHILL, *Representation and approximation of surfaces,* in Mathematical Software III, J. R. Rice, ed., Academic Press, New York, 1977, pp. 69–120.

[2]   ———, *Surfaces in computer aided geometric design: A survey with new results,* Computer Aided Geometric Design, 2 (1985), pp. 1–17.

[3]   R. E. BARNHILL, R. P. DUBE AND F. F. LITTLE, *Properties of Shepard's surfaces,* Rocky Mountain J. Math., 13 (1983), pp. 365–382.

[4]   R. E. BARNHILL, B. R. PIPER AND S. E. STEAD, *Surface representation for the graphical display of structured data,* NASA Interchange Final Report, Computer Aided Geometric Design, 2 (1985), pp. 185–187. A later form appears in The Visual Computer, 1 (1985), pp. 108–111.

[5]   R. E. BARNHILL, B. R. PIPER AND K. L. RESCORLA, *Surface Representation for the Graphical Display of Structured Data,* NASA-Ames Research Center, Computational Research and Technology, Moffett Field, CA, 1985.

[6]   R. E. BARNHILL, G. T. MAKATURA AND S. E. STEAD, *A new look at higher dimensional surfaces through computer graphics,* this Volume, 1987.

[7]   A. BOWYER, *Computing Dirichlet tessellations,* Computer J., 24 (1981), pp. 162–166.

[8]   J. D. FOLEY AND A. VAN DAM, *Fundamentals of Interactive Computer Graphics,* Addison-Wesley, Reading, MA, 1982.

[9]   R. FRANKE AND G. NIELSON, *Smooth interpolation of large sets of scattered data,* Internat. J. Numer. Meth. Engng., 15 (1980), pp. 1691–1704.

[10] R. L. HARDY, *Multiquadratic equations of topography and other irregular surfaces,* J. Geophys. Res., 76 (1971), pp. 1905–1915.

[11] C. L. LAWSON, C^1 *surface interpolation for scattered data on a sphere,* Rocky Mountain J. Math., 14 (1984), pp. 159–176.

[12] F. F. LITTLE, *Convex combination surfaces,* in Surfaces in Computer Aided Geometric Design, R. E. Barnhill and W. Boehm, eds., North-Holland, Amsterdam, 1983, pp. 99–107.

[13] D. SHEPARD, *A two-dimensional interpolation function for irregularly-spaced data,* Proceedings of the 1968 ACM National Conference, 1968.

[14] S. E. STEAD, *Smooth multistage multivariate approximation,* Ph.D. thesis, Div. Applied Mathematics, Brown Univ., Providence, RI, 1983, and CAGD Report, Dept. Mathematics, Univ. Utah, Salt Lake City, UT.

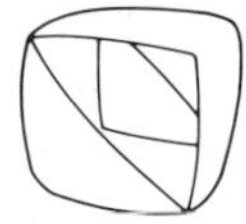

Shape Interrogation:
A Case Study

FREDERICK MUNCHMEYER

Abstract. A reference surface is interrogated for its shape. The surface is a double sinusoid on the unit square. The half-amplitude of the surface is one. Classical measures of shape are selected for this case study. They are: Gaussian curvature, mean curvature, principal curvatures, principal directions and asymptotic directions. A graph of each measure is shown and we observe that not one of these graphs gives a complete picture of the shape of the sinusoid. Graphs of selected pairs of the classic measures are also shown. The increased amount of information thus presented in one figure is still not a complete definition of shape and the analyst must choose a set of graphs that will give sufficient information for the application which is at hand.

The shape of the reference surface is rich, as indicated by graphs of the classic measures. In particular, we observe at least 35 unique singular points on the graphs. Nineteen of the singularities are intuitive. The remaining singularities are unexpected and the importance of each of them to engineering analysis or to appearance remains to be investigated.

1. Introduction. Our concerns are the shapes of sculptured surfaces, and in particular engineering surfaces. The surfaces of interest may be designed for flow control. They may be structural and may also be judged for appearance. Some examples are the hulls of ships, the skins of aircraft, auto bodies, and inlet ducts. Other examples exist, but these will suffice as they meet our use criteria which are flow control, structure, and appearance. We wish to interrogate surfaces to gain ideas about shape and we begin with curves in the surface.

It is very practical to discuss the shape of one plane curve in an engineering surface. In some design disciplines, a wire frame of plane curves supports the surface. That is, the designer constructs a wire frame first and then lays the final surface over the frame. His surface may contain the full arc of each curve. In such an instance, the designer pays close attention to the shape of each plane curve that he creates. It is practical to do so by displaying a graph of curvature versus arc length. The curvature at a point is the local shape of the plane curve and the graph shows the distribution of its shape.

It is also practical to discuss the shape of a space curve, but it is more difficult. The designer may encounter space curves at the geometric boundaries of his work. The deck edge of a ship's hull is one example. The curvature distribution of one plane projection of a deck edge does not describe shape fully. If the designer works with mechanical tools, he attempts to describe the shape of a space curve by graphing curvature distributions only. He does so for two or three of its plane projections. With new tools he can describe shape fully by graphing both curvature and torsion. Nevertheless, to visualize the shape of a space curve using two graphs is more difficult than to picture the shape of a plane curve from one graph.

When we move up from curves to surfaces, the number of quantitative shape measures grows and the number of ways to display shape grows more rapidly. The classical measures of shape of a surface at a point are:

Gaussian curvature: K,

mean curvature: H,

principal curvatures: κ_1 and κ_2,

principal directions, and

asymptotic directions.

To show the shape of a region, we can plot the distribution of K, which means either level curves (contours) or color. See, for example, [1]–[4]. We can also plot contours of H, contours of maximum and minimum principal curvatures, fields of principal directions, or fields of asymptotic directions. Further, we can overlay two of these by plotting contours of K over a color distribution of H, for example.

Classical measures yield both quantitative and qualitative information. Additional quantitative measures are proposed as in [5], [6]. We gain other qualitative views of the shape from a wire frame drawing, a shaded rendering, an artist's sketch or a physical model. Many ideas about shape are important but we shall not deal with all of them.

Rather than choose an engineering sculptured surface for this case study, we choose a double sinusoid. It is defined by

$$z = \sin 2\pi x * \sin \pi y$$

and is indicated by a line drawing in Fig. 1. It is not remarkable and probably would not be used for flow control. It resembles a solitary short-crested ocean wave consisting of one crest and one trough. The shape is rich.

One of the riches in classical shape measure is the notion of a geometric singularity. We will show that the double sinusoid has at least 35 geometric singularities on $[0, 1] \times [0, 1]$. The several singularities in the graphs of its shape measures suggest that real engineering surfaces also contain singular points.

If we want to measure the shape of an engineering surface then it must have continuous derivatives. Some shape measures require continuity of the second derivatives and some require third derivative continuity. All the derivatives of the double sinusoid are continuous.

The example is easy to construct, easy to analyze, to manufacture, and to visualize. Most of our work is on $0 \le x \le 1$, and $0 \le y \le 1$. Some is confined to the

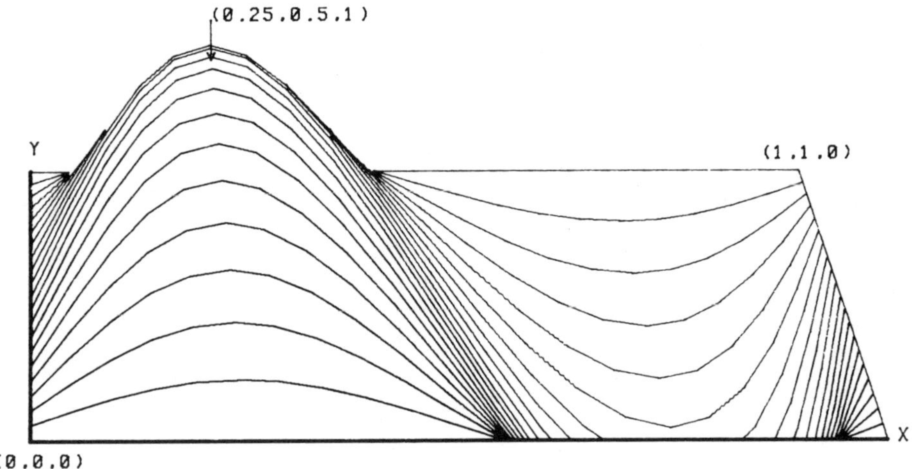

Figure 1. Line drawing of a double sinusoid. The equation for this surface is: $x = \sin 2\pi x * \sin \pi y$; $0 \leq x \leq 1$, $0 \leq y \leq 1$.

lower left octant, $[0, 0.25] \times [0, 0.5]$. That shape is found four times in the wave crest. Four additional reflections of the lower left octant appear in the trough.

In one of the first shape interrogations, Felix Klein showed level curves of Gaussian curvature. We follow his lead.

2.   The Distribution of Gaussian Curvature.   Curvature κ indicates the shape of a plane curve at a point and that idea can be extended to a surface by making plane cuts through it. At a point on the surface, erect a normal vector and cut the surface with planes that contain the normal. Next examine the curvature of every normal section curve at the point. With exceptions, one of those curves has a greater signed curvature than any other, and one has the least signed curvature. Euler observed this geometric property in 1760 [7]. The maximum and minimum values are the principal curvatures κ_1 and κ_2. (The exceptional surfaces are locally planar and spherical.) Euler also showed how to calculate the curvature of any normal section curve at this point from the two principal curvatures. Given the shape of every normal section curve through a point, we then have the local shape of the surface. Gaussian curvature is the product of those two scalars:

$$K = \kappa_1 * \kappa_2.$$

We use K to compare the shape of an engineering surface with the shapes of common surfaces. If $K > 0$, then the surface is elliptic. If $K < 0$, it is hyperbolic. If $K = 0$, then the surface is either locally flat or it can be flattened without stretching; it is developable.

The attractiveness of K as a measure of shape extends from points to regions. If a graph of Gauss contours shows that $K > 0$ everywhere over a continuously curving

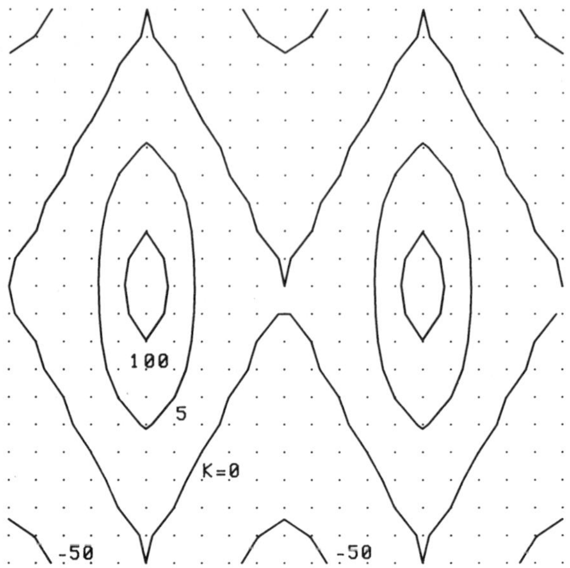

Figure 2. Four contours of Gaussian curvature K over a double sinusoid on the unit square. See text regarding the imperfections in this trace of the contour $K = 0$.

region, then the entire region is elliptic. It is either full or hollow (convex or concave); K does not distinguish.

If $K < 0$ everywhere then the region is hyperbolic or saddle-shaped.

If $K = 0$ throughout, craftsmen can fabricate the entire region by rolling a flat sheet of stock. In practice, if K is small over a region then the fabricator may treat it as developable. That is, he may approximate the designer's surface with rolled plates.

Figure 2, in black and white, shows the distribution of K over the double sinusoid. To generate this figure, we first calculate K on a 21×21 mesh. Stoker [9] and related texts provide formulas for K and the other geometric properties which we examine.

Next we estimate locations of contour points by linear interpolation and trace the contours through them. (The current version of the tracing algorithm may skip points that belong to some of the contours.)

Observe the parabolic contour, $K = 0$, in Fig. 2. In the projection onto the $x - y$ plane, this contour divides the surface into diamonds and triangles. The two diamonds enclose elliptic regions and the triangles bound hyperbolic regions.

In this case, the wiggles in the parabolic contour of Fig. 2 are due to the interpolation scheme and the scale; they disappear when we examine one octant of the surface, as in Fig. 3. If the wiggles are true then the surface is not considered to be fair; see [2], [3], [4] for discussions of fairness and Gauss contours. The shape of the double sinusoid will satisfy any criterion of fairness.

In Fig. 4 (see color insert), the distribution of K is shown in color. Red indicates $K > 0$, an elliptic region, and blue indicates a hyperbolic region. Brighter red and

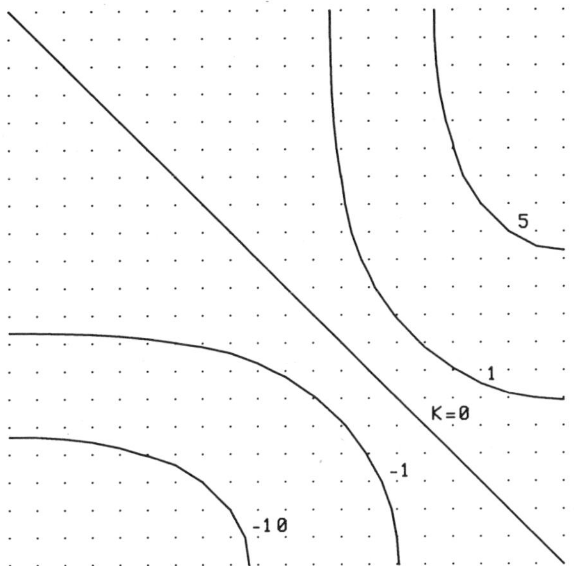

Figure 3.   Four contours of Gaussian curvature over a double sinusoid. The region shown is the lower left octant of the unit square: $0 \le x \le 0.25$, and $0 \le y \le 0.5$. Compare the $K = 0$ contour here and in Fig. 2. The straight contour is a correct projection.

brighter blue indicate greater magnitudes. We used 15 intensity levels in these early trials. In Fig. 5 (see color insert), we combine contours and color.

The idea of plotting Gauss contours is not new. Felix Klein did it on a marble surface, as reported by Hilbert and Cohn-Vossen [8]. Klein speculated that the artistic beauty of the statue Apollo Belvedere "$\cdots$was based on certain mathematical relations $\cdots$. But the [parabolic contours drawn on the bust] did not possess a particularly simple form, nor did they follow any general law that could be discerned."

Klein's experience not withstanding, Gauss contours plotted on an engineering surface do give a designer insight regarding shape. At the least, they point out the existence of unwanted humps, hollows, and possible flats, features which detract from good flow, good structure and good appearance. However, we do not tout Gaussian curvature as a sufficient measure of shape for all cases. For example, it will not distinguish a flat from a developable region.

3.   Mean Curvature.   Mean curvature H is just the arithmetic mean of κ_1 and κ_2. Figure 6 in black and white and Fig. 7 (see color insert), in color with contours, show its distribution over the double sinusoid. Again, red shows $H > 0$, blue means $H < 0$ and brighter colors indicate greater magnitudes.

Compare Figs. 4 and 7. Gaussian curvature tells whether a region is elliptic, hyperbolic or parabolic. Mean curvature indicates whether the region is full or hollow. Neither measure, alone, gives a sufficient indication of shape.

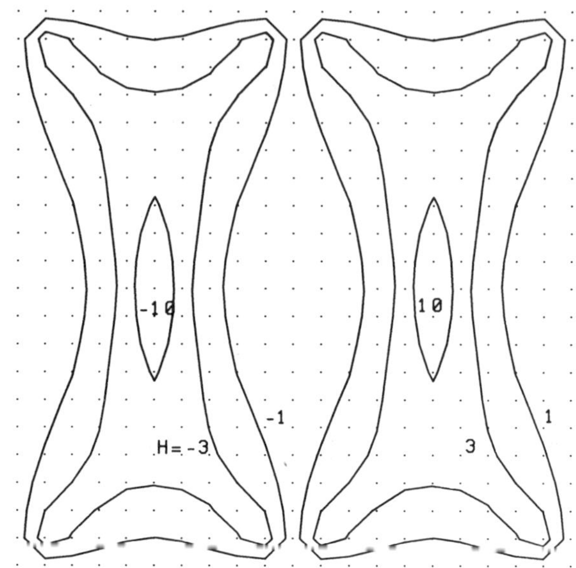

Figure 6.   Six contours of mean curvature H over a double sinusoid on the unit square.

In Fig. 7, note the contours $H = 3$ and $H = -3$. They begin to pinch. In Fig. 8, we zoom in to examine one pinch. It shows two unexpected stationary points in the lower left octant. One of these lies in the center of the H-contour loop; it is a local maximum. The other is the pinch point itself. It marks a saddle point in H.

4.   Principal Curvatures and Principal Directions.   Principal curvatures may interest engineering analysts and fabricators more than do K and H. Principal directions may interest them equally. For example, if a sheet of stock is to be shaped by rolling, then a principal direction tells how to feed the sheet into the rolls and the corresponding principal curvature indicates the set of the rolls. Figures 9a and 9b show contours of maximum curvature and a field of directions of maximum curvature.

5.   Asymptotic Directions.   The Dupin indicatrix is useful for examining shape at a point [9]. It is the intersection of another plane with the surface. This one is parallel to the tangent plane and infinitesimally distant from it. At a hyperbolic point, the indicatrix is a hyperbola. The asymptotes of the hyperbola are the asymptotic directions of the surface. They are the directions in which the surface does not curve or the directions in which the normal section is a straight line.

When we apply Dupin at many points on a hyperbolic surface the resulting asymptotes produce a direction field. Figure 10 shows those two direction fields for the entire sinusoidal surface. Evidently the direction field does not exist inside the diamonds where the surface is elliptic.

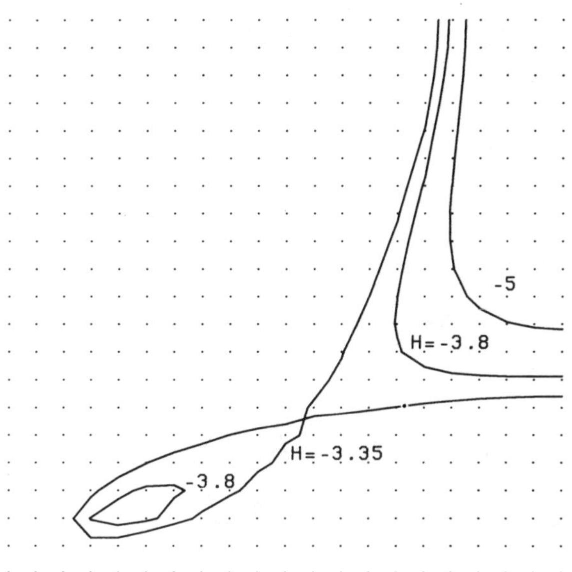

Figure 8.   Three contours of mean curvature over the double sinusoid on a lower left octant of the unit square: $0 \le x \le 0.25$, and $0 \le y \le 0.5$. This is one of the pinch points seen in Figs. 6 and 7. The detail indicates that the -3.35 contour is self intersecting at $x = 0.135$, $y = 0.135$. However, caution is required in this interpretation. The -3.35 contour could be drawn to show two arcs kissing rather than crossing. The -3.8 contour indicates a local minimum in H.

The two asymptotic directions at a point should be orthogonal wherever the mean curvature H is zero. Further, they should coalesce in the vicinity of an asymptotic line. Figure 10 exemplifies these geometric properties of a surface.

Compare Fig. 9b and Fig. 10 at the point $[0.3, 0.1]$, for example. When the asymptotic directions coalesce, the result lies in a principal direction.

6.   Combined Measures of Shape.   Each of the individual measures presented here gives a different view of shape. It is useful to combine two or more graphs to gain new insight. Figure 11 (see color insert) shows contours of K combined with colors of H. In Fig. 12 the directions of maximum curvature trace over contours of κ_1.

Evidently, visualizing the shape of a free-form surface is more difficult than visualizing the shape of a plane curve. The classic measures permit many useful views of shape. The choice of views depends on the application. For example, if Gaussian contours are important, mean curvature will distinguish convex regions from concavities. Similarly, H contours combine with K to distinguish flat regions from developables. The fields of principal directions combined with contours of κ_1 and κ_2 find uses in fluid flow, structural analysis and fabrication.

7.   Singularities.   At least 35 unique singular points exist on the double sinusoid on $[0, 1] \times [0, 1]$. Some belong to vector fields and the others to direction fields. To count them, we look first for the most intuitive geometry. The first two are the top of

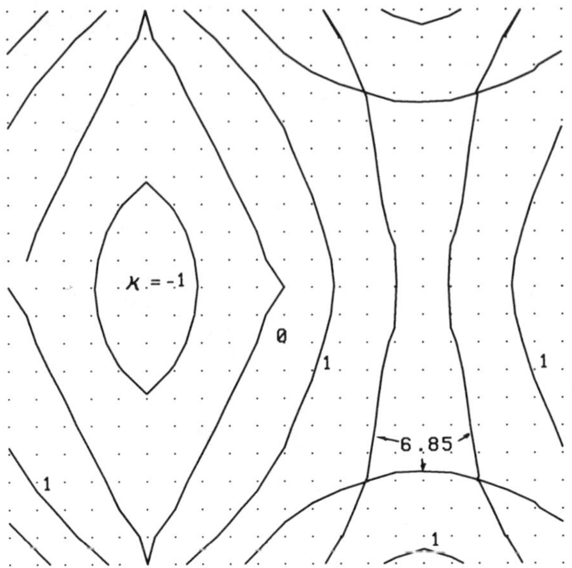

Figure 9a. Four contours of maximum principal curvature over a double sinusoid on the unit square. Note the self-intersections of the 6.85 contour. Note also the warning given with Fig. 8: at these singular points, the arcs may kiss, not cross.

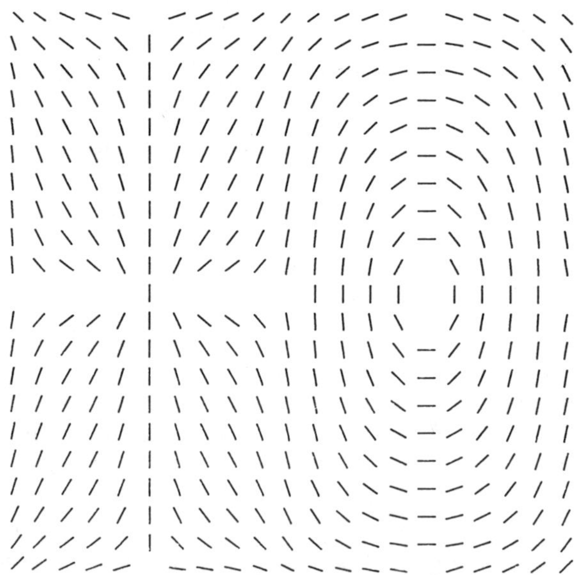

Figure 9b. The directions of maximum curvature over a double sinusoid on the unit square. The projection of this field onto an x-y plane is true only at points where the tangent plane is also an x-y plane. There are six such points: the four corners, the peak where $z = 1$ and the trough or pit where $z = -1$. The four umbilics, all on the line $y = 0.5$, are singularities in this field.

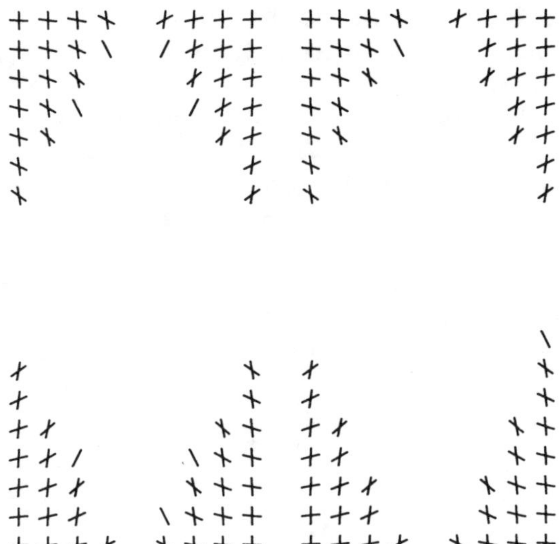

Figure 10.   The fields of asymptotic directions over a double sinusoid on the unit square.

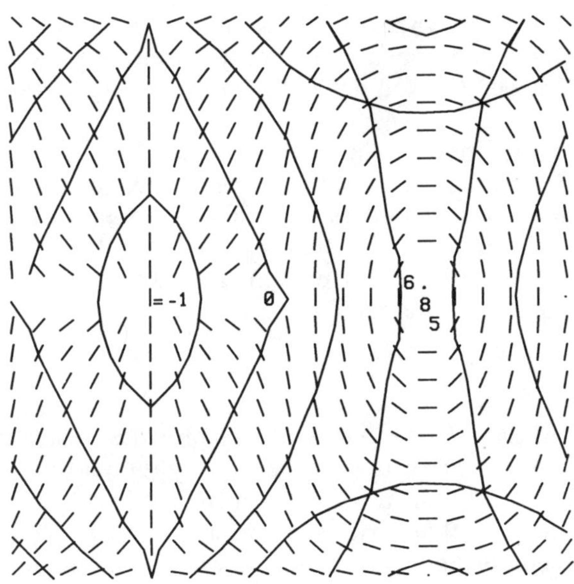

Figure 12.   Four contours of maximum principal curvature and the corresponding field of principal directions. The surface is the double sinusoid on the unit square and the projection is onto the plane $z = 0$.

the wave crest at the point $[0.25, 0.5]$ and the bottom of the trough at $[0.75, 0.5]$. The scalar fields of K, H and κ are stationary at those points. Their gradient vector fields are singular.

Next look at the saddles centered at the four corners and at the points $[0.5, 0]$ and $[0.5, 1]$. The gradient vector fields of K, H and κ are singular at these six points also.

Mean curvature, H, is zero on the perimeter of the unit square and on the line $x = 0.5$. When both K and H are zero, the surface is flat. This occurs at the seven points of the K-diamonds.

The final four intuitive singularities are the umbilics. All four are on the line $y = 0.5$ and are symmetric with respect to the crest and the trough. The first is found at $x = 0.2055$. The fields of principal directions are singular at the umbilics.

To summarize the intuitive singularities on $[0, 1] \times [0, 1]$, we have:

$$
\begin{array}{rl}
2 & \text{peaks} \\
6 & \text{saddles} \\
7 & \text{flats} \\
\underline{4} & \underline{\text{umbilics}} \\
\end{array}
$$

Subtotal: 19 intuitive singularities.

Sixteen definitely remain and eight others are probable. The 16 are stationary points in H and the eight are stationaries in κ. Figure 8 shows two of the stationary points in H, which are also singularities in gradient fields. They are in the lower left octant at the point $[0.06, 0.06]$ where $H < -3.8$, a local minimum, and at $[0.135, 0.135]$ where the contour $H = -3.35$ appears to intersect itself. We are not certain whether the contour intersects itself at a well defined point, or whether two arcs of the contour kiss.

Figure 9a shows four of the singularities in κ. They are the crossings of the contour $\kappa \max = 6.85$. Contours of $\kappa \min$ are symmetric. In the lower left octant, the contour $\kappa \min = -6.85$ appears to intersect itself at the point $[0.14, 0.15]$.

We estimate the locations of the crossing points of H and κ graphically. They appear to be distinct but we have not proven that matter conclusively. Thus we list eight probable singularities. The final total is either 35 or 43 singular points on $[0, 1] \times [0, 1]$.

8. Summary. Classic differential geometry provides means for shape interrogation. The shape of a plane curve is described in one graph and two graphs give the shape of a space curve, but six or more views of surface shape are possible when using classical analyses. Some of those views allow us to assign standard descriptors of surface shape such as elliptic and hyperbolic. Other views yield numbers that are useful in engineering analyses of fluid flow, structural behavior and fabrication. Finally, as in this case study, some views may exhibit unexpected singularities in shape. We have not yet determined the importance of geometric singularities to engineering analyses or to appearance.

REFERENCES

[1] A. R. FORREST, *On the Rendering of Surfaces,* Proc. Siggraph 79, in Computer Graphics, 13 (1979), pp. 253–259.

[2] J. C. DILL, *An Application of Color Graphics to the Display of Surface Curvature,* Proc. Siggraph 81, in Computer Graphics, 15 (1981), pp. 153–161.

[3] F. MUNCHMEYER, *Mathematical ship lines and surfaces,* Marine Tech., 19 (1982), pp. 219–227.

[4] H. NOWACKI AND D. REESE, *Design and fairing of ship surfaces,* in Surfaces in Computer Aided Geometric Design, R. E. Barnhill and W. Boehm, eds., North-Holland, Amsterdam, 1983, pp. 121–134.

[5] R. KLASS, *Correction of local surface irregularities using reflection lines,* Computer Aided Design, 12 (1980), pp. 73–76.

[6] T. POESCHL, *Detecting surface irregularities using isophotes,* Computer Aided Geometric Design, 1 (1984), pp. 163–168.

[7] L. EULER, *Résearches sur la courbure des surfâces,* Memories de l'Academie des Sciences de Berlin, 16 (1767), pp. 119–143.

[8] D. HILBERT AND S. COHN-VOSSEN, *Geometry and the Imagination,* Chelsea, New York, 1952.

[9] J. J. STOKER, *Differential Geometry,* John Wiley, New York, 1969.

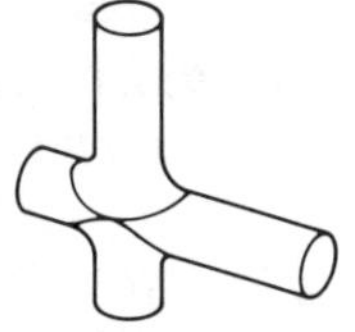

Part IV:
Intersection Algorithms

The geometric algorithms employed by a CAD system can be classified into geometry generation *and* geometry processing *algorithms. Geometry generation is concerned with the construction (the design) of curves, surfaces, and solids from lower level data: curves are generated from points, surfaces from points or curves, etc. (Many papers in the previous sections deal with generation algorithms.) Geometry processing uses data thus generated to produce "secondary" data: this could be the volume of a solid, the cutter path for machining a surface, or the offset curve of a given curve.*

Historically, the development of CAD methods was characterized by the geometry generation methods: Bézier and B-spline curves and surfaces and Coons patches are all in this category. Generating geometry is, however, only a first step in the production of an object. In order to arrive at the desired final result—this may be part of a car body or a sophisticated graphical rendition—it is necessary to process the generated geometry. An essential building block for many processing algorithms is the determination of intersections. *Some examples:*

• The theoretical offset curve to a given curve will usually have self-intersections. In order to find a physically meaningful offset curve (for example to generate a cutter path), one must first determine those self-intersections. The situation is the same for offset surfaces, although much more complicated.

• "Ray tracing" is a standard technique for the realistic rendering of objects. One of the major components in that method is the computation of intersections between straight lines (rays) and the objects to be rendered.

• *The Boolean operations for solids require accurate intersection algorithms (see, for example the papers by D. Field and R. Goldman in the second section). In fact, the absence of such algorithms had a severe impact on solid modelers: they were typically restricted to plane and quadric surface faces simply to ensure that intersections could be found reliably.*

• *The construction of fillet surfaces, i.e., surfaces that round off sharp corners or edges between surfaces, typically requires the knowledge of that edge—very often the intersection line between two surfaces. Fillet surfaces (also called blending surfaces) are considered by A. Rockwood and J. Owen and by C. Hoffmann and J. Hopcroft.*

• *In scientific computing, when graphical results are needed, the contour lines of a given function are often of importance: an example are lines of constant temperature in meteorology. These "contour lines" are the intersections between a (constant) plane and the graph of the given function. A generalization to the contouring of functions of three variables is discussed by C. Petersen, B. Piper and A. Worsey.*

Algorithms for the intersection of geometric entities depend to a high degree on the internal object representation. Since there are two standard forms for the representation of free-form surfaces—parametric and implicit (see the paper by T. Sederberg for a comparison)—there are three classes of intersection algorithms. Each case is addressed by an article in this section: parametric/parametric by V. Chandru and B. Kochar; parametric/implicit by R. Farouki; and implicit/implicit by J. Owen and A. Rockwood.

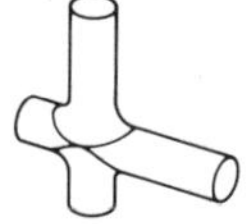

Analytic Techniques for
Geometric Intersection Problems

VIJAYA CHANDRU AND BIPIN S. KOCHAR

Abstract. This paper addresses the problem of computing the intersection of rational parametric surface patches. Given the modern trends in geometric modeling, techniques for computing the intersection of bicubic surface patches would be particularly useful. Numerical methods for tracing out the intersection curve(s) can be made far more reliable if an initial point on or close to the intersection curve is given. Sederberg showed how simple constructs of Elimination Theory in Algebraic Geometry could be used to find such an initial intersection point. These constructs, however, did not yield practical algorithms except for very special biquadratic Steiner surfaces. In the first part of the paper we explore several other special cases for which intersection point(s) can be found with reasonable computational effort. In the second part we develop efficient procedures for tracing completely the intersection of a plane with rational parametric surfaces. These are related to a particular type of surface section evaluation. Such methods are also discussed by Farouki in his paper entitled *Direct surface section evaluation* which appears elsewhere in this volume.

Introduction. Computational geometry of planar and three-dimensional objects is undergoing a resurrection of sorts. Much of the impetus for this resurgence comes from the availability of powerful geometric modeling systems and thence from the perceived applications in automated manufacturing, computer graphics, robotics, etc. Computational geometers today are faced with the task of devising efficient algorithms for the analysis of geometric models. Quick progress has been made in combinatorial computational geometry. The theoretical computer scientists have been the main contributors in this effort. An excellent survey paper by Lee and Preparata [11] has already appeared as have some monographs (Mehlhorn [13], Preparata and Shamos [16]). The analysis of "free-form" geometric models has also made some progress although much of the basic work still remains to be done.

In this paper we focus on a particular aspect of the analysis of free-form geometric models, namely intersection problems. The fundamental nature of the intersection problem is attested to by noting its direct application in hidden-line elimination, collision-avoidance, NC-Program verification and constructive solid geometry representation. Research on this problem has proceeded in three different directions. The problem of intersecting two objects composed of quadric surface patches has been well studied and for the most part well solved (Blinn [3], Levin [12], Ocken et al. [15]). For more general parametric surface patches several iterative numerical techniques, including subdivision methods, have been given (Timmer [18], Faux and Pratt [7] and Mortenson [14]). These techniques can be made failsafe but only with extensive computations. A third direction was proposed recently in a remarkable doctoral dissertation by Sederberg [17]. In this work Sederberg shows how some simple constructs of elimination theory in algebraic geometry can be used to "implicitize" any rational polynomial parametric surface. This removed a commonly held misconception of geometric modelers (see Faux and Pratt [7]) that implicitization was not possible in general. Sederberg also showed that this elimination process could in principle be applied towards the analytic solution of intersection problems.

The main idea in the elimination approach is to boil down the computation of an intersection point of a curve and a surface to the computation of real roots of a univariate rational polynomial. Since numerical analysts have developed extremely efficient algorithms for the latter problem a solution is at hand (in principle). However, Sederberg also pointed out the quick growth of the degree of the resulting polynomial when general free-form surfaces are involved. For example two bicubic surfaces yield a resultant polynomial of degree 324. Although this leaves the general problem unsolved for practical purposes, Sederberg showed that for some special cases the "state of the art" could be expanded by his techniques. In particular he discussed Steiner surface patches which strictly generalize quadric surfaces and gave efficient algorithms for this case.

In a similar spirit we examine various special cases in § 1 of the paper. Our approach is different from Sederberg's in that we consider each special case uniquely and apply an elimination technique based on consistency conditions to it. In developing these techniques for rational cubic curves we found a characterization of space cubic curves as the intersection of three quadric surfaces. The details are given in § 1.4. An iterative numerical method may be used for tracing the intersection curve of two surfaces based on Gauss' least squared error method (Beightler [2], Chandru and Kochar [5]). We have carried out some preliminary computational tests with this method and are encouraged by the results. This method is of course only locally convergent but can be made reliable by having a start point exactly on the intersection curve or "close" to it. Such a start point would be generated by the elimination techniques developed in § 1.

In § 2 we consider the special case of a plane intersecting a parametric surface and describe a technique for tracing the intersection curve(s) based on the quadratic root and cubic root formulae. These methods may be viewed as solutions to a particular kind of surface sectioning problem. Farouki [6] discusses the more general sectioning problem where instead of a plane, we may have an arbitrary (low degree) algebraic surface.

1. Elimination Techniques for Intersections. This section contains some results on intersection problems. The common thread in all these results is the use of elimination techniques to boil down the intersection question to the computation of real roots of a univariate polynomial (for a curve and surface intersection). This approach was pioneered by Sederberg [17] in his dissertation via the so-called resultant methods. Our approach generates several univariate polynomials whose common roots define the intersection points. This allows the possibility of using the polynomial g.c.d. algorithm (cf. Knuth [10]) to decrease the degree of the resultant polynomial. Thus even if the degree of the individual polynomials is not minimal, there is a computationally inexpensive way of possibly reducing the degree of the polynomial whose roots are to be computed numerically.

1.1. Point on a curve.

PROBLEM 1.1. *Given a point $P(x_0, y_0, z_0)$ in three-dimensional Euclidean space and a rational cubic curve in parametric form, determine whether P lies on $\vec{r}(u)$.*

$$\vec{r}(u) = \left\{ \frac{f_x(u)}{f_w(u)}, \frac{f_y(u)}{f_w(u)}, \frac{f_z(u)}{f_w(u)} \right\}, \qquad (u \in \mathbb{R}).$$

Clearly, P lies on $\vec{r}(u)$ if and only if the system

$$(1.2) \qquad \left. \begin{array}{l} x_0 f_w(u) = f_x(u) \\ y_0 f_w(u) = f_y(u) \\ z_0 f_w(u) = f_z(u) \end{array} \right\} \text{ has a solution for some } u \in \mathbb{R}.$$

Since $\vec{r}(u)$ is assumed to be of cubic degree, we can reformulate (1.2) to be of the form

$$(1.3) \qquad A\bar{u} = b \quad \text{where } \bar{u} = (u^3, u^2, u)^T \text{ and } A \text{ is } (3 \times 3), b \text{ is } (3 \times 1).$$

The following algorithm solves Problem 1.1.

ALGORITHM 1.4.   (Point on a Cubic Curve).
(0) Row reduce A to an upper triangular form A' using Gaussian elimination. The new form of the linear system (1.2) is given by $A' \bar{u} = b'$.
(1) If $\det A' \neq 0$, then compute $(y_1, y_2, y_3)^T = (A')^{-1}b'$ else execute step (2). If in addition $y_1 = y_2 y_3$ and $y_2 = y_3^2$, then return "P is on $\vec{r}(u)$." Else return "P is not on $\vec{r}(u)$."
(2) Pick the lowest degree polynomial amongst the three rows of $(A'\bar{u} - b')$ and find the real roots of this polynomial. Verify if one of the roots is common to all three polynomials. If so, output "P is on $\vec{r}(u)$" else output "P is not on $\vec{r}(u)$."

We note that in the worst case, the algorithm has to solve a polynomial equation of degree 3 in step (2). The elimination technique used here is straightforward Gaussian elimination since the matrix A is made up of all constants.

1.2. Point on a surface.

PROBLEM 1.5. *Given a point $P(x_0, y_0, z_0)$ and a rational bicubic parametric surface*

$$\vec{r}(u, v) = \left\{ \frac{f_x(u, v)}{f_w(u, v)}, \frac{f_y(u, v)}{f_w(u, v)}, \frac{f_z(u, v)}{f_w(u, v)} \right\},$$

determine whether P lies on $\vec{r}(u, v)$.

Again, P is on $\vec{r}(u, v)$ if and only if the system

$$(1.6) \qquad \left. \begin{array}{l} x_0 f_w(u, v) = f_x(u, v) \\ y_0 f_w(u, v) = f_y(u, v) \\ z_0 f_w(u, v) = f_z(u, v) \end{array} \right\} \text{ has a solution for some } u \in \mathbb{R}, \, v \in \mathbb{R}.$$

Simplifying (1.6) we obtain the homogeneous system

$$(1.7) \qquad f_i(u, v) = 0 \quad \text{for } i = 1, 2, 3.$$

Since $\vec{r}(u, v)$ was assumed bicubic, we have

$$(1.8) \qquad f_i(u, v) = f_{i0}(v) + f_{i1}(v)u + f_{i2}(v)u^2 + f_{i3}(v)u^3$$

with (1.7) given by

$$(1.9) \qquad -\bar{f}_0(v) = [\bar{f}_3(v)\bar{f}_2(v)\bar{f}_1(v)][u^3 u^2 u]^T \quad \text{where } \bar{f}_i(v) = [f_{1i}(v)f_{2i}(v)f_{3i}(v)]^T.$$

Solving (1.9) by Cramer's Rule we get

$$(1.10) \qquad \begin{bmatrix} u^3 \\ u^2 \\ u \end{bmatrix} = \frac{-1}{\bar{f}_3 \cdot (\bar{f}_2 \times \bar{f}_1)} \begin{bmatrix} \bar{f}_0 \cdot (\bar{f}_2 \times \bar{f}_1) \\ \bar{f}_0 \cdot (\bar{f}_1 \times \bar{f}_3) \\ \bar{f}_0 \cdot (\bar{f}_3 \times \bar{f}_2) \end{bmatrix} \triangleq \frac{1}{D(v)} \begin{bmatrix} F_3(v) \\ F_2(v) \\ F_1(v) \end{bmatrix}.$$

For this solution to be consistent, we require the checks $u^3 = (u^2)(u)$ and $u^2 = (u)(u)$ to hold.

$$(1.11) \qquad F_3(v)D(v) = F_2(v)F_1(v), \qquad F_2(v)D(v) = F_1(v)F_1(v).$$

Note that (1.11) gives us two 18th degree univariate polynomial equations. If there is at least one common (real-valued) root of these polynomials then P is on $\vec{r}(u, v)$ else P is not on $\vec{r}(u, v)$. Hence Problem 1.5 can be solved by one application of the g.c.d. algorithm followed by numerical solution for the roots of the g.c.d. polynomial. Note also that we have implicitly assumed that the parameters u, v are ranged over the entire real line. The reader should verify that bounded ranges for u and v cause no difficulty to the solution procedure given above.

1.3. A line and a surface.

PROBLEM 1.12. *Given a straight line $\vec{r}(t)$ and a rational, bicubic surface $\vec{r}(u, v)$ report all points in their intersection.*

Let $\vec{r}(u, v)$ be defined as in the previous section and let $\vec{r}(t) = \vec{r}_0 + t\vec{n}$. Furthermore, let $\bar{n}_1$ and $\bar{n}_2$ be two distinct vectors in $\mathbb{R}^3$ orthogonal to $\bar{n}$. The points of intersection must satisfy the conditions:

(1.13) $$\bar{n}_i^T \vec{r}(u, v) = \bar{n}_i^T \bar{r}_0 = c_i \quad \text{a constant for } i = 1, 2.$$

Hence we have two bicubic equations

(1.14) $$f_i(u, v) = f_{i0}(v) + f_{i1}(v)u + f_{i2}(v)u^2 + f_{i3}(v)u^3 = 0 \quad \text{for } i = 1, 2$$

where $f_i(u, v) = c_i f_w(u, v) - \bar{n}_i^T f(u, v)$ and $f(u, v) = (f_x(u, v), f_y(u, v), f_z(u, v))^T$.

Multiplying the first equation by $f_{23}(v)$ and the second by $f_{13}(v)$, then taking their difference and finally scaling the difference by u yields the third bivariate polynomial equation

(1.15)
$$f_3(u, v) = f_{31}(v)u + f_{32}(v)u^2 + f_{33}(v)u^3 = 0, \text{ with}$$
$$f_{31}(v) = f_{23}(v)f_{10}(v) - f_{13}(v)f_{20}(v),$$
$$f_{32}(v) = f_{23}(v)f_{11}(v) - f_{13}(v)f_{21}(v),$$
$$f_{33}(v) = f_{23}(v)f_{12}(v) - f_{13}(v)f_{22}(v).$$

Now as in the previous section we can apply Cramer's rule to (1.14) and (1.15) to obtain

$$(u^3 u^2 u)^T = (1/D(v))(F_3(v), F_2(v), F_1(v))^T$$

where $D(v)$ and $F_i(v)$ for $i = 1, 2, 3$ are degree 12 polynomials.

Enforcing the consistency conditions $(u^3) = (u^2)(u)$ and $(u^2) = (u)^2$ yields the following degree 24 polynomial equations in v alone:

(1.16) $$F_3(v)D(v) - F_2(v)F_1(v) = 0, \quad F_2(v)D(v) - F_1(v)F_1(v) = 0.$$

The intersection points of the line and surface must come from the common real roots of (1.16). Hence we may apply the g.c.d. algorithm here. The resultant methods (Kajiya [9]) yield a single polynomial of degree 18 for this problem.

1.4. Intersection of two curves.[1]

PROBLEM 1.17. *Given two rational cubic curves $\vec{r}_1(t)$ and $\vec{r}_2(S)$ report all the points of intersection.*

For this problem we shall first analyze rational cubic curves. This is necessary because there are substantive differences between different types of such curves (for example between space curves and planar curves). For the moment we shall deal with only one cubic curve $\vec{r}(t)$ (given in homogeneous coordinates)

(1.18) $$\vec{r}(t) = C[1\, t\, t^2\, t^3]^T = \bar{c}_0 + \bar{c}_1 t + \bar{c}_2 t^2 + \bar{c}_3 t^3.$$

---

[1] We have recently learned that Goldman [8] has independently found many of the results of this section. In a private communication, Dr. Goldman informed us that Theorem 1.19 is a "folklore" result in algebraic geometry.

We consider the following classification of rational cubic curves:

(a) $\vec{r}(t)$ is a *Space Curve* if rank $C = 4$,
(b) $\vec{r}(t)$ is a *Plane Curve* if rank $C = 3$,
(c) $\vec{r}(t)$ is a *Straight Line* if rank $C = 2$,
(d) $\vec{r}(t)$ is a *Point* if rank $C = 1$.

We prove the following "structure" theorem constructively.

THEOREM 1.19.   *A rational cubic space curve $\vec{r}(t)$ is always representable as the intersection of three quadric surfaces.*

Proof.   From the definitions we have $[w\,x\,y\,z]^T \triangleq \vec{r} = C[1\,t\,t^2\,t^3]^T \triangleq \vec{r}(t)$ with C nonsingular, i.e., $[1\,t\,t^2\,t^3]^T = C^{-1}\vec{r} = [\bar{d}_0 \cdot \vec{r}, \ \bar{d}_1 \cdot \vec{r}, \ \bar{d}_2 \cdot \vec{r}, \ \bar{d}_3 \cdot \vec{r}]^T$.

The consistency conditions $(t^2)^2 = (t)(t^3)$; $(1)(t^2) = (t)^2$ and $(1)(t^3) = (t)(t^2)$ give the following *necessary* conditions:

$$(\bar{d}_2 \cdot \vec{r})(\bar{d}_2 \cdot \vec{r}) = (\bar{d}_1 \cdot \vec{r})(\bar{d}_3 \cdot \vec{r}),$$

(1.20)
$$(\bar{d}_0 \cdot \vec{r})(\bar{d}_2 \cdot \vec{r}) = (\bar{d}_1 \cdot \vec{r})(\bar{d}_1 \cdot \vec{r}),$$

$$(\bar{d}_0 \cdot \vec{r})(\bar{d}_3 \cdot \vec{r}) = (\bar{d}_1 \cdot \vec{r})(\bar{d}_2 \cdot \vec{r}).$$

Now (1.20) clearly represents three quadric surfaces. After reformulation we obtain

$$\vec{r}^T M_i \vec{r} = 0 \quad \text{for } i = 1, 2, 3,$$

$$M_1 = 2\bar{d}_2\bar{d}_2^T - \bar{d}_1\bar{d}_3^T - \bar{d}_3\bar{d}_1^T,$$

(1.21)
$$M_2 = 2\bar{d}_1\bar{d}_1^T - \bar{d}_0\bar{d}_2^T - \bar{d}_2\bar{d}_0^T,$$

$$M_3 = \bar{d}_0\bar{d}_3^T + \bar{d}_3\bar{d}_0^T - \bar{d}_1\bar{d}_2^T - \bar{d}_2\bar{d}_1^T.$$

On further simplification,

(1.22)
$$M_i = (C^{-1})^T B_i (C^{-1}), \quad i = 1, 2, 3,$$

$$B_1 = \begin{bmatrix} 0 & 0 & 0 & 0 \\ 0 & 0 & 0 & -1 \\ 0 & 0 & 2 & 0 \\ 0 & -1 & 0 & 0 \end{bmatrix}, \quad B_2 = \begin{bmatrix} 0 & 0 & -1 & 0 \\ 0 & 2 & 0 & 0 \\ -1 & 0 & 0 & 0 \\ 0 & 0 & 0 & 0 \end{bmatrix}, \quad B_3 = \begin{bmatrix} 0 & 0 & 0 & 1 \\ 0 & 0 & -1 & 0 \\ 0 & -1 & 0 & 0 \\ 1 & 0 & 0 & 0 \end{bmatrix}.$$

Now to prove that the three quadric surfaces given by (1.21) are sufficient, let us suppose there is an arbitrary point $\vec{r}$ which lies in their intersection. The nonsingularity of C implies that $\vec{r} = C(s_0 s_1 s_2 s_3)^T$ for some choice of $s_0 s_1 s_2$ and s_3. By assumption then we have

(1.23)
$$s^T B_i s = 0 \quad \text{for } i = 1, 2, 3 \text{ and } s = (s_0 s_1 s_2 s_3)^T,$$

or equivalently,

$$s_1 s_3 = s_2^2, \quad s_2 s_0 = s_1^2, \quad s_0 s_3 = s_1 s_2.$$

First suppose s_0 and s_1 are both nonzero. Then (1.23) implies that s_2 and s_3 are also nonzero and that $\vec{r}$ is given by $(s_0)\, C(1, t, t^2, t^3)^T$ where $t = s_1/s_0$. Since homogeneous coordinates are scale invariant, this implies that $\vec{r}$ lies on $\vec{r}(t)$ the cubic curve. Next suppose s_0 is nonzero and that s_1 is zero. Then (1.23) implies that

$s = (s_0, 0, 0, 0)^T$. In this case $\vec{r}$ equals $\vec{r}(0)$ and therefore is again on the cubic curve. Finally, suppose that s_0 is zero. Then (1.23) implies that $s = (0, 0, 0, s_3)^T$ and hence $\vec{r} = s_3 \vec{c}_3$. This means that $\vec{r}$ equals $\lim_{t \to \infty} \vec{r}(t)$ and is therefore on the cubic curve. $\square$

The construction of the three quadric surfaces in the proof of the theorem can be further analyzed to show the following consequences. We omit the proof details for the sake of brevity.

COROLLARY 1.24. *The three quadric surfaces given by* (1.17) *form a basis for the space of all quadric surfaces that contain the cubic curve* $\vec{r}(t)$.

COROLLARY 1.25. *An integral cubic space curve* $\vec{r}_i(t)$ *is always representable as the intersection of two quadric surfaces.* (*Note that in this case* $C_1 = (1, 0, 0, 0)$.)

COROLLARY 1.26. *Two rational cubic space curves* $\vec{r}_1(t)$ *and* $\vec{r}_2(s)$ *intersect in no more than five points.*

Armed with these results we can now present an efficient algorithmic scheme for computing the intersection of two rational cubic curves.

ALGORITHM 1.27.   (Two Rational Cubic Curves).

We assume that the two cubic curves $\vec{r}_1(t)$ and $\vec{r}_2(s)$ are either space curves or plane curves (i.e., rank C_1, $C_2 = 3$ or 4), since otherwise the problem is trivial. There are still four combinations to consider.

Case 1. (Two Space Curves). In this case we have $\vec{r}_1(t) = C_1(1\ t\ t^2\ t^3)^T = C_2(1\ s\ s^2\ s^3)^T = \vec{r}_2(s)$ with rank $C_1 = $ rank $C_2 = 4$. Therefore,

$$(1\ t\ t^2\ t^3)^T = C_1^{-1} C_2 (1\ s\ s^2\ s^3)^T.$$

The consistency conditions

$$(1)(t^2) = (t)^2, \quad (t)(t^3) = (t^2)^2, \quad (1)(t^3) = (t)(t^2),$$

yield three degree 6 polynomials in s above whose common real root represents the intersection points. A linear combination of these polynomials yields a resultant polynomial of degree 5 or less. Note that this constructively proves Corollary 1.23.

Case 2.   ($\vec{r}_1(t)$ a Space Curve and $\vec{r}_2(s)$ Planar).   In this case we have C_2 of rank three. Hence we can find a vector b such that,

$$b^T \vec{r}_2(s) = 0.$$

Therefore, the intersection points must satisfy

$$b^T \vec{r}_1(t) = 0$$

which is a degree three polynomial equation to solve.

Case 3. (Two Planar Curves on Distinct Planes).   This case is analogous to Case 2 above.

Case 4.   (Two Planar Curves on the Same Plane).   This case is the worst case as the resultant polynomial is of degree 9. The standard resultant techniques described by Sederberg [17] solves this case.

1.5. Intersection of a curve and a surface.

PROBLEM 1.28. *Given a rational quadratic curve $\vec{r}(t)$ and a rational biquadratic surface $\vec{r}(u, v)$ report all the points in their intersection.*

We use the notation

$$\vec{r}(t) \triangleq \left[\frac{g_x(t)}{g_w(t)}, \frac{g_y(t)}{g_w(t)}, \frac{g_z(t)}{g_w(t)} \right]^T,$$

$$\vec{r}(u, v) \triangleq \left[\frac{f_x(u, v)}{f_w(u, v)}, \frac{f_y(u, v)}{f_w(u, v)}, \frac{f_z(u, v)}{f_w(u, v)} \right]^T.$$

Thus every intersection point must satisfy the following equations for some value of the constant k:

$$\begin{bmatrix} g_{x0} & g_{x1} & g_{x2} \\ g_{y0} & g_{y1} & g_{y2} \\ g_{z0} & g_{z1} & g_{z2} \\ g_{w0} & g_{w1} & g_{w2} \end{bmatrix} \begin{bmatrix} 1 \\ t \\ t^2 \end{bmatrix} = k \begin{bmatrix} f_x(u, v) \\ f_y(u, v) \\ f_z(u, v) \\ f_w(u, v) \end{bmatrix}$$

or more compactly,

$$(1.29) \qquad \bar{g}_0 + \bar{g}_1 t + \bar{g}_2 t^2 = k\vec{f}(u, v).$$

Since $\bar{g}_i (i = 0, 1, 2)$ are (4×1) vectors there is a $\bar{g}_3$ orthogonal to every $\bar{g}_i (i = 0, 1, 2)$. Hence we obtain the following biquadratic equation:

$$(1.30) \qquad \bar{g}_3^T \vec{f}(u, v) = 0.$$

Case 1. ($\bar{g}_0, \bar{g}_1$ and $\bar{g}_2$ are linearly independent). Since $\bar{g}_i (i = 0, 1, 2, 3)$ are (4×1) vectors we can find $\bar{h}_1$, $\bar{h}_2$ and $\bar{h}_3$ such that

$$\bar{h}_1 \text{ is orthogonal to } \bar{g}_1, \bar{g}_2 \text{ and } \bar{g}_3,$$
$$\bar{h}_2 \text{ is orthogonal to } \bar{g}_0, \bar{g}_2 \text{ and } \bar{g}_3,$$
$$\bar{h}_3 \text{ is orthogonal to } \bar{g}_0, \bar{g}_1 \text{ and } \bar{g}_3.$$

Hence we obtain from (1.29) that

$$\bar{h}_1^T \bar{g}_0 = k \bar{h}_1^T \vec{f}(u, v),$$
$$(\bar{h}_2^T \bar{g}_1)t = k \bar{h}_2^T \vec{f}(u, v),$$
$$(\bar{h}_3^T \bar{g}_2)t^2 = k \bar{h}_3^T \vec{f}(u, v).$$

The consistency condition $(t^2/k)(1/k) = (t/k)^2$ yields the second equation which is biquartic in u and v.

$$(1.31) \qquad (\bar{h}_2^T \bar{g}_1)^2 (\bar{h}_3^T \vec{f}(u, v))(\bar{h}_1^T \vec{f}(u, v)) = (\bar{h}_1^T \bar{g}_0)(\bar{h}_3^T \bar{g}_2)(\bar{h}_2^T \vec{f}(u, v))^2,$$

i.e., $d_0(u) + d_1(u)v + d_2(u)v^2 + d_3(u)v^3 + d_4(u)v^4 = 0$ where $d_i(u)$ are quartic polynomials for $i = 0, 1, 2, 3, 4$.

Now we can obtain two more equations in u and v by multiplying (1.30) by v and v^2, respectively. The four equations together give us the square system

$$\begin{bmatrix} d_1(u) & d_2(u) & d_3(u) & d_4(u) \\ e_1(u) & e_2(u) & 0 & 0 \\ e_0(u) & e_1(u) & e_2(u) & 0 \\ 0 & e_0(u) & e_1(u) & e_2(u) \end{bmatrix}\begin{bmatrix} v \\ v^2 \\ v^3 \\ v^4 \end{bmatrix} = \begin{bmatrix} -d_0(u) \\ -e_0(u) \\ 0 \\ 0 \end{bmatrix}.$$

Applying Cramer's rule to the system above we obtain

$$(v, v^2, v^3, v^4)^T = \left(\frac{1}{D(u)}\right)(b_1(u), b_2(u), b_3(u), b_4(u))^T,$$

where $D(u)$ and $b_i(u)$ are degree 10 polynomials.

Finally, the consistency conditions $(v^4) = (v^2)^2$, $(v^4) = (v)(v^3)$, $(v^3) = (v)(v^2)$ and $(v^2) = (v)^2$ give four degree 20 polynomials whose common real roots are the required intersection points. This is another case where the individual polynomials are not of minimal degree (degree 16 is achievable by resultant methods). However, the polynomial g.c.d. of the four polynomials may be of comparable (or lower) degree.

Case 2. ($\bar{g}_0$, $\bar{g}_1$ and $\bar{g}_2$ are linearly dependent). In this case we can find $\bar{g}_4$ orthogonal to $\bar{g}_0$, $\bar{g}_1$ and $\bar{g}_2$ and distinct from $\bar{g}_3$. Thus a second biquadratic equation in u and v is easily obtained from

$$(1.32) \qquad\qquad \bar{g}_4^T \vec{f}(u, v) = 0,$$

i.e., $a_0(u) + a_1(u)v + a_2(u)v^2 = 0$.

Combining (1.30) and (1.32) we get the simple square system

$$(1.33) \qquad\qquad \begin{bmatrix} a_1(u) & a_2(u) \\ e_1(u) & e_2(u) \end{bmatrix}\begin{bmatrix} v \\ v^2 \end{bmatrix} = \begin{bmatrix} -a_0(u) \\ -e_0(u) \end{bmatrix}.$$

Now the consistency condition $v^2 = (v)^2$ yields a degree 8 polynomial in u alone. Thus this case is easier than the previous one. This can be gleaned from the following reparameterization of $\vec{r}(t)$.

Let $\bar{g}_2 = \alpha_0 \bar{g}_0 + \alpha_1 \bar{g}_1$; then defining

$$s \triangleq \left(\frac{t + \alpha_1 t^2}{1 + \alpha_0 t^2}\right),$$

we get

$$\vec{r}(t) = \vec{r}(s) = \begin{pmatrix} (g_{x0} + g_{x1}s)/(g_{w0} + g_{w1}s) \\ (g_{y0} + g_{y1}s)/(g_{w0} + g_{w1}s) \\ (g_{z0} + g_{z1}s)/(g_{w0} + g_{w1}s) \end{pmatrix},$$

i.e., $\vec{r}(s)$ is a (rational) straight line.

1.6. A related interference problem. In this section we shall solve a geometric interference problem by using some of the techniques developed earlier in the paper. By an interference problem we mean the problem of *detecting* when two geometric objects intersect. Thus the solution to such a problem consists of a mere "Yes" or "No."

PROBLEM 1.34.   *Given a plane $P = \{\bar{r} : \bar{n}^T \bar{r} = d\}$ and an integral bicubic surface $\bar{r}(u, v)$, do the two have a nonnull intersection?*

Consider the two constrained optimization problems:

$$(1.35) \qquad \min(\max) \{\bar{n}^T \bar{r} : \bar{r} = \bar{r}(u, v), 0 \le u \le 1, 0 \le v \le 1\}.$$

We have assumed bounds on the parameters u and v. The unbounded case is simpler to analyze. Let us call a point $\bar{r}$ *extremal* in (1.35) if it is an optimal solution to either the minimization or the maximization problems. It is self-evident then that P avoids $\bar{r}(u, v)$ if and only if all extremal points lie strictly on one side of the plane P. To use this idea we will first find the extremal points using various necessary conditions for optimal solutions. First we consider the following categorization:

$A = \{\bar{r} : \bar{r}$ is in the interior of $\bar{r}(u, v)$ and is extremal$\}$,

$B = \{\bar{r} : \bar{r}$ is on a boundary curve of $\bar{r}(u, v)$ and is extremal$\}$,

$C = \{\bar{r} : \bar{r}$ is a corner point of $\bar{r}(u, v)$ and is extremal$\}$.

Obviously $A \cup B \cup C$ gives a complete list of the extremal points. We show now how to find supersets $A' \supseteq A$, $B' \supseteq B$ and $C' \supseteq C$. Thence by simple enumerative checking of function values $(\bar{n}^T \bar{r})$ for $\bar{r} \in A' \cup B' \cup C'$ we can determine on which side of the plane the extremal points lie.

Computing A'. We know from the calculus of optimization that all interior extremal solutions must satisfy the following stationarity condition:

$$(1.36) \quad f_u(u, v) = \frac{\partial f(u, v)}{\partial u} = 0, \qquad f_v(u, v) = \frac{\partial f(u, v)}{\partial v} = 0 \quad \text{where } f(u, v) = \bar{n}^T \bar{r}(u, v).$$

Combining (1.36) with the condition $v f_v(u, v) = 0$, we obtain a (3×3) bicubic system which we can solve as we did in § 1.2 (point on a surface). The resulting polynomial whose real roots have to be computed will have degree no larger than 16.

Computing B'. For each of the four boundary curves, we may again apply the stationarity conditions. Thus for example when $f(0, v)$ is considered we have the necessary condition

$$(1.37) \qquad\qquad f_v(0, v) = 0$$

which can be solved trivially since $f_v(0, v)$ is a quadratic polynomial.

Computing C'. Actually we need no computation for this case as we can set $C' \triangleq \{\bar{r}(0, 1), \bar{r}(1, 0), \bar{r}(1, 1), \bar{r}(0, 0)\}$.

This completes the discussion of our solution to Problem 1.34. In § 2 of this paper we will show another scheme for solving a generalization of this problem, which requires the computation of the real roots of a degree 12 polynomial.

2. Covering Disjoint Intersections.[2] Iterative methods for tracing the intersection curve of two surfaces would miss disjoint curve segments unless starting points

---

[2] An anonymous referee has pointed out that the techniques developed in this part of the paper may be viewed as specializations of the cylindrical decomposition method of Arnon [1].

along each disjoint curve segment were guaranteed. In the following sections we describe an approach for the case that one surface is taken to be a plane. The second surface is taken to be either Steiner, biquadratic or bicubic. Farouki [6] considers a generalization of this problem where the first surface is an arbitrary algebraic surface of low degree (plane, quadric, etc.). The generic problem that we shall look at is as follows:

PROBLEM 2.1.   *Given a plane $P = \{\bar{r} : \bar{n}^T \bar{r} = d\}$ and a rational surface $\vec{r}(u, v)$ for $0 \le u, v \le 1$, trace the intersection curves(s).*

Clearly the curve(s) of intersection must satisfy the condition

$$(2.2) \qquad \bar{n}^T \vec{r}(u, v) = d.$$

The main idea is to use the formulae for real roots of quadratic and cubic polynomials in conjunction with (2.2) to both narrow the ranges of search for parameter values and to ensure the covering of disjoint intersection curve segments.

2.1. A plane and a Steiner patch.   In this case we have the second surface defined by a parameterization of the form

$$(2.3) \qquad \vec{r}(u, v) = \left(\frac{f_x(u, v)}{f_w(u, v)}, \frac{f_y(u, v)}{f_w(u, v)}, \frac{f_z(u, v)}{f_w(u, v)} \right)^T$$

where f_x, f_y, f_z and f_2 are all of the form

$$f.(u, v) = a_0 + a_1 u + a_2 v + a_3 uv + a_4 u^2 + a_5 v^2.$$

In addition, since the Steiner Surface Patch is a triangular patch the boundary curves are defined by $u = 0$, $v = 0$ and $u + v = 1$. From the definition (2.3) it should be apparent that Steiner surfaces are special cases of biquadratic surfaces. Sederberg [17] discusses Steiner surfaces in some detail and gives elegant proofs of various properties of these patches. The intersection curve of a plane and a Steiner surface is a degree two, rational polynomial curve in parameter space. This may be seen quite easily by noting that (2.2) for the case of Steiner surfaces $\vec{r}(u, v)$ is of the quadric form

$$(2.4) \qquad a_{11}u^2 + 2a_{12}uv + a_{22}v^2 + 2a_{13}u + 2a_{23}v + a_{33} = 0,$$

i.e.,

$$[u \, v \, 1] \begin{bmatrix} a_{11} & a_{12} & a_{13} \\ a_{12} & a_{22} & a_{23} \\ a_{13} & a_{23} & a_{33} \end{bmatrix} \begin{bmatrix} u \\ v \\ 1 \end{bmatrix} = 0$$

or, more compactly,

$$p_0(u) + p_1(u)v + p_2 v^2 = 0.$$

Note that $p_0(u)$ is quadratic in u, $p_1(u)$ is linear in u and p_2 is independent of u. All intersection points correspond with real solutions of u and v in (2.4). Hence a necessary condition that these intersection points must meet is that

$$(2.5) \qquad K(u) \triangleq p_1^2(u) - 4p_0(u)p_2 \ge 0.$$

Since $K(u)$ is quadratic in u, we can find its roots analytically and hence identify the ranges within $[0, 1]$ for u within which $K(u)$ is nonnegative. Clearly there may be 0, 1, or 2 of such disjoint ranges. Each value of u satisfying $K(u)$ nonnegative gives rise to at most two distinct values of v such that (u, v) is an intersection point. This follows from the quadratic root formula solution of (2.4) which reads:

$$(2.6)\quad v_1(u) = (-p_1(u) + (K(u))^{1/2})(2p_2)^{-1}, \quad v_2(u) = (-p_1(u) - (K(u))^{1/2})(2p_2)^{-1}.$$

Now we can use the boundary conditions $v = 0$ and $u + v = 1$ to obtain additional pruning of possible u values defining intersection curves. This can be done by substituting $v = 0$ and $v = 1 - u$ in (2.6) and solving the resulting quadratic polynomial equations to obtain at most four critical values of u to consider for each $v_i(u)$, $(i = 1, 2)$.

DEFINITION 2.7.

$$I_1(u) \triangleq \{u : 0 \le u \le 1 - v_1(u), \; K(u) \ge 0 \text{ and } v_1(u) \ge 0\},$$
$$I_2(u) \triangleq \{u : 0 \le u \le 1 - v_2(u), \; K(u) \ge 0 \text{ and } v_2(u) \ge 0\}.$$

From the discussion above it follows that $I_1(u)$ and $I_2(u)$ can be computed analytically, i.e., no iterative procedures are needed. This constitutes the first step of the algorithm given below for tracing the intersection of a plane and a Steiner surface patch.

ALGORITHM 2.8.   (STEINER-PLANE).
(0)   Compute $I_1(u)$ and $I_2(u)$. For $i = 1, 2$ approximate $I_i(u)$ with a finite set of possible values of u.
(1)   For each value of $u \in I_i(u)$ computed above we use (2.6) to find the corresponding $v_i(u)$ and plot $\vec{r}(u, v_i(u))$.

2.2. A plane and a biquadratic surface. Since $\vec{r}(u, v)$ is biquadratic in this case we have (2.2) of the form

$$(2.9) \qquad\qquad p_0(u) + p_1(u)v + p_2(u)v^2 = 0.$$

Unlike the case of Steiner surfaces this time we have $p_0(u)$, $p_1(u)$ and $p_2(u)$ all of quadratic degree. A necessary condition that all intersection points must meet is that

$$(2.10) \qquad\qquad K(u) \triangleq p_1^2(u) - 4p_0(u)p_2(u) \ge 0.$$

Analogous to the previous section then we can define $v_i(u)$ and $I_i(u)$ for $i = 1, 2$. Note that there are typically four boundary curves $(\vec{r}(0, v), \vec{r}(1, v), \vec{r}(u, 0), \vec{r}(u, 1))$ for the "rectangular" biquadratic patches. The algorithm for this case is also completely analogous to STEINER-PLANE except that solving for the critical values of u arising from (2.10) involves solving for the real roots of a quartic polynomial.

2.3. A plane and a bicubic surface. A bicubic surface patch $\vec{r}(u, v)$ with $0 \le u$, $v \le 1$ gives rise to the following necessary condition for intersection points with a plane:

$$(2.11) \qquad\qquad a_0(u) + a_1(u)v + a_2(u)v^2 + a_3(u)v^3 = 0.$$

In the previous two sections we used the simple condition for roots of a quadratic polynomial to be real to derive bounds on the values of the parameter u. The analogy to cubic polynomials does not go through. However, we do have the following fact that we can apply to this case.

(2.12)   Cardan's method for cubic equations (cf. Carr [4]) implies that (2.11) has three real solutions v if and only if $K(u) \triangleq [-4a_0a_2^3 + a_1^2a_2^2 + 18a_0a_1a_2a_3 - 4a_1^3a_3 - 27a_0^2a_3^2] \geq 0$. Otherwise (2.11) has exactly one real solution v for a given real value of u.

The derivation of (2.12) is a straightforward application of Cardan's method. We note also that (2.11) always has at least one real solution (u, v) since any plane in $\mathbb{R}^3$ and any (unbounded) bicubic surface $\vec{r}(u, v)$ have a nonnull intersection. Since $K(u)$ in (2.12) is a degree 12 polynomial we would use a numerical root-solver to find the intervals of u for which there are three v solutions.

(2.13)
$$I_1(u) \triangleq \{u : 0 \leq u \leq 1,\ 0 \leq v_1(u) \leq 1\},$$
$$I_i(u) \triangleq \{u : 0 \leq u \leq 1,\ K(u) \geq 0,\ 0 \leq v_i(u) \leq 1\} \quad \text{for } i = 2, 3.$$

ALGORITHM 2.14.   (BICUBIC-PLANE).
(0)   Compute $I_i(u)$ for $i = 1, 2, 3$. Approximate $I_i(u)$ for $i = 1, 2, 3$ with a finite set of possible values of v.
(1)   For each value of $u \in I_i(u)$ computed above we use (2.11) to find the corresponding v value(s) and plot $\vec{r}(u, v_i(u))$.

Step (0) above of course requires the solution of a degree 12 polynomial equation which can be done quite efficiently by standard numerical procedures. This preprocessing may well be worth the effort since many unnecessary evaluations of $v_2(u)$ and $v_3(u)$ can be avoided.

3. Conclusions.   In this paper we have presented analytic techniques for finding intersection points. These can be used in conjunction with numerical methods for tracing the intersection curve. It is our belief that such an integrated (or semi-analytic) approach is essential for a general purpose and fail-safe algorithm to be developed in practice. We have not fully implemented such a scheme although our experience with the individual pieces leads us to believe that the overall scheme is practicable. In § 2 of the paper we showed that for the special case of surface and plane intersection, the analytic methods could solve the intersection problem by themselves.

As to future research, aside from the implementation issues, we suggest that several other special cases of curve and surface intersection can be studied from the perspectives we developed in § 1 of the paper. For example, it can be easily shown that the line and biquadratic intersection problem can be reduced to solving for the roots of a degree 8 polynomial. There seems to be a number of applications in the automatic generation of machine tool paths where the techniques of this paper can be fruitfully applied.

Acknowledgments.   The research of V. Chandru was partially supported by a grant from Hughes Aircraft Company. The research of B. S. Kochar was partially supported by CIDMAC, Purdue University.

REFERENCES

[1] D. S. ARNON, *Topologically reliable display of algebraic curves*, ACM Computer Graphics, Proc. Siggraph 83, 17 (1983), pp. 219–227.

[2] C. S. BEIGHTLER, D. T. PHILLIPS AND D. J. WILDE, *Foundations of Optimization*, 2nd Edition, Prentice-Hall, Englewood Cliffs, NJ, 1979.

[3] J. F. BLINN, *The algebraic properties of homogeneous second order surfaces*, course notes 15, Mathematics of Computer Graphics, Siggraph 84, 1984.

[4] G. S. CARR, *Formulas and Theorems in Mathematics*, 2nd Edition, Chelsea, New York, 1970.

[5] V. CHANDRU AND B. S. KOCHAR, *Semi-analytic techniques for geometric intersection problems*, Research Memorandum Series, Report No. 85–12, School of Industrial Engineering, Purdue Univ., West Lafayette, IN, 1985.

[6] R. T. FAROUKI, *Direct surface section evaluation*, this Volume, 1987.

[7] I. D. FAUX AND M. A. PRATT, *Computational Geometry for Design and Manufacture*, Ellis Horwood, 1981.

[8] R. N. GOLDMAN, *The method of resolvents: A technique for the implicitization, inversion and intersection of non-planar, parametric rational cubic curves*, Control Data Corporation, Arden Hills, MN, 1984.

[9] J. T. KAJIYA, *Ray tracing parametric patches*, ACM Computer Graphics, Proc. Siggraph 82, 16 (1982), pp. 245–254.

[10] D. E. KNUTH, *The Art of Computer Programming, Volume 2: Seminumerical Algorithms*, Addison-Wesley, Reading, MA, 1969.

[11] D. T. LEE AND F. P. PREPARATA, *Computational geometry: A survey*, IEEE Trans. Comput., 33 (1984), pp. 1072–1101.

[12] J. Z. LEVIN, *A parametric algorithm for drawing pictures of solid objects bounded by quadric surfaces*, CACM, November 1976.

[13] K. MEHLHORN, *Multi-Dimensional Searching and Computational Geometry*, EATCS Monographs on Theoretical Computer Science, Springer-Verlag, Berlin, 1984.

[14] M. E. MORTENSON, *Geometric Modeling*, John Wiley, New York, 1985.

[15] S. OCKEN, J. T. SCHWARTZ AND M. SHARIR, *Precise implementation of CAD primitives using rational parametrizations of standard surfaces*, Technical Report No. 67, New York Univ., Dept. Computer Science, New York, NY, 1983.

[16] F. P. PREPARATA AND M. I. SHAMOS, *Computational Geometry*, Springer-Verlag, New York, 1985.

[17] T. W. SEDERBERG, *Implicit and parametric curves and surfaces for computer aided geometric design*, Ph.D. thesis, Purdue Univ., Dept. Mechanical Engineering, West Lafayette, IN, 1983.

[18] H. G. TIMMER, *Analytic background for computation of surface intersections*, Douglas Aircraft Company Technical Memo. C1-250-CAT-77-036, April 1977.

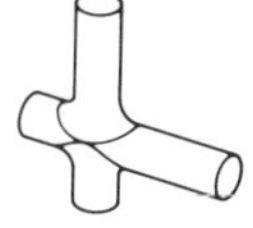

Direct Surface Section
Evaluation

RIDA T. FAROUKI

Abstract. The problem of sectioning a finite parametric rational polynomial patch $\mathbf{r} = \mathbf{r}(u, v)$ by an unbounded implicit algebraic surface $f(x, y, z) = 0$ is considered. The section curve is described precisely by a high order algebraic curve $F(u, v) = 0$ in the parameter space of the patch. Algebraic methods are employed to characterize the section curve, i.e., to identify its essential features: the number of open segments and closed loops, and any singular points such as self-intersections, cusps and isolated points.

A set of characteristic points for the section curve is proposed. These fall into three categories: (1) border points, where the curve crosses the boundary of the finite parameter domain; (2) turning points, where the curve tangent is parallel to the coordinate axes $u = 0$ or $v = 0$; and (3) singular points, where the curve possesses no unique tangent. For singular points, the number and directions of the real, distinct tangents are also determined.

The characteristic points dissect the section curve into a set of monotonic branches. Each characteristic point is assigned a definite link multiplicity. The number of monotonic branches of the curve is uniquely determined by the sum of the link multiplicities. To identify the branches and generate ordered sequences of points along each, power series expansions are employed to step along the curve between pairs of characteristic points.

1. Introduction. Recent years have witnessed burgeoning interest and activity in *solid modeling* as a paradigm for computer aided design in three dimensions (Requicha and Voelcker [19], [20]). Several different formulations have been proposed, but in many respects the *constructive solid geometry* (CSG) approach captures the most versatile ideal of solid modeling.

In a CSG modeler, the regularized (Tilove and Requicha [29]) Boolean operations of union, difference and intersection are performed on elementary solid volumes or *primitives* to generate solids of greater complexity. The set of primitives includes at minimum the familiar "natural" solids: polyhedra, cylinders, cones, spheres and tori, and in principal is extensible to volumes bounded by general surface elements.

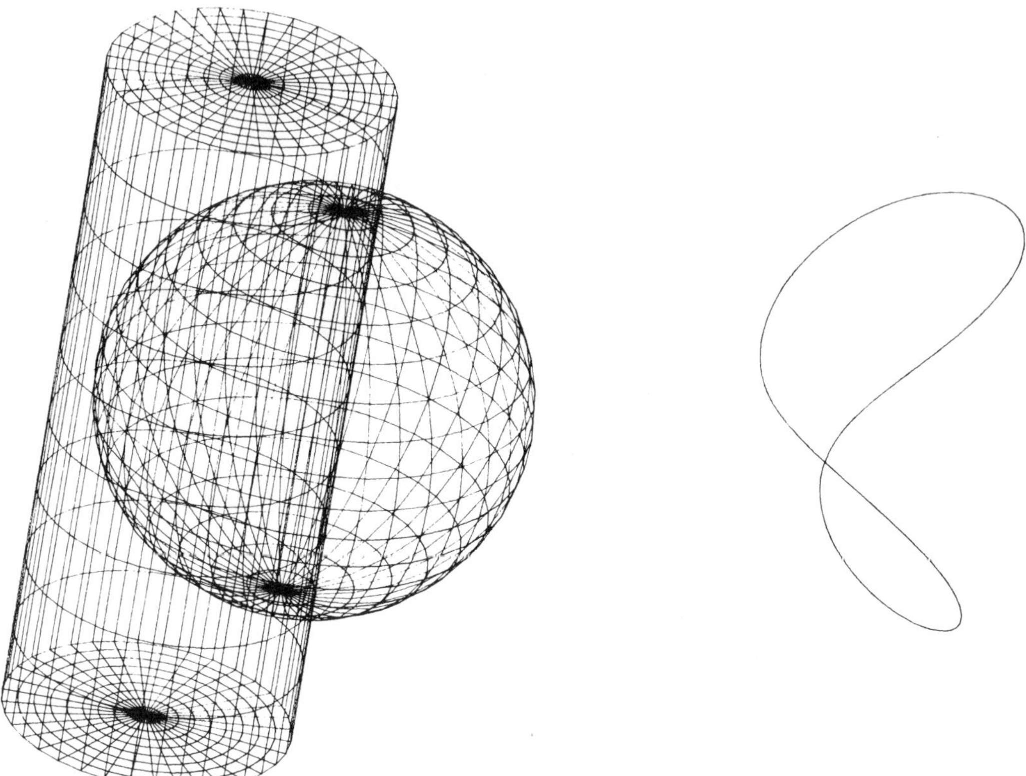

Figure 1a.   The curve of intersection of two overlapping solid primitives.

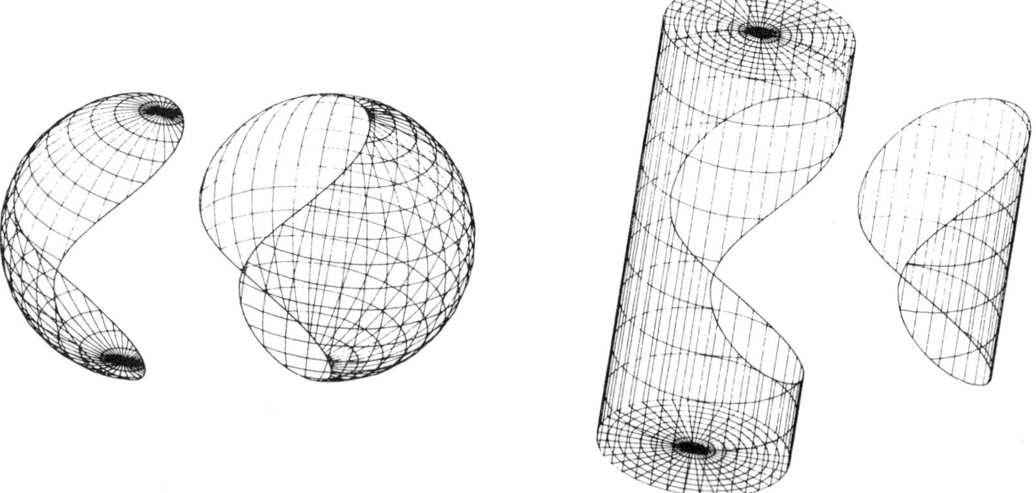

Figure 1b.   The boundary surfaces of the primitives trimmed by their mutual curve of intersection.

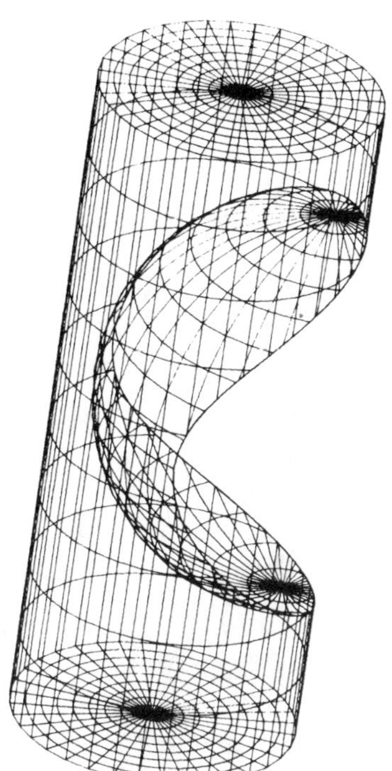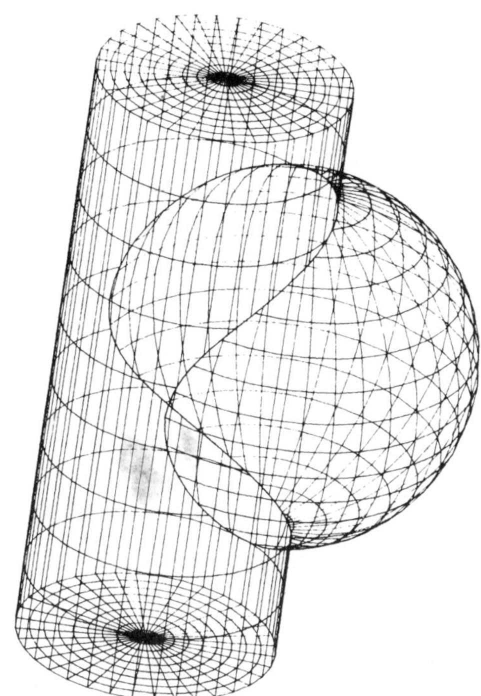

Figure 1c. Boundary representations for Boolean solid combinations formed from the trimmed surface elements.

Figure 1 illustrates the logical computational stages in a CSG modeler. Given two nondisjoint primitives A and B, the mutual curve of intersection is determined. Except in the most trivial cases, this curve possesses no simple analytical representation. The intersection curve *trims* the surface of A into segments lying inside and outside the volume of B, respectively, and vice versa. The boundary surfaces of the four Boolean solid combinations,

$$(1) \qquad A \cup B, \quad A - B, \quad B - A, \quad A \cap B,$$

are obtained by combining appropriate elements from the trimmed surfaces of A and B. In practice, many primitives are involved and primitive boundary surfaces must be successively trimmed while traversing a *Boolean tree*.

Unfortunately, this conceptual description of the CSG process encounters severe difficulties in practical implementation: the related problems of surface intersection and surface trimming are still something of a black art.

The published literature on surface intersections is extremely sparse, and that on trimmed surfaces practically nonexistent. The difficulties reflect both the intrinsic geometric complexity of the problem and a prevailing ignorance of the basic nature of its solutions, although recent revival of classical algebraic methods (Sederberg [24], [25]) has yielded important insight.

Table 1.   Surface intersections.

Surface types	Nature of solution	Level of difficulty
1. two planes	straight line	trivial
2. plane section of quadric	planar conic curve	simple
3. two quadric surfaces	quartic space curve	semi-analytic
4. plane section of bicubic patch	degree 18 plane curve	numerical
5. two bicubic patches	degree 324 space curve	heuristic

Table 1 illustrates a sampling of the spectrum of surface intersection problems at five representative cases of increasing complexity, as encountered in practical modeling environments. Case 1 is trivial, and forms the point of departure for numerous systems which render all curved surfaces by polygonal tesselations, thereby approximating intersections in a piecewise-linear fashion. This includes the B-spline subdivision algorithms (Cohen, Lyche and Riesenfeld [3]; Lane and Riesenfeld [13]). Case 2 is resolved by a coordinate transformation which maps the section plane into $z = 0$.

The intersections of quadric surfaces, case 3, represent current state of the art in solid modeling, insofar as rigorous and robust methods are available for computing these intersections (Levin [16]; Morgan [17]; Sarraga [23]). The resulting quartic space curves are complex entities which may suffer degeneracies such as cusps and self-intersections when the surfaces touch with tangency contact (Sommerville [28]).

In this paper we study a class of intersections typified by case 4, the plane section of a bicubic patch, which we term *parametric surface sections*. A parametric surface section is a cut through a parametric surface patch, defined on a finite parameter domain, by an unbounded implicit algebraic surface of the form $f(x, y, z) = 0$—typically a plane, quadric or other low degree surface.

Such surface sections play an important role in solid modeling, but do not answer all requirements. Ideally, robust solutions to the surface intersection problems characterized by case 5 must be found. Current practical algorithms for addressing such problems have a strong numerical or heuristic flavor (e.g. Houghton et al. [12]; Owen and Rockwood [18]).

Computational methods for sectioning have improved steadily in recent years (cf. Blinn [2]; Geisow [9]; Hoitsma and Roche [11]; Varady [30]; Lee and Fredericks [15]). Here we outline new algorithms based on a direct analysis of the algebraic curve properties in parameter space. A more detailed description of the method can be found in Farouki [7].

The remainder of this paper is structured as follows. Section 2 describes some prerequisite algebraic methods for bivariate polynomials. In §3 we derive the section as an algebraic curve in parameter space, and in §4 we obtain characteristic points which dissect it into monotonic branches. Section 5 describes procedures for tracing the curve between these points. Some simple examples are given in §6. Finally, §7 assesses the practical and theoretical significance of our results, and indicates directions for future work.

2. Elimination Methods for Bivariate Polynomials. A requirement for the section curve evaluation procedure is the ability to determine the roots of two simultaneous, bivariate polynomials of the form

$$(2) \qquad \sum_{i=0}^{m} \sum_{j=0}^{n} a_{ij} u^i v^j = 0, \qquad \sum_{k=0}^{p} \sum_{l=0}^{q} b_{kl} u^k v^l = 0,$$

in the unit square $[0, 1] \times [0, 1]$. These roots correspond to the intersections of two algebraic curves within the unit square.

We employ elimination techniques to identify the number and location of the roots in a precise, deterministic manner. Iterative numerical schemes, such as the two-dimensional Newton–Raphson method (Dahlquist and Bjorck [5]) are not sufficiently reliable for the present purposes.

We choose to eliminate v from (2) to obtain a single univariate polynomial in u—the *resultant* of (2) with respect to v (symmetric arguments are employed to form the resultant with respect to u). This is accomplished by Sylvester's expansion (Salmon [22]; cf. also Sederberg, Anderson and Goldman [26]; Goldman and Sederberg [10]).

Writing the coefficients of each power of v in (2) as polynomials in u:

$$(3a) \qquad \sum_{j=0}^{n} \alpha_j(u) v^j = 0, \qquad \alpha_j(u) = \sum_{i=0}^{m} a_{ij} u^i,$$

$$(3b) \qquad \sum_{l=0}^{q} \beta_l(u) v^l = 0, \qquad \beta_l(u) = \sum_{k=0}^{p} b_{kl} u^k,$$

we regard each monomial $1, v, v^2, \cdots$ as an independent "variable" and generate a system of linear equations in these variables whose coefficients are the polynomials α_j and β_l in u.

Equation (3a) contains terms to order v^n, equation (3b) to order v^q. We generate q equations from (3a) by multiplying by factors $1, v, \cdots, v^{q-1}$, and n equations from (3b) by multiplying by factors $1, v, \cdots, v^{n-1}$. This yields a total of $n + q$ homogeneous linear equations in the variables $1, v, \cdots, v^{n+q-1}$. The coefficient matrix for this linear system has the following form:

$$(4) \qquad \begin{bmatrix} \alpha_0 & \alpha_1 & \cdot & \cdot & \alpha_n & & & & \\ & \alpha_0 & \alpha_1 & \cdot & \cdot & \alpha_n & & & \\ & & \cdot & \cdot & \cdot & \cdot & \cdot & & \\ & & & \alpha_0 & \alpha_1 & \cdot & \cdot & \alpha_n & \\ \beta_0 & \beta_1 & \cdot & \cdot & \beta_q & & & & \\ & \beta_0 & \beta_1 & \cdot & \cdot & \beta_q & & & \\ & & \cdot & \cdot & \cdot & \cdot & \cdot & & \\ & & & \beta_0 & \beta_1 & \cdot & \cdot & \beta_q & \end{bmatrix}.$$

When the above linear system possesses a nontrivial solution, i.e., when the determinant of matrix (4) is zero, then equations (2) are satisfied by a common value for v. With q rows of degree-m polynomials α_j and n rows of degree-p polynomials β_l, the order of this matrix is $n + q$, and its determinant, the resultant of (2) with respect to v, is of degree $mq + np$ in u.

The roots of the resultant must generally be determined by means of numerical methods. Conversion to the Bernstein basis provides a stable and efficient algorithm (Lane and Riesenfeld [14]) for determining real roots in the unit interval, although convergence problems arise with multiple roots. These may be remedied by successive differentiation of the polynomial.

To determine the unique value of v corresponding to each root of the resultant for u, we substitute the root into the matrix of polynomials (4) and recall from the theory of homogeneous linear equations that the ratio $v = v^j/v^{j-1}$ of any two "variables" v^j and v^{j-1} is simply the ratio of the cofactors $C_{i,j+1}/C_{ij}$ of (4) for the corresponding column indices $j + 1$ and j, and arbitrary row index i.

This method fails, however, at multiple roots of the resultant, since the rank of matrix (4) is reduced by the root multiplicity. A foolproof alternative in such situations is to simply substitute the resultant root into the original bivariate polynomials (2) and compare their roots for common values.

If we chose to eliminate u instead of v, the resultant polynomial in v would have been of the same degree, $mq + np$. However, the order of the matrix would be $m + p$. It is advantageous to eliminate that variable which results in a more compact matrix.

3. Section Curve Representation in Parameter Space. We consider sectioning parametric surface patches which may be expressed in the rational polynomial form

$$(5) \qquad \mathbf{r}(u, v) = \sum_{j=0}^{p} \sum_{k=0}^{q} w_{jk} \mathbf{r}_{jk} u^j v^k \bigg/ \sum_{j=0}^{p} \sum_{k=0}^{q} w_{jk} u^j v^k,$$

defined on the unit parameter square $0 \le u$, $v \le 1$. These forms encompass the quadric surfaces and a variety of swept surfaces and spline surfaces (Farouki and Hinds [8]). The sectioning surface is taken to be a degree-n algebraic surface, defined by an inhomogeneous polynomial of the form

$$(6) \qquad \sum_{i=0}^{n} \sum_{j=0}^{n-i} \sum_{k=0}^{n-i-j} c_{ijk} x^i y^j z^k = 0,$$

including terms $x^i y^j z^k$ of total degree $i + j + k$ less than or equal to n. The number of such terms, i.e., the number of distinct real coefficients c_{ijk} defining the surface, is $\frac{1}{6}(n + 1)(n + 2)(n + 3)$.

Equation (6) defines an *unbounded* surface, in the sense that all real coordinates x, y, z satisfying it represent points lying on that surface. Such surfaces extend to infinity or close upon themselves, but possess no sharp boundaries. Thus, in sectioning patch (5) by surface (6), the extent of the section curve is determined entirely by the parameter domain of the patch.

We are concerned here principally with bicubic patches ($p = q = 3$) of the form (5) and planar ($n = 1$) or quadric ($n = 2$) sectioning surfaces of the form (6). The arguments apply to higher degree surfaces, but the polynomial degrees encountered during intermediate processing may be very high.

Substituting the parametric patch equations (5) directly into the algebraic surface definition (6) yields an exact representation

$$(7) \qquad F(u, v) = \sum_{r=0}^{M} \sum_{s=0}^{N} a_{rs} u^r v^s = 0$$

for the section curve as an algebraic curve in the parameter space (u, v) of the patch. The degree in u is $M = np$ and the degree in v is $N = nq$. Such a curve is a special instance of a plane algebraic curve of total degree $M + N$.

The coefficients a_{rs} in (7) are easily obtained by multiplying through by the nth power of the denominator polynomial in (5) after substituting in (6) and collecting like terms. This is straightforward for planar or quadric sections, but for higher order surfaces, computation of the coefficients is cumbersome.

The section curve evaluation procedure now reduces to identifying the fundamental characteristics of the algebraic curve (7), and mapping it in smooth segments or as ordered sequences of points to three dimensions. This is accomplished by identifying a set of characteristic points for the curve and traversing each curve branch between pairs of these points.

Before proceeding, however, we consider the possibility of performing a *transfinite* mapping of (7), i.e., an exact mapping of the entire continuum of points of the curve. Such a mapping is possible if $F(u, v) = 0$ can be rewritten in the parametric form $\{u = u(t), v = v(t)\}$, since substitution into (5) gives

$$(8) \qquad \mathbf{r}(t) = \mathbf{r}[u(t), v(t)]$$

as an exact three-dimensional representation of the section curve.

However, a degree-n algebraic curve $F(u, v) = 0$ possesses a parametric representation $\{u = u(t), v = v(t)\}$ in terms of elementary rational functions only if its coefficients satisfy stringent conditions. Specifically, the curve must have zero genus g as defined by

$$(9) \qquad g = \frac{1}{2}\left[(n - 1)(n - 2) - \sum_{i} h_i(h_i - 1) \right],$$

where the h_i are the multiplicities of the singular points of the curve.

For example, all conic section curves and plane cubic curves with one double point are rational. Section curves of the form (7), however, are rational only under exceptional circumstances and therefore possess no exact three-dimensional parametric representations. They must therefore be approximated by smooth segments or sequences of points.

4. Characteristic Points of the Section Curve. We now determine a set of characteristic points for the section curve (7) defined on the unit parameter square. We employ without proof concepts from the theory of algebraic curves. This has largely been neglected in current literature on computational geometry (see, however, Arnon [1]). The reader should consult the references (Salmon [21]; Walker [31]; Coolidge [4]; Semple and Kneebone [27]) for further details.

The set of characteristic points must include the following: the start and endpoints of all open segments; at least one point on each closed loop; all singular points; and a "sufficient" number of points in the vicinity of all curve features, such as regions of high curvature.

The tangent direction for the curve $F(u, v) = 0$ is $\pm(F_v, -F_u)$. The curve tangent is thus parallel to the coordinate axes $u = 0$ and $v = 0$ when $F_v = 0$ and $F_u = 0$, respectively, and indeterminate when $F_u = F_v = 0$. This behavior of the tangent forms the basis for our choice of characteristic points.

The characteristic points occur at the roots of univariate polynomials whose coefficients may be determined by precise algebraic methods. They fall into three basic categories, and dissect the section curve into a set of *monotonic branches*. Along each branch, the tangent direction varies by at most 90°, and u may be regarded as a monotonic function of v or vice versa.

The characteristic points provide a minimum base for "filling in" the section curve, i.e., establishing the identity of each branch between characteristic point pairs and generating as many intermediate points as required. This process is called *tracing* the section curve, and will be discussed below.

4.1. Border points. The border points of the section curve (7) for the unit parameter square are the sets of points

$$(10) \qquad (0, v_{0i}), \quad (1, v_{1i}), \quad (u_{0i}, 0), \quad (u_{1i}, 1),$$

where the v_{0i}, v_{1i}, u_{0i}, u_{1i} are the roots on $[0, 1]$ of the univariate polynomials

$$(11) \qquad F(0, v) = 0, \quad F(1, v) = 0, \quad F(u, 0) = 0, \quad F(u, 1) = 0,$$

respectively. If $F(u, v)$ is of degree (M, N) in (u, v), there can be at most N border points along $u = 0$ and $u = 1$, and M border points along $v = 0$ and $v = 1$.

4.2. Turning points. The u- and v-turning points of the section curve (7) are the points where the curve tangent is parallel to the axes $u = 0$ and $v = 0$, respectively. They occur at the roots of the simultaneous equations

$$(12) \qquad F = F_v = 0, \qquad F = F_u = 0.$$

These simultaneous equations are solved by the methods described in § 2. Only the real roots within the unit parameter square are desired.

If F is of degree (M, N) in (u, v), then F_u and F_v are of degree $(M - 1, N)$ and $(M, N - 1)$, respectively. Thus, there can be at most $2MN - M$ u-turning points and $2MN - N$ v-turning points.

For example, for a plane section of a bicubic patch $(M = N = 3)$ there can be up to 15 turning points for each direction; for a quadric section of a bicubic patch $(M = N = 6)$ there can be up to 66 turning points for each direction.

4.3. Singular points. The singular points occur where the curve tangent is indeterminate, at the simultaneous roots of the *three* equations

$$(13) \qquad F = F_u = F_v = 0.$$

Since the roots of these three equations have already been found pairwise in determining the u- and v-turning points, no additional computation is expended in locating the singular points.

If any (u, v) pairs within the unit parameter square are indicated as both u- and v-turning points (to some specified resolution), they are identified as singular points rather than turning points.

Since the singular points must satisfy three simultaneous equations in two variables, they do not occur in general curves. The curve coefficients must satisfy rigorous criteria for singular points to exist. These may be expressed as the vanishing of the *resultant* of equations (13) with respect to u and v.

Since F_u is of degree $(M - 1, N)$ and F_v is of degree $(M, N - 1)$, there can be at most $2MN - M - N + 1$ singular points. For plane and quadric sections of a bicubic patch there can be up to 13 and 61 singular points, respectively.

The *multiplicity* of these singular points must be determined. The simplest singular points are *double points* where $F_u = F_v = 0$, but at least one of (F_{uu}, F_{uv}, F_{vv}) is nonzero. In general, a singular point has multiplicity h if F and all its partial derivatives to order $h - 1$ are zero, but at least one partial derivative of order h is nonzero.

It was noted above that at a singular point F_u and F_v vanish and the tangent becomes indeterminate. By this we mean that no *unique* tangent to the curve exists at such a point. It is possible, however, for *several* tangents to be defined at singular points. For a singular point of multiplicity h, there may be up to h tangent directions, the real roots of the homogeneous polynomial

$$(14) \qquad \sum_{k=0}^{h} \frac{h!}{(h-k)!\, k!} \lambda^k \mu^{h-k} \frac{\partial^h F}{\partial u^k\, \partial v^{h-k}} = 0$$

in the direction cosines (λ, μ) (cf. Walker [31] and references cited above).

For the case of a double point, the direction cosines of the tangents are the roots of the homogeneous quadratic equation

$$(15) \qquad \lambda^2 F_{uu} + 2\lambda\mu F_{uv} + \mu^2 F_{vv} = 0,$$

which may be real and distinct, real and coincident, or complex conjugates, corresponding to the three possible morphologies of a double point: self-intersection, cusp and isolated point, respectively.

For the section curve characterization procedure, it is necessary to record the multiplicity of each singular point and the number and directions of its real tangents, as determined from (14).

5. Section Curve Tracing.   The border points, turning points and singular points dissect the section curve into a set of smooth, monotonic branches. We now determine which points are actually linked by a branch, employing a local power series expansion to traverse each branch. This branch identification procedure also generates ordered sequences of points along each branch, which may be used for spline or other approximations of that branch.

5.1. Link multiplicities for characteristic points. We assign a link multiplicity m_i to each characteristic point, indicating the number of curve branches entering or leaving that point. For border points, the link multiplicity is 1. For turning points it is 2.

For singular points, the link multiplicity is $2k$, where k ($\leq h$) is the number of real, distinct tangent directions at the singular point. At double points, for example, the link multiplicity is 4 for self-intersections, 2 for cusps and 0 for isolated points (which are usually discarded).

It is assumed that no singular points or turning points lie exactly on the parameter domain boundary. If this occurs, the link multiplicity must be modified accordingly. The number of monotonic branches N_b in the section curve is determined uniquely by the link multiplicities:

$$(16) \qquad N_b = \frac{1}{2} \sum_i m_i,$$

where the summation is over all characteristic points.

5.2. Power series expansion along curve branches. The "implicit function theorem" states that in the vicinity of any point satisfying an implicit equation $F(u, v) = 0$ for which $F_v \neq 0$, a function $v = v(u)$ exists, analytic in some neighborhood, such that $F[u, v(u)] = 0$. Conversely, if $F_u \neq 0$, a function $u = u(v)$ exists such that $F[u(v), v] = 0$.

The method for curve tracing between characteristic points described here is based on approximating $v = v(u)$ and $u = u(v)$ by Taylor series expansions. Any monotonic curve branch may be developed by successive applications of the Taylor series expansion in a single variable, u or v.

Along the curve $F(u, v) = 0$, the change in F corresponding to steps (du, dv) is $dF = F_u du + F_v dv \equiv 0$ and hence the expressions

$$(17) \qquad \frac{dv}{du} = -\frac{F_u}{F_v}, \qquad \frac{du}{dv} = -\frac{F_v}{F_u},$$

hold for the derivatives dv/du and du/dv. Note that dv/du and du/dv diverge as $F_v \to 0$ and $F_u \to 0$, respectively. Higher order derivatives are obtained by repeated application of the total derivative operators

$$(18) \qquad \frac{d}{du} = \frac{\partial}{\partial u} - \left(\frac{F_u}{F_v}\right)\frac{\partial}{\partial v}, \qquad \frac{d}{dv} = -\left(\frac{F_v}{F_u}\right)\frac{\partial}{\partial u} + \frac{\partial}{\partial v}$$

along $F(u, v) = 0$ to (17). The expressions for these derivatives become rather cumbersome, but in general we employ small steps and truncate the series after only three or four terms.

Alternative methods for deriving the power series coefficients expressing a monotonic neighborhood of a polynomial implicit function are given by de Montaudouin and Tiller [6], who obtain an exact recursion relation for these coefficients by algebraic methods.

Expanding in a power series in u, the increment Δv corresponding to any step Δu from a base point (u_0, v_0) is then

$$(19) \qquad \Delta v = \frac{dv}{du}\Delta u + \frac{1}{2!}\frac{d^2v}{du^2}(\Delta u)^2 + \frac{1}{3!}\frac{d^3v}{du^3}(\Delta u)^3 + \cdots,$$

where the derivatives are evaluated at (u_0, v_0).

Such a power series describes the neighborhood of any point of the curve where $F_v \neq 0$. If F_v is zero but $F_u \neq 0$, an analogous power series in increments Δv may be used. At a singular point, however, F_u and F_v are simultaneously zero, and the power series (19) fails. We therefore preclude singular points as base points for Taylor expansions.

5.3. Some implementation considerations. We now briefly sketch the basic features of a curve tracing procedure implemented on the above principles. The procedure is frankly numerical in nature, and requires judicious choice of tolerances for optimum robustness. A more detailed exposition is given in Farouki [7]. Some simple examples in the next section provide concrete illustrations of the method.

Starting at border points or turning points, curve branches are traced in small steps until a second characteristic point is encountered to within specified tolerance. The link counts of the starting and terminating points are incremented. The directions of departure and approach are also recorded. When the link multiplicities and tangent directions of all characteristic points have been realized, the curve characterization is complete. The number of branches determined is checked against the expected value N_b.

Singular points are not employed as start points for curve tracing since the power series (19) fails. The exceptional case of a curve branch between two singular points thus requires special treatment (Farouki [7]). The intermediate branch points generated in the stepping procedure may or may not be satisfactory in accuracy and frequency for subsequent purposes. If not, new points may be generated now that all branches are unambiguously identified.

The choice of step sizes Δu and Δv is critical, since in commencing any branch development at a given point, it is not known a priori in which point the branch terminates. The residual of $F(u, v)$ provides a simple but not optimal measure of deviation from the true curve at each step. An adaptive method which checks the residual and rejects unsatisfactory steps in favor of smaller ones is recommended. This also obviates concern about the radius of convergence of the power series (de Montaudouin and Tiller [6]).

A fourth order series provides reasonable balance between step size and computational effort per step for a given accuracy. A single Newton–Raphson iteration after each power series step significantly damps the cumulative error. Note that at u- and v-turning points, power series in increments Δv and Δu, respectively, must be employed. The linear term vanishes identically, so the higher order terms must be accurate to avoid large initial errors.

The power-series tracing method has provided satisfactory reliability in plane sectioning bicubic patches, but is by no means foolproof. The issue of reliably traversing an algebraic curve segment in a stable, convergent manner with strict control of deviation is of fundamental importance and deserves further attention.

6. Some Illustrative Examples. A few elementary examples illustrate the above principles in practice.

6.1. Example 1. Consider the section curve given by

$$(20) \quad F(u, v) = 64u^3 + 1024uv^2 - 48u^2 - 1024uv - 1280v^2 + 268u + 1280v - 321 = 0.$$

Substituting 0 and 1 for u and v in (20), we obtain the following two border points:

$$(21) \qquad \left(1, \frac{1}{2} \pm \frac{3\sqrt{3}}{16}\right).$$

The resultant of F and F_v with respect to v is

$$(22) \qquad 1024u^5 - 3328u^4 + 3712u^3 - 1696u^2 + 340u - 25.$$

This has the root $u = \frac{1}{4}$ in the unit interval; the corresponding value of v is $\frac{1}{2}$. This point is a candidate u-turning point. The resultant of F and F_u with respect to v is:

$$(23) \qquad 4096u^6 - 18432u^5 + 28416u^4 - 18176u^3 + 5616u^2 - 840u + 49.$$

This has the root $u = \frac{1}{4}$ in the unit interval; the corresponding value of v is $\frac{1}{2}$. This point is a candidate v-turning point.

Since $(\frac{1}{4}, \frac{1}{2})$ is flagged as both a u- and v-turning point, we identify it as a singular point. At this point, $F_{uu} = 0$, $F_{uv} = 0$ and $F_{vv} = -2048$. The singular point is therefore a double point, with tangents (λ, μ) given by

$$(24) \qquad 0\lambda^2 + 0\lambda\mu - 2048\mu^2 = 0.$$

There is only one linearly independent tangent: $(1, 0)$, and the double point is therefore a cusp.

The two border points are assigned a link multiplicity of 1 each, the cusp a link multiplicity of 2. The number of monotonic section curve branches is thus $N_b = 2$. Marching in small negative steps Δu we reach the cusp from each of the border points, satisfying all link multiplicities and completing the section curve characterization (Fig. 2).

6.2. Example 2. Consider the section curve given by

$$(25) \qquad F(u, v) = 729v^3 + 1152u^2 - 1134v^2 - 1536u + 396v + 472 = 0.$$

Substituting 0 and 1 for u and v in (25), we obtain the four border points

$$(26) \qquad \left(\frac{2}{3} \pm \frac{\sqrt{5}}{12}, 0\right), \qquad \left(\frac{2}{3} \pm \frac{7\sqrt{2}}{48}, 1\right).$$

Since F_v is independent of u, we may take the resultant of F and F_v with respect to u as simply F_v:

$$(27) \qquad 2187v^2 - 2268v + 396.$$

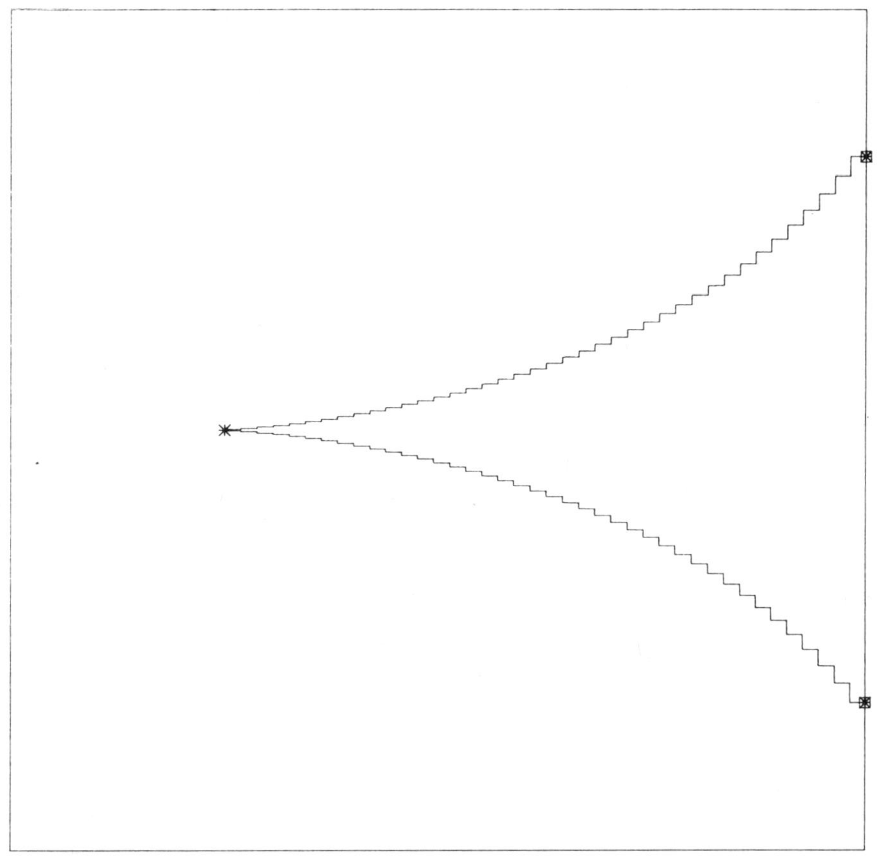

Figure 2. The characterization of the section curve analyzed in Example 1.

This has the roots $v = 2/9$ and $v = 22/27$ in the unit interval; corresponding values for u are 2/3 and $2/3 \pm 4\sqrt{3}/27$, respectively, giving three candidate u-turning points. Since F_u is independent of v, we may take the resultant of F and F_u with respect to v as simply F_u:

$$(28) \qquad\qquad\qquad 2304u - 1536.$$

This has the root $u = 2/3$ in the unit interval; the corresponding value for v is 2/9. This is a candidate v-turning point.

Since $(2/3, 2/9)$ is flagged as both a u- and v-turning point, we identify it as a singular point. At this point, $F_{uu} = 2304$, $F_{uv} = 0$ and $F_{vv} = -1296$. It is therefore a double point, with tangents (λ, μ) given by:

$$(29) \qquad\qquad\qquad 2304\lambda^2 + 0\lambda\mu - 1296\mu^2 = 0.$$

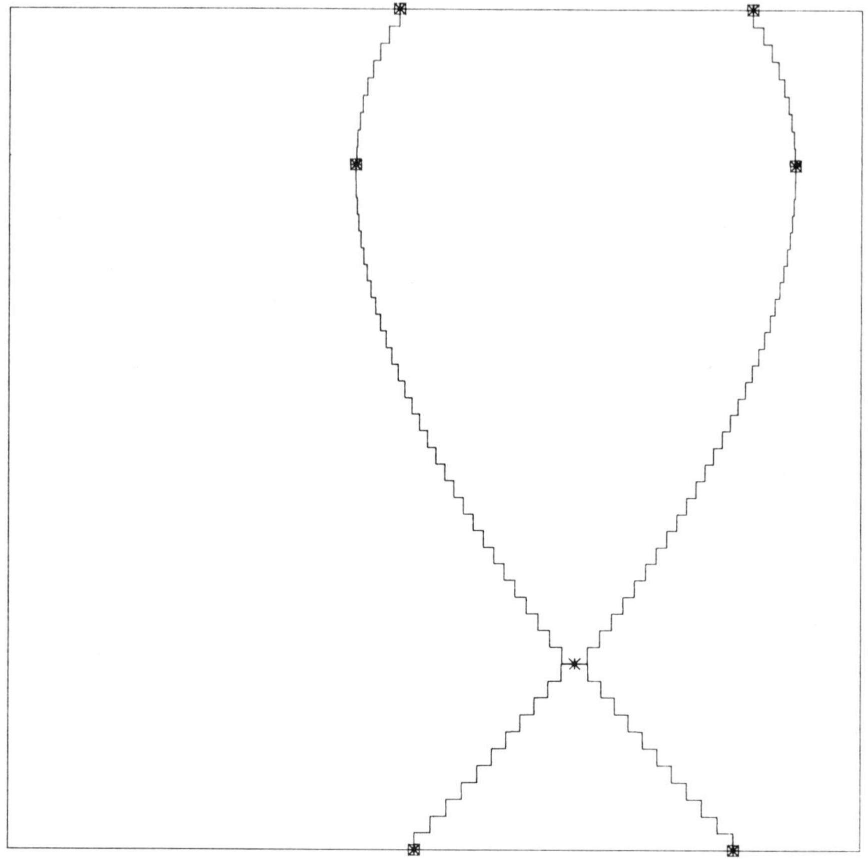

Figure 3. The characterization of the section curve analyzed in Example 2.

There are two linearly independent tangents: $(1, \pm 4/3)$, and the double point is therefore a self-intersection.

The following link multiplicities are assigned: 1 for each of the four border points; 2 for each of the two remaining u-turning points; and 4 for the self-intersection. The number of monotonic branches is thus $N_b = 6$.

Marching in small positive steps Δv from the two border points along $v = 0$, we reach the singular point; and marching from those along $v = 1$ in small negative steps Δv, we reach the two u-turning points. Finally we march from the two u-turning points in small negative steps Δv (i.e., the currently unliked directions) and reach the self-intersection twice. All link multiplicities are thus satisfied, and the curve characterization is complete (Fig. 3).

7. Concluding Remarks. While the elementary examples above are verifiable by hand, the viability of the techniques proposed here must be established in the context of floating point computations. For plane sections of bicubic patches, we find that characteristic points can be determined to 12 significant digits or better with double precision arithmetic. This is an acceptable level of accuracy, even for situations where the results are channeled to subsequent computational stages.

For plane sections of bicubic patches, the resultant (4) for determining the characteristic points is of degree 15. For quadric sections, the degree jumps to 66 and the stability and efficiency of the method is more questionable.

Consider next the intersection of two bicubic patches. If one patch is implicitized, it yields a degree 18 surface (Sederberg [24]). Substituting the parametric form for the other patch gives an algebraic curve $u^{54}v^{54} + \cdots$. The resultant for determining characteristic points is then a degree 5778 polynomial!

The current method is thus obviously inappropriate for practical implementation in the context of free-form parametric surface intersections. Numerical schemes, such as the B-spline subdivision methods, provide a more direct and rapid approximate solution in such instances.

It should be noted, however, that the high polynomial degrees encountered above are not simply artifacts of our formulation. They reflect the intrinsic potential complexity of the intersection and must be addressed at some stage if *guaranteed* accuracy and reliability are required. The notion that the intersection of two bicubic patches may possess up to 5778 turning points in each direction is a sobering theoretical insight.

For the family of simpler surfaces arising frequently in solid modeling (the quadrics, ruled surfaces, surfaces of revolution, and other elementary swept surfaces), however, the present method holds promise for systematic, reliable evaluation of intersections and forms the basis for a rigorous trimmed surface formalism:

Sectioning a finite parametric patch by an unbounded implicit surface provides a definite trimmed surface representation as the two distinct regions $F(u, v) > 0$ and $F(u, v) < 0$ within the given parameter domain. When two finite patches intersect the intersection curve may be obtained by implicitizing either one and substituting the parametric form for the other. However, this curve is not defined on an unambiguous parameter domain and does not alone furnish a trimmed surface representation.

When all patch-patch intersections for two solid primitives are considered, a trimmed surface patch may be formulated as those regions enclosed by a set of *piecewise algebraic loops*—closed loops in the parameter space of the patch which are composed piecewise of general monotonic algebraic curve segments. The formalization of these ideas and the provision of interrogation algorithms for such trimmed parametric surface entities offers scope for much future work.

REFERENCES

[1] D. S. ARNON, *Topologically reliable display of algebraic curves,* ACM Computer Graphics, 17 (1983), pp. 219–227.

[2] J. BLINN, *Computer display of curved surfaces,* Ph.D. thesis, Univ. Utah, Salt Lake City, UT, 1978.

[3] E. COHEN, T. LYCHE AND R. RIESENFELD, *Discrete B-splines and subdivision techniques in computer aided geometric design and computer graphics,* Computer Graphics and Image Processing, 14 (1980), pp. 87–111.

[4] J. L. COOLIDGE, *A Treatise on Algebraic Plane Curves*, Dover, New York, 1959.

[5] G. DAHLQUIST AND A. BJORCK, (trans. N. Anderson), *Numerical Methods*, Prentice-Hall, Englewood Cliffs, NJ, 1974.

[6] Y. DE MONTAUDOUIN AND W. TILLER, *Applications of power series in computational geometry*, preprint.

[7] R. T. FAROUKI, *The characterization of parametric surface sections*, Computer Vision, Graphics and Image Processing, 33 (1986), pp. 209–236.

[8] R. T. FAROUKI AND J. K. HINDS, *A hierarchy of geometric forms*, IEEE Computer Graphics and Applications, 5 (1985), pp. 51–78.

[9] A. D. GEISOW, *Surface interrogations*, Ph.D. thesis, University of East Anglia, Norwich, UK, 1983.

[10] R. N. GOLDMAN AND T. W. SEDERBERG, *Some applications of resultants to problems in computational geometry*, preprint.

[11] D. H. HOITSMA AND M. ROCHE, *The computation of all plane-surface intersections for* CAD/CAM *applications*, in Computer Aided Geometry Modeling, NASA C.P. 2272, 1983, pp. 15–18.

[12] E. G. HOUGHTON, R. F. EMNETT, J. D. FACTOR AND C. L. SABHARWAL, *Implementation of a divide-and-conquer method for intersection of parametric surfaces*, Computer Aided Geometric Design, 2 (1985), pp. 173–183.

[13] J. M. LANE AND R. F. RIESENFELD, *A theoretical development for the computer generation and display of piecewise polynomial surfaces*, IEEE Transactions on Pattern Analysis and Machine Intelligence, 2 (1980), pp. 35–46.

[14] ——, *Bounds on a polynomial*, Nordisk tidskrift for Informationsbehandling (D.I.T.), 21 (1981), pp. 112–117.

[15] R. B. LEE AND D. A. FREDERICKS, *Intersection of parametric surfaces and a plane*, IEEE Computer Graphics and Applications, 4 (1984), pp. 48–51.

[16] J. Z. LEVIN, *Mathematical models for determining the intersections of quadric surfaces*, Computer Graphics and Image Processing, 11 (1979), pp. 73–87.

[17] A. P. MORGAN, *A method for computing all solutions to systems of polynomial equations*, ACM Trans. Math. Software, 9 (1983), pp. 1–17.

[18] J. C. OWEN AND A. P. ROCKWOOD, *Intersection of general implicit surfaces*, this Volume, 1987.

[19] A. A. G. REQUICHA AND H. B. VOELCKER, *Solid modeling: A historical summary and contemporary assessment*, IEEE Computer Graphics and Applications, 2 (1982), pp. 9–24.

[20] ——, *Solid modeling: Current status and research directions*, IEEE Computer Graphics and Applications, 3 (1983), pp. 25–37.

[21] G. SALMON, *A Treatise on the Higher Plane Curves*, Hodges, Smith and Co., Dublin, 1879 (reprinted by Chelsea, New York).

[22] ——, *Lessons Introductory to the Modern Higher Algebra*, Hodges, Smith and Co., Dublin, 1885 (reprinted by Chelsea, New York).

[23] R. F. SARRAGA, *Algebraic methods for intersections of quadric surfaces in* GMSOLID, Computer Vision, Graphics and Image Processing, 22 (1983), pp. 222–238.

[24] T. W. SEDERBERG, *Implicit and parametric curves and surfaces for computer aided geometric design*, Ph.D. thesis, Purdue Univ., West Lafayette, IN, 1983.

[25] ——, *Algebraic geometry for surface and solid modeling*, this Volume, 1987.

[26] T. W. SEDERBERG, D. C. ANDERSON AND R. N. GOLDMAN, *Implicit representation of parametric curves and surfaces*, Computer Vision, Graphics and Image Processing, 28 (1984), pp. 72–84.

[27] J. G. SEMPLE AND G. T. KNEEBONE, *Algebraic Curves*, Oxford University Press, Oxford, 1959.

[28] D. M. Y. SOMMERVILLE, *Analytical Geometry of Three Dimensions*, Cambridge University Press, Cambridge, 1951.

[29] R. B. TILOVE AND A. A. G. REQUICHA, *Closure of Boolean operations on geometric entities*, Computer Aided Design, 12 (1980), pp. 219–220.

[30] T. VARADY, *Surface-surface intersections for double-quadratic parametric patches in a solid modeler*, Proc. U.K.–Hungarian Seminar, Computational Geometry for CAD/CAM, Cambridge University, 1983.

[31] R. J. WALKER, *Algebraic Curves*, Princeton University Press, Princeton, NJ, 1950 (reprinted by Dover Publications, New York).

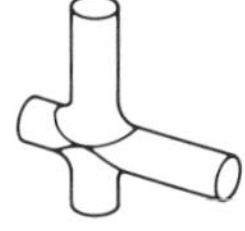

Intersection of General
Implicit Surfaces

JOHN C. OWEN AND ALYN P. ROCKWOOD

Abstract. An algorithm is described for obtaining all of the intersection points and curves between two general implicit surfaces in a region of space. The algorithm pays special attention to points and curves of tangency and to the question of numerical resolution. The algorithm has been implemented in the ROMULUS solid modeling system.

1. Introduction. In this paper we describe a computer algorithm that will find the intersections between two surfaces, which are defined in the implicit form

$$(1.1) \qquad\qquad f_1(\mathbf{r}) = 0, \qquad f_2(\mathbf{r}) = 0$$

where $\mathbf{r}$ is a point in a region of three-space.

Our interest in this problem arises from our work on introducing blending surfaces into the ROMULUS solid modeler (Rockwood and Owen [1]). The definition of the blending surfaces can result in fairly complicated, nonpolynomial forms for the implicit functions f_1 and f_2 and so we have been led to develop an algorithm which does not depend on the analytic form of these functions. Rather, the algorithm is cast in the form of "evaluators" so that the user of the algorithm is required to provide certain specified evaluations of the surface functions at points in space, but not their functional form.

A solid modeling program imposes exacting requirements on its surface-surface intersection algorithm. Foremost among these is the reliable handling of numerical accuracy and resolution particularly close to singular points and singular curves of the intersection. This is important in order to ensure that the topological structure of a solid body is always consistent with the structure implied by the geometry of its surfaces. We shall assume that we have a small resolution length scale, called RESABS, which is much smaller than the region in which the intersections are to be found but which is much larger than the precision to which real arithmetic can be performed. For example, in our implementation RESABS is 1.E-10 when the size of the surfaces is about unity and arithmetic precision is about 1.E-16. The requirements of the intersection algorithm can be summarized as follows:

(1)   Locate within RESABS all of the distinct intersection curves between the two surfaces in a given region of space. We will assume that this region is defined by a rectangular box which will be called the intersection box.

(2)   Locate curves and points at which the surfaces are tangent. These are the singular curves and singular points of the intersection. A singular point or curve should also be counted if the surfaces can be moved by less than RESABS and thereby give rise to the singular case. This allows the classification of singular intersections without appealing to arithmetic precision.

(3)   The algorithm must work in interactive time. It is difficult to be precise about what this means, except that a single surface intersection must take seconds rather than hours of computation time.

The algorithm described here will locate and classify all points on the intersection for which

(1)   The surface normals are not parallel and cannot be made parallel by moving the surfaces less than RESABS, i.e., nonsingular points

(2)   The surface normals are parallel but there is at least one direction along which the surface curvatures are not the same. These are points of first order contact. The algorithm will fail for points at which the surfaces are tangent and have the same curvatures in all directions.

We stress that the ability to handle first order tangencies is vitally important in solid modeling. Any pair of surfaces can always be oriented so that they touch and thus have points of first order contact. Designers seem to be attracted to these special configurations! On the other hand, higher order contacts can only be generated by specially chosen pairs of surfaces. In our application for blending surfaces, for example, higher order tangencies occur mostly in symmetric configurations which we treat by special case analysis.

The intersection algorithm falls into two parts. In the first part, three space is subdivided successively into rectangular boxes down to a level which is determined by the minimum of a characteristic length for each of the surfaces. This length must be supplied by the user of the algorithm. Its specification is discussed further below but might typically be the smallest radius of curvature on the surface. At each level of subdivision, only boxes which are known or believed to contain both surfaces are divided further. The user is required to make this determination. In this way a tree is built whose leaf nodes are boxes which contain, or are close to, intersections of the surfaces.

In the second part of the algorithm the exact location and structure of the intersections is obtained by making local expansions of the surface functions starting from the centers of the leaf node boxes. If the surface function gradients are not close to parallel then linear expansion and iteration is sufficient. If the surface gradients approach parallel, then the local expansion is carried up to quadratic order. This is the novel part of the algorithm. It gives a picture of the intersection curves as conics in the mean tangent plane of the two surfaces and allows the location of first order singular points and curves. It also allows the resolution of pairs of closely spaced curves due to near tangency of the two surfaces, thereby allowing the resolution of separate curves which are much more closely spaced than the resolution of the subdivision. This part of the algorithm requires the user to evaluate the surface function, its gradient vector and its second derivative matrix for given points in space.

In the next section we describe our local expansion method. In § 3 we discuss the subdivision method and give a description of the overall algorithm.

2. Description of First Order Singularities by Quadratic Expansion. Suppose that $\mathbf{x}$ is a point on or close to the intersection of the two surfaces which are described by $f_1 = 0$ and $f_2 = 0$. Let us suppose also that the surfaces are nonsingular everywhere in the region of interest so that the gradients of f_1 and f_2 are not of zero length.

Now consider a small displacement $\mathbf{dx}$ to a second point which is on the intersection of the two surfaces. Expanding the surfaces up to second order gives the following conditions for $\mathbf{x} + \mathbf{dx}$ to be on the intersection:

$$(2.0) \quad \begin{aligned} f_1 + \mathbf{dx} \cdot \nabla(f_1) + 0.5 * \mathbf{dx} \cdot \nabla^2(f_1) \cdot \mathbf{dx} + o(\mathbf{dx}^3) = 0, \\ f_2 + \mathbf{dx} \cdot \nabla(f_2) + 0.5 * \mathbf{dx} \cdot \nabla^2(f_2) \cdot \mathbf{dx} + o(\mathbf{dx}^3) = 0 \end{aligned}$$

where $\nabla(f_1)$ is the first derivative vector and $\nabla^2(f_1)$ is the second derivative matrix of f_1. Both are evaluated at the point $\mathbf{x}$.

If $\nabla(f_1)$ and $\nabla(f_2)$ are linearly independent, then the quadratic terms can be neglected and we have the simple local picture of a straight line intersection curve. Iteration to the intersection curve is a simple Newton–Raphson scheme.

On the other hand, if $\nabla(f_1)$ and $\nabla(f_2)$ are linearly dependent, or nearly so, then $\mathbf{x}$ is close to a singular point or singular curve of the intersection and the quadratic terms are important.

The singularity in the system can be isolated into one equation by dividing each equation by the length of the gradient of its function and then forming the sum and difference equations

$$(2.1) \quad \begin{aligned} (f_1/Gf_1 + f_2/Gf_2) + \mathbf{dx} \cdot (\nabla(f_1)/Gf_1 + \nabla(f_2)/Gf_2) \\ + 0.5 * \mathbf{dx} \cdot (\nabla^2(f_1)/Gf_1 + \nabla^2(f_2)/Gf_2) \cdot \mathbf{dx} = 0, \end{aligned}$$

$$(2.2) \quad \begin{aligned} (f_1/Gf_1 - f_2/Gf_2) + \mathbf{dx} \cdot (\nabla(f_1)/Gf_1 - \nabla(f_2)/Gf_2) \\ + 0.5 * \mathbf{dx} \cdot (\nabla^2(f_1)/Gf_1 - \nabla^2(f_2)/Gf_2) \cdot \mathbf{dx} = 0 \end{aligned}$$

where Gf_1 and Gf_2 are the lengths of the vectors $\nabla(f_1)$ and $\nabla(f_2)$.

Now suppose that $\nabla(f_1)$ and $\nabla(f_2)$ are in the same direction. Then the linear term in (2.1) has a nonzero coefficient and the quadratic terms can be neglected. In (2.2) however, the linear term becomes zero as $\nabla(f_1)$ and $\nabla(f_2)$ tend to parallel and the quadratic terms give the leading order behavior. Thus close to singularities the intersection curve is described by the following equations:

$$(2.3) \quad (f_1/Gf_1 + f_2/Gf_2) + \mathbf{dx} \cdot (\nabla(f_1)/Gf_1 + \nabla(f_2)/Gf_2) = 0,$$

$$(2.4) \quad \begin{aligned} (f_1/Gf_1 - f_2/Gf_2) + \mathbf{dx} \cdot (\nabla(f_1)/Gf_1 - \nabla(f_2)/Gf_2) \\ + 0.5 * \mathbf{dx} \cdot (\nabla^2(f_1)/Gf_1 - \nabla^2(f_2)/Gf_2) \cdot \mathbf{dx} = 0. \end{aligned}$$

The vector $(\nabla(f_1)/Gf_1 + \nabla(f_2)/Gf_2)$ is the normal to the mean of the tangent planes to the two surfaces. The component of $\mathbf{dx}$ in this direction is determined by (2.3). The vector $(\nabla(f_1)/Gf_1 - \nabla(f_2)/Gf_2)$ is orthogonal to $(\nabla(f_1)/Gf_1 + \nabla(f_2)/Gf_2)$ and so (2.4) describes the structure of the intersection as a conic in the mean tangent plane.

The singular points and curves of the intersection are identified with the degenerate conic cases of coincident lines (tangent curve) and intersecting lines (singular point).

Before discussing the various conics in more detail we make three remarks:

(1) Equations (2.3) and (2.4) are derived by an expansion in $\mathbf{dx}$. Thus the conic representation of the intersection given by (2.4) is valid only in the region where the length of $\mathbf{dx}$ is small enough that the quadratic term in (2.1) and the neglected cubic and higher order terms in the original expansion can be neglected. It is also necessary that both the linear and quadratic terms in (2.4) are not zero. The neglected higher order terms are small provided $\mathbf{dx}$ is small compared with the curvatures of both the surfaces. The quadratic term in (2.4) is nonzero provided that the surface curvatures are not the same in all directions.

(2) The expression f/Gf gives the first order approximation to the distance of the point $\mathbf{x}$ from the surface. When the surfaces become tangent, the constant term $(f_1/Gf_1 - f_2/Gf_2)$ is approximately the distance between the two surfaces. This means that the value of this term can be changed by RESABS by moving either of the surfaces by RESABS. Thus, configurations in the surface-surface intersection which are within RESABS of being singular are identified by checking if the constant term in the conic equation is within RESABS of a singular case.

The constant term in (2.4) can be interpreted as the distance between the surfaces because we took a particular linear combination of the surface equations. The classical method of forming pencils of intersection curves between implicit surfaces by taking arbitrary linear combinations (Sommerville [2]; Levin [3]) does not provide a basis for reliable computer algorithms because this distance measure is thereby lost.

(3) The conic approximation may show two disconnected pieces of curve or two pieces which could be disconnected inside the intersection box, such as a very long thin ellipse. Separate pieces of curve which are much closer than the initial bisection resolution can be resolved by using the conic expansion to provide a "help point" on a nearby piece of curve.

We now return to a more detailed discussion of the conic solutions to (2.3) and (2.4). Expand $\mathbf{dx}$ in three orthogonal vectors

$$(2.5) \qquad\qquad \mathbf{dx} = n\mathbf{N} + u\mathbf{U} + v\mathbf{V}$$

where $\mathbf{N}$ is the normal to the mean tangent plane

$$(2.6) \qquad\qquad \mathbf{N} = (\nabla(f_1)/Gf_1 + \nabla(f_2)/Gf_2).$$

The two remaining orthogonal vectors $\mathbf{U}$ and $\mathbf{V}$ lie in the mean tangent plane. They can always be chosen so that there is no term from (2.4) which couples $\mathbf{U}$ and $\mathbf{V}$. That is, they are chosen so that $\mathbf{U} \cdot (\nabla^2)(f_1)/Gf_1 - \nabla^2(f_2)/Gf_2) \cdot \mathbf{V}$ is zero.

The coefficient n is determined by (2.3) as

(2.7)
$$n = -(f_1/Gf_1 + f_2/Gf_2)/(2 + 2\mathbf{\nabla}(f_1) \cdot \mathbf{\nabla}(f_2)/(Gf_1 Gf_2))$$

and (2.4) gives a conic equation for u and v as

$$au^2 + bv^2 + cu + dv + e = 0$$

where

(2.8)
$$\begin{aligned}
e &= f_1/Gf_1 - f_2/Gf_2 + 0.5n^2 \mathbf{N} \cdot C \cdot \mathbf{N}, \\
d &= (\mathbf{\nabla}(f_1)/Gf_1 - \mathbf{\nabla}(f_2)/Gf_2) \cdot \mathbf{V} + n\mathbf{N} \cdot C \cdot \mathbf{V}, \\
c &= (\mathbf{\nabla}(f_1)/Gf_1 - \mathbf{\nabla}(f_2)/Gf_2) \cdot \mathbf{U} + n\mathbf{N} \cdot C \cdot \mathbf{U}, \\
b &= (\mathbf{V} \cdot C \cdot \mathbf{V})/2, \\
a &= (\mathbf{U} \cdot C \cdot \mathbf{U})/2,
\end{aligned}$$

and

$$C = \mathbf{\nabla}^2(f_1)/Gf_1 - \mathbf{\nabla}^2(f_2)/Gf_2.$$

Of course, if the point $\mathbf{x}$ is on the intersection then f_1, f_2, n and e are all zero. The standard theory of conics (Sommerville [2]) now permits the classification of the local structure of the intersection.

(1)   *A single straight line.* For example two intersecting planes. Both the quadratic terms are zero, but at least one of the linear terms is not.

(2)   *A pair of distinct straight lines.* For example two intersecting cylinders with parallel axes. Both a and c (or b and d) are zero and

$$\text{abs}\,(e - (d^2/b)/4) > \text{RESABS}.$$

If $\mathbf{x}$ is on one of the lines, then it is easy to obtain a help point on the other line.

(3)   *A pair of coincident straight lines.* For example two cylinders with parallel axes separated by the sum of their radii. Both a and c (or b and d) are zero and

$$\text{abs}\,(e - (d^2/b)/4 < \text{RESABS}.$$

This means that there is an intersection along a line tangency.

(4)   *A nondegenerate hyperbola.* For example, the configuration shown in Fig. 1. Both a and b are nonzero with opposite signs and

$$\text{abs}\,(e - (c^2/a + d^2/b)/4) > \text{RESABS}.$$

If the two branches of the hyperbola come close then we can obtain a help point on the second branch of the hyperbola.

(5)   *A degenerate hyperbola.*   For example the configuration shown in Fig. 2. This occurs when a and b have opposite signs and

$$\text{abs}\,(e - (c^2/a + d^2/b)/4) < \text{RESABS}.$$

The point where the two lines cross is a singular point. The precise location of the singular point of the intersection is found by iterating the expansion discussed here until the lines cross at the origin in the (u, v) plane. Notice that the conditions for

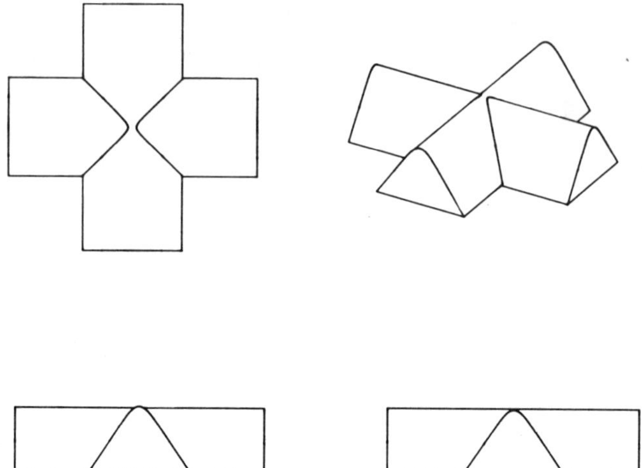

Figure 1. A near singular intersection of two blending surfaces. The figure shows orthographic views of a solid body containing the surfaces with hidden lines removed.

singularity is, with respect to RESABS, the given length resolution, and not zero. Help points on other segments of the degenerate hyperbola can be recorded if need be.

(6) *An ellipse or circle. a* and *b* have the same sign and

$$\text{Sign } (a)(e - (c^2/a + d^2/b)/4) < \text{RESABS}.$$

Although the intersection looks locally like a single closed curve, it may still be useful to obtain a help point on the opposite side of the curve. For example, a curve which looks

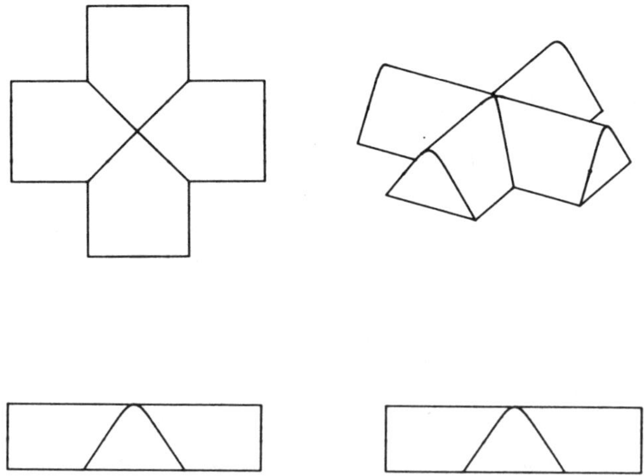

Figure 2. A singular intersection of the same surfaces as in Fig. 1.

locally like a long thin ellipse may be split by the intersection box or may not be closed as a global curve.

(7) *A parabola.* The same remarks apply as for the ellipse if the parabola is long and thin.

(8) *A single point.* a and b have the same sign and

$$\text{abs} \, (e - (c^2/a + d^2/b)/4) < \text{RESABS}.$$

Notice again that the distinction between point contact and a small elliptical intersection curve involves a comparison with RESABS. The precise location of the point contact can be determined by iterating the expansion onto the point.

(9) *An imaginary conic.* a and b have the same sign and

$$\text{Sign} \, (a)(e - (c^2/a + d^2/b)/4) > \text{RESABS}.$$

There is no intersection near the point of the expansion.

(10) *Undefined.* a, b, c and d are all zero. The intersection is not defined in this order of expansion. The surface normals are parallel and the curvatures are identical. The intersection cannot be located accurately.

3. Subdivision of Space.   Help points for the precise location of the intersection curves are obtained by successive bisection of three-space down to some predetermined size which will be called DELTA. The algorithm is supplied with a rectangular box in which the intersections are to be found. The longest side of the box which results at any level in the bisection tree is bisected to get the successor level and this is continued until the longest box side is smaller than DELTA. This method means that the boxes tend towards a cubic shape. For our purposes this was found to be preferable to an octree approach where all three box axes are divided simultaneously and the leaf node boxes always have the same shape as the original box.

The bisection is also terminated if, at any level, it can be determined with certainty that the box does not contain the intersection of the two surfaces. A sufficient (though not necessary) condition is that the box is known with certainty not to contain at least one of the surfaces.

The user of the algorithm is required to provide a test which will determine whether a surface passes through a box with the following properties:

(1) If the surface does pass through the box, the test always returns true.

(2) If the surface does not pass through the box, the test should, where possible, return false. The likelihood of the test returning false should tend to certainty as the longest box size shrinks to zero about a fixed center.

The size of the final bisection tree and efficiency of the algorithm is dependent on how accurately the surface through box test is performed. If the size of the original intersection box is L the number of leaf nodes for nontangent intersections will be of order (L/DELTA). Without the surface through box test the number would be order $(L/\text{DELTA})^3$.

In our current implementation, the surface through box test is made by estimating a lower bound to the distance from the center of the box to the surface and comparing this with the longest diagonal of the box.

It is possible that a much more general and efficient method can be obtained using the method of interval arithmetic (Moore [4]). The test box is regarded as a vector interval and the functions f_1 and f_2 are evaluated using interval arithmetic to yield scalar intervals. If these intervals do not contain zero then the surface does not pass through the box. The methods of interval arithmetic can be applied to a wide class of functions (not just polynomials) and can be shown to converge in the sense of point 2 above. This would give a very general algorithm in which the user is required only to make function evaluations for points and vector intervals, to provide a bisection resolution DELTA and nothing else. We are currently experimenting with the use of interval arithmetic.

Once the bisection tree has been built, the algorithm proceeds in the following stages:

(1) Select any leaf node of the tree. Relax the center of the box onto the intersection curve, using the conic expansion method discussed above if necessary.

(2) Form a list of secondary help points which may lie on disjoint but closely spaced pieces of intersection curve using the conic expansion method.

(3) Starting at the relaxed point "march out" the segment of intersection curve in a forward then a backward direction until it leaves the intersection box or returns to the start point. "Marching" means generating an ordered sequence of points on the curve by successively moving a short distance along the tangent to the curve and then relaxing back onto the curve. If necessary the relaxation can be done using the conic expansion. The conic expansion is also used to locate any singular points which may be encountered on the curve.

(4) All of the leaf nodes through which the intersection curve passes (or may possibly pass) are marked in the bisection tree. Any of the secondary help points which are encountered on the march are removed from the list.

(5) The intersection curves of any remaining secondary help points are marched out as in the preceding step.

(6) Any remaining unmarked leaf node in the bisection tree is selected and its center relaxed onto the intersection. If the relaxed point is also in an unmarked leaf node a new intersection curve has been found and it is treated as from step (2) above. Otherwise another unmarked cell is selected until there are none left.

This algorithm will locate each disjoint segment of curve of the intersection of two surfaces exactly once.

3.1. Specification of DELTA. The only part of the algorithm remaining to be discussed is the definition of DELTA, the resolution for the bisection. The conic expansion allows the resolution of closely spaced curves, but only if they are close because the surfaces are nearly tangent. We shall call this "quadratically related." The bisection must be taken to such a level that, for each disjoint piece of intersection (or quadratically related pair of intersections) there is at least one leaf node that contains only that intersection and whose center will relax onto that intersection.

In our implementation we associate a length scale DELTA separately with each surface and take the bisection resolution DELTA to be the minimum of the two. The DELTA for each surface is the smallest length scale over which the surface deviates

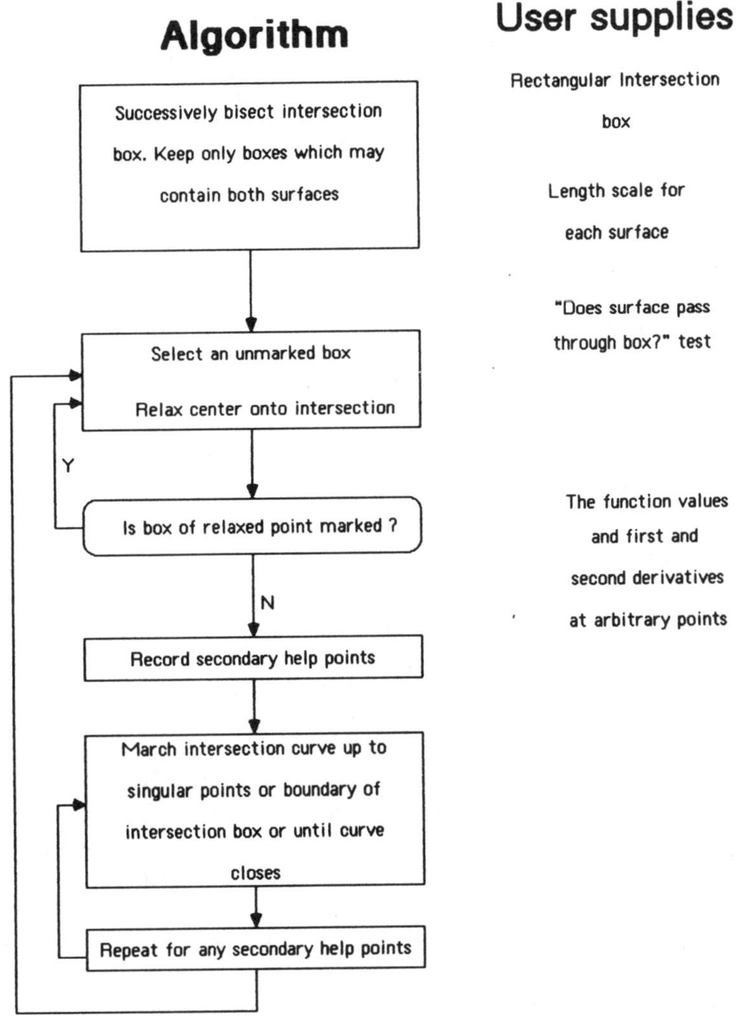

Figure 3.   A flow chart of the intersection algorithm.

significantly from planarity. This is obviously a global property of the surface and can only be determined from a knowledge of the global structure of the surface. For example, for a simple surface such as a sphere or cylinder it is a fraction of the radius of curvature. For a piece of corrugated iron it would be a fraction of the distance between the corrugations. For blending surfaces we find it satisfactory to use half of the range or radius of the blend.

Figure 3 shows a flow chart of the algorithm.

4.   Conclusion.   The algorithm described here has been implemented to intersect the blending surfaces described elsewhere (Rockwood and Owen [1]) in the

Figure 4.   The intersection of solid models made from blending surfaces in ROMULUS.

ROMULUS solid modeler. It has proved to be reliable and efficient. Figure 4 shows a hidden line picture of a wine glass built from blending surfaces intersected with a rotated copy of itself. Notice the singular points where intersection curves cross.

The time taken to do an intersection depends upon the type of surface, the size of the intersection box and the value of the bisection resolution DELTA. It does not depend significantly on the value of RESABS since all of the relaxations (even near singular points) are quadratically convergent. Typically a surface-surface intersection will take a few seconds on a VAX 11/780. Most of this time is spent in marching out the intersections once the bisection tree has been built. We find that when doing a Boolean operation between bodies with blending surfaces, only a small fraction of the computation time is spent in the surface-surface intersection routine.

There are a number of obvious extensions of our algorithm. We use it to find silhouette curves on blending surfaces simply by expressing the silhouette as the intersection of the surface with its silhouette surface.

The algorithm can also be used to intersect a parametric with an implicit surface by substituting the parametric definition for a point into the implicit surface equation and doing both the bisection and the conic expansion in the parameter space. The algorithm can also be extended to intersect two parametric surfaces.

The main weakness of the algorithm is that it will not handle surface intersections with more than simple tangencies. One way to get round this limitation is to continue the bisection down to RESABS when a point of high order contact is detected. This can be expected to leave a sea of boxes clustered around the singular point. The singular point can then be located at the center of the sea. We are currently experimenting with this idea.

REFERENCES

[1] A. P. ROCKWOOD AND J. C. OWEN, *Blending surfaces in solid modeling,* this Volume, 1987.

[2] D. M. Y. SOMMERVILLE, *Analytic Geometry of Three Dimensions,* Cambridge University Press, Cambridge, 1934.

[3] J. Z. LEVIN, *A parametric algorithm for drawing pictures of solid objects composed of quadric surfaces,* Commun. ACM, 19 (1976), pp. 555–563.

[4] R. E. MOORE, *Interval Analysis,* Prentice-Hall, Englewood Cliffs, NJ, 1966.

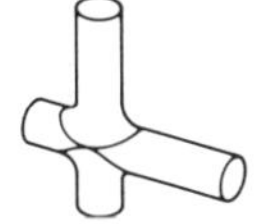

The Potential Method for
Blending Surfaces and Corners

CHRISTOPH HOFFMANN AND JOHN HOPCROFT

Abstract. The potential method for blending implicit algebraic surfaces is summarized, summarizing and extending work previously reported. The method is capable of deriving blends for pairs of algebraic surfaces, and is guaranteed to produce blending surfaces of lowest possible degree for two quadrics in general position.

Two paradigms are given by which to understand the method. The first paradigm views the blends as surfaces swept out by a family of space curves. The second, more general paradigm considers the surfaces as a result of deformation of a parameter space effected by substitution. The method has a general formulation based on projective parameter spaces.

The deformation by substitution paradigm is extended to blend blending surfaces at solid vertices without a degree penalty, under the assumption that the vertex valence has been reduced to three. It may also lead to a general solution for blending patches of algebraic surfaces that meet tangentially. A special case of this problem is solved and illustrated.

1. Introduction. Mechanical parts have primary surfaces whose shapes are functionally important and secondary surfaces whose purpose it is to smoothly connect the functional, primary surfaces. Usually, the secondary surfaces are only approximately specified in the engineering drawings of the part, and their exact shape, within bounds, is irrelevant. Surfaces of this type are called *blending surfaces* or *blends,* and we consider here how to derive them in a simple and systematic manner.

Specifying a blend for a computer based geometric modeling system should be simple; after all, the surfaces need to conform only approximately to the designer's shape requirements. Yet this is not the case. The principal difficulty is in shaping and positioning these surfaces so as to achieve tangency to the primary surfaces. As a consequence, blending surfaces have a higher algebraic degree, and are mathematically more complicated to derive than the primary surfaces they connect. Thus, a long term goal of our research into the existence and properties of blending surfaces is automating the derivation of blending surfaces, by computer, from the adjacent primary surfaces and a few, spatially intuitive parameters such as width and approximate curvature.

This paper surveys the mathematical aspects of deriving blending surfaces, given the primary surfaces to which tangency is to be achieved. Throughout, the surfaces are assumed to be algebraic, and are specified by their implicit equation, i.e., by an equation $F = 0$, where F is a polynomial in x, y, and z. Specifically, we address the *potential* method, introduced in Hoffmann and Hopcroft [1], [2], which has the advantage of deriving, in a very simple manner, a rather extensive and flexible class of blending surfaces of low degree. Other work on blending implicit algebraic surfaces is described in Middleditch and Sears [3] and in Rockwood and Owen [4], and these approaches are related to the potential method in § 3.

Although the blending problem has been formulated in terms of surfaces, the techniques can be integrated into a constructive solid geometry (CSG) based modeling system. Even Rockwood and Owen [4], reporting on work intended for a new version of the boundary representation based modeling system ROMULUS, was first experimentally tested in a CSG based modeler according to A. Rockwood.

Throughout our work, we stress the importance of obtaining blending surfaces of low degree. This is a practical consideration. Both the size of the surface representation, as well as the difficulty encountered by e.g., root finding algorithms, grow quickly with increasing degree. Fortunately, we not only derive low degree blends, we can also show rigorously that they are of the lowest degree possible for quadratic surfaces in general position.

This paper describes work in progress. Consequently, a number of results have a preliminary character, and topics in need of further investigation are indicated throughout. We have attempted to portray the material in as intuitive a way as possible, hoping to make it accessible to a large audience. The Uniqueness Theorem has a very technical proof that draws on algebraic geometry and the theory of ideals, and is omitted here. The interested reader is referred to Hoffmann and Hopcroft [2] for a complete derivation. However, many of the technical aspects are readily accessible to the nonspecialist, and where this is the case, we did not hesitate to go into full detail.

Sections 2–4 deal with blending two surfaces. Although written for intersecting surfaces, one can also blend nonintersecting surfaces. In the case of quadrics, the corollary to the Uniqueness Theorem (§ 4) explains how this can be done. Sections 5–7 deal with blending corners and patched algebraic surfaces, a comparatively less developed area. Here the main point is that special geometric properties of quadrics can be lifted to higher degree algebraic surfaces, by a simple intuitive approach.

2. Blending Two Intersecting Surfaces.   We explain a method for blending two intersecting algebraic surfaces, initially developed in Hoffmann and Hopcroft [1], called the *potential method*. All polynomials are assumed to be in x, y and z, with real coefficients, unless stated otherwise. To avoid confusion, we distinguish between the polynomial F and the surface $S(F)$ whose implicit equation is $F = 0$. In general, the intersection of two surfaces $S(G)$ and $S(H)$ is a space curve denoted $S(G, H)$, and the intersection of three surfaces is a set of points denoted $S(G, H, K)$. Note that the point $(1, 2, 3)$ can also be written as the intersection of three planes $S(x - 1, y - 2, z - 3)$.

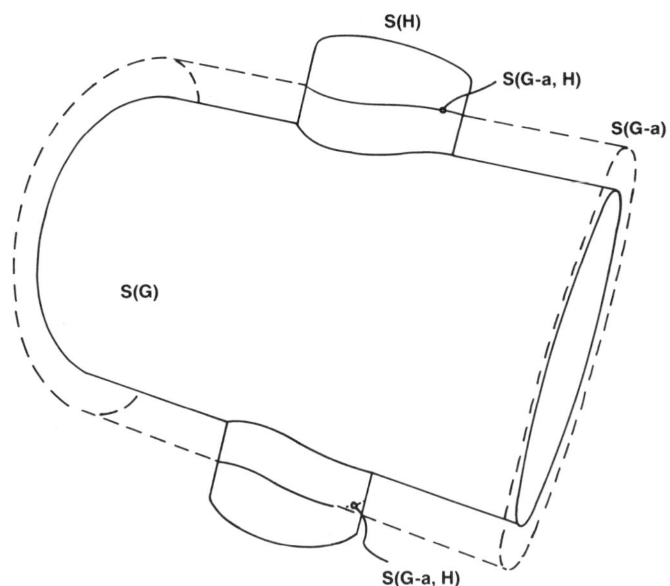

Figure 2.1.   The curve $S(G - a, H)$ where the blending surface meets $S(H)$ tangentially.

A surface $S(F)$ partitions space into three sets: all points (a, b, c) such that $F(a, b, c) > 0$, called the *outside* of $S(F)$; all points such that $F(a, b, c) < 0$, called the *inside* of $S(F)$; and all points such that $F(a, b, c) = 0$, i.e., $S(F)$. Note that in general the inside of $S(F)$ is equal to the outside of $S(-F)$, and vice versa.

2.1.   The affine method.   Given the surface $S(G)$, we define a family of surfaces, parameterized by s, via $G - s = 0$. For a particular value of $s \neq 0$, $S(G - s)$ is always entirely on the inside or the outside of $S(G)$, depending on the sign of s. For example, consider a circular cylinder $S(G)$. Then G may always be chosen such that for positive s, $S(G - s)$ is a circular cylinder of larger radius. In a like manner the family $S(H - t)$ of surfaces based on the surface $S(H)$ is defined. For our example we pick $S(H)$ as another cylinder. A maximum radius is fixed by picking a constant a and intersecting $S(G - a)$ with $S(H)$, as shown in Fig. 2.1. We choose a such that $S(G - a)$ intersects $S(H)$ in a nondegenerate space curve $S(G - a, H)$. The space curve $S(G - a, H)$ is associated with the point $(a, 0)$ in s-t parameter space. Now reduce the value of s from a to 0, while simultaneously increasing the value of t from 0 to some other constant b. Each intermediate pair (u, v) of s-t values corresponds to a space curve $S(G - u, H - v)$. If the values for s and t lie on a curve $f(s, t) = 0$, then the corresponding space curves lie on a surface whose equation is $F(x, y, z) = f(G, H) = 0$. Figure 2.2 shows the correspondence of an arc of the curve f to a segment of the surface $S(F)$. Here the radius of $S(G)$ is 8, and the radius of $S(H)$ is 4.

As shown in Hoffmann and Hopcroft [1], if $f = 0$ is tangent to the s-axis at $(a, 0)$ then $S(F)$ is tangent to $S(H)$ in the curve $S(G - a, H)$. Likewise, tangency of $f = 0$ to

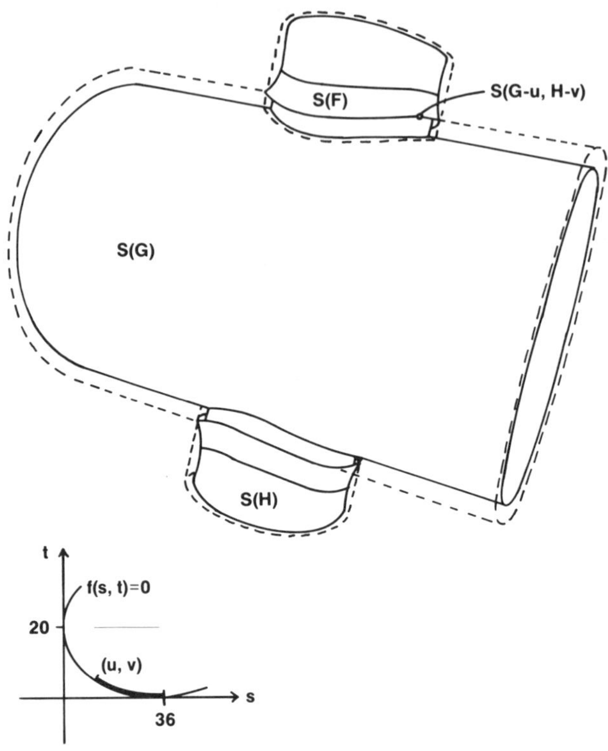

Figure 2.2. Correspondence of an arc of the parameter curve to a part of the blending surface.

the t-axis at $(0, b)$ implies tangency of $S(F)$ to $S(G)$ in $S(G, H - b)$. Higher order continuity of f with the coordinate axes gives higher order continuity of the surfaces.

Essentially, the procedure just described is the simple version of the potential method and we call it the *affine* potential method. It is not fully general, as we shall see below.

Although all subsequent examples concentrate on blending quadrics, the method applies to blending arbitrary algebraic surfaces. However, for higher degree surfaces the intrinsic surface geometry is more complicated. For example, $S(G - a)$ may split into components. While the method is robust and very intuitive for many surfaces, and especially for all quadrics, much exploration of its general behavior needs to be done.

2.2. Significance of the parameters when blending with a conic. The above procedure does not depend on the degrees of G, H and f, but as we are interested in low degree blending surfaces, we choose f of as low a degree as possible, i.e., as a conic. With the required tangency conditions f can be written as

$$f(s, t) = b^2s^2 + a^2t^2 + a^2b^2 - 2ab^2s - 2a^2bt + 2\lambda st.$$

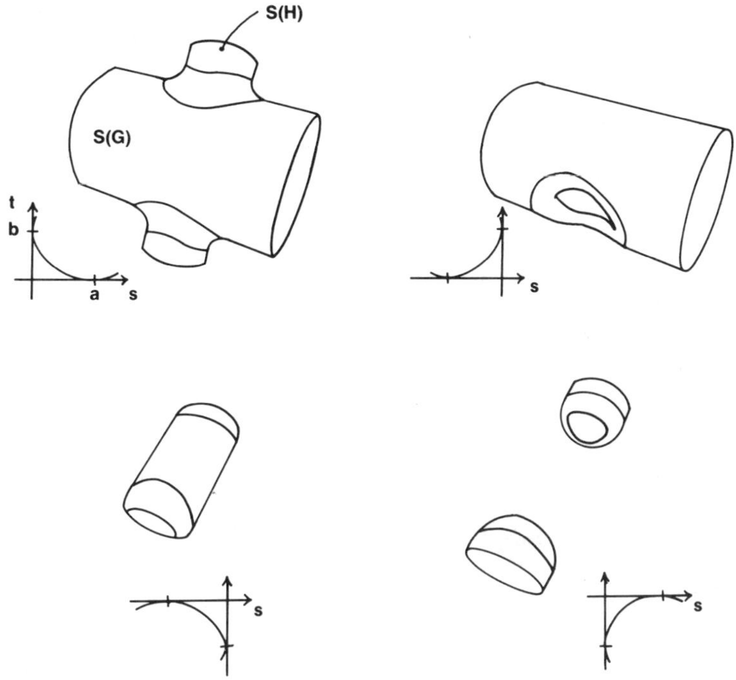

Figure 2.3.   Positional correspondence of the blending surface and the parameter curve.

The corresponding blending surface $S(F)$ is seen to possess degree $\max(2m, 2n)$, assuming that G has degree m and H has degree n. For quadrics, therefore, we obtain quartic blending surfaces.

In using a conic for blending, the points of tangency may be positioned by choosing a and b. Note that the signs of a and b determine in which quadrant the conic lies. The choice of λ determines the actual conic. We now explain how these choices affect the blend using as an example two circular cylinders whose axes intersect at right angles.

If a is positive, then $S(G - s)$ is on the outside of $S(G)$. Accordingly, the surface $S(F)$ is on the outside of $S(G)$. If a is negative, however, $S(F)$ must lie on the inside of $S(G)$. Similarly, $S(F)$ must lie on the outside or the inside of $S(H)$ depending on whether b is positive or negative. The quadrant positions of f, and the corresponding blending surface positions are illustrated in Fig. 2.3.

Moving $(a, 0)$ further away from the origin moves the curve of tangency $S(G - a, H)$ further away from the intersection curve $S(G, H)$ of the surfaces being blended. Similarly, the curves $S(G, H)$ and $S(G, H - b)$ are further apart when the magnitude of b is enlarged. There is no simple relationship between the magnitude of, say, a and the (mean) Euclidean distance of $S(G - a, H)$ to $S(G, H)$. In the case of quadrics, Middleditch and Sears [3] give a method for the affine potential method.

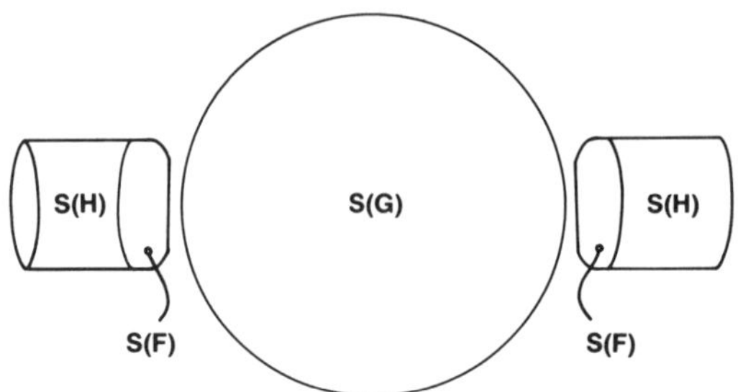

Figure 2.4.   Failure to blend due to improper choice of a and b.

For certain (s, t)-values the space curve $S(G - s, H - t)$ may degenerate to a point or vanish in real three-space. Accordingly, $S(F)$ may *miss* one of the surfaces or be disconnected in real three-space, although the latter phenomenon does not happen for intersecting quadrics. For example, consider the two cylinders $G = x^2 + y^2 - 8^2 = 0$ and $H = y^2 + z^2 - 4^2 = 0$. We choose $a = 57$ and $b = -34$, and blend with $\lambda = 0$. The resulting surface is shown in Fig. 2.4, where the gap between the blending surface and $S(G)$ is approximately 0.5. The problem here is that $S(H - b)$ does not have real points, and so does not intersect $S(G)$ in a space curve. In the case of blending quadrics it is easy to avoid this situation, but little is known about it in the case of blending higher order surfaces.

The type of conic chosen for f is determined by λ. The important values and the resulting curve shapes are summarized below and illustrated in Fig. 2.5:

$$\lambda = -\infty, \qquad \text{a pair of lines, } s = 0 \text{ and } t = 0,$$
$$-\infty < \lambda < -ab, \qquad \text{hyperbola,}$$
$$\lambda = -ab, \qquad \text{parabola,}$$
$$-ab < \lambda < ab, \qquad \text{ellipse; a circle if } a = b \text{ and } \lambda = 0,$$
$$\lambda = ab, \qquad \text{the line } bs + at - ab, \text{ counted double.}$$

Figures 2.6, 2.7 and 2.8 (see color insert) show the shape of the resulting blends for $G = y^2 + z^2 - 9$ and $H = x^2 + y^2 - 1$ with $a = 7$ and $b = 3$. In Fig. 2.6, λ is 20, just a little under the critical value of ab at which the surface would degenerate into the (ellipsoid) $S(bG + aH - ab)$. In Fig. 2.7, λ is 0 and in Fig. 2.8, λ is -750 and thus its magnitude is large compared to ab. The blending surface visibly begins to approximate the other degeneracy, namely the union of the two cylinders. We see that λ controls the distribution of curvature of the cross-sections of the blend.

Only a portion of the surface $S(F)$ is used in blending; the portion of $S(F)$ that corresponds to the arc of $f = 0$ lying on the inside of the line $bs + at - ab = 0$. This is the part of $S(F)$ that lies on the inside of the surface $S(bG + aH - ab)$, an ellipsoid in our example. In the illustrations all blending surfaces have been clipped accordingly.

2.3.   Parameter space.   We have conceptualized the blending surface as swept out by curves of intersection of two families of surfaces, controlled by a curve in

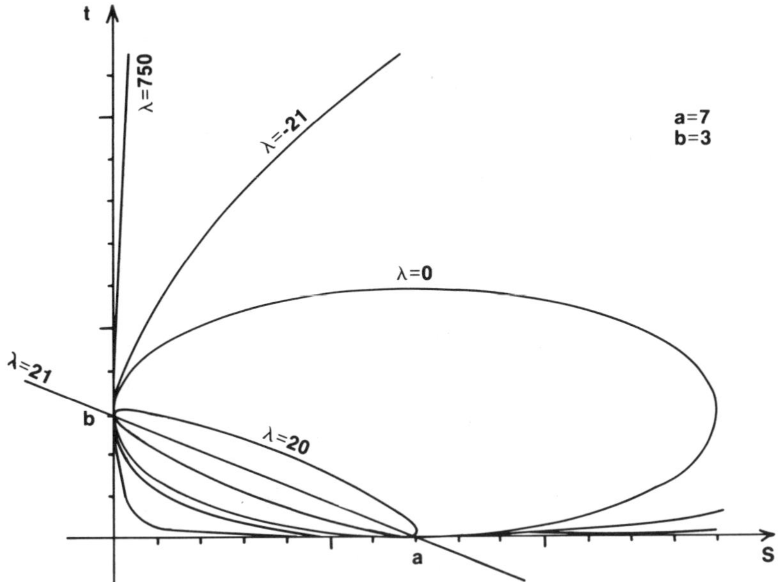

Figure 2.5.   Family of quadric parameter curves for controlling the curvature of the blending surface.

two-dimensional parameter space. A different conceptualization is possible by considering a three-dimensional parameter space. The curve f is now replaced by a conic cylinder. Note that this cylinder is a blending surface for the planes $S(s)$ and $S(t)$, as shown in Fig. 2.9.

The Cartesian coordinate system of this parameter space is based on the three principal planes, $S(r)$, $S(s)$, and $S(t)$, that intersect pairwise in the three coordinate axes. Every line parallel to the plane $S(r)$ is the intersection of the planes $S(s - u)$ and

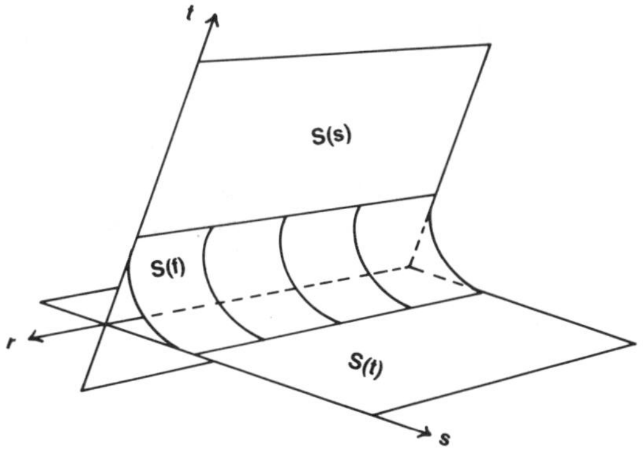

Figure 2.9.   Three-dimensional parameter space configuration for blending two surfaces.

$S(t - v)$. Suppose we replace the planes $S(s)$ and $S(t)$ with the curved surfaces $S(G)$ and $S(H)$, where G and H are polynomials in x, y, and z. We view these surfaces as two of three principal surfaces defining a curved coordinate system of xyz-space, without regard to whether in this system there is a one-to-one correspondence between coordinate values and points in Euclidean space. Then to the line $S(s - u, t - v)$ corresponds in general the space curve $S(G - u, H - v)$. Moreover, if the line $S(s - u, t - v)$ lies on the cylinder $S(f(s, t))$ in parameter space, then the curve $S(G - u, H - v)$ lies on the surface $S(f(G, H))$ in xyz-space.

Just as the surface $S(f)$ is tangent to $S(s)$ and $S(t)$ in parameter space, so $S(F)$ is tangent to $S(G)$ and $S(H)$ in xyz-space. Specifically, since $S(f)$ is tangent to $S(s)$ in the curve $S(s, t - b)$, so $S(F)$ is tangent to $S(G)$ in $S(G, H - b)$. Likewise, $S(F)$ is tangent to $S(H)$ in $S(G - a, H)$ since $S(f)$ is tangent to $S(t)$ in $S(s - a, t)$. In parameter space, only the portion on the inside of the plane $S(bs + at - ab)$ is of interest as blend. Similarly, in xyz-space, only the portion inside $S(bG + aH - ab)$ is of interest.

We have just described an equivalent blending procedure in which, by substitution of G for s and H for t, the rst coordinate system is replaced by a curved coordinate system with two of its principal surfaces being $S(G)$ and $S(H)$. We can think of this process as a warping of space in which the blend $S(f)$ is deformed into the corresponding blend $S(F)$. Note that this notion of deformation cannot be thought of as a continuous process as in, e.g., topology, since $S(f)$ and $S(F)$ may have different genera. Nevertheless, the paradigm is useful when blending corners of solids, and may lead to a successful procedure for blending patches of implicit algebraic surfaces. In particular, it can be thought of as reducing the blending of algebraic surfaces to blending planes. From now on we shall drop the sweep paradigm in favor of the substitution paradigm.

2.4. The projective method.   The affine formulation of the potential method given above is not fully general. Consider blending a circular cylinder $S(G)$ with a sphere $S(H)$, say $G = x^2 + z^2 - 4$ and $H = x^2 + y^2 + (z - 3)^2 - 1$. No matter how a is chosen, the curve of tangency $S(G - a, H)$ lies on $S(G - a)$, a concentric cylinder. Therefore, the oblique blend shown in Fig. 2.10 (see color insert) cannot be derived by the affine method. It is, however, a quartic surface obtained from f by substitution, using the *projective* potential method.

In the projective potential method, the intersecting families of surfaces are defined as $G - sW = 0$ and $H - tW = 0$, where W is a polynomial that may be chosen arbitrarily, but must not be the zero polynomial. Again, for degree consideration, we will choose W to have at most degree 2 when blending quadrics. The difference between the two methods is merely that in the affine method W is chosen as 1.

The blend of Fig. 2.10 is obtained by substituting G/W for s and H/W for t in

$$f(r, s, t) = b^2 s^2 + a^2 t^2 + a^2 b^2 - 2ab^2 s - 2a^2 bt + 2\lambda st,$$

where $a = b = 1$, $\lambda = \frac{1}{2}$ and $W = x^2 + z^2 + 2y/3 - 2z + 8/9$. Note that the procedure is equivalent to substituting G for s, H for t, and W for w in the homogeneous form

$$f(r, s, t, w) = b^2 s^2 + a^2 t^2 + a^2 b^2 w^2 - 2ab^2 sw - 2a^2 btw + 2\lambda st.$$

Certain projective blending surfaces derived in this way are the *projective transform* (see, e.g., Snyder and Sisam [7]) of affine blending surfaces for different surfaces G and H. Every blending surface derived from f using $W = U^2$, where U is some linear form, is the projective transform of another blending surface derived with the affine potential method.

3. A Uniqueness Theorem.   We have outlined a method for deriving blending surfaces which for quadric surfaces obtains degree 4 blending surfaces. It is natural to ask how general this method is, and how it relates to other blending methods proposed in the literature, e.g., Middleditch and Sears [3], Rockwood and Owen [4] and Rossignac and Requicha [5]. The Uniqueness Theorem provides a comprehensive, but not complete, answer to this question. We develop the theorem informally, since an exact formulation requires a fair number of concepts from algebraic geometry. The interested reader is referred to Hoffmann and Hopcroft [2] for complete details and the proof of the theorem.

UNIQUENESS THEOREM.   *Given two quadrics, $S(G)$ and $S(H)$, mark on each a space curve by intersecting $S(G)$ with an auxiliary quadric surface $S(H')$, and by intersecting $S(H)$ with a second auxiliary quadric surface $S(G')$. All degree 4 surfaces $S(F)$ that are tangent to $S(G)$ in the curve $S(G, H')$, and are tangent to $S(H)$ in the curve $S(H, G')$ may be derived from f using the (projective) potential method.*

In the case of the affine potential method the auxiliary surfaces are given by $G' = G - a$ and $H' = H - b$. In § 4, we replace these two auxiliary surfaces with a single, common one.

The Uniqueness Theorem has a number of hypotheses that must be satisfied. Expressed in intuitive terms, these are:

(1)   The curves $S(G, H)$, $S(G, H')$ and $S(H, G')$ are not the union of algebraic curves of lower degree and are all distinct;

(2) $S(F)$ is not the union of algebraic surfaces of lower degree;

(3)   the quadratic terms in G and H do not possess a common factor.

A thorough discussion of these hypotheses can be found in Hoffmann and Hopcroft [2].

In Middleditch and Sears [3] a blending method has been proposed that blends two quadrics with a degree 4 surface. Because of the Uniqueness Theorem, we know that the method is no more powerful than the potential method. In fact, it is a formulation of the affine potential method.

In Rockwood and Owen [4] a blending method has been proposed which derives blending surfaces as a function of G, H, and their gradient functions. For arbitrary quadrics, blending surfaces of degree 8 are obtained, but when blending cylinders and spheres term cancellation takes place and surfaces of degree 4 are obtained. Because of the above theorem, those degree 4 surfaces could equally well have been derived with the potential method, i.e., the gradient functions are not used in an essential way for those surfaces.

Suppose a blending method is sought that is to deliver degree 4 surfaces of constant curvature for blending quadrics. Because the surfaces obtained by the potential method do not possess constant curvature, the theorem states that this project must fail. Higher algebraic degrees are needed. Note, however, that such surfaces can be approximated in various ways, i.e., Rossignac and Requicha [5] and Rockwood and Owen [4].

In Middleditch and Sears [3] a blend is shown for two axially intersecting circular cylinders of equal radius. The blending surface shows a bulge. By the sweep paradigm of the affine potential method, the bulge seems unavoidable. Here it is not possible to draw conclusions from the theorem: The curve of intersection of the cylinders is reducible to two ellipses, a violation of hypothesis (1) above. Indeed, in Warren [8] a degree 4 blend without a bulge is given for precisely this case, consisting of the reducible surface $S(F)$ that is the union of two one-sheeted hyperboloids. The surface may also be derived with the projective potential method.

4. The Projective Potential Method for Quadrics. In the projective form of the potential method, the constants a and b no longer have the direct interpretation given to them in § 2.2. A different approach to controlling the blending surface is needed and is provided by the following corollary that is a consequence of the proof of the Uniqueness Theorem:

COROLLARY. *Given two quadrics $S(G)$ and $S(H)$. There is a degree 4 blending surface $S(F)$ tangent to both $S(G)$ and $S(H)$ if and only if the respective curves of tangency lie on a common quadric $S(\bar{G})$. Moreover, every such surface is derivable from the potential method.*

What is this surface $S(\bar{G})$? Recall that the curves of tangency of $S(f)$ to $S(s)$ and $S(t)$ are $S(s - a, t)$ and $S(s, t - b)$, in parameter space. The plane through these two lines is given by $bs + at - ab = 0$. Hence the surface $\bar{G}$ is just $S(bG + aH - ab)$ in the affine formulation, and $S(bG + aH - abW)$ in the projective formulation of the method. In § 2.2 this surface was used for clipping the unwanted parts of the blending surface.

As it were, the constants a and b may be replaced by 1, as W assumes their role. The projective quadric into which to substitute is then given by

$$f(r, s, t, w) = (s - w)^2 + (t - w)^2 - w^2 + 2\lambda st.$$

Given the quadrics $S(G)$ and $S(H)$ to be blended, we pick a quadric surface $S(\bar{G})$ such that it intersects $S(G)$ and $S(H)$ in the desired curves of tangency. We determine W from $\bar{G} = G + H - W$ and so obtain, by substitution into f above, the one-parameter family of blending surfaces given by

$$F = \mu GH + \bar{G}^2$$

where $\mu = 2\lambda - 2$. Here λ retains its previous interpretation as the parameter controlling the curvature distribution.

While this procedure is satisfactory mathematically, it does pose difficulties for automating the choice of blends, because the determination of $\bar{G}$ is not simple. More work is needed to give this method the practicality that its flexibility deserves.

5. Corner Blending. Edges between two faces of an object end at vertices. If one or more of the incident edges have to be smoothed by blends, then one must terminate a blend or smoothly combine several blends meeting at the vertex. Both problems have received attention in the literature, but much work remains to be done. Middleditch and Sears [3] and Rossignac and Requicha [5] can provide an entry into this problem.

Consider the problem of combining three joining blends at a vertex. In principle, a solution to this problem can be adapted to an arbitrary number of meeting blends, after reducing the vertex valence to three with the help of an auxiliary surface snubbing the vertex. Suppose we combine three blends as follows: first, blend two of them with a new blending surface; then, blend this new blending surface with the remaining third blend. The difficulty with this approach is that in each step the surface degree is doubled. Hence, if the three edge blends are degree 4 each, the two additional surfaces have degree 8 and 16, respectively. We seek alternatives which do not drive up the degree of the combining blending surfaces.

The general approach taken to obtain low-degree corner blends first solves the problem in parameter space with planes as primary surfaces and quadrics as their blending surfaces. Then this solution is lifted to the vertex of the solid at hand, by substitution. In this manner, any corner of three quadrics will be blended entirely by degree 4 surfaces. Of course, the method is not limited to quadrics as primary surfaces.

Recall the interpretation of quadrant position of the parametric base curve, as shown in Fig. 2.3. The blending surface is on the outside of $S(G)$ when $a > 0$, and on the inside of $S(G)$ when $a < 0$. Similarly, it is on the outside of $S(H)$ when $b > 0$, and on its inside when $b < 0$. In blending a three edge corner, we first examine on which side of the adjacent faces the edge blending surfaces are. Two generic cases arise:

(1)   For every face of the vertex, the two adjacent edge blending surfaces are always on the same side, i.e., always on the outside or always on the inside.

(2)   There is one face whose two adjacent edge blends are on the same side, and two faces such that the adjacent edge blends are on opposite sides of the same face. No other cases are possible at vertices with three edges.

Throughout this section, we use only the affine potential method. Additional work is required to extend the techniques given here to the general potential method.

5.1.   Adjacent blends always on the same face side.   Assume that the corner is formed by the surfaces $S(G)$, $S(H)$, and $S(K)$. The generic situation is shown in Fig. 5.1. In parameter space, the three faces meeting at the vertex are modeled by the three principal coordinate planes, $S(r)$, $S(s)$ and $S(t)$, and the vertex is represented by the origin. In the Fig. 5.1 we have assumed that the edge blends are on the outside of every face. If this is not so, i.e., if the edge blends adjacent to the face $S(G)$ are on the inside of $S(G)$, we simply substitute $-G$ for r. This is equivalent to reformulating Fig. 5.1 in the second octant.

As an example, we blend the edges parametrically with circular cylinders, and combine them at the vertex with a sphere. The respective equations are the following:

$$B1: \quad (r-1)^2 + (s-1)^2 - 1 = 0,$$
$$B2: \quad (s-1)^2 + (t-1)^2 - 1 = 0,$$
$$B3: \quad (t-1)^2 + (r-1)^2 - 1 = 0,$$
$$B4: \quad (r-1)^2 + (s-1)^2 + (t-1)^2 - 1 = 0.$$

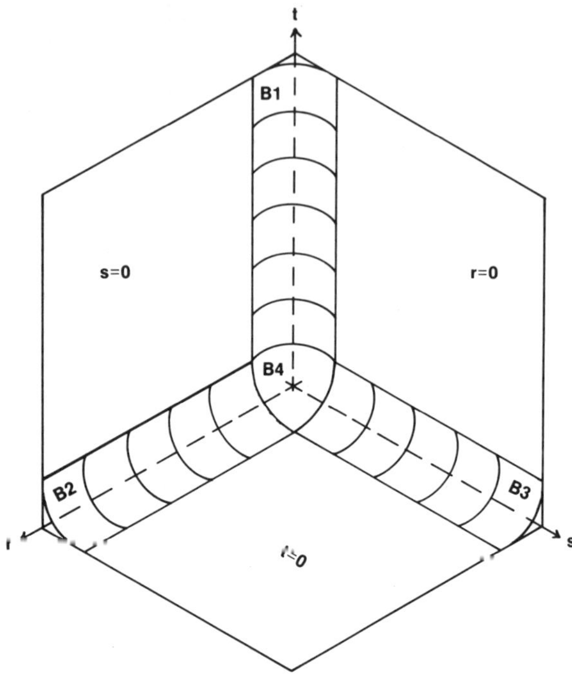

Figure 5.1.   Parameter space configuration for vertex blending when adjacent blends are all on the same surface side.

Clipping the unwanted parts is accomplished for each prototype blend by retaining those parts that are on the *inside* of the following planes:

$$\text{C1:}\ r + s - 1 = 0 \quad \text{and} \quad 1 - t = 0,$$

$$\text{C2:}\ s + t - 1 = 0 \quad \text{and} \quad 1 - r = 0,$$

$$\text{C3:}\ t + r - 1 = 0 \quad \text{and} \quad 1 - s = 0,$$

$$\text{C4:}\ r - 1 = 0 \quad \text{and} \quad s - 1 = 0 \quad \text{and} \quad t - 1 = 0.$$

In order to blend three intersecting cylinders, given by $G = x^2 + y^2 - 1$, $H = y^2 + z^2 - 1$, and $K = z^2 + x^2 - 1$, we substitute G, H, and K for r, s, and t, respectively, and obtain

$$\text{B1':}\ (G - 1)^2 + (H - 1)^2 - 1 = 0,$$

$$\text{B2':}\ (H - 1)^2 + (K - 1)^2 - 1 = 0,$$

$$\text{B3':}\ (K - 1)^2 + (G - 1)^2 - 1 = 0,$$

$$\text{B4':}\ (G - 1)^2 + (H - 1)^2 + (K - 1)^2 - 1 = 0.$$

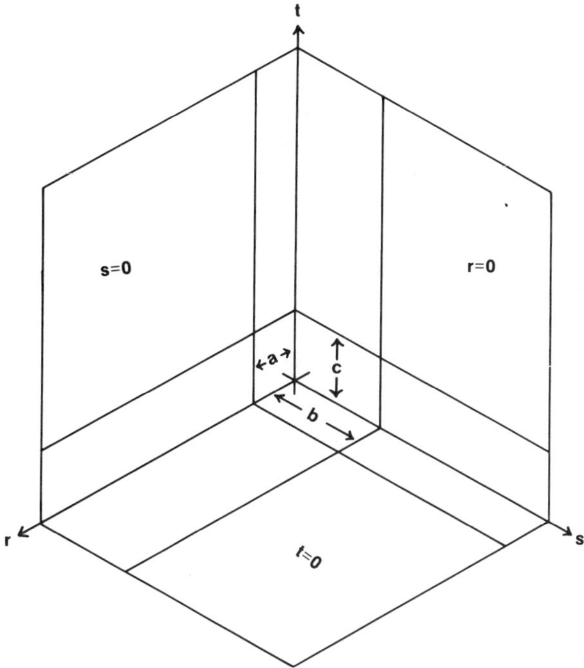

Figure 5.4.   Parameters for controlling tangency position when blending vertices of valence three.

These surfaces need to be clipped. This is accomplished just as in parameter space, by retaining those parts which are on the *inside* of the following surfaces, given by:

$$C1': \quad G + H - 1 = 0 \quad \text{and} \quad 1 - K = 0,$$
$$C2': \quad H + K - 1 = 0 \quad \text{and} \quad 1 - G = 0,$$
$$C3': \quad K + G - 1 = 0 \quad \text{and} \quad 1 - H = 0,$$
$$C4': \quad G - 1 = 0 \quad \text{and} \quad H - 1 = 0 \quad \text{and} \quad K - 1 = 0.$$

The resulting object is shown in Fig. 5.2 (see color insert). Here, the blends B1′ through B3′ are shown in light blue, and the blend B4′ is shown in purple. All surfaces meet C^1-continuously. Since three quadrics in general intersect in eight points, the surface B4′ has eight real components, as seen in Fig. 5.3 (see color insert). An interesting aspect of the method is that if the edge blends B1–B3 do not intersect after substitution, then the surface B4′ is either clipped away entirely or becomes imaginary.

What flexibility does this method offer? With the affine potential method, the width of the three blends is controlled by six constants a_k and b_k, but three are now dependent, so there remain exactly three independent constants, a, b, and c. Figure 5.4 shows the lines of tangency positioned in parameter space as a function of a, b, and c. Moreover, there is only one free parameter λ controlling the curvature distribution of all three edge blends simultaneously. The generic formulas are

$$B1: \quad b^2r^2 + 2ab\lambda rs + a^2s^2 - 2b^2ar - 2a^2bs + a^2b^2 = 0,$$
$$B2: \quad c^2r^2 + 2ac\lambda rt + a^2t^2 - 2c^2ar - 2a^2ct + a^2c^2 = 0,$$
$$B3: \quad c^2s^2 + 2bc\lambda st + b^2t^2 - 2c^2bs - 2b^2ct + b^2c^2 = 0,$$
$$B4: \quad a^2b^2c^2(r^2/a^2 + s^2/b^2 + t^2/c^2 + 2(1+\lambda)(1 - r/a - s/b - t/c))$$
$$+ 2\lambda abc(rsc + rbt + ast) = 0$$

and the clipping planes are given by

$$C1: \quad br + as - ab = 0 \quad \text{and} \quad c - t = 0,$$
$$C2: \quad cr + at - ac = 0 \quad \text{and} \quad b - s = 0,$$
$$C3: \quad cs + bt - bc = 0 \quad \text{and} \quad a - r = 0,$$
$$C4: \quad r - a = 0 \quad \text{and} \quad s - b = 0 \quad \text{and} \quad t - c = 0.$$

To blend a corner with a surface having the same degree as the edge blending surfaces, the shaping parameters must be coordinated in this way. More work is required to study if the projective potential method offers greater flexibility and permits, for example, to control edge blend curvature independently.

5.2. Edge blends on opposite sides. The other case to consider is a vertex, two of whose faces have their adjacent edge blending surfaces on opposite sides. Again, we may have to substitute negated face equations depending on the position of the edge blends. This will be necessary in the example below.

The generic situation is shown in Fig. 5.5. Again, the vertex is the origin. Note that $S(r)$ and $S(s)$ have the adjacent blending surfaces on opposite sides. In parameter space, we take two of the edge blends, B1 and B2, as circular cylinders of equal radius. Since they are axially intersecting cylinders of equal radius, there is a hyperboloid of one sheet tangent to both which may be used to join B1 and B2. In the figure the hyperboloid is shown as B4. With a cylinder radius 1, we may take a hyperboloid whose major axes are $m = \sqrt{3}$, $n = 1$ and 1. Here m and n may be chosen differently, but must satisfy $m^2 - n^2 = 2$, so that the hyperboloid remains tangent to the cylinders. Finally, B3 is a hyperbolic cylinder matching the hyperboloid's cross-section in the plane $t = 1$. The exact equations are

$$B1: \quad (r-1)^2 + (t-1)^2 - 1 = 0,$$
$$B2: \quad (s-1)^2 + (t-1)^2 - 1 = 0,$$
$$B3: \quad (r+1)^2 + (s+1)^2 - 1 - 4rs = 0,$$
$$B4: \quad 3(t-1)^2 - (r-1)^2 - (s-1)^2 + 4(r-1)(s-1) - 3 = 0.$$

The planes in which the hyperboloid is tangent to B1 and B2 are $s - 2r + 1 = 0$ and $r - 2s + 1 = 0$, and are used for clipping. The respective clipping equations, adjusted

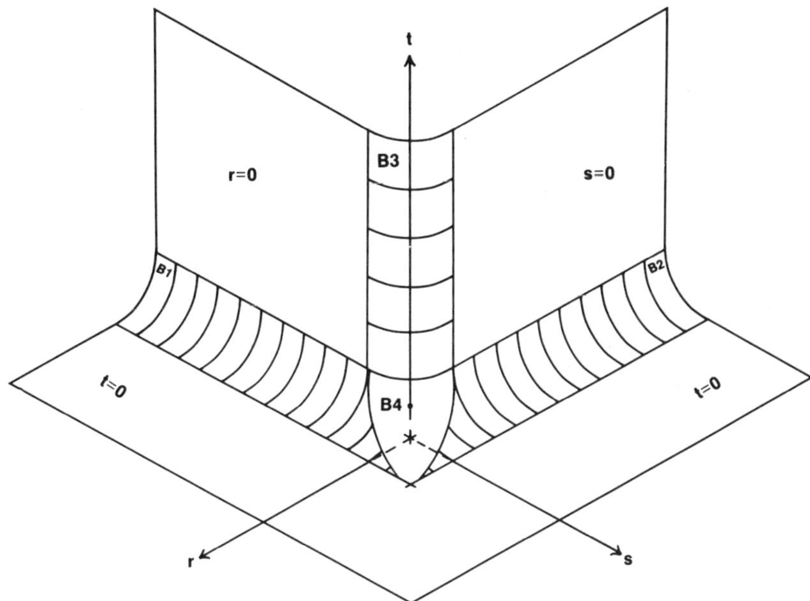

Figure 5.5.   Parameter space configuration for vertex blending when not all adjacent blends are on the same surface side.

such that the wanted portion is on the surface inside, are given by

$$C1: \quad r + t - 1 = 0 \quad \text{and} \quad s - 2r + 1 = 0,$$

$$C2: \quad s + t - 1 = 0 \quad \text{and} \quad r - 2s + 1 = 0,$$

$$C3: \quad -r - s - 1 = 0 \quad \text{and} \quad 1 - t = 0 \quad \text{and} \quad r + s = 0,$$

$$C4: \quad 2r - s - 1 = 0 \quad \text{and} \quad 2s - r - 1 = 0 \quad \text{and} \quad t - 1 = 0.$$

Here the third constraint on B3 is needed to remove the second branch of the hyperbolic cylinder. As previously, we substitute for r, s and t the surfaces intersecting in the vertex, observing on which side of the face the adjacent edge blends lie.

Consider blending the cylinder configuration shown in Fig. 5.6. Here $S(G)$ and $S(H)$ are two intersecting cylinders with radius $\sqrt{2}$, and $S(K)$ is the cylinder of radius 1, removed from the other two cylinders. Note that we blend the outside of $S(K)$ to the inside of both $S(G)$ and $S(H)$, whereas the outside of $S(G)$ is blended to the outside of $S(H)$. Consequently, we substituted $-G$ for r, $-H$ for s, and K for t. The result is shown in Fig. 5.7 (see color insert). Note that all surfaces meet C^1-continuously.

In parameter space, we may replace the circular cylinders B1 and B2 with elliptic ones, but their intersection must remain a pair of intersecting conics, so that the corner remains a quadric. The width of the third edge blend is controlled by the eccentricities of the corner hyperboloid. This case is more awkward than the previous case since the major axes of the hyperboloid do not lie parallel to the principal coordinate axes. As before, general formulas can be worked out.

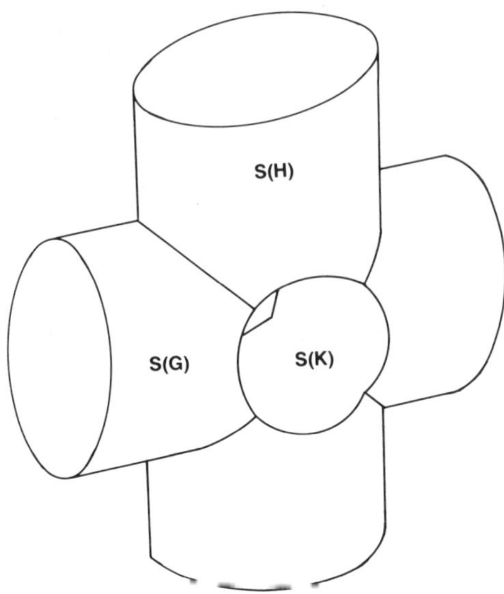

Figure 5.6. Object requiring the parameter configuration of Fig. 5.3.

6. Patched Algebraic Surfaces. The surface of an object consists, in general, of patches of algebraic surfaces. When two patches intersect transversally at an edge, then the blending methods outlined above apply. But suppose the two surfaces $S(G)$ and $S(H)$ meet tangentially in an edge and both intersect a common third surface $S(K)$ transversally in another edge that we wish to smooth, as shown in Fig. 6.1.

We may blend $S(G)$ with $S(K)$ and separately $S(H)$ with $S(K)$. Even though the curves of tangency may be correctly lined up, it is very likely that the two blending surfaces do not meet along the seam, as shown in Fig. 6.2. Again, it is our wish to provide solutions that do not raise the degree of the blending surfaces unnecessarily. For instance, the situation depicted in Fig. 6.2 may be solved with one blending surface of degree 4, the other of degree 8, but such a solution seems unsatisfactory unless we can prove that there are no lower degree surfaces with the necessary properties.

While the general problem remains unsolved, there is a situation in which degree 4 surfaces automatically match: Assume that $S(G)$ and $S(H)$ intersect tangentially in an edge e, and that both intersect a surface $S(K)$. If there is a plane h such that the edge e is the complete intersection of $S(G)$ with h and also the complete intersection of $S(H)$ with h, then the curves of intersection of the blending surfaces $S(f(G, K))$ and $S(f(H, K))$ with that plane are equal.

The situation is illustrated by Fig. 6.3 (see color insert). Here the green ellipsoid, given by $G = x^2/25 + y^2/9 + z^2/16 - 1$, intersects tangentially the yellow hyperboloid, given by $H = -x^2/25 + y^2/9 + z^2/16 - 1$. Both surfaces, in turn, transversally intersect a cylinder given by $K = x^2 + y^2 - 1$ and shown in red. Both transversal intersections are blended with the same parametric cylinder $f(r, s, t) = (s - 4)^2 + 4(t - 1)^2 - 4 = 0$. The

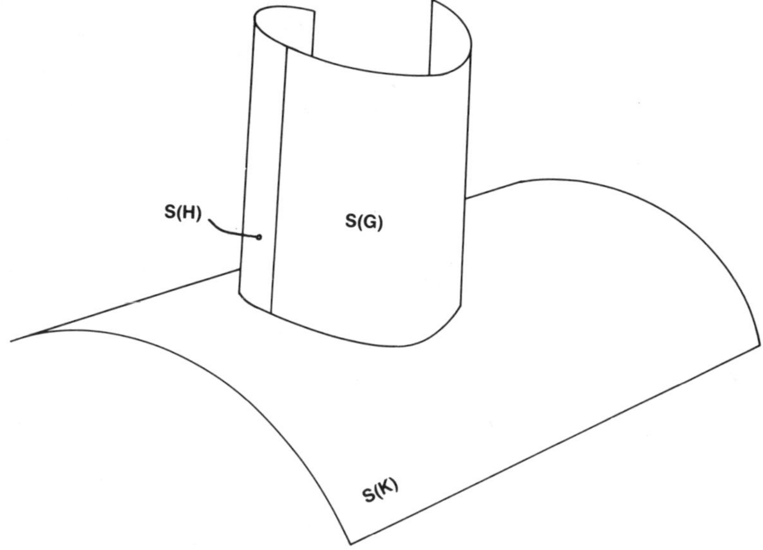

Figure 6.1.   Patches of algebraic surfaces.

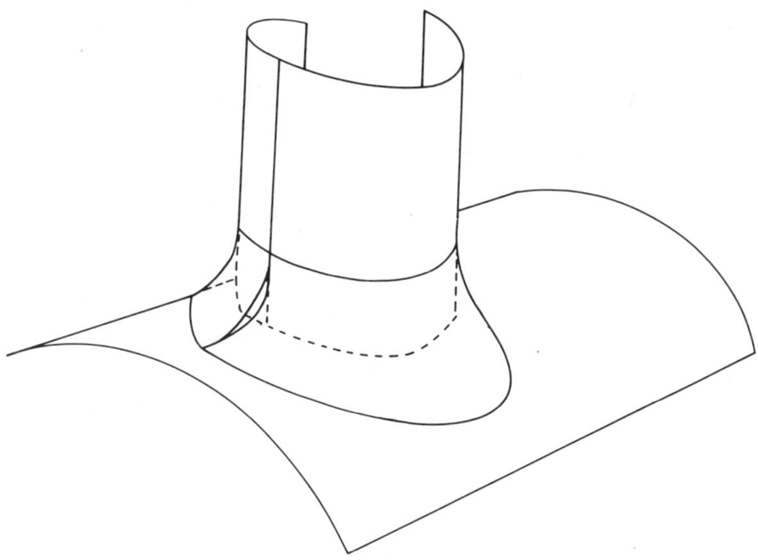

Figure 6.2.   Blend discontinuity for patches of algebraic surfaces.

resulting (light blue) blending surface $S(f(G, K))$ matches the (purple) blend $S(f(H, K))$ C^1-continuously.

7. Dimensionality of Parameter Space. In our approach to low degree blending surface derivation, we have derived the actual blends from simple blends to planar surfaces in three-dimensional parameter space. Essentially we have advocated reducing the problem of blending algebraic surfaces to the problem of blending planar surfaces. This works quite well, since the deformation effected by substitution for the principal coordinate surfaces usually is not too drastic. That is, the surfaces $G - s = 0$ are usually very similar in shape to the surface $S(G)$.

In all cases considered here, the parameter space has at most three dimensions. This seems artificial, and we believe that the investigation of parameter space configurations of higher dimensionality can lead to better ways to blend complex corners than reducing vertex valence to three. It may also help to localize the shape control of edge blends at a vertex.

Blending algebraic patches is a more difficult matter. The major problem is that when two patches $S(G)$ and $S(H)$ are C^1-continuous along an edge, there is no guarantee that the surface families $G - s = 0$ and $H - t = 0$ are related in a deep way. It is possible that an approach working in higher-dimensional parameter space can provide results, or that the projective method yields the necessary tools.

Sederberg [6] works with implicit surfaces that possess rational parameterizations. This approach is interesting since it aims at a spatially intuitive procedure for deriving and placing free form surfaces. In fact, some of our blending surfaces are known to possess rational parameterization. For instance, the Steiner surface advocated by Sederberg is also a blending surface. Consider, for example, the Steiner surface $x^2y^2 + y^2z^2 + z^2x^2 - 2xyz = 0$. Since its equation may be written as

$$(x^2 + y^2 - 1)(z^2) + (xy - z)^2 = 0,$$

it is a blending surface where the quadrics blended are the cylinder $x^2 + y^2 - 1 = 0$ and the double plane $z^2 = 0$. The common quadric defining the curve of intersection is $xy - z = 0$, a hyperbolic paraboloid. The exact relationship between the class of all quartics having a rational parameterization on the one hand, and the degree 4 blending surfaces for two quadrics on the other, is not understood at this time.

Acknowledgments. We are indebted to Dean Krafft who wrote the display algorithm with which our pictures have been generated. Joe Warren carefully read an earlier version of this paper and made useful suggestions. Our discussions with Alyn Rockwood, Jim Owen, Alan Middleditch, and Ken Sears in Albany were also useful.

This work was supported in part by National Science Foundation grants ECS 83-12096 and MCS 82-17966.

REFERENCES

[1] C. M. HOFFMANN AND J. E. HOPCROFT, *Automatic surface generation in computer aided design*, The Visual Computer, 1 (1985), pp. 95–100.

[2] C. M. HOFFMANN AND J. E. HOPCROFT, *Quadratic blending surfaces,* Tech. Rept. 85–674, Computer Science Dept., Cornell Univ., Ithaca, NY, 1985.

[3] A. MIDDLEDITCH AND K. SEARS, *Blend surfaces for set theoretic volume modeling systems,* Computer Graphics, 19 (1985), pp. 161–170.

[4] A. ROCKWOOD AND J. OWEN, *Blending surfaces in solid modeling,* this Volume, 1987.

[5] J. ROSSIGNAC AND A. REQUICHA, *Constant-radius blending in solid modeling,* Comput. Mech. Engrg., 3 (1984), pp. 65–73.

[6] T. W. SEDERBERG, *Algebraic geometry for surface and solid modeling,* this Volume, 1987.

[7] V. SNYDER AND C. H. SISAM, *Analtyic Geometry of Space,* H. Holt and Co., New York, 1914.

[8] J. WARREN, Ph.D. thesis, Dept. Computer Science, Cornell Univ., Ithaca, NY, 1986.

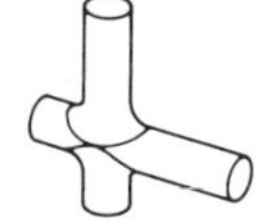

Blending Surfaces in
Solid Modeling

ALYN P. ROCKWOOD AND JOHN C. OWEN

Abstract. Solid geometric modelers typically employ only a restricted set of primitive surfaces. Blending surfaces can significantly extend the range of objects which can be modeled. A pattern is established and examples are given for devising implicitly defined blending surfaces. One form, the super elliptic blend, is described in detail. It is shown to integrate nicely into a "solid" modeling environment and to have intuitive inputs, viz., ranges and thumbweight. A modification of the super-elliptic blend allows blending on blends, a facility which increases the flexibility and the applicability of the blending surfaces. Characteristics of this blending surface, such as smoothness, continuity class and convexity, are investigated mathematically.

1. Introduction. The term blending surface usually means either a surface which forms a smooth transition between different, but intersecting surfaces, or a surface which smoothly joins two or more disconnected surfaces. The blending surfaces described in this paper were developed primarily with the first task in mind; that is, they are intended to replace the kinks and creases, where the surfaces come together, with smooth transitions. The filleting operation is an example. The technique does have applications to the second problem, but they are not pursued here.

Such blending surfaces are required in mechanical design for several reasons, including dissipating stress concentrations, enhancing fluid flow and improving aesthetics. Furthermore, some manufacturing processes are incompatible with sharp edges and corners. High pressure injection molding is an example of one.

Solid modelers handle only a restricted set of surface types such as quadrics and tori. Introduction of blending surfaces to these solid modelers can significantly extend the range of objects which can be represented by the modeler. The stimulus for this work was provided by the need to expand the solid geometric modeler ROMULUS to include general blending surfaces. From the outset three goals were postulated for the development of blending surfaces in ROMULUS:

(1)   The surfaces must integrate into a "solid" modeling system.

367

(2)   The surfaces should be easy to define; their specification should have intuitive advantages.

(3)   The surfaces should have the broadest possible application to the blending problem, e.g., blending multiple surfaces, possessing shaping controls, etc.

Goal (1) suggests the use of implicit surfaces if possible; that is, surfaces which are defined by $f(\mathbf{x}) = 0$ for $f: \mathbf{R}^3 \to \mathbf{R}$, $\mathbf{x} \in \mathbf{R}^3$. Implicit functions divide space into two halfspaces, namely, $f(\mathbf{x}) > 0$ and $f(\mathbf{x}) \le 0$. Testing whether a point is inside or outside an implicitly defined object, or on the surface of the object, reduces to a simple substitution of the point into the function and subsequent evaluation. This contrasts with parametrically defined surfaces, for instance, for which inside and outside testing is difficult. Such an "inside-outside" test is essential for solid modeling. Furthermore, the surface normal and curvature of an implicitly defined surface are typically easier to evaluate than for a parametrically defined surface. These are important for many solid modeling operations such as finding intersection curves with other surfaces (Owen and Rockwood [6]).

It was also found that implicit surfaces were very convenient for defining blends. The simplicity of definition contributed towards satisfying goal (2). Several examples of this are given later.

Parametric surface patch methods exhibit considerable flexibility for shaping and design. Creating an implicit surface form which exhibits some of this flexibility without infringing upon goals (2) or (1) posed the greatest challenge in developing a form for blending surfaces. The majority of this paper addresses the problem of developing an implicit form with sufficient flexibility to satisfy goal (3).

2. The Pattern for Defining Blends.   We consider the problem of blending between two surfaces defined by $P_1(\mathbf{x}) = 0$ and $P_2(\mathbf{x}) = 0$. Initially, we assume the blending surface (also called the blend) will be contained in the regions $P_1(\mathbf{x}) \ge 0$ and $P_2(\mathbf{x}) \ge 0$. This assumption will be relaxed in § 5.

The first step in the general method for defining blends is to define a blend between two intersecting lines in $\mathbf{R}^2$. The blend is a kernel set of some function $F: \mathbf{R}^2 \to \mathbf{R}$ which joins the lines, i.e., $F^{-1}(0)$ is the blend between $P_1^{-1}(0)$ and $P_2^{-1}(0)$.

Example 2.1.   Let P_1 and P_2 be linear functions as in Fig. 2.1. We seek a blend which results in a circular transition between $P_1^{-1}(0)$ and $P_2^{-1}(0)$, regardless of the angle between the lines.

$F(\mathbf{x})$ is the distance from a point $\mathbf{x}$ to a circle of radius r which touches both lines. θ is the angle between ∇P_1 and ∇P_2, α is the angle between the bisector (dashed) of the lines and the line through both the center of the circle to $\mathbf{x}$. From the figure,

$$P_1 = r - (r - F) \cos\left(\frac{\theta}{2} - \alpha\right) \quad \text{and} \quad P_2 = r - (r - F) \cos\left(\frac{\theta}{2} + \alpha\right).$$

From this we obtain

$$(2.1) \qquad F = r - \frac{1}{\sin\theta}[(r - P_1)^2 + (r - P_2)^2 - 2(r - P_1)(r - P_2)\cos\theta]^{1/2}.$$

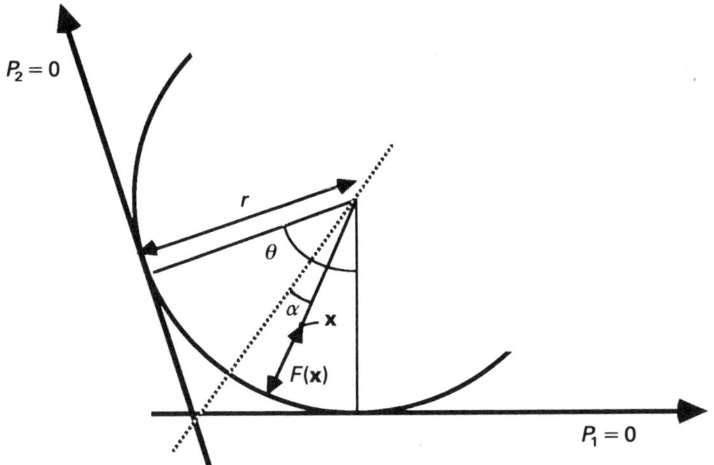

Figure 2.1.   Defining a circular blend.

The arc from $F = P_1$ to $F = P_2$ is the circular part of the blend, implying a region of definition

(2.2) $\mathcal{R} \equiv \{\mathbf{x} \mid P_1(\mathbf{x}) < r - (r - P_2(\mathbf{x})) \cos \theta \text{ and } P_2(\mathbf{x}) < r - (r - P_1(\mathbf{x})) \cos \theta\}.$

The function defining the blend is

(2.3) $F = \begin{cases} r - \dfrac{1}{\sin \theta}[(r - P_1)^2 - (r - P_2)^2 - 2(r - P_1)(r - P_2)\cos \theta]^{1/2} & \text{on } \mathcal{R}, \\ \min(P_1, P_2) & \text{on } \tilde{\mathcal{R}}. \end{cases}$

$F^{-1}(0)$ will consist of a circular arc connecting two pieces of straight lines as long as the lines are not parallel, otherwise there is no circular part.

The second step in the pattern is to generalize P_1 and P_2 to be functions from $\mathbf{R}^n$ to $\mathbf{R}$ and to define the *primitive deformation* $P = (P_1, P_2): \mathbf{R}^n \to \mathbf{R}^2$.

The blend between $P_1^{-1}(0)$ and $P_2^{-1}(0)$ is the kernel of F compose P; that is, the blend is $B_{12}^{-1}(0)$ where $B_{12} = F \circ P$. $P_1^{-1}(0)$ and $P_2^{-1}(0)$ are called *primitives*.

Example 2.2.   In (2.3) let $n = 3$ and let

$$P_1(x, y, z) = 5 - (x^2 + y^2)^{1/2} \quad \text{and} \quad P_2(x, y, z) = 5 - (x^2 + z^2)^{1/2}.$$

P_1 and P_2 define crossing cylinders of equal radius. The blending surface depicted in Fig. 2.2 is for $r = 3$.

Note that the blend dissipates at the *points of tangency*; that is, where $P_1(\mathbf{x}) = P_2(\mathbf{x})$ and $\nabla P_1(\mathbf{x}) = \alpha \nabla P_2(\mathbf{x})$, $\alpha \in \mathbf{R}\backslash\{0\}$. This can be understood by considering how the region $\tilde{\mathcal{R}}$ is defined. For points of tangency, $\cos \theta = 1$ which implies that all points of tangency are in $\tilde{\mathcal{R}}$. Such behavior for a blend is often desirable because it simulates the so-called "rolling ball" blend.

The rolling ball blend consists of part of a *tubular surface* which is the envelope of a sphere of constant radius which contacts two primitives. More precisely, if $\mathbf{x}$ is a point

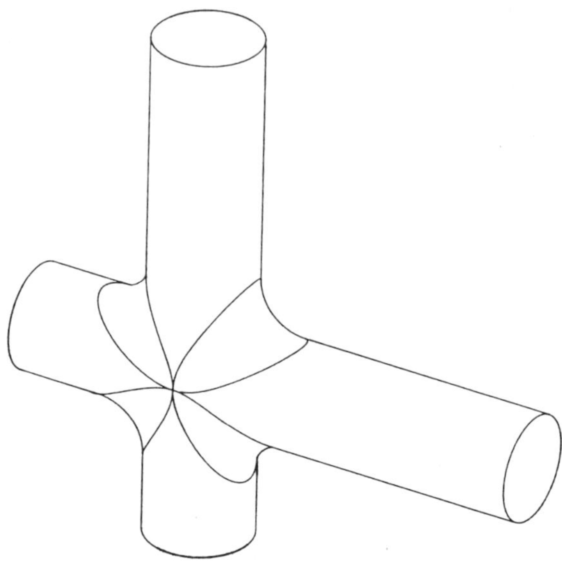

Figure 2.2.   A rolling ball blend between two cylinders.

on a tubular surface, then there exists a $\mathbf{y}$ corresponding to the centre of a sphere, such that

$$\|\mathbf{x} - \mathbf{y}\| = r, \qquad P_1(\mathbf{y}) - r = 0,$$
$$P_2(\mathbf{y}) - r = 0 \quad \text{and} \quad (\mathbf{x} - \mathbf{y}) \cdot [\nabla P_1(\mathbf{y}) \times \nabla P_2(\mathbf{y})] = 0,$$

where r is the radius of the sphere and P_1 and P_2 measure Euclidean distance from the primitives. It can be shown that (2.1) for general primitives is a first order linearization of the tubular surfaces, i.e., for $\mathbf{y} = \mathbf{x} + \delta\mathbf{y}$ and $|\nabla P_1| = |\nabla P_2| = 1$, F satisfies

$$F = r - \|\mathbf{x} - \mathbf{y}\|, \qquad P_1(\mathbf{x}) + \delta\mathbf{y} \cdot \nabla P_1(\mathbf{x}) = r,$$
$$P_2(\mathbf{x}) + \delta\mathbf{y} \cdot \nabla P_2(\mathbf{x}) = r \quad \text{and} \quad \delta\mathbf{y} \cdot [\nabla P_1(\mathbf{y}) \times \nabla P_2(\mathbf{y})] = 0.$$

The proof requires linear independence of ∇P_1 and ∇P_2 and solves for $\delta\mathbf{y}$ in the last three equations which is then substituted into $F = r - \|\mathbf{x} - \mathbf{y}\| = r - \|\delta\mathbf{y}\|$.

This blending form gains utility from the fact that the exact definition of a tubular surface produces highly unwieldy solutions. For crossing cylinders the simplest form has 969 terms, for instance. In practice, the exact shape of the blend is not often important. $B_{12}^{-1}(0)$ as an approximation to the rolling ball blend seems to work well as long as r in (2.2) is less than half the smallest radius of curvature at the primitives.

3.   The Super-Elliptic Blend.   The blending surface in the previous example is intuitive and useful, due in part to its simple input (radius of the ball) and also to its familiarity in the manufacturing industry. We expect, however, to be able to construct more versatile and "flexible" forms, using the pattern established. The first step in this

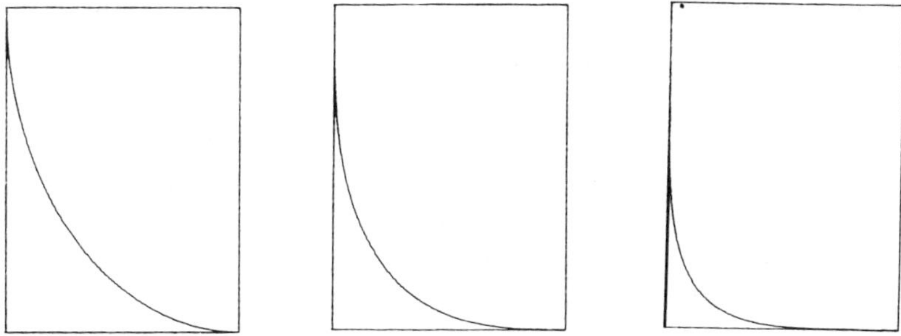

Figure 3.1.   Super-ellipses with $t = 2$, 3 and 5.

process is to consider choosing a more versatile form for F, the function which defines the blend in $\mathbf{R}^2$. An immediate generalization would be to choose F so that its kernel is an appropriate ellipse which rests between the lines. Even more general is the so-called super-ellipse.

Assume that $P_1^{-1}(0)$ and $P_2^{-1}(0)$ are, respectively, the x and y axes in the plane. A super-elliptic arc which touches both axes is the zero contour of

$$(3.1) \qquad S(x, y) = 1 - \left[1 - \frac{x}{r_1}\right]_+^t - \left[1 - \frac{y}{r_2}\right]_+^t ,$$

where r_1, r_2, $t \in \mathbf{R}^+$ are constants and where $[a]_+ = \max (0, a)$.

The effects of the three parameters are illustrated in Fig. 3.1. r_1 measures the distance along the x axis where the arc heels to the axis; r_2 does the same on the y axis. These two parameters are referred to as *ranges*. The parameter t is used to change the shape of the blend, drawing it closer to the primitive as t increases. It is called the *thumbweight*.

Following the pattern, the blending surface defined for primitives P_1 and P_2 from $\mathbf{R}^n$ to $\mathbf{R}$ is defined by

$$(3.2) \qquad S_{12} = 1 - \left[1 - \frac{P_1}{r_1}\right]_+^t - \left[1 - \frac{P_2}{r_2}\right]_+^t .$$

Typically, we let $n = 3$.

3.1. Properties of $S_{12}^{-1}(0)$.   Given $\mathbf{x} \in \mathbf{R}^n$, if $P_1(\mathbf{x}) < 0$, then $[1 - P_1/r_1]_+^t > 1$ which implies $S_{12}(\mathbf{x}) < 0$. Similarly $P_2(\mathbf{x}) < 0$ implies $S_{12}(\mathbf{x}) < 0$. Therefore $S_{12}^{-1}(0)$ is contained in the region given by $P_1 \geqq 0$ and $P_2 \geqq 0$.

Define a region

$$\mathcal{R} \equiv \{\mathbf{x} \in \mathbf{R}^n \mid P_1(\mathbf{x}) < r_1 \text{ and } P_2(\mathbf{x}) < r_2\},$$

which is called the *region of support*. The blend is disjoint from the primitives in $\mathcal{R}$, i.e.,

$$(\mathcal{R} \cap S_{12}^{-1}(0)) \cap (P_1^{-1}(0) \cup P_2^{-1}(0)) = \varnothing.$$

The complementary region $\tilde{\mathscr{R}}$ is called the *region of identity*, and it satisfies

$$\tilde{\mathscr{R}} \cap S_{12}^{-1}(0) \equiv \tilde{\mathscr{R}} \cap (P_1^{-1}(0) \cup P_2^{-1}(0)).$$

If P_1 and P_2 are C^k continuous, then for $t > k$, S_{12} is C^k continuous. (Note that $\max[0, x]^t$ is C^k continuous if $t > k$.) For $k \geq 1$, C^k continuity of S_{12} is necessary, but not sufficient to consider $S_{12}^{-1}(0)$ a smooth set, according to the notion of smoothness which we subscribe to, which is that the blending surface is smooth if it is manifold. The smoothness of $S_{12}^{-1}(0)$ is investigated in § 4 as a special case of an alternative form, for which $S_{12}^{-1}(0)$ is the kernel.

PROPOSITION 3.1.   *If in* (3.2) $P_1(\mathbf{x}) = x$ *and* $P_2(\mathbf{x}) = y$ *and* $t > 1$, *then the set*

$$S_{12}^+ \equiv \{\mathbf{x} \in \mathbf{R}^2 \mid S_{12}(\mathbf{x}) > 0\}$$

is convex.

Proof.   The functions $\{[1 - x/r_1]_+^t; x \in \mathbf{R}\}$ and $\{[1 - y/r_2]_+^t; y \in \mathbf{R}\}$ are convex because their first derivatives increase monotonically when $t > 1$. Hence $\{S_{12}(\mathbf{x}); \mathbf{x} \in \mathbf{R}^2\}$ is concave, which implies the required result. $\square$

The proposition implies that any straight line in $\mathbf{R}^2$ will intersect $S_{12}^{-1}(0)$ at most twice. It is a variation diminishing property which implies that the blending surface does not contain any extraneous wiggles. Letting P_1 and P_2 be general functions, $\mathbf{R}^n \to \mathbf{R}$ can introduce nonconvexities into S_{12}^+; in other words, the primitive deformation does not necessarily preserve convexity. However, convexity of a set is preserved by linear maps, including for example, a linear injection $\mathbf{R}^2 \to \mathbf{R}^n$ from which the following is obtained.

COROLLARY 3.2.   *If in* (3.2) P_1 *and* P_2 *are linear functions and* $t > 1$, *then* S_{12}^+ *is convex.*

If P_1 and P_2 are polynomial and t is an integer, then S_{12} is piecewise polynomial. For computational reasons, it is often suggested that the order of the polynomial be as low as possible (Hoffmann and Hopcroft [4]). There are several ostensible reasons for this. One is the proliferation of terms which results from a polynomial of high degree in several variables. Another is the variation of the surface which can be typified by looking at the intersections of a straight line with the surface. The problem reduces to finding roots of a single variable along the line. The complexity of the root structure usually corresponds to the degree of the polynomial. Low degree is, at best, only a sufficient condition, not a necessary condition. S_{12} provides examples of this. First, the cost of evaluating S_{12} at a point does not depend on how large t is.

Second, for P_1 and P_2 linear, Corollary 3.2 implies that the number of intersections of $S_{12}^{-1}(0)$ with a straight line will be two, at most. This is independent of how t is chosen. If t is a positive integer, the degree of the polynomial pieces of S_{12} can be arbitrarily high, but does not affect the number of intersections between $S_{12}^{-1}(0)$ and the line.

3.2.   Variable range.   A useful modification of (3.2) is obtained by replacing the constants r_i by positive functions of position $R_i : \mathbf{R}^n \to \mathbf{R}$, thus

$$(3.3) \qquad S_{12} = 1 - \left[1 - \frac{P_1}{R_1}\right]_+^t - \left[1 - \frac{P_2}{R_2}\right]_+^t.$$

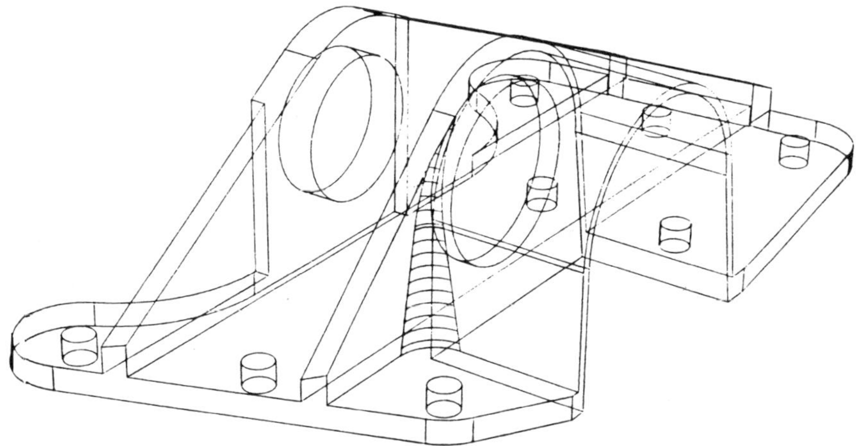

Figure 3.2.   Blend (ribbed) with variable range.

Example 3.3.   Let P_1 and P_2 be defined as in Example 2.2. Let $R_1 = r - (r - P_2)\cos\theta$ and $R_2 = r - (r - P_1)\cos\theta$, where θ is the angle between the gradients of P_1 and P_2 at a point and where r is a positive constant. The resulting blend for $t = 2$ and $r = 3$ would produce a figure very similar to Fig. 2.2. This is because the region of support for $S_{12}^{-1}(0)$ is defined by the R_i which were obtained from (2.2). The cross-sectional shape of the blend would differ. In the case where P_1 and P_2 are chosen to be linear in $\mathbf{R}^2$, for instance, the blend will contain part of an elliptic arc, instead of a circular one. In general, however $S_{12}^{-1}(0)$ behaves like a rolling ball blend in that it contains points of tangency. Thus the super-elliptic form with variable range has a pseudo-rolling ball form when the R_i are defined as above.

Example 3.4.   In $\mathbf{R}^3$ two points are chosen on the intersection of the surfaces to be blended. Let E_1 and E_2 be the planes normal to the intersection track at these points. Ranges are given for the blend on each of these normal planes; thus a_1e_1 and a_2e_1 are the ranges given for E_1.

The blend between the two planes is then given by (3.3) where

$$(3.4) \qquad R_i = \sum_{j=1}^{2} \frac{a_ie_j * E_j^s}{(E_1^s + E_2^s)}$$

where the exponent s is a shaping factor which determines the manner in which the cross-sectional shape on one plane changes to the shape on the other blend.

Figure 3.2 shows a blend (ribbed) in which the range diminishes from one plane (at the base of the blend) to zero on another plane (at the top of the blend).

4.   Blending with Blends.   There are several advantages to be gained from treating a blend as a primitive surface which is then subsequently used in another blend. We mention three. They are (1) increased flexibility in blending, (2) overlapping blends without concern, and (3) the blending of more than two primitives. Figures

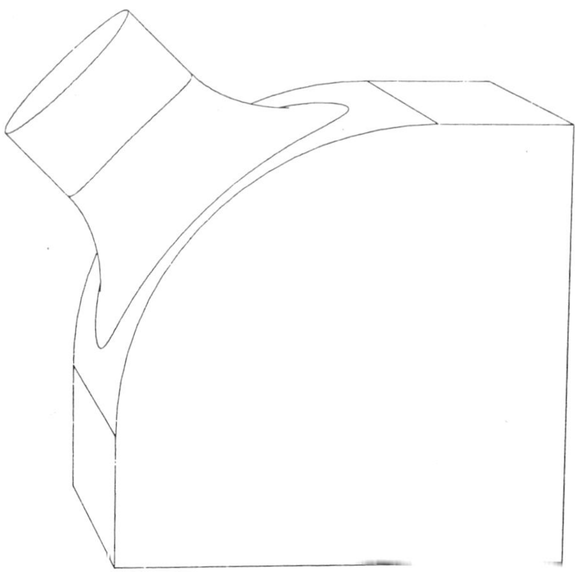

Figure 4.1. Subsequent blending.

4.1–4.3 give examples. Figure 4.1 shows a cube with one edge blended which is subsequently blended to a cylinder. Figure 4.2 shows overlapping blends and Fig. 4.3 depicts a blend between three planes. All figures involve blending with blends. The methods for achieving blends on blends is described in this and the next sections.

An important consideration in blending is the way in which $P_1(\mathbf{x})$ acts as a measure of distance from the point $\mathbf{x}$ in $\mathbf{R}^n$ to the primitive $P_1^{-1}(0)$. Middleditch and Sears [5] restrict their primitives to ones which can be easily reconstructed so that $P_1(\mathbf{x})$ is a measure of Euclidean distance. We do not make such restrictions but we will want to avoid primitives in which the function as a distance measure may produce unintuitive or surprising results in the blend. S_{12} is an example of an undesirable form for a primitive, even assuming that the primitive functions P_1 and P_2 of S_{12} are acceptable.

4.1. The problems with S_{12} for defining primitives. Given the primitive functions and ranges as in (3.2) divide $\mathbf{R}^n$ into four regions (see Fig. 4.4):

(I) $P_1 < r_1$ and $P_2 \geq r_2$,

(II) $P_1 < r_1$ and $P_2 < r_2$,

(III) $P_1 \geq r_1$ and $P_2 < r_2$,

(IV) $P_1 \geq r_1$ and $P_2 \geq r_2$.

S_{12} as given by (3.2) is identically equal to 1 everywhere on region IV, and thus $\nabla S_{12} = 0$ on IV. If S_{12} were the primitive function in the definition of another blend, then throughout IV the blend would remain simply a constant offset of the other primitive; the blend would never heel to the primitive in IV regardless of how far

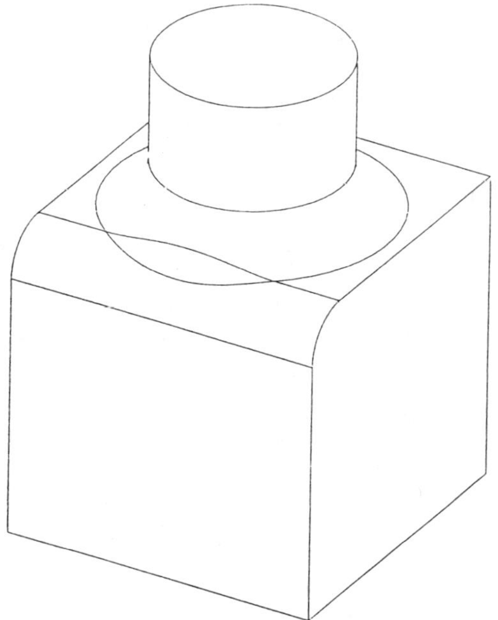

Figure 4.2.   Overlapping blends.

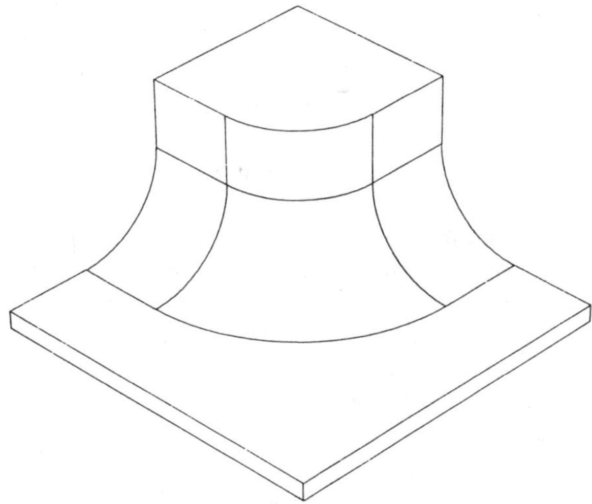

Figure 4.3.   Blending multiple surfaces.

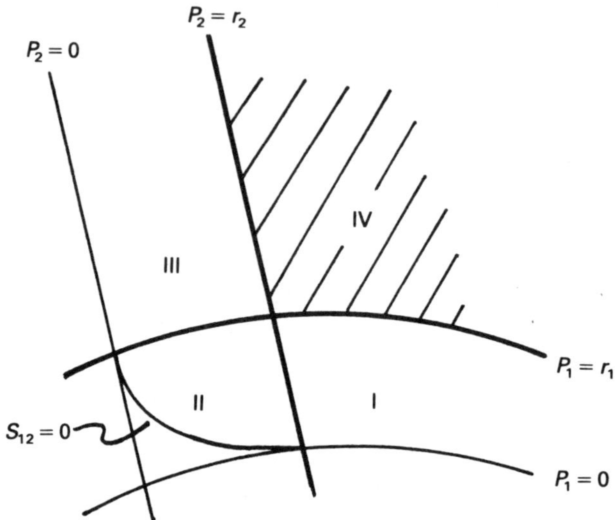

Figure 4.4. Regions of S_{12}.

away, in a Euclidean sense, it was from $S_{12}^{-1}(0)$. This is one example of the sort of behavior which should be avoided in choosing a primitive function.

A second problem with S_{12} is that in the regions where $S_{12}^{-1}(0)$ and $P_1^{-1}(0)$ (or $P_2^{-1}(0)$) are identical we would like the distant measures to correspond, that is on I we would like $S_{12} = P_1$ and on III, $S_{12} = P_2$. This is not the case, except for $S_{12} = P_1 = 0$ on I and $S_{12} = P_2 = 0$ on III.

4.2. A variation on S_{12}. We seek a function G such that $G^{-1}(0) \equiv S_{12}^{-1}(0)$, but such that G has better properties as a distance measure. In fact, the discovery of G may help us understand what properties are generally useful for primitives.

One possibility for G is to let it measure Euclidean distance, i.e.,

$$G(\mathbf{x}) = \inf_{\mathbf{y} \in S_{12}^{-1}(0)} \|\mathbf{x} - \mathbf{y}\|.$$

The main problem with this definition is that it can be computationally intractable, depending on S_{12}.

Another problem is illustrated by example.

Example 4.1. Let E be the function from $\mathbf{R}^2$ to $\mathbf{R}$ which is the Euclidean distance from the solid curve of Fig. 4.5. $E^{-1}(0)$ consists of parts of the axes and the quarter part of the circle. The gradient, ∇E, is continuous, except where $x = y \geq 1$ as illustrated by the double line in the figure. Given a second primitive and blending parameters such that the blend crossed the double line, it is possible that the discontinuity would be reflected in the blend.

A new form for the blend is defined by the equation

$$(4.1a) \qquad \left(1 - \frac{G_{12}(\mathbf{x})}{r(\mathbf{x})}\right)^t = \left[1 - \frac{P_1(\mathbf{x})}{r_1(\mathbf{x})}\right]_+^t + \left[1 - \frac{P_2(\mathbf{x})}{r_2(\mathbf{x})}\right]_+^t$$

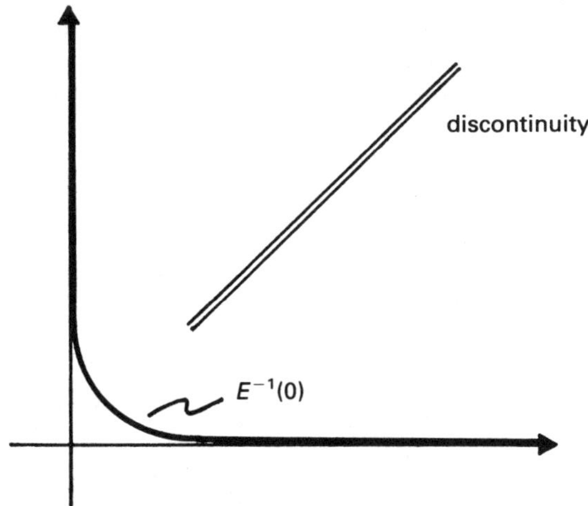

Figure 4.5. C^1 discontinuity in Euclidean distance.

where $r(\mathbf{x})$ is the convex combination of the r_i, $i = 1, 2$;

$$(4.1b) \qquad r(\mathbf{x}) = \sum_{i=1}^{2} [Q_i(\mathbf{x})/(Q_1(\mathbf{x}) + Q_2(\mathbf{x}))]r_i(\mathbf{x})$$

where $Q_i(\mathbf{x}) = [1 - P_i(\mathbf{x})/r_i(\mathbf{x})]^t_+$, $i = 1, 2$. The function $r(\mathbf{x})$ is somewhat arbitrary. We desire $r(\mathbf{x}) = r_1(\mathbf{x})$ in region I, $r(\mathbf{x}) = r_2(\mathbf{x})$ in region III and $r(\mathbf{x})$ to be C^1 in I $\cup$ II $\cup$ III. The convex combination of (4.1b) is suitable as will be shown later.

Observe that $r(\mathbf{x})$ is not defined in IV. The definition given in (4.1a, b), which is valid outside IV, is extended to $\mathbf{R}^n$ by

$$(4.2) \qquad G_{12}(\mathbf{x}) = \begin{cases} r(\mathbf{x})(1 - Q_1(\mathbf{x}) + Q_2(\mathbf{x}))^{1/t} & \text{if } \mathbf{x} \in \text{II}, \\ P_1(\mathbf{x}) & \text{if } \mathbf{x} \in \text{I}, \\ P_2(\mathbf{x}) & \text{if } \mathbf{x} \in \text{III}, \\ \min(P_1(\mathbf{x}), P_2(\mathbf{x})) & \text{if } \mathbf{x} \in \text{IV} \end{cases}$$

Because $G_{12}^{-1}(0) \equiv S_{12}^{-1}(0)$, all the properties of the original blend are retained by G_{12}. The region of support $\mathcal{R}$ corresponds to region II.

Regarding G_{12} as a distance measure, note that $G_{12} = P_1$ in I and $G_{12} = P_2$ in III. In IV there is a set of points for which G_{12} may not be C^0 or C^1 continuous. The situation is similar to Example 4.1 in which the distance measure was Euclidean. Outside IV, the conditions for C^1 continuity are not unexpected.

PROPOSITION 4.2.   *Let P_1 and P_2 be C^1 and $t > 1$; then G_{12} is C^1 continuous outside* IV.

Proof. We check the continuity across the boundary of III and II, i.e., where $P_1 = r_1$. By differentiating (4.1a) we obtain

$$t\left(1 - \frac{G}{r}\right)^{t-1} (\nabla rG - r\nabla G)/r^2$$

$$= t\left(1 - \frac{P_1}{r_1}\right)^{t-1} (\nabla r_1 P_1 - r_1 \nabla P_1)/r_1^2 + t\left(1 - \frac{P_2}{r_2}\right)^{t-1} (\nabla r_2 P_2 - r_2 \nabla P_2)/r_2^2.$$

Substituting $P_1 = r_1$, $r = r_2$ and $G = P_2$ gives

$$\nabla rP_2 - r_2\nabla G = \nabla r_2 P_2 - r_2\nabla P_2$$

on the boundary. Now

$$P_1 = r_1 \Rightarrow Q_1 = 0 \quad \text{and} \quad \nabla Q = 0 \quad \text{so} \quad \nabla r = r_2\nabla\left(\frac{Q_2}{Q_1+Q_2}\right) + \left(\frac{Q_2}{Q_1+Q_2}\right)\nabla r_2$$

$$\Rightarrow \nabla r = \frac{\nabla Q_2(Q_1+Q_2) - Q_2(\nabla Q_1 + \nabla Q_2)}{(Q_1+Q_2)^2} r_2 + \nabla r_2 = \nabla r_2 \quad \text{since } Q_2 \neq 0 \text{ in II.}$$

Therefore, $\nabla G = \nabla P_2$. Similarly, at $P_2 = r_2$, $\nabla G = \nabla P_1$. $\square$

The C^1 continuity of G_{12} is not enough to guarantee smoothness of the blend (see, e.g., Brocker [1, p. 24, Whitney's Theorem]). A stronger condition is necessary.

4.3. Conditions for $S_{12}^{-1}(0)$ to be a manifold. Digressing briefly, a C^k m-manifold in $\mathbf{R}^n$ is a set for which there is an open neighborhood about each point, which is C^k differentiably homeomorphic, i.e., C^k diffeomorphic, to an m-dimensional plane for some $m < n$ (see, e.g., Spivak [10]). In other words, a C^k m-manifold implies the existence of a C^k differentiable change of coordinates which takes an m-plane onto the manifold locally. This means that a C^1 manifold, for instance, will not kink, tear or self-intersect; it embodies a notion of a smooth set of points.

A set of points $F^{-1}(0)$ for $F : \mathbf{R}^n \to \mathbf{R}$ is a $C^1(n-1)$-manifold if F is C^1 continuous on an open set containing $F^{-1}(0)$ and $\nabla F \neq 0$ on $F^{-1}(0)$ (see Bruce and Giblin [2, p. 56]). These are sufficient but not necessary conditions.

With respect to $S_{12}^{-1}(0)$, Proposition 4.2 provides the continuity condition, i.e., $S_{12}^{-1}(0) \equiv G_{12}^{-1}(0)$ is contained in an open set on which G_{12} is C^1 continuous.

In $\tilde{\mathcal{R}}$ (outside II), $\nabla G_{12} = \nabla P_1$, $\nabla G_{12} = \nabla P_2$ or, where $P_1 = P_2$, the gradient does not exist, therefore the question of where $\nabla G = 0$ depends on the primitive functions only.

In $\mathcal{R}$, let $D = Q_1 + Q_2$ and assume $t > 1$, then

$$\nabla G_{12} = \nabla r(1 - D^{1/t}) - \frac{r}{t} D^{(1/t)-1}\nabla D$$

and

$$\nabla r = \sum_{i=1}^{2} \nabla\left(\frac{Q_i}{D}\right)r_i + \frac{\nabla r_i Q_i}{D} = \sum_{\substack{i=1 \\ j=3-i}}^{2} (\nabla Q_i Q_j - \nabla Q_j Q_i)r_i/D^2 + \left(\frac{Q_i}{D}\right)\nabla r_i.$$

Hence $\nabla G_{12} = 0$ implies

$$\left(\frac{R}{t}\right)^{(1/t)-1}\nabla D = \left(\sum (\nabla Q_i Q_j - \nabla Q_j Q_i)r_i/D^2 + \left(\frac{Q_i}{D}\right)\nabla r_i\right)(1 - D^{1/t}).$$

Case 1. $D = 1$. The set $D^{-1}(1)$ is just the set $G_{12}^{-1}(0)$ and then $\nabla D = 0$ because $(R/t)D^{(1/t)-1} > 0$. Expanding ∇D, collecting terms into positive multipliers α_i, β_i, $i = 1,$ 2 yields

$$\sum \alpha_i \nabla P_i r_i - \beta_i \nabla r_i P_i = 0.$$

Case 2. $D \neq 1$. Absorbing constants into $\alpha \neq 0$ gives

$$\alpha \nabla D = \sum (\nabla Q_j Q_i - \nabla Q_i Q_j)\frac{r_i}{D} + Q_i \nabla r_i$$

$$= \nabla Q_1 Q_2 (r_1 - r_2)/D + \nabla Q_2 Q_1 (r_2 - r_1)/D + \sum Q_i \nabla r_i.$$

If $r_2 = r_1$, then $\alpha \nabla D = Q_1 \nabla r_1 + Q_2 \nabla r_2$; otherwise the factors $Q_2(r_1 - r_2)/D$ and $Q_1(r_2 - r_1)/D$ can be written as nonzero factors, therefore $\alpha \nabla Q_1 + \beta \nabla Q_2 = \sum Q_i \nabla r_i$.

In both cases, if the r_i are constant, the condition $\nabla G_{12} = 0$ implies that ∇P_1 and ∇P_2 are linearly dependent. When the r_i are not constant the situation is less obvious, though some sufficient criteria can be stated (see Rockwood [8]).

In summary, G_{12} as a distance measure from $S_{12}^{-1}(0)$ is an improvement on S_{12}. Outside $\mathcal{R}$, $G_{12} = P_1$ or $G_{12} = P_2$, its C^1 discontinuities are similar to that of Euclidean distance measure and $\nabla G_{12} = 0$ at understandable points; in particular, when the r_i are constant, then the surfaces $G_{12}^{-1}(k)$ are C^1 $(n-1)$-manifolds for k such that $\mathbf{x}$ is not in IV and $\nabla P_1(\mathbf{x})$ is linearly independent of $\nabla P_2(\mathbf{x})$. We are now prepared to extend the concept of blending to solid models in general.

5. Blending a Solid Model. Given a set of primitive functions $P_1 \cdots P_k$ define a solid model as any set of points satisfying $M(\mathbf{x}) \leq 0$, $\mathbf{x} \in \mathbf{R}^n$, where M is any composition of max or min functions with the P_i or $-P_i$ as arguments.

This definition corresponds to the familiar constructive solid geometry (CSG) model in which intersection, union and difference operators are applied to primitive sets $\bar{P}_i \equiv \{\mathbf{x} \in \mathbf{R}^n \mid P_i(\mathbf{x}) \leq 0\}$ or $-\bar{P}_i = \{\mathbf{x} \in \mathbf{R}^n \mid P_i(\mathbf{x}) \geq 0\}$ [7].

Example 5.1. In $\mathbf{R}^3$, let $P_1(\mathbf{x}) = x$, $P_2(\mathbf{x}) = y$, $P_3(\mathbf{x}) = z$ and $P_4(\mathbf{x}) = (x^2 + y^2 + z^2)^{1/2} - 1$; then $M(\mathbf{x}) = \max(\min(P_1, P_2, P_3), P_4) \leq 0$ corresponds to the intersection of the set $\bar{P}_4$ with $\bigcup_{i=1}^{3} \bar{P}_i$. The solid model is depicted in Fig. 5.1, along with the associated CSG tree.

Next, we introduce the notation $G_{12} = G\langle P_1, P_2 \rangle$, and then define $G_{123} = G\langle P_1, G\langle P_2, P_3 \rangle \rangle$. Let r_{23} be the range for the $G\langle P_2, P_3 \rangle$ term and $r(2, 3)$ be the convex combination of r_2 and r_3 and so forth; then

$$G_{123} = r(1, 2, 3)\left(1 - \left\{\left[1 - \frac{P_1}{r_1}\right]_+^t + [1 - G\langle P_2, P_3 \rangle / r_{23}]_+^t\right\}^{1/t}\right)$$

$$(5.1) \qquad = r(1, 2, 3)\left(1 - \left\{\left[1 - \frac{P_1}{r_1}\right]_+^t - \left[1 - r(2, 3)\left(1 - \left\{\left[1 - \frac{P_2}{r_2}\right]_+^t\right.\right.\right.\right.$$

$$\left.\left.\left.\left. - \left[1 - \frac{P_3}{r_3}\right]_+^t\right\}^{1/t}\right)\middle/ r_{23}\right]_+^t\right\}^{1/t}\right).$$

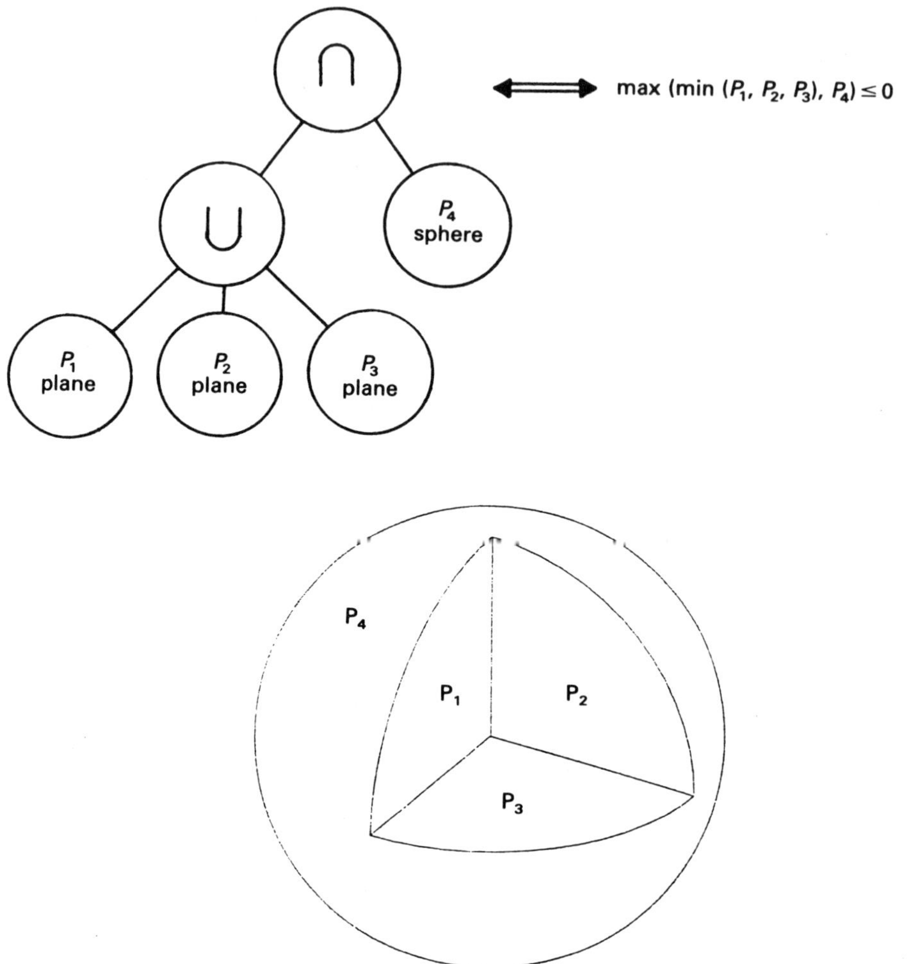

Figure 5.1. A CSG model.

If $r_{23} = r(2, 3)$, then

$$(5.2) \qquad G_{123} = r(1, 2, 3)\left(1 - \left\{\sum_{i=1}^{3}\left[1 - \frac{P_i}{r_i}\right]_+^t\right\}^{1/t}\right),$$

where $r(1, 2, 3)$ is the convex combination of r_1, r_2 and r_3, an extension of (4.1b). Expression (5.2), although a special case of a blend on a blend, is symmetric with respect to the primitives; that is, there is no implied hierarchical evaluation of primitives as in (5.1). Extending (5.2) we define

$$(5.3) \qquad G\langle P_1, P_2 \cdots P_k\rangle = r(1, 2, \cdots k)\left(1 - \left\{\sum_{i=1}^{k}\left[1 - \frac{P_i}{r_i}\right]_+^t\right\}^{1/t}\right)$$

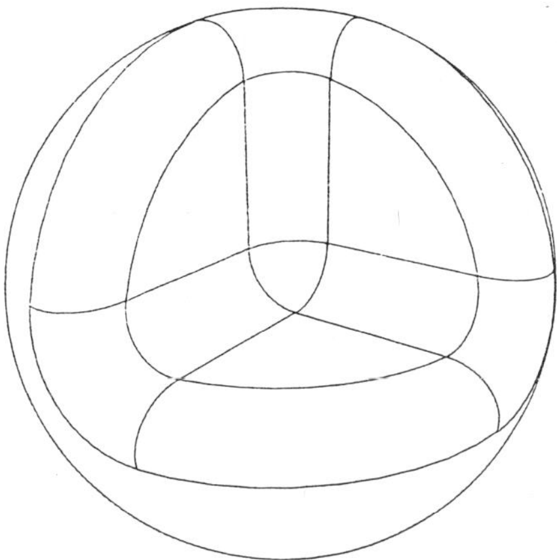

Figure 5.2. A blended CSG model.

as a special case of a blend on a blend $\cdots$ etc., which, like (5.2), is without need to evaluate primitives hierarchically.

We apply $G\langle P_1, P_2 \cdots P_k \rangle$ to blend any union of solid primitives $\bigcup_{i=1}^k \bar{P}_i$, that is $G\langle P_1 \cdots P_k \rangle \leq 0$ replaces $\min (P_1 \cdots P_k) \leq 0$ as the defining inequality.

The blend of $\bigcap_{i=1}^k \bar{P}_i \leq 0$ can be defined as well by replacing $\max (P_1 \cdots P_k) \leq 0$ with $\hat{G}\langle P_1 \cdots P_k \rangle \leq 0$, where

$$(5.4) \qquad \hat{G}\langle P_1, P_2 \cdots P_k \rangle = -G\langle -P_1, -P_2, \cdots, -P_k \rangle.$$

Any solid model is now blended by replacing the min and max functions in the defining inequality by the appropriate blending function, given respectively in (5.3) and (5.4).

Example 5.2. The defining inequality in Example 5.1, when "blended," results in $\hat{G}\langle G\langle P_1, P_2, P_3 \rangle, P_4 \rangle \leq 0$. Figure 5.2 depicts the blended object.

6. Conclusion. We have established a method for developing blends between two implicitly defined primitives in $\mathbf{R}^n$. The method has been demonstrated by developing the super-elliptic blend $S_{12}^{-1}(0)$, which was shown to have desirable characteristics such as intuitive specification and reasonable computability. S_{12} was altered to G_{12} for which $S_{12}^{-1}(0) \equiv G_{12}^{-1}(0)$, but such that G_{12} had better characteristics for blending with blends as primitives. A special form of blending on blends was derived from G_{12} which was then applied to blend solid models defined by a CSG tree. It should here be noted that the CSG approach was chosen because it provided the most concise and easily understood application of blending to a solid model. There are other definitions of solid models (see Requicha and Voelcker [7]) to which the concepts of blending can be applied. In fact, ROMULUS is not a CSG modeler, but a boundary representation modeler. The description of blending in such a context, however, would

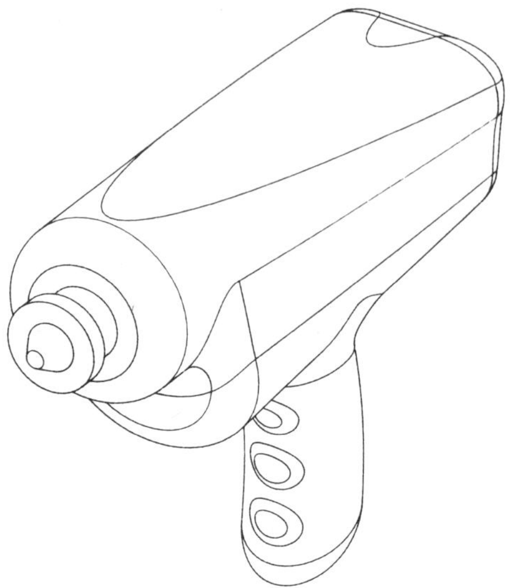

Figure 6.1.   Drill.

have been too lengthy for this paper, although most of the basic principles remain the same as with CSG models.

As an example of the times required in blending we offer the drill depicted in Fig. 6.1. It is a reasonably sophisticated object which demonstrates the aesthetic aspect of designing with blends. Using ROMULUS, it took 417.27 CPU seconds on a VAX 11/780 to produce the hidden line picture. Furthermore, there is room for considerable optimization in this task.

Figure 6.2.   Distributor cap.

The possible C^1 discontinuities in G_{12} have not limited the models thus far, but could potentially become a design constraint. An area of useful research is, therefore, to find another form for which $S_{12}^{-1}(0)$ remains the kernel. One choice which is being considered is to define a mapping $H_{12} = k$ where $S\langle P_1 - k, P_2 - k \rangle = 0$, using the notation $S_{12} = S\langle P_1, P_2 \rangle$. H_{12} can be shown to be a function and to possess better continuity characteristics than G_{12}, but it requires numerical evaluation.

Another area of research is in considering other blending types which derive from the established pattern in § 2.

Related issues involve intersecting the blends with other surfaces and display of the blends. Figure 6.2 shows a shaded picture of a distributor cap modeled with blending surfaces.

The topic of blending in solid modeling has recently received a burst of attention, including the following papers: Hoffmann and Hopcroft [4], Middleditch and Sears [5] and Rossignac and Requicha [9]. It is a useful endeavor for the interested reader to consider the similarities and differences in the approaches given. All were developed independently.

Acknowledgments. We are pleased to acknowledge the help of several individuals. Peter Veenman, Managing Director of Shape Data Ltd., has given encouragement and logistical support. Professor M. J. D. Powell, University of Cambridge, has made many technical comments and observations. Mike Atkins, Ian Jones, Peter Charrot and Stewart Newlove have contributed towards the implementation in ROMULUS.

REFERENCES

[1] Th. Bröcker, *Differentiable Germs and Catastrophes*, Cambridge Univ. Press, Cambridge, 1975.

[2] J. W. Bruce and P. J. Giblin, *Curves and Singularities*, Cambridge Univ. Press, Cambridge, 1984.

[3] I. D. Faux and M. J. Pratt, *Computational Geometry for Design and Manufacture*, Ellis Harwood, Chichester, UK, 1979.

[4] C. Hoffmann and J. Hopcroft, *The potential method for blending surfaces and corners*, this Volume, 1987.

[5] A. E. Middleditch and K. H. Sears, *Blend surfaces for set theoretic volume modeling systems*, Computer Graphics, 19 (1985), pp. 161–170.

[6] J. C. Owen and A. P. Rockwood, *Intersection of general implicit surfaces*, this Volume, 1987.

[7] A. A. G. Requicha and H. B. Voelcker, *Solid modelling: A history, summary and contemporary assessment*, IEEE Computer Graphics and Applications, 2 (1982), pp. 9–14.

[8] A. P. Rockwood, Ph.D. thesis, Univ. Cambridge, Cambridge, UK, in preparation.

[9] J. R. Rossignac and A. A. G. Requicha, *Constant-radius blending in solid modelling*, Comput. Mech. Engrg., July 1984, p. 65.

[10] M. Spivak, *Calculus on Manifolds*, W. A. Benjamin, New York, 1965.

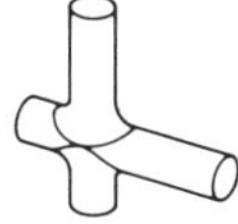

Adaptive Contouring of a
Trivariate Interpolant

CARL S. PETERSEN, BRUCE R. PIPER AND ANDREW J. WORSEY

Abstract. An adaptive method for contouring a trivariate interpolant to scattered data is presented. The method uses a degree reduction scheme coupled with domain subdivision and computes piecewise linear approximations to contours to within a given error tolerance. The adaptive nature of the scheme ensures that to within tolerance no contours are missed.

In general, the interpolant we contour is constructed from positional data only. It is piecewise cubic and globally C^1 continuous. Using the interpolation scheme in a preprocessing stage, we can approximate contours of a general trivariate function.

1. Introduction. Scientific data sometimes correspond to functions of three variables, as, for example, in the cases of

(1)   temperature as a function of three spatial variables,

(2)   soil characteristics as a function of latitude, longitude and depth.

There is, therefore, a need for trivariate interpolants to scattered data and because of computational and physical considerations, it is often desirable that they be local, piecewise polynomial and C^1 continuous.

There are presently only two interpolants with these features. The first is due to Alfeld [1]. It is piecewise quintic and requires derivative information up through second order to be supplied as input data. The second is the three-dimensional case of the general interpolation scheme due to Worsey and Farin [9]. The interpolant is piecewise cubic and, in comparison, requires approximately half as much input data since no second order derivative information is needed.

In order to analyze and graphically display such interpolants, it is necessary to examine three-dimensional "slices." This is most conveniently done by considering contour levels since they convey a good deal of information about the behavior of the surface.

The purpose of this paper is to present an adaptive contouring algorithm for the interpolant of Worsey and Farin [9]. In order to develop a complete package, we first of all consider the problem of constructing the interpolant. To ensure greatest generality we assume only positional data are given. This means that a certain amount of preprocessing is necessary before the interpolant which represents these data can be defined. We address this preprocessing aspect of the contouring package in the next section. The rest of the paper is devoted to a discussion of the contouring method.

In § 3 we outline the basic steps of our method to give the reader a general overview. This also serves to provide a framework for the discussion in § 4, where we consider precise computational details of our implementation of the algorithm. We conclude the main part of the paper by presenting the results of applying our method to certain test problems.

The contouring method is an extension of that developed by Petersen [8] for contouring bivariate functions, in that given a user specified error tolerance ε, we systematically reduce the original piecewise cubic surface to a piecewise linear approximation. The error in this approximation is governed by ε and can be made arbitrarily small. Contouring the approximation exactly then gives a continuous piecewise linear approximation to the contour of the original surface. In the bivariate case, this approach to the contouring problem has certain computational advantages. Petersen [8] highlights these by comparing his method to other contouring schemes.

The piecewise linear approximation is obtained by implementing a method for polynomial degree reduction coupled with domain subdivision. This is done adaptively, according to local behavior of the surface. This means that no contours are missed (to within tolerance) and by using the Bernstein–Bézier form for a polynomial throughout, we implement these steps easily and efficiently. Unnecessary computations are avoided by computing the approximation in a neighborhood of the contour only.

2. Data Preprocessing. We assume we are given only the positional data

$$(2.1) \qquad\qquad w_i = w(x_i, y_i, z_i), \qquad i = 1, \cdots, n$$

where the points $(x_i, y_i, z_i) \equiv V_i$ are scattered in $\mathbb{R}^3$ and w is some underlying primitive function. These are actually the only data supplied to a contouring package in many cases and w is generally unknown. Therefore, as a first step, we need to construct a surface which represents this data set. We do so using a condensed form of the trivariate piecewise cubic, globally C^1 interpolation scheme due to Worsey and Farin [9].

In cases where w is actually known, we can always approximate it using the same approach. Of course, in this case the quality of the approximation will depend on the precise nature of the data set. However, by sampling w sufficiently finely we can obtain a "good" approximation to it and hence, ultimately, its contours.

In the condensed form of the interpolant we use, the cross boundary derivatives are linear; see Worsey and Farin [9]. In order to construct this interpolant P to the data (2.1) some preprocessing is necessary, namely:

(1) The points V_i; $i = 1, \cdots, n,$ must be tessellated into tetrahedra to form a domain.

(2) Gradient data at the vertices V_i; $i = 1, \cdots, n,$ must be obtained.

To tessellate the points, we implement the algorithm due to Bowyer [3]. This produces the Delaunay tessellation, T. We approximate the gradient at a vertex V_i in this tessellation by

$$(2.2) \qquad \frac{\displaystyle\sum_{t \in T_i} \left(\frac{\nabla L_t}{d_{ij}^2 d_{ik}^2 d_{il}^2} \right)}{\displaystyle\sum_{t \in T_i} \left(\frac{1}{d_{ij}^2 d_{ik}^2 d_{il}^2} \right)}.$$

In this expression, T_i is the set of all tetrahedra in T with V_i as a vertex and $t \in T_i$ has vertices V_i, V_j, V_k, and V_l. L_t is the linear interpolant to the given positional data on such a generic tetrahedron, ∇ is the gradient operator and

$$(2.3) \qquad \begin{aligned} d_{ir}^2 &= \|V_i - V_r\|^2 \\ &= (x_i - x_r)^2 + (y_i - y_r)^2 + (z_i - z_r)^2, \qquad r = j, k, l. \end{aligned}$$

Of course, other gradient estimation schemes exist and could be used instead. However, the scheme presented here has the advantages of being local and simple. It is the natural extension to three variables of the triangular Shepard method due to Little [6]. If exact gradient data are available, they should be used rather than these approximations when constructing the interpolant. In this case the second preprocessing step is unnecessary.

At this point, with the preprocessing completed, the scheme of Worsey and Farin [9] may be implemented. Each tetrahedron in the domain T is subdivided into twelve subtetrahedra which we now consider as forming the domain. A trivariate cubic polynomial is defined over each tetrahedron, using the Bernstein–Bézier representation, to obtain a globally C^1 interpolant P to the data (2.1). The Bézier ordinates defining the surface are stored and used to compute contours of the interpolant. These ordinates are simply the coefficients in the particular representation we use for the surface and for a generic tetrahedron t only their w-components are stored. The x, y, and z components are calculated from the vertices of t as and when they are needed.

3. Computing Contours. The method we propose for computing a level surface of the interpolant P involves 5 basic steps. In this section we give a general overview of these steps rather than dwelling upon details of the implementation. We hope that this provides the reader with an understanding of the essential structure of the algorithm. Specific details of the computational aspects of the method are considered in the next section.

(a) Find a tetrahedron t in the domain which contains the given contour level.

(b) Compute an $(n-1)$st degree approximation to the nth degree surface patch.

(c) Compare the $(n-1)$st degree approximation of the surface with the nth degree representation. If they are within some error tolerance either the approximation is linear, in which case go to (e), or it is quadratic so that we need to replace n by $(n-1)$ and return to (b). If they are not within the error tolerance then:

(d) Subdivide this surface patch and any neighboring patches as necessary, in order to maintain a tessellation in the domain. Perform steps (b) and (c) for each new patch which contains the contour level.

(e) Draw the contour for the patch and return to (a).

The contouring method revolves around the schemes for degree reduction and surface subdivision, involved in (b), (c) and (d). They are both local schemes and easy to implement. Since together they form the backbone of the contouring algorithm the method is quite fast. Further, the error tolerances we build into them are adaptive. This guarantees that to within tolerance no contours are missed.

4. Implementation. We now address the details and computational aspects of our method by describing the global procedure for computing the level surface

$$(4.1) \qquad \{(x, y, z) : P(x, y, z) = w_c\}$$

of the interpolant P, where w_c is a given number. In some sense, the steps outlined in the previous section may be thought of as forming a framework for the following discussion. However, they do not, and indeed are not intended to, agree directly with the steps involved in the precise implementation of the algorithm that we now consider.

Step 1. For reasons of computational efficiency, we implement a simple test which eliminates from further consideration any tetrahedra which cannot possibly contain the contour. This test is based on the fact that the value of a Bézier polynomial lies within the convex hull of the ordinates defining it. We compare the maximum and minimum w-components of the Bézier ordinates, w_{max} and w_{min}, respectively, for each tetrahedron in the domain with the function value w_c. If

$$w_c < w_{min} \quad \text{or} \quad w_c > w_{max},$$

the tetrahedron may be eliminated. However, if

$$w_{min} \leq w_c \leq w_{max},$$

the tetrahedron may or may not contain part of the contour and thus cannot be eliminated at this stage.

Typically, this simple test eliminates a large number of tetrahedra in the domain from any further consideration in computing the contour.

Step 2. We compute the volumes of the tetrahedra that remain by evaluating the appropriate 4 by 4 determinant and form a list L indexed by decreasing volume. The tetrahedra in this list are then systematically processed until we find one which contains the contour. This is done by implementing Steps 3 and 4. Ultimately, the list L will be exhausted so that to within tolerance no contours are missed. Listing the tetrahedra by volume was found to produce the most stable approach. We begin with the first tetrahedron in the list.

Step 3. Over this tetrahedron, we compute a quadratic approximation to the cubic Bézier surface, using the method of degree reduction due to Petersen [7]. (A complete discussion of the method together with all the relevant formulae may be found in the Appendix to this paper. We choose not to present the details at this point so as to avoid the possibility of sidetracking the reader and confusing the structure of the algorithm. We simply note here that the method derives from the formulae used to express an nth degree polynomial in Bernstein–Bézier form as one of degree $(n + 1)$.)

The quadratic approximation Q is then compared to the original cubic C in the following way.

We represent Q exactly as a cubic $\tilde{C}$ in Bézier form and calculate the l_2 norm of the difference between the ordinates of $\tilde{C}$ and those of C. The result, γ, is then compared with η which is the product of the user specified tolerance ε and the minimum of the l_2 norm of the gradients at the vertices of the tetrahedron. If

$$(4.2) \qquad\qquad \gamma < \eta,$$

we check whether the contour level is contained within the convex hull of the ordinates defining Q, since if not we may eliminate the tetrahedron from further consideration. If so, the degree reduction process is applied to Q to obtain a linear approximation. We use the same test to judge the quality of this approximation.

The whole point to the degree reduction is to establish whether or not the linear polynomial determined by the vertex values is "close enough" to the given cubic. This then begs the question as to why the intermediate step of degree reducing to a quadratic is implemented. This important point is addressed in Step 4, since the justification is centered around the subdivision process.

In the event that the degree reduction is successful, we compute and draw the contour for the linear approximation, flag that the tetrahedron under consideration has been processed and move to Step 5. If not, or if (4.2) was violated, we move to Step 4.

Step 4.   If the degree reduction fails, that is, if at any stage the $(n-1)$st degree polynomial is not a sufficiently good approximation to the nth degree representation, then we subdivide the surface patch. Over the new, smaller patches the degree reduction process is more likely to be successful since a lower degree polynomial may better approximate the original.

The degree reduction of the original cubic to a quadratic is likely to produce an acceptable approximation. (This claim is based on experimental observations.) However, this is not the case when degree reducing the quadratic to the linear polynomial, so that typically we need to subdivide the quadratic patch. This is computationally more efficient than subdividing the original cubic, which we would have to do if we degree reduce the cubic to the linear approximation directly. Since the subdivision process is a major computational expense in the algorithm, this motivates the intermediate step of degree reducing to a quadratic.

To subdivide the surface over a tetrahedron t in the domain, we first of all subdivide t. There are many ways in which this could be done. However we are constrained by the following considerations. Firstly, we must maintain a tessellation and, secondly, in order to maintain numerical stability we want to avoid constructing nearly planar tetrahedra. Extending the results of Petersen [8] for subdividing a triangle, we subdivide t by joining the midpoint of the longest edge to the opposite vertices of all tetrahedra in the domain sharing that edge. This splits t into two new tetrahedra t_1 and t_2. Next we subdivide the surface defined over t accordingly. Since we are using the Bernstein–Bézier representation throughout this can be done easily and efficiently; see Farin [4].

We examine the tetrahedra t_1 and t_2, using the simple test outlined in Step 1, to see if they contain the contour. If they do not, we eliminate them from further consideration, consider the next tetrahedron in the list L and return to Step 3. If t_1 (resp. t_2) is not eliminated on the basis of this test, we apply the degree reduction process to t_1 (resp. t_2) by returning to Step 3.

Examining the domain in this systematic and exhaustive fashion, we are able to find a tetrahedron which contains the contour level.

Step 5.   We consider next a tetrahedron, t, which neighbors one that has been processed and contains the contour. If t has already been processed, we continue looking until we find a tetrahedron which has not yet been processed and which neighbors one containing the contour. Then we return to Step 3 and continue.

Although there may be disjoint sections of the level surface for the given function value, by searching the data base for other tetrahedra which contain the contour we ensure that all of them are found. Also, with the particular data structure used we identify the occurrence of closed contour surfaces. As a final remark on the implementation, the convex hull property of Bézier polynomials used in Step 1 is also used to find contours which are wholly contained within a single tetrahedral surface patch.

Using this implementation of the algorithm we are able to follow a contour in a continuous fashion. However, there are some degenerate cases that may arise and which require further detailed comment.

When a tetrahedron is subdivided, we must ensure that we do not inadvertently skip a tetrahedron which contains the contour. We do this by referring back to the previous tetrahedron to identify which tetrahedron is to be processed. Also, during the subdivision process, a tetrahedron which has already been contoured may be subdivided. In this case we must recompute the contour for each new section and find its new representation.

Having successfully reduced the surface to a linear representation over a tetrahedron we have to draw the contour. This is done at Step 3 and there are four cases to consider when doing so.

(1)   The contour intersects the tetrahedron at one vertex only.
(2)   The contour intersects the tetrahedron along an edge only.
(3)   The intersection of the contour with the tetrahedron is a triangle.
(4)   The intersection of the contour with the tetrahedron is a quadrilateral.

These cases are illustrated in Figs. 4.1–4.4.

In the first two, we simply eliminate the tetrahedron from further consideration and proceed. In the third case, the three points are used to draw a planar section of the contour. Neighboring tetrahedra which share a common face crossed by this section are identified and each one is added to the list of tetrahedra to process, providing it has not already been processed.

The last case is the most difficult to handle since the four points do not necessarily lie on a plane. We therefore take the average of these points to determine a fifth and then render the contour as four planar sections. In this way we approximate the behavior of the contour over the tetrahedron symmetrically. As before, the neighboring tetrahedra which need to be examined are identified and added to the list for later processing.

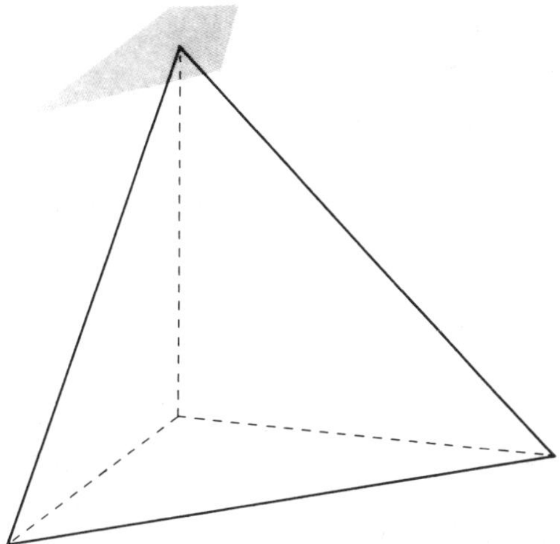

Figure 4.1. The contour passes through only one vertex.

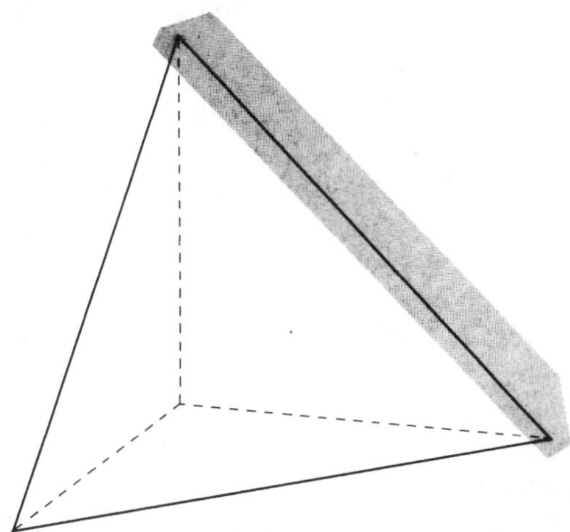

Figure 4.2. The contour passes through two vertices.

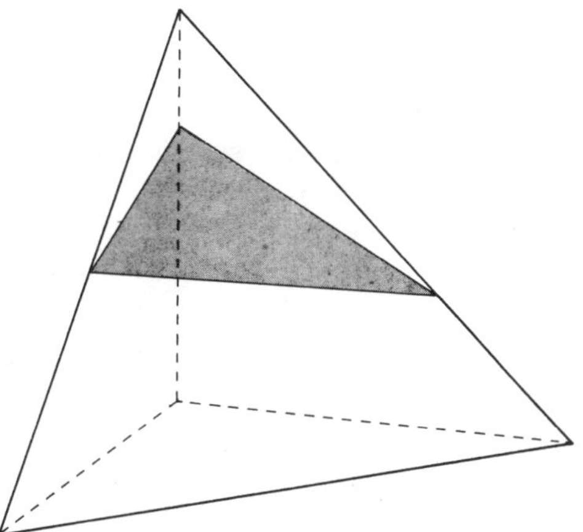

Figure 4.3. The contour crosses three edges.

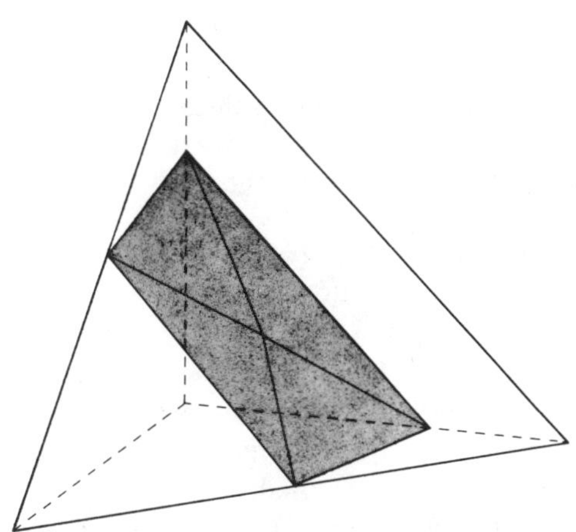

Figure 4.4. The contour crosses four edges.

5.   Examples.   In this section we present the results of applying our contouring method to two test problems. Although the functions we use may appear simple, their contours are not trivial to compute and provide a good test for our algorithm. In the first example we illustrate that the algorithm is able to deal with closed contours while the function used in the second example has contours with disjoint sections.

Example 1.   We consider the function

$$(5.1) \qquad\qquad\qquad w(x, y, z) = x^2 + y^2 + z^2$$

and take the positional data generated by evaluating w at the domain points

$$(5.2) \qquad (-1+ih, -1+jh, -1+kh), \qquad i, j, k \in \{0, 1, \cdots, 5\}, \quad h = 0.4.$$

These points are tessellated to form the Delaunay tessellation and the gradient data needed to construct the interpolant are determined exactly from the primitive function w, rather than approximated. The interpolant is defined over the three-cube

$$(5.3) \qquad\qquad [-1, 1] \times [-1, 1] \times [-1, 1],$$

and since the interpolation scheme is quadratically precise, the primitive function (5.1) is actually reproduced. Therefore, contours of the interpolant are spheres or sections of same.

In Fig. 5.1 (see color insert) we show two contour surfaces generated by our scheme, together with the boundary edges of the domain (5.3). The blue sphere is the computed contour for the function value 0.8, while the green surface is the 1.2 contour. The picture is rendered using three point light sources and Phong shading; see Foley and Van Dam [5, pp. 577–584].

Example 2. We use the underlying primitive function

$$(5.4) \qquad\qquad w(x, y, z) = xyz$$

and take the positional data generated by evaluating w at the points (5.2). As in the previous example, we form the Delaunay tessellation of these points and use exact gradient data in forming the interpolant over the three-cube (5.3).

Figure 5.2 (see color insert) shows two computed contour surfaces for the interpolant. Each of these is comprised of four disjoint sections. The contour corresponding to the function value 0.1 is shown in red, while the purple surfaces are the sections of the -0.1 contour. The boundary edges of the domain for the interpolant are also shown (they appear as white lines), and, as in the previous example, the picture is rendered using three point light sources and Phong shading.

6. Appendix—Degree Reduction. We present here the formulae used in our degree reduction scheme for approximating an nth degree polynomial over a tetrahedron by one of degree $(n-1)$. Although for the contouring method presented in this paper n can only be either three or two, we begin by considering the general case in order to try and motivate our approach.

An nth degree polynomial, B_n, defined over a tetrahedron T, with vertices p_1, p_2, p_3, and p_4, is written in Bernstein–Bézier form as

$$(6.1) \quad B_n(\bar{V}; b_1, b_2, b_3, b_4) = \sum_{i+j+k+l=n} \frac{n!}{i!\, j!\, k!\, l!} b_1^i b_2^j b_3^k b_4^l V_{i,j,k,l}, \qquad \bar{V} = (V_{i,j,k,l})$$

where (b_1, b_2, b_3, b_4) are the barycentric coordinates of a point with respect to the vertices of T and the $V_{i,j,k,l}$ are the Bézier ordinates.

We want to approximate the polynomial $B_n(\bar{V}; b_1, b_2, b_3, b_4)$ by an $(n-1)$st degree polynomial $B_{n-1}(\bar{W}; b_1, b_2, b_3, b_4)$, with the $W_{i,j,k,l}$ given in terms of the original $V_{i,j,k,l}$. This is the essence of our degree reduction scheme. The derivation of the formulae for doing this follows from the fact that we can always write an nth degree Bézier polynomial as one of degree $(n + 1)$. A discussion of this degree raising process is also given by de Boor [2].

Naturally, if we raise the degree of an nth degree Bézier polynomial and then lower the result, we get back the coefficients of the original polynomial. Symbolically this may be written as

$$L(R(\bar{V})) = \bar{V}$$

where the operator R raises the degree of the polynomial and L lowers the degree.

We raise the degree of $B_n(\bar{V}; b_1, b_2, b_3, b_4)$ in equation (6.1) by multiplying by $(b_1 + b_2 + b_3 + b_4) \equiv 1$. This gives

(6.2)
$$
\begin{aligned}
B_n(\bar{V}; b_1, b_2, b_3, b_4) \times (b_1 + b_2 + b_3 + b_4) =\ & \sum_{i+j+k+l=n} \frac{n!}{i!\,j!\,k!\,l!} b_1^{i+1} b_2^{j} b_3^{k} b_4^{l} V_{i,j,k,l} \\
&+ \sum_{i+j+k+l=n} \frac{n!}{i!\,j!\,k!\,l!} b_1^{i} b_2^{j+1} b_3^{k} b_4^{l} V_{i,j,k,l} \\
&+ \sum_{i+j+k+l=n} \frac{n!}{i!\,j!\,k!\,l!} b_1^{i} b_2^{j} b_3^{k+1} b_4^{l} V_{i,j,k,l} \\
&+ \sum_{i+j+k+l=n} \frac{n!}{i!\,j!\,k!\,l!} b_1^{i} b_2^{j} b_3^{k} b_4^{l+1} V_{i,j,k,l}.
\end{aligned}
$$

In terms of $(n + 1)$, this may be rewritten as

(6.3)
$$
\begin{aligned}
B_{n+1}(\bar{V}; b_1, b_2, b_3, b_4) =\ & \sum_{i+j+k+l=n+1} \frac{n!}{(i-1)!\,j!\,k!\,l!} b_1^{i} b_2^{j} b_3^{k} b_4^{l} V_{i-1,j,k,l} \\
&+ \sum_{i+j+k+l=n+1} \frac{n!}{i!\,(j-1)!\,k!\,l!} b_1^{i} b_2^{j} b_3^{k} b_4^{l} V_{i,j-1,k,l} \\
&+ \sum_{i+j+k+l=n+1} \frac{n!}{i!\,j!\,(k-1)!\,l!} b_1^{i} b_2^{j} b_3^{k} b_4^{l} V_{i,j,k-1,l} \\
&+ \sum_{i+j+k+l=n+1} \frac{n!}{i!\,j!\,k!\,(l-1)!} b_1^{i} b_2^{j} b_3^{k} b_4^{l} V_{i,j,k,l-1}.
\end{aligned}
$$

We can express this nth degree polynomial as one of degree $(n + 1)$ using new Bézier ordinates $P_{i,j,k,l}$. The Bézier polynomial defined by this set of ordinates is

(6.4) $\quad B_{n+1}(\bar{P}; b_1, b_2, b_3, b_4) = \sum_{i+j+k+l=n+1} \frac{(n+1)!}{i!\,j!\,k!\,l!} b_1^{i} b_2^{j} b_3^{k} b_4^{l} P_{i,j,k,l}, \qquad \bar{P} = (P_{i,j,k,l}).$

The formulae for the ordinates $P_{i,j,k,l}$ in terms of the $V_{i,j,k,l}$ now follow from setting

$$B_{n+1}(\bar{P}; b_1, b_2, b_3, b_4) = B_{n+1}(\bar{V}; b_1, b_2, b_3, b_4)$$

and equating coefficients. This gives

(6.5) $\quad P_{i,j,k,l} = \dfrac{(i \times V_{i-1,j,k,l} + j \times V_{i,j-1,k,l} + k \times V_{i,j,k-1,l} + l \times V_{i,j,k,l-1})}{(n+1)}$

where

(6.5a)
$$P_{n+1,0,0,0} = V_{n,0,0,0}, \qquad P_{0,n+1,0,0} = V_{0,n,0,0},$$
$$P_{0,0,n+1,0} = V_{0,0,n,0}, \qquad P_{0,0,0,n+1} = V_{0,0,0,n}.$$

Hence, we have defined the Bézier ordinates for an $(n+1)$st degree representation of an nth degree Bézier polynomial in terms of the original ordinates.

We can rewrite (6.5) and (6.5a) for $P_{i,j,k,l}$, to obtain the Bézier ordinates of our $(n-1)$st degree approximation $B_{n-1}(\bar{W}; b_1, b_2, b_3, b_4)$, to the nth degree polynomial $B_n(\bar{V}; b_1, b_2, b_3, b_4)$. In general this gives us four possible values for these ordinates, namely:

$$(6.6) \quad W^1_{i-1,j,k,l} = \frac{(n \times V_{i,j,k,l} - j \times W^1_{i,j-1,k,l} - k \times W^1_{i,j,k-1,l} - l \times W^1_{i,j,k,l-1})}{i}, \quad i \neq 1,$$

$$(6.7) \quad W^2_{i,j-1,k,l} = \frac{(n \times V_{i,j,k,l} - i \times W^2_{i-1,j,k,l} - k \times W^2_{i,j,k-1,l} - l \times W^2_{i,j,k,l-1})}{j}, \quad j \neq 1,$$

$$(6.8) \quad W^3_{i,j,k-1,l} = \frac{(n \times V_{i,j,k,l} - i \times W^3_{i-1,j,k,l} - j \times W^3_{i,j-1,k,l} - l \times W^3_{i,j,k,l-1})}{k}, \quad k \neq 1,$$

$$(6.9) \quad W^4_{i,j,k,l-1} = \frac{(n \times V_{i,j,k,l} - i \times W^4_{i-1,j,k,l} - j \times W^4_{i,j-1,k,l} - k \times W^4_{i,j,k-1,l})}{l}, \quad l \neq 1.$$

The ordinates at the vertices of the tetrahedron T are unaltered in the degree reduction process and hence are given by

(6.10)
$$W_{n-1,0,0,0} = V_{n,0,0,0}, \qquad W_{0,n-1,0,0} = V_{0,n,0,0},$$
$$W_{0,0,n-1,0} = V_{0,0,n,0}, \qquad W_{0,0,0,n-1} = V_{0,0,0,n}.$$

The Bézier ordinates on the left-hand sides of (6.6)–(6.9) are written in terms of known values, since if negative subscripts appear for the ordinates on the right we set those terms to zero. Now we examine the problem that arises from the multiple definitions for the ordinates given by (6.6)–(6.9). In the particular instances where we use the degree reduction in the contouring scheme n is either three or two, so these are the only cases we consider.

When we reduce a quadratic to a linear polynomial, that is in the case when $n = 2$, the new ordinates for the approximation are completely determined by (6.10). It is only when $n = 3$ where we reduce a cubic to a quadratic that (6.6)–(6.9) are relevant. In this case we resolve the ambiguities in the definitions of the ordinates for the quadratic by taking averages.

More precisely, the ordinates associated with the midpoints of the edges of the tetrahedron T are taken to be

(6.11)
$$W_{1100} = (W^1_{1100} + W^2_{1100})/2,$$
$$W_{1010} = (W^1_{1010} + W^3_{1010})/2,$$
$$W_{1001} = (W^1_{1001} + W^4_{1001})/2,$$
$$W_{0110} = (W^2_{0110} + W^3_{0110})/2,$$
$$W_{0101} = (W^2_{0101} + W^4_{0101})/2,$$
$$W_{0011} = (W^3_{0011} + W^4_{0011})/2$$

where the terms in the right-hand sides of these equations are calculated from (6.6)–(6.9) as and when necessary. The remaining ordinates which define the quadratic approximation, namely those associated with the vertices of the tetrahedron T, are unaltered by the degree reduction process. They are the same as the corresponding ordinates of the original cubic and are given by (6.10) with $n = 3$.

Acknowledgments. This research was supported in part by the Department of Energy under contract DE-AC02-85ER12046 to the University of Utah, Salt Lake City, Utah. The authors would like to thank the referees for their extremely valuable comments on an earlier draft of this paper. Their efforts are much appreciated.

REFERENCES

[1] P. ALFELD, *A trivariate Clough-Tocher scheme for tetrahedral data,* Computer Aided Geometric Design, 1 (1984), pp. 169–181.

[2] C. DE BOOR, *B-form basics,* this Volume, 1987.

[3] A. BOWYER, *Computing Dirichlet tessellations,* Computer J., 24 (1981), pp. 162–166.

[4] G. FARIN, *Triangular Bernstein-Bézier patches,* Computer Aided Geometric Design, to appear.

[5] J. D. FOLEY AND A. VAN DAM, *Fundamentals of Interactive Computer Graphics,* Addison-Wesley, Reading, MA, 1982.

[6] F. F. LITTLE, *Convex combination surfaces,* in Surfaces in Computer Aided Geometric Design, R. E. Barnhill and W. Boehm, eds., North-Holland, Amsterdam, 1983.

[7] C. S. PETERSEN, *Contours of three and four dimensional surfaces,* Masters thesis, Dept. Mathematics, Univ. Utah, Salt Lake City, UT, 1983.

[8] ———, *Adaptive contouring of three-dimensional surfaces,* Computer Aided Geometric Design, 1 (1984), pp. 61–74.

[9] A. J. WORSEY AND G. FARIN, *An n-dimensional Clough–Tocher interpolant,* Constructive Approximation, to appear.

Index